Drug and Biological Development

Drug and Biological Development
From Molecule to Product and Beyond

Editor

Ronald P. Evens

Ronald P. Evens, Pharm D., FCCP
President, MAPS 4 Biotech, Inc.
Clinical Professor, University of Florida
Jacksonville, FL
Maps4biotech@aol.com

Library of Congress Control Number: 2007925368

ISBN: 978-0-387-32978-9   e-ISBN: 978-0-387-69094-0

Printed on acid-free paper.

© 2007 Springer Science+Business Media, LLC

All rights reserved. This work may not be translated or copied in whole or in part without the written permission of the publisher (Springer Science+Business Media, LLC, 233 Spring Street, New York, NY 10013, USA), except for brief excerpts in connection with reviews or scholarly analysis. Use in connection with any form of information storage and retrieval, electronic adaptation, computer software, or by similar or dissimilar methodology now known or hereafter developed is forbidden.
The use in the publication of trade names, trademarks, service marks, and similar terms, even if they are not identified as such, is not to be taken as an expression of opinion as to whether or not they are subject to proprietary rights.

Printed in the United States of America.

9 8 7 6 5 4 3 2 1

springer.com

Acknowledgments

The members of the editorial board of this book have been professional associates in health care and research in the industry and academia during the editor's career, as well as friends. Each brings his own base of experience and talents to this scholarly work. Their advice and engagement as editorial advisors and authors have enriched the content and were indispensable to me as the editor. The authors comprise 24 talented group of researchers, clinicians, and educators in health care, both industry and academia, who offer the readers of our book their extensive, varied knowledge and broad experiences. Over a 35-year career in health care, academia, and the industry (both the biotechnology and pharmaceutical worlds), I have been blessed with the unique opportunity to work with many thought leaders and creative scientists, leading clinicians, bright students, and top-flight business people, many of which are editors and authors in this book.

For the editor, my friends, colleagues, co-workers, department heads, staff, and bosses, and students as well, have all served me well as advisors and mentors. They have taught me much about science, research, education, business, human nature, patient care, and professionalism. And most important to me in my life, these associates have fostered in me the desire to care for, assist, and lead people in all walks and all levels of life, to learn, succeed, and evolve. Even at the risk of leaving someone out, I will enumerate some of these special people; my high school teachers, Mr. Felsinger and Ms. Scime in Buffalo, New York, and professors, Drs. Paul Parker, Ann Amerson, and Charles Walton in Kentucky; my residency comrades, Drs. Robert Talbert, Rusty Ryan, John Kerege, and Bob Anderson; my Pharm.D. fellows that I trained, Brock Guernsey, David Magnuson, Donald Fraser, and William Clementi; my good friends and business and academic colleagues, Edward Kinney, Bill Ashton, Craig Books, Kathy Wiltsey, Tommy Beard, Philip vonHolle, Thomas Lytle, John Perry, and Stephen Allen, Drs. Robert Elenbaas; Cliff Littlefield, Joel Covinsky, Elaine Waller, Thomas Foster, Lewis Smith, and Robin Campbell; my bosses, Drs. James Doluisio, Bill Miller in academia, and Ed Fox, Paul Dawson, Stanley Benson, and Kevin Sharer in industry; my management team, Drs. Donna Mapes, Gary Hubler, Jeanne Flynn, August Salvado, Homa Yeganegi, Alan Rosman, Stanfell Boone, David McFadden, Larry Green, Billwelch, Bill Carter, and Grant Lawless. My family has been more than instrumental in supporting me and serving as a guiding light; my parents, Dorthea and William Evens, my children, Andrew Evens, Julie Evens, and Alexander Evens, and my wife, Sally Ann Evens. I have devoted much of my last 25 years of professional life to drug development, people development, and professional education; this book is the major effort in my career to pull all these learning opportunities together and share them with young clinicians, students, and future researchers, and their mentors.

The chapter on commercialization reflects Tom Lytle's thoughts and experiences, as well as the thoughts of many others with whom he has worked for more than 30 years of experience in sales, marketing, and launching new products in the pharmaceutical and biotech industry. It is not possible to thank or recognize each person in a chapter like this, so the author takes this opportunity to recognize that what is presented here reflects my own ideas on the subject, as well as the collective thinking and best practices of many past and current industry leaders. Their contribution to my training, thinking and knowledge on this topic is recognized and deeply appreciated.

In the preparation of this book and educational materials, special thanks are in order to Amgen Inc, who provided an educational grant to the University of Florida Foundation to cover some of the book's production costs. No Amgen input was sought or given in any way in writing this scholarly endeavor regarding the content or style. No review or editing was permitted. The editor, editorial board, authors, and publisher had full and independent control of the content, examples, and formats.

As a guide for the course directors and users of the book and CD-ROM, several references are provided that offer information (statistics and reports) about the pharmaceutical and biotechnology industries and also disease statistics useful for product development. The title and publisher along with Web

- BIO: Biotechnology Industry Organization; Washington, D.C.
  www.bio.org
- Center watch; Boston MA 02210
  www.centerwatch.com
- CMS: Centers for Medicare & Medicaid, HHS ; Washington, D.C.
  www.cms.hhs.gov/
- CSDD: Tufts Center for Drug Development ; Boston, MA 02111
  www.tufts.edu
- E&Y: Ernst & Young ; San Francisco, CA
  www.ernst&young.com
- FDA: Food & Drug Administration; Rockville, MD
  www.fda.gov
- IMS: IMS Health; Plymouth Meeting, PA 19462
  www.IMShealth.com
- PhRMA: Pharmaceutical Research & Manufacturers Association, Washington, D.C.
  www.phrma.org
- R&D Directions; Med Ad News: Engel Publishing; Newtown, PA 18940
  www.editorial@engelpub.com

Figure. Information Sources

sites are listed below for 10 such resources, among the literally hundreds of health Web sites. Industry statistics and drug development issues in the industry can be found in Web sites from the government, industry organizations, industry service organizations, journal publications, and a university-based service. In the text below, I will list altogether 20 such sites.

The industry sites (3) come from the Biotechnology Industry Organization, Pharmaceutical Research and Manufacturing Association, and various individual company Web sites. Government sites (5) include the Food and Drug Administration and the Centers for Medicare and Medicaid Services in the United States; in addition, three other government sites will contain useful information from the U.S. Centers for Disease Control and Prevention, the U.S. National Institutes of Health (various institutes based on disease areas such as the National Cancer Institute and National Institute for Mental Health), and the European Medicines Evaluations Agency. Health care information is available from foundations and related health organizations that support research and patient care for specific diseases (e.g., National Headache Foundation, American Diabetes Association, American Society for Clinical Oncology, Alzheimer's Foundation, Osteoporosis Foundation). Service organizations (4) that support and evaluate the industry and related health care community for drug research are also available with Web sites and reports (e.g., Thomson/CenterWatch, IMS Health, Ernst and Young, Parexel International Corporation, and Boston Consulting Group). The leading university-based organization for industry and related health care information is the Tuft's University Center for the Study of Drug Development in Boston, Massachusetts. Journal publications (5) are yet another choice for health care and industry information and statistics (e.g., *R&D Directions* and *Med Ad News* from Engel Publications, *Pharmaceutical Executive* from Advanstar Communications, and *Nature Reviews Drug Discovery* and *Nature Biotechnology* from the Nature Publishing Group). These resources were used in the writing of this book.

Contents

Introduction 1

1. Health Care and Industry Overview and Challenges 5
   Ronald P. Evens
   Health Care Issues 5
   Industry Statistics (The Pharma and Biotech Industries) 11
   Research Activities and Costs 14
   R&D Productivity 16
   Organization of a Pharmaceutical Company 19
   R&D Challenges and Issues 24
   Summary of Research and Development Approaches 29
   References 31

2. R&D Planning and Governance 33
   Ronald P. Evens and Joel Covinsky
   Development Schemas and Leadership 34
   Clinical Development Strategy 36
   Leadership 38
   Product Development Paradigm 39
   Portfolio Planning Management (PPM) Description 43
   PPM Process Components 47
   PPM Analyses (Examples) 54
   Project Management 60
   Summary 62
   References 64

3. R&D Outcomes 66
   Ronald P. Evens
   Public/Patient Outcomes 66
   Product Outcomes 70
   Research Outcomes 73
   Data/Information Outcomes 76
   Company/Business Outcomes 79
   References 82

4. Discovery and Nonclinical Development 84
   Stephen F. Carroll
   The Discovery Process 85
   Targets 93

       Products 95
       Nonclinical Development and Testing 98
       IND-Enabling Studies 102
       Added Discovery Work 103
       References 106

5. Types of Clinical Studies 107
   Lewis J. Smith
   Introduction 107
   Phase 1 Studies 107
   Phase 2 Studies 108
   Phase 3 Studies 109
   Phase 4 Studies 111
   Special Studies 112
   Summary 120
   References 121

6. Metabolism and Pharmacokinetics 123
   Jun Shi, Shashank Rohatagi, and Vijay O. Bhargava
   Introduction 123
   Value of PK/PD 124
   PK/PD Concepts 126
   Drug Development Value Chain 132
   References 147

7. Laws and Regulations: The Discipline of Regulatory Affairs 148
   Elaine S. Waller and Nancy L. Kercher
   Regulatory Authorities and the Laws 149
       Protecting the Public Health 149
       Advancing the Public Health 150
       Helping the Public Get Accurate, Science-Based Information 150
   Regulatory Development Strategies 157
       Fast-Track Program 157
       Priority Review 157
       Rolling Submission 157
       Accelerated Approval 158
       Orphan Drugs 158
       Rx to OTC Switch 159
   Submissions to Regulatory Authorities 160
   Product Review 165
       Package Insert 167
       Advertising 167
   Postapproval Maintenance 168
   Compliance/Quality Assurance 171
   General References 177

8. Clinical Trial Operations 178
   Carl L. Roland and Paul Litka
   Introduction 178
   Clinical Trial Conduct 181
   Contract Research Organizations (CROs) 193
   Selected Issues in Clinical Development 197
   Summary 200
   References 200

9. Formulation and Manufacturing 202
   Leo Pavliv and James F. Cahill
   Dosage Form Decisions 202
   Formulation Development 205
   Early Manufacturing 214
   Process Development 216
   General References 221

10. Commercial Division 222
    Thomas Lytle
    Challenges 222
    Framework 224
    Commercialization Process 228
    Practical Approaches 235
    Commercial Responsibilities 237
    References 239

11. Medical Affairs and Professional Services 240
    Ronald P. Evens
    Introduction 240
    Departmental Issues 241
    Medical Information Services 245
    "Standard Letter" Generation 246
    Product Package Insert 246
    Patient Package Insert 246
    Pipeline Product Information 247
    FDAMA "Safe Harbor" 247
    Medical Inquiries (Off-label) to Sales Representatives 247
    Scientific Publication Tracking 248
    Formulary Material Development 248
    Scientific Meeting Support 248
    Disease State Management 248
    Patient Assistance Programs 249
    Research Questions to be Addressed 249
    Special Patient Registries 249
    Communication Triage 249
    Industry Requirements for Postmarket Safety Surveillance 250
    MedWatch 251
    HIPPA 252
    Medical Science Liaisons 252
    Medical Communications 256
        "The Orange Book" 258
        United States Pharmacopeia Drug Information® 258
        United States Pharmacopeia 258
        Martindale: The Extra Pharmacopoeia 258
        AHFS–Drug Information Essentials 258
        Risk management 261
        Product Defects 263
        Known Side Effects 263
        Avoidable Side Effects 263
        Unavoidable Side Effects 263
        Medication or Device Errors 263
        Unexpected Side Effects 264
        Long-term Effects 264
        Effects of Off-label Uses 264

  Effects in Populations Not Studied ... 265
   Remaining Uncertainties ... 265
 Outcome and Pharmacoeconomics Research ... 265
 Late-Phase Clinical Trials ... 267
 References ... 272

12. Special Considerations in Research ... 275

12.1. Cardiovascular Drug Development ... 275
C. Michael White, Jessica Song, and Jeffrey Kluger
 State of Cardiovascular Research ... 275
 New Modalities ... 278
 Cardiovascular Drug Development ... 279
 Standard of Care ... 281
 Clinical Trial Types ... 281
 Acute and Chronic Therapy Trials ... 284
 Special Populations Studies ... 286
 Conclusions ... 290
 References ... 290

12.2. Infectious Diseases ... 294
Ralph H. Raasch
 "Bad Bugs, NO Drugs"—The Diminishing Antibiotic Pipeline ... 294
 Problem of Antimicrobial Resistance ... 297
 Rational Antibacterial Drug Development ... 300
 Proposed Approaches to Address "Bad Bugs, NO Drugs" ... 306
 References ... 307

12.3. Oncology ... 308
Suzanne F. Jones and Howard A. Burris III
 Background Issues ... 309
 Phase 1 Clinical Trials ... 310
 Phase 2 Clinical Trials ... 313
 Other Design Issues in Oncology ... 316
 Era of Targeted Therapy ... 317
 References ... 319

12.4. Pediatrics ... 321
Philip D. Walson
 The Case for Pediatric Research with Drugs ... 322
 American Academy of Pediatrics and Research ... 325
 Legislative and Government Initiatives in Pediatric Research ... 327
 The Science in Pediatric and Clinical Studies ... 333
 References ... 337

12.5. Psychiatry ... 338
Michael W. Jann, John J. Brennan, and Roland Garritsen VanderHoop
 Introduction ... 338
 Patient Populations in Psychiatry ... 339
 Acute versus Chronic Trials—Design and Conduct Issues ... 340
 Pharmacology ... 344

Contents   xi

    Research Challenges 346
       Dos 347
       Don'ts 347
    Summary 348
    References 349

Appendix 1 351

Appendix 2 366

Index 374

Introduction

Ronald P. Evens

Editors and Authors 2
   Editor 2
   Editorial Board 2
   Chapter/Section Authors 3

This book and CD-ROM contain an extensive discussion of product development in the pharmaceutical and biotechnology industries from discovery, to product launch, and through life cycle management for the new researcher in academia or industry. The primary goal is the education of new researchers in the academic medical center and industry environments about industry-based research and product development. The perspective is product development (drugs and biologicals) especially from the industry situation, along with collaboration with medical center scientists. References are quite extensive to support the work, numbering more than 500. The authors collectively have several hundred years of experience at senior levels in product development in the industry or research experience in the academic or clinical setting. The book has many tables of data and information, illustrations, and examples to elaborate on the issues, problems, challenges, and successes in product development.

The collaboration of industry scientists and marketers with their university and medical center colleagues, as advisors and key investigators, is an indispensable key to successful drug development and an important part of this book's discussion as well. Drug research by the pharmaceutical and biotechnology industry has been a success story over the past 40 years with hundreds of new products advancing the state of medicine, plus some products each year for previously untreatable or poorly managed diseases. Four constituencies are engaged: (1) patients in meeting unmet medical needs, (2) companies in both financial success and research advances, (3) universities, who obtain research grants, create drug discoveries as well, and conduct much of the clinical and other research leading to product approval, and (4) government regulators, responsible for both public safety and health, industry regulation, and new product approvals. The needs, challenges, and controversies in the industry are also addressed throughout the chapters. This book shares how this success and the challenges are accomplished by the various groups of specialized people, with all the organization requirements, in compliance with the many laws and regulations, and with the many processes and outcomes necessary from each contributing industry department.

This preface and introduction to the book provides a discussion on the needs and use for the book, brief biographies of the editorial board, a brief description of each of the authors, acknowledgments, and a list of key information sources about the industry and related information.

The format is optimized for the education and training of health care professionals, especially fellows (M.D., Pharm.D., and Ph.D.) in training at universities and other new researchers. The format of the book is uniquely geared for the training setting with PowerPoint style slides to summarize the information and give illustrations (that is, tables, lists, and diagrams) and accompanied by detailed narrative descriptions for explanation and elaboration. Industry and research experts (multidisciplinary: M.D., Pharm.D., and Ph.D.) are the editors and authors. An added CD-ROM is available to enhance the utility of the book for course directors, providing them with highly sought after slides to deliver lectures. No single book employs such an educational format for fellows or new researchers and covers the full scope of drug development from discovery through a product's life cycle.

Drug research is a major mission at all medical and pharmacy colleges and medical centers. Collaboration between a drug company and universities in drug research is the typical and indispensable arrangement for drug development, wherein the university provides the patients, research staff, investigators, and expert advisors. At these health science campuses, fellows and residents in medicine (M.D., D.O.), pharmacy fellows (Pharm.D.), and graduate candidates (Ph.D., M.S.N., M.P.H.) strive to understand the drug development process and increase their related research skills. They also desire to improve their collaboration with industry scientists, both in seeking research funding and conducting drug studies. Their university faculty often has limited practical expertise in the breadth and details of work and the nuances of industry-based research, such that they seek outside assistance from the industry to help train the fellows. Also, R&D departments in the industry, as well as their marketing divisions, have new staff entering the industry without formal training in drug development. Job effectiveness is needed as soon as possible; this book and CD-ROM can be part of their education.

The book covers in 12 chapters all the steps in drug and biological product development by a company from discovery to marketing and later life cycle management: Health Care and Industry Overview, Planning and Governance, R&D Outcomes, Discovery, Types of Clinical Studies, Metabolism and Pharmacokinetics, Regulations and Laws, Clinical Operations, Manufacturing and Formulations, Commercialization Division, Medical Affairs and Professional Services, and Special Considerations for Research in five selected therapeutic areas (cardiovascular, infectious diseases, oncology, pediatrics, and psychiatry). Two appendices are included to assist the reader in the jargon common to the industry.

The book includes four components for each chapter: (1) brief introductions for each chapter and major section in a chapter; (2) copies of PowerPoint type figures and tables, including any necessary illustrations, lists, graphics, compilation of terms, and data; (3) narratives accompanying and following each slide that explain and elaborate upon the content on the slides, cover real-life company examples and industry controversies, and include graphs, tables, and figures for illustration of key points; (4) references for further study and resources. A few slides and concepts were repeated in multiple chapters wherein the topic could be addressed from a different and useful perspective for the reader.

The content can be used at two levels. In each chapter, a subset of slides (about 10) can offer an overview on the subject, which is used for more general audiences. Detailed slides (up to 20 to 30 more) are available to fully elucidate the subject for fellows or staff needing more education for their jobs or coursework. In educational terms, the overview would be equivalent to 1 credit hour course (10–12 lecture hours); the total material would be equivalent to a 3–4 credit hour course (30–40 hours of lecture/discussions).

Individual courses in drug development often cover one or two semesters (10–60 lectures or workshops) at many health science universities, for example, Northwestern University, University of Chicago, Campbell University, University of North Carolina, Duke University, University of California at San Diego; New York University, Drexel University, Boston University, Tufts University, Georgetown University, University of Texas, University of Kentucky, and University of Florida. Master's degree or postgraduate programs addressing drug development also are common in fellowship training at medical and pharmacy schools. We hope that this book will support these educational initiatives.

Editors and Authors

The book is authored by a multidisciplinary team of researchers, clinicians, and marketers from industry and the medical/pharmaceutical community. They collectively have done extensive work in drug development, measured by the many product applications prepared and ushered through regulatory authorities, by scope and quality of institutions and companies at which they worked, by number of years devoted to research and product development, and by hundreds of publications. The experience base in research and product development is several hundred years collectively at 30 pharmaceutical and biotechnology companies, plus professorships, teaching positions, and research directorships at 20 health science centers or universities.

Editor

Ronald P. Evens, Pharm.D., FCCP

The editor is Ronald P. Evens, Pharm.D., FCCP, who currently is President of MAPS 4 Biotec, Inc, and Clinical Professor, University of Florida, Jacksonville, FL. Dr. Evens has more than 20 years of industry experience (clinical research, product development, marketing support, and leadership) at Amgen Inc. and Bristol-Myers Company, plus more than 20 years of academic/health science center experience as a professor at seven medical/pharmacy universities in Buffalo, NY; Lexington, KY; San Antonio, TX; Memphis, TN; Los Angeles, CA; Chicago, IL; and Gainesville, FL. His professional record also includes 12 national health care advisory boards, more than 100 publications, 14 book chapters, and many lectures on biotechnology, drug development, and related topics at national professional organizations and universities. Dr. Evens received a B.S. in pharmacy at University of Buffalo, Pharm.D. at University of Kentucky; internship at E.J. Meyer Memorial Hospital; clinical residency at A.B. Chandler Medical Center in Lexington, KY; plus a strategic leadership certificate at Center for Creative Leadership and a marketing certificate at University of Southern California, Marshall School of Business; Fellowship with American College of Clinical Pharmacy.

Editorial Board

The multidisciplinary editorial board has extensive experience (collectively more than 100 years) in industry at major pharmaceutical companies and/or leading universities in research, education, and product development. Their publications number in the hundreds, and the training of fellows and young researchers hallmarks their careers.

Robin Campbell, Ph.D.

CEO, SinusPharma, Thousand Oaks, CA (previously, Vice-president, Oncology, and product team leader at Amgen; manager at Ciba-Geigy; B.S. at University of North Carolina, Chapel Hill, Ph.D. in microbiology and immunology at Wake Forest University).

Joel Covinsky, Pharm.D.

Consultant, Kansas City, MO (previously, Vice-president at Aventis, Hoescht-Marion-Roussell, and Marion-Merrell-Dow in clinical research and product planning; associate professor at University of Missouri at Kansas City Medical School; B.S. in pharmacy at Philadelphia College of Pharmacy and Sciences, Pharm.D. at University of Kentucky, clinical residency at A.B. Chandler Medical Center, Lexington, KY).

Thomas S. Foster, Pharm.D.

Professor, Pharmacy Practice and Anesthesiology, Colleges of Pharmacy and Medicine, University of Kentucky, Lexington, KY (previously, B.S. in pharmacy from University of Buffalo, Pharm.D. at University of Kentucky, and clinical residency at A.B. Chandler Medical Center, Lexington, KY).

Lewis J. Smith, M.D.

Professor of Medicine, Associate Vice-President of Research, College of Medicine, Northwestern University, Chicago, IL (previously, B.S. at City College of New York, M.D. at University of Rochester, and pulmonary fellowship at Boston University).

Salomon A. Stavchansky, Ph.D.

Professor of Pharmaceutics, Alcon Centennial Professor of Pharmacy, College of Pharmacy, The University of Texas at Austin, Austin, TX (previously, B.S. in pharmacy at National University of Mexico, Ph.D. at University of Kentucky).

Chapter/Section Authors

The 24 chapter authors are experienced in the industry as researchers, marketers, and clinicians and/or are experienced university faculty in research, clinical practice, and education. These authors average more than 20 years of experience and many serve as faculty at major universities, which includes fellowship training in drug development. The pharmaceutical and biotechnology companies represented by the authors are substantial in number, diverse, small and large, well-established old guard and start-ups, and influential players in research and product advances (Abbott, Agouron, Amgen, Armour, Aventis, Bristol-Myers, Cato, Ciba-Geigy, Cognetix, Cumberland, Elkins-Sinn, Endo, Immunex, Lederle, Marion-Merrell-Dow, Novartis, Ortho, Pfizer, ProCyte, Radiant, Sankyo Pharma, Schein, Solvay, Squibb, Upjohn, Warner-Lambert, and Xoma). The university experience base includes faculty appointments at 16 prestigious institutions over their careers (alphabetical order), for example, University of Cincinnati, University of Connecticut, Harvard University, University of Kentucky, Mercer University, University of Missouri at Kansas City, University of Nebraska, New Jersey College of Medicine and Dentistry, University of North Carolina at Chapel Hill, Northwestern University, Ohio State University, University of the Pacific, University of Southern California, University of Tennessee, University of Texas, and University of Utah.

Vijay Bhargava, Ph.D. (Metabolism and Pharmacokinetics Chapter)

Vice-president and Global Head of Metabolism and Pharmacokinetics at Novartis Pharmaceutical, East Hanover, NJ (previously, senior director at Aventis and Hoescht-Marion-Roussel; assistant professor at University of Nebraska; B.S. in pharmacy at Bombay University, Ph.D. in Pharmacokinetics and Biopharmaceutics at Virginia Commonwealth University).

John J. Brennan, Ph.D. (Psychiatry)

Group director, Solvay Pharmaceuticals for Clinical development, Medical Affairs, Clinical Pharmacology, and Bioanalytical Chemistry for Men's and Women's Health, Marietta, GA (previously, Squibb Institute for Medical Research, Ortho Pharmaceuticals; B.A. in chemistry at Temple University, Ph.D. in pharmaceutical sciences at Philadelphia College of Pharmacy and Sciences).

Howard A. Burris, III M.D. (Oncology)

Director of Drug Development, The Sarah Cannon Cancer Center, and associate, Tennessee Oncology PLLC, Nashville, TN (previously, B.S. at U.S. Military Academy at West Point, M.D. at University of South Alabama, residency and fellowship at Brooke Army Medical Center).

James F. Cahill, M.S., Ph.D. (Manufacturing and Formulations)

Vice-president, Biopharmaceutical Sciences, Cato Research, Ltd., Richboro, PA (previously, Armour Pharmaceuticals, Rorer Pharmaceuticals, Elkins-Sinn, Inc, International Diagnostics Inc, Princeton Labs; B.S. at Capitol University, M.S. and Ph.D. in microbiology and immunology at University of Utah).

Stephen F. Carroll, Ph.D. (Discovery)

President, Altair BioConsulting, Seattle, WA (previously, Vice-president, Scientific and Product Development, Xoma, LLC; Vice-president, Preclinical Research, Xoma; assistant professor, Harvard University; B.A. in biology at University of California at San Diego, Ph.D. in microbiology at University of California at Los Angeles).

Joel Covinsky, Pharm.D. (Governance and Planning)

Ronald P. Evens, Pharm.D., FCCP (Health Care and Industry Overview; Governance and Planning; R&D Outcomes; Medical Affairs and Professional Services).

Michael W. Jann, Pharm.D. (Psychiatry)

Professor and Chairman, Pharmacy Practice and Pharmaceutical Sciences, Research Director, Mercer University, Atlanta, GA (Previously, B.S. at California State University, Los Angeles, Pharm.D. at University of Southern California, psychiatry residency at University of Tennessee).

Suzanne F. Jones, Pharm.D. (Oncology)

Associate Director, Drug Development, The Sarah Cannon Research Center, associate, Tennessee Oncology, Nashville,

TN (previously, Pharm.D. at University of North Carolina at Chapel Hill, residency and oncology fellowship at University of Texas Health Science Center at San Antonio).

Nancy L. Kercher, M.B.A. (Regulatory Affairs)

President, Strategic Biotechnology Consulting, Seattle, WA (previously, Vice-president, Regulatory, Immunex; positions at Abbott and Upjohn; B.S. in chemistry and M.B.A. at Western Michigan University).

Jeffrey Kluger, M.D. (Cardiovascular Diseases)

Clinical Professor of Medicine, University of Connecticut, and Director, Heart Rhythm Service, Hartford, CT (previously, M.D. at New York Medical College, residency at Beth Israel Hospital of New York, cardiology fellowship at Cornell University Medical Center).

Paul Litka, M.D. (Operations)

Consultant in pharmaceutical industry and clinical research organizations, Salt Lake City, UT (previously, Vice-president, Clinical Drug Development, Anesta Corporation, Senior Vice-president, Clinical Research and Regulatory Affairs, Magainin Pharmaceuticals, and also senior positions at Institute for Biological Research and Development, Inc., SmithKlineBeecham Laboratories, and SmithKlineFrench Labs; B.S. in languages at New York University, M.D. at College of Medicine and Dentistry of New Jersey, residency at St. Michaels Medical Center).

Thomas Lytle, BBA, MBA (Commercial Division)

Senior vice-president, Cytogen, Morrisville, NJ (previously, Vice-president, New Product Marketing and Strategic Planning, Amgen Inc.; vice-president, Lederle Labs and Pfizer, Inc; B.A in marketing at Western Michigan University, M.B.A. at LaSalle University).

Leo Pavliv, R.Ph., M.B.A. (Manufacturing and Formulations)

Vice-president, Operations, Cumberland Pharmaceuticals, Nashville, TN (previously, Cato Research, LLC, Agouron Pharmaceuticals, ProCyte Corporation, Interferon Sciences, Parke-Davis/Warner-Lambert; B.S. in pharmacy and M.B.A. at Rutgers University).

Ralph H. Raasch, Pharm.D., FCCP (Infectious Diseases)

Associate Professor, Department of Pharmacy Practice, School of Pharmacy, Clinical Associate Professor, School of Medicine, University of North Carolina, Chapel Hill, NC (previously, B.S. at University of California at Davis, Pharm.D. and residency at University of California at San Francisco).

Shashank Rohatagi, Ph.D., M.B.A. (Metabolism and Pharmacokinetics)

Director, Sankyo Pharma, Parsippany, NJ (previously, Director in Pharmacokinetics at Aventis; B.S. in pharmacy at Birle Institute of Technology and Sciences, Ph.D. in pharmacokinetics at University of Florida, M.B.A at St. Joseph's University).

Carl L. Roland, Pharm.D. (Study Operations)

Director of Clinical Research, King Pharmaceuticals, Cary, NC (previously, assistant professor at University of Utah, Pharmacotherapy and Pharmacoeconomics Center; Vice-president, Clinical Development, Cognetix, Inc., research positions at Glaxo-Welcome and Anesta Pharmaceuticals; Pharm.D. at University of Illinois, Chicago, IL).

Jun Shi, M.D., M.Sc., FCP (Metabolism and Pharmacokinetics)

Director, Clinical Pharmacology and Drug Dynamics, Forest Laboratories, St. Louis, MO (previously, Director at Aventis; M.D. at Shanghai Medical University, M.Sc. at Southern Methodist University, postdoctorate in pharmacokinetics at University of California at San Francisco, fellow in Clinical Pharmacology).

Lewis Smith, M.D. (Types of Studies)

Jessica Song, M.A., Pharm.D. (Cardiovascular Diseases)

Clinical assistant professor, University of Pacific, and clinician at Santa Clara Valley Medical Center, San Jose, CA (Previously, B.A. in chemistry at University of Washington, M.A. in chemistry at Johns Hopkins University, Pharm.D. at University of California at San Francisco, residency and fellowship in cardiology at University of Connecticut Medical Center).

Roland Garritsen VanderHoop, M.D., Ph.D. (Psychiatry)

Endo Pharmaceuticals, Senior Vice-president, Research & Development & Regulatory Affairs, Chadds Ford, PA (previously, SVP, R&D & Chief Medical Officer, Serologicals Corp.; VP R&D & Chief Scientific Officer, Solvay Pharmaceuticals; B.S., Ph.D. and M.D. at University of Utrecht, The Netherlands).

Elaine S. Waller, Pharm.D., M.B.A. (Regulatory Affairs)

Vice-president, Regulatory Affairs & Quality Assurance, Sonus Pharmaceuticals, Seattle, WA (previously, Chief Operating Officer, Radiant Research; senior positions at Hoechst-Marion-Roussel; associate professor, University of Texas at Austin, College of Pharmacy, and assistant director, Drug Dynamics Institute; B.S. in pharmacy and Pharm.D. at University of Missouri at Kansas City).

Philip D. Walson, M.D. (Pediatrics)

Professor of Pediatrics & Pharmacology, University of Cincinnati, Director of Clinical Pharmacology Division, Cincinnati Children's Medical Center and Chief Scientific Officer of Pediatric Clinical Trials International, Cincinnati, OH (previously, Professor of pediatrics at Ohio State University and Chief of Clinical Pharmacology and Toxicology; Medical Director of Clinical Pharmacology at University of Arizona; B.S. at University of California at Berkley, M.D. at University of California at San Francisco, pediatrics residency at University of Chicago, NIH fellowship in cardiology at University of California at San Francisco).

C. Michael White, Pharm.D, FCCP. (Cardiovascular Diseases)

Associate professor, University of Connecticut, School of Pharmacy, Co-director, Arrythmia & Cardiac Pharmacology & Research Group, Hartford Hospital, Hartford, CT (previously, B.S. in pharmacy and Pharm.D. at Albany College of Pharmacy, fellowship at Hartford Hospital; fellowship with American College of Clinical Pharmacy).

1
Health Care and Industry Overview and Challenges

Ronald P. Evens

Health Care Issues 5
Industry Statistics (The Pharma and Biotech Industries) 11
Research Activities and Costs 14
R&D Productivity 16
Organization of a Pharmaceutical Company 19
R&D Challenges and Issues 24
Summary of Research and Development Approaches 29
References 31

Drug and biological product development is a global, massive, complex enterprise that entails health care systems, disease knowledge, drug knowledge, research experiences (basic and clinical research with many disciplines, technologies, and processes), personnel/professional affairs, business and marketing practices, public relations, legal and regulatory issues, and global business, cultural, and medical factors. This chapter is intended to provide some background context for product development regarding applicable general health care issues, a description of the industry and key statistics, the organization of a pharmaceutical company, and drug and biological product development challenges. Thus, a framework is provided for the following 11 chapters that will discuss all the people, processes, systems, and outcomes for drug and biological development applicable in the United States and in Europe as well.

The four areas covered in this chapter include (1) health care issues (spending, changes over time, utilization, causes of disability, improvements with drugs) that serve as a broad context for new product development; (2) industry statistics regarding drug sales, drug costs, research and development (R&D) costs, research activity, Investigational New Drug applications (INDs), New Drug Approvals (NDAs), new molecular entities (NMEs), time frames and speed to market; (3) the organization of a typical company in the industry (FIPCO; fully integrated pharmaceutical company), especially describing seven divisions (research, clinical development, marketing, sales, medical, manufacturing, and global operations); and (4) research and development (R&D) issues that are challenges to drug development by the industry (e.g., complex milleau of phases and content of R&D, major disease, business and clinical challenges, technology issues, blockbuster discussion, collaborations, and culture).

Health Care Issues

In preparation for the discussion of drug development, an important context is health care delivery and its related costs and benefits, which are key health issues potentially influencing drug development. This section discusses national health care incomes and expenditures, changes over time in them, causes of existing disabilities that create opportunities for improved health care, factors impacting health care utilization (increases and decreases), improvements in health care over the past couple of decades, along with the reasons for these changes, and specific health care factors impacting drug development.

The cost of health care in the United States of America (USA) rose to $1.6 trillion (National Health Expenditures; NHE) by 2002, or about $5,440 per person (Fig. 1.1). This amount consumed about 14.9% of the gross domestic product (GDP) of the economy at that time. The total NHE has been growing by about 5–10% per year since 1995. The NHE

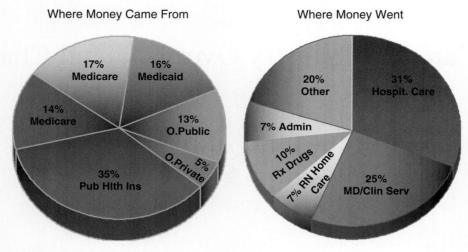

FIG. 1.1. Nation's Health Dollar ($1.6 Trillion, 2002)

increase is driven by changes to two factors: (1) utilization of services and products and (2) medical prices. The percentage contribution of these two factors varies annually. Each factor contributed about 50% to the growth in NHE from 2000 to 2002. Utilization is comprised of usage (e.g., number of visits or products), new technologies, and the mix of services. Figure 1.1 demonstrates that the income sources for the U.S.A. health dollar were private health insurance (36%), major government programs (33%, that is, Medicare, 17%, Medicaid, 16%), other public income, (13%), out-of-pocket payments by patients (14%), and other private sources (5%) in 2002. Other public resources include worker's compensation, Department of Defense, Veterans Administration, Indian Health Service, state and local subsidies, school health, and state children's health insurance program (SCHIP), a Medicaid supplement. SCHIP covered 19.8% of all children in the USA in 1998 for physician, emergency room, and hospital visits plus immunization, and surprisingly 15% of children had no health insurance. Other private sources include philanthropy, private construction, and in-plant industrial construction. Figure 1.1 also displays that the nation's health dollar expenditures included hospital care (31%), physician and clinical services (25%), nursing home care (7%), prescription drugs (10%), program administration and net costs (7%), and other costs (20%) in 2002. Other costs includes dental services, home health, durable medical products, over-the-counter medicines and sundries, public health, research, and construction [1–3].

In 2002, the 10% of U.S. NHE for prescription drug purchases totaled about $162.4 million per CMS and $208 million per industry, which increased to about $248 billion in 2004. Payments for prescription drugs include private health insurance (48%), out-of-pocket payments (OOP) by patients (30%), and government programs (22%). Prescription drugs costs have grown by about 15–16% per year (2000 to 2004), but not at an increasing rate in the past 4 years. The measures that have slowed the increase in spending on prescription drugs include more generic drug use, fewer new drugs in the marketplace, formulary controls (lists of approved products for use), prior authorization policies, special high-technology budgets, and higher tiered copayment growth for patients. Managed care organizations, prescription-based managers, hospital systems, and health insurers have used the aforementioned tools to help control costs [1–3].

Over the last decade (1990 to 2000), health care has changed dramatically in a variety of significant ways that has impact on drug utilization and needs for drug development (Fig. 1.2). The type of health care spending and site of care evolved, with the percentage contribution of hospital care costs falling from 36.5% to 31.7%. For the most part, this change was based on inpatient care as the site of care falling from 76% to 63%, whereas outpatient care conversely rose from 24% to 37%. Hospital admissions fell from 122 per 1,000 in population to 114, while outpatient hospital visits rose about 29% from the 1990s to 2000. Also, the average hospital stay went down from 7 to 5 days, also having a major impact on the percentage of care in hospitals versus outpatients. Prescription drug costs rose from 5.8% to 9.4% of NHE over this decade, whereas physician and clinic percentage contribution to costs were stable at 25.2% to 25%, as a percentage of NHE. Please be reminded, however, that actual health care costs in all three segments rose substantially over this decade (e.g., hospital costs rising by 3–10% per year). Furthermore, the type of payer changed a great deal as well, with private health insurance moving from only 24% to 46% and out-of-pocket payments by patients falling from 60% to 31%. Where the care occurs and who pays for services will be major influences on the type of drugs necessary to meet disease needs, health care system needs, payer coverage, and patient preferences [1–6].

1. Health Care and Industry Overview and Challenges

	1990	2000			
	% NHE				
Hospital care	36.5%	31.7%	Rx Drug Spending by Payer		
Prescription drugs:	5.8%	9.4%		1990	2000
Physicians	25.2%	25%	Insurance	24% to 47%	
Inpatient care:	76%	63%	Public	16% to 22%	
Outpatient care:	24%	37%	OOP	60% to 31%	
Hospital stay :	7	5 days			

FIG. 1.2. U.S. Health Care Spend; Changes '90 to '00
Source: CMS, Office of the Actuary, National Health Statistics Group, June 2002 Ed.

Decreases in HCU:
- Public health advances
- Lower risk factors for disease
- More treatment cures
- Shift in sites of care
- Payer pressure for less cost

Increases in HCU :
- Population growth
- Aging population
- New technology
- New drugs
- More use of existing drugs
- Health policy changes

Either an Increase or Decrease in HCU:
- Supply of health services or products
- Guidelines or consensus documents
- Health insurance coverage changes
- Change in practice patterns
- Change in consumer preferences
- Public or political pressure for change

FIG. 1.3. Health Care Utilization – Forces for Change

Health care utilization is a complex subject influenced by a wide breadth of factors related to disease, health systems, preferences of patients and providers, society, and government. Drug utilization can change health care utilization; however, the converse is true that health care utilization can impact drug use. Some factors can either increase or decrease the use of health services (Fig. 1.3). For example, the supply of health services can go up or down (e.g., number of physicians, hospital beds, or surgery centers), which can change the type or mix of drugs needed. Second, new guideline or consensus documents from health agencies or medical societies may recommend a new standard for more or less services or drugs (e.g., National Institutes of Health [NIH] consensus panel in 1994 recommending both antibiotic and antisecretory drugs for peptic ulcer disease with *H. pylori* infection; consensus conferences in 2003–2004 recommending less use of hormone replacement therapy at menopause because of cancer risks). Third, practice patterns may change (e.g., new diet advances, more outpatient surgery, new care for the elderly, or drug use [antisecretory drugs] instead of surgery [gastric resection] for peptic ulcer care). Fourth, health insurance coverage is a major driver in health-seeking behavior. Workers with family coverage will receive more health care and drug use than the families without health insurance. Fifth, consumer preferences for services may change, perhaps influenced by direct to consumer advertising, or major new advances in therapy, or cultural changes (e.g., cosmetic surgery is much more commonplace, treatment for male impotence, or treatment of anemia in cancer patients). Sixth, public or political pressure may force change in health utilization (e.g., the acquired immune deficiency [AIDs] crisis requiring aggressive novel treatments to be available at an accelerated pace, and public desire for alternative, nonprescription, self-help medicines [herbals, vitamins, and mineral products]) [5].

Forces for change that can decrease health care utilization (Fig. 1.3) are suggested in five areas: (1) public health advances such as immunization or sanitation improvement, reducing infectious diseases (e.g., flu vaccine program, children vaccinations, sewer systems in third world countries); (2) lowering risk factors or more prevention measures, such as smoking cessation programs associated with less lung cancer, cholesterol lowering drug use with aggressive cholesterol targets and less heart disease; (3) new treatments that are cures for disease or radically reduce disease (e.g., new cancer drugs for lymphoma, leukemias, and breast cancer); (4) a shift in the site of patient care (e.g., more assisted living with better access to health services, or more outpatient surgery); and (5) added payer pressures to reduce costs of care through various means (e.g., formularies [fewer and less costly drugs being used], physician referral systems [less specialist care], and negotiated provider and drug acquisition fees [less cost per service or product]).

On the reverse side, reasons for increases in health care utilization (Fig. 1.3) encompass at least six areas: First, population growth, simply more people need more services and products. Second, new technologies for diagnosis or treatment (e.g., MRI for cancer or heart disease detection, cardiac stents for heart blockage). Third, major new policy initiatives (e.g., Medicare coverage for the catastrophic illness of kidney failure and then coverage for anemia care with epoietin products in 1989; Medicare Modernization Act [2004] for prescription drugs for elderly). Fourth, age is another major factor increasing health care utilization in several ways. The over-65-year-old age group has grown from 31 million in 1990 to 34 million in 2000. The aging of the population creates the attendant increases in multiple and more chronic type diseases in patients associated with aging. Another aging factor is the U.S. cultural tendency to use all possible measures to treat older relatives, with dramatic increases of cost of care with age. Therefore, you observe the following health care costs based on age: 30–50 years old, about $1,000 to 3,000 per capita NHE; vs. 60 year olds, $5,000; vs. 70 year olds, $7,500; vs. 80 year olds, $14,000 [3–5].

The fifth and sixth reasons for increased health care utilization deal with more drug use, which was well documented by the Centers for Disease Control 2004 report on health statistics. This report documents visits to physician offices or hospital clinics, and mentions prescription drugs that were ordered, provided, or renewed. Fifth, a greater number of novel drugs available that usually are more costly and used chronically for many diseases (e.g., newer biological products for rheumatoid arthritis, colitis, psoriasis, asthma, and cancers); and more asthma drug use, especially new agents, such as Advair®, combination therapy, and leukotriene inhibitors (e.g., Singulair®). Sixth, a greater use of existing drugs for better disease control (e.g., anticholesterol drugs [Lipitor®, Zocor®, and Crestor®]) being used routinely with standardized guidelines for all providers and insurers. Such medication use in physician visits rose from 16 per 100 population from 1995–1996 to 44 per 100 in 2001–2002, a 175% increase, especially statin drugs. Also for example, more antidepressant drug use occurred with 3% of population receiving a drug in the past month in 1988–1994 increasing to 7% in 1999–2000, and a 75% increase in office visits wherein an antidepressant was mentioned from 1995 to 2000, especially selective serotonin reuptake inhibitors. For physician visits, 62% resulted in at least one drug association. Number of drugs recorded for the visits rose from 1.1 billion in 1995–1996 to 1.5 billion in 2001–2002 [3–5].

Opportunities to create new and better drugs or alternative choices are major drivers for drug development for the many chronic diseases that continue to be treated with only symptom resolution and/or partial control. Such chronic diseases without cures lead to disabilities and higher costs of direct care and indirect costs such as lost wages and productivity. This situation (chronicity and disability) is the norm for most diseases, except for most infectious diseases. Figure 1.4 lists such chronic conditions. The first column is the disease prevalence from the 2002 National Health Interview Survey in U.S. from the Centers for Disease Control. The second and third columns look at the U.S. population over 65 years old; chronic diseases per 100 persons (%) and then hospitalization

Disease	% Pop.^	> 65 yrs % **	> 65 Hospitaliztn*
Back pain	26.4	-	-
Hypertension	21.2	40.3	-
Arthritis	20.9	48.3	8.6
Sinusitis	14.1	-	-
Migraine	15.0	-	-
Heart disease	11.2	30.8	80.4
Asthma	10.6	13.8	-
Ulcers	7.6	-	-
Cancer	7.1	7.4	21.8
Diabetes	6.6	12.6	5.8
Bronchitis, chronic	4.4	(in asthma#)	9.4
Pneumonia	-	-	20.7
Stroke / Cerebrovascular	2.4	7.1	21.5

^ % in National Health Survey for U.S. adults in 2002, CDC, NCHS July 2004 ; ** % in Surveillance for morbidity & mortality among older adults, 1995 - 96, CDC, MMWR 1999; 48 (s508):7-25; * same in 65 y.o., but per 100,000 population

FIG. 1.4. Leading Causes of Disease and Disability

rates per 100,000 in this U.S. population. Middle age group (45–54 years old) also has a major increase in limitations of activities from disease over younger adults (e.g., twofold increase in limited activity with arthritis, lung disease, and mental illness; 4–5 fold increase with diabetes and heart/circulatory problems). It is obvious that disease and disability substantially increases with age. Much need exists for better disease control and improved treatments [6, 7].

Furthermore, science is advancing in identifying additional mechanisms for chronic diseases, creating new targets for drug development (e.g., tumor necrosis factor and interleukins in 2000+ vs. prostaglandins in 1980s in rheumatoid arthritis). Also, chronic diseases are exceptionally prevalent in the population. The top 10 diseases causing significant disability account for about 30 million cases in 1992–1993, and more than 90 million Americans live with chronic illness. About 70% of deaths are attributable to chronic disease. Chronic illness is responsible for 75% of NHE, more than $1.2 trillion. Asthma episodes (annual prevalence) occur at about 40 per 1,000 in population. Prevalence of depression is 6.6% of population in the 2000–2001 time period. High cholesterol (over 240 mg/dL) occurred in 17% of population in the 1999–2000 time frame. Arthritis costs $22 billion in health care costs and $60 billion in lost productivity. Diabetes costs $132 billion in direct and indirect costs. Figure 1.4 presents the 13 leading causes of disability with an approximation of the frequency in 2002 [3, 6, 7].

In two of the top three disability areas, arthritis and back problems, we are treating the signs and symptoms of the diseases (generally without cures), and they are chronic and often progressive in their pathophysiology. Although many treatments already are available, they are all only partially successful in controlling the acute or even chronic manifestations, such that new treatments are desired by providers and patients. Cardiovascular diseases that lead to commonly occurring disability include heart trouble, high blood pressure, and stroke (collectively, millions of cases). Sensory problems of the eyes and ears are common disabilities, 15.3% and 9.3% of the population, respectively. Diabetes mellitus is very common as noted and is increasing in frequency in the population in all age groups significantly, and especially in minorities, African American, Hispanic, and American Indian. The death rate from diabetes has risen from 18.2 to 25.2 per 100,000 population [3, 6, 7].

Progress is being made dramatically in improved health care with a variety of diseases, especially heart diseases, stroke, and cancer, lowering death rates, and increasing life expectancy (Fig. 1.5). A combination of factors is producing these favorable results, such as (1) better public awareness and health-seeking behavior (e.g., immunization programs), (2) earlier diagnosis (e.g., mammography done at earlier age), (3) improved diagnostic techniques (e.g., magnetic resonance imaging [MRIs] for cancer or heart attacks, better simpler blood glucose diagnostic kits), (4) new drug availability (e.g., clot busting agents and new antiplatelet drugs), (5) greater use

	1980	1990	2000	% Change
Disability Rate	26.2	23.5	19.7	25% Decr
Death Rates - Overall (per 100,000)	1039.1	938.7	872.0	16% Decr
Heart Attack →	345.2	249.6	186.9	46% Decr
Stroke →	96.2	65.3	60.8	37% Decr
Breast Cancer →	32.3	31.8	25.4	21% Decr
Life Expectancy (years)	73.7	75.3	76.8	4% Incr

FIG. 1.5. Health Care Improvements

of existing drugs (e.g., statin drugs for cholesterol), (6) new technology (devices) (e.g., implantable defibrillators), and (7) improved surgical techniques (e.g., in stroke, carotid endaterectomy, or early aneurysm surgery) [3, 5, 7, 8].

Health care costs have risen significantly by $2,254 (102%) per person per year over the past 20 years, but major health gains have been achieved during this timeframe. Figure 1.5 suggests that improved health care has benefited several diseases, with overall disability rates decreasing over the past 20 years by about 25%. The overall number of hospital days fell 56% from 129.7 to 56.6 per 100 persons, suggesting, to some extent, better population health. Death rates for heart attacks, stroke, and breast cancer have improved 46%, 37%, and 21%, respectively. All cancer deaths have been reduced by 10% in past 12 years. Overall death rates have been reduced by 16%, and life expectancy has risen by about 4% (3.2 years). Another way to look at health gains is to document the financial benefits (in dollars gained) from disease improvement for each $1 invested in health care. Each dollar invested in therapy of breast cancer is estimated to result in a $4.50 health gain; for stroke, $1.55 gain; for diabetes mellitus, $1.49; and for heart attacks, $1.10—besides the benefits of less trauma and family disruption. These health gains likely relate to improved diagnosis and care, better drug therapies, and better health awareness and preventative care [9].

Science is advancing at an ever increasing pace and changing the face of both the diagnosis and therapy of disease with dramatic new findings, and hence health care advances with it. The best example perhaps is cancer with the associated new benefits of extended life measured in years, but the benefits add major new costs to the health care system. In the 1950s and 1960s, a cancer diagnosis was the death sentence for patients in nearly all diseases, and the therapies were limited primarily to cell poisons, such as antimetabolites and alkylating agents with very debilitating and major life-threatening toxicities. Biology and drug research improved to a point in the 1970s and 1980s to create some new classes of life-extending drugs (e.g., taxanes, platins, topoisomerase inhibitors, and aromatase inhibitors). Now in the 1990s and the dawn of the 21st century,

Targets:
- Oncogenes
- Lymphocyte CD-20
- Tyrosine kinase receptors
- Proteosome inhibitors
- EGFR inhibitors
- VEGF inhibitors/Angiogenesis
- VEGF trap
- Aurora kinase inhibitors
- Cancer cell
- Anti-sense inhibitor
- Apoptosis enhancing
- Ribonuclease (RNA enzyme)
- Endothelin A antagonist

Company:
- Genetech (Herceptin)
- Idec-Biogen (Rituxan)
- Astra-Zeneca (Iressa)
- Millennium (Velcade)
- Imclone (Erbitux), OSI (Tarceva)
- Genentech (Avastin)
- Regeneron (VEGF trap)
- Vertex (VX-680)
- Dendreon (Vaccine)
- Isis (Affinitak)
- OSI (Exisulind)
- Alfacell (Onconase)
- Abbott (Xinlay)

FIG. 1.6. Biotechnology & Cancer – New Science

Patients & Disease:
- Aging population
- Advances in science (Disease biology, Genomics)
- Oncology – chronic disease
- Pharmacoeconomic data demands
- Medication safety focus (IOM, Businesses)

Health Care Delivery:
- Patient care from hospital to ambulatory
- Guidelines for disease & treatment
- Government regulations (R&D & S&M)
- Payers for drugs (Insurance & Medicare)
- Costs for health care & drugs

FIG. 1.7. Trends (2000+) & R & D Impacts

biotechnology is unlocking many secrets of genetics, proteomics, and especially intracellular function, such that new targets and new drugs, as well as whole new categories of therapeutics, are available to support and treat the cancer patient (Fig. 1.6). The complications of cancer and its drug therapies, that is, anemia, neutropenia, and mucositis, can be controlled with protein growth factors. Monoclonal antibodies have been humanized and conjugated; they are now available to attack oncogenes or cellular nucleotide polymorphisms and carry risks of less toxicity than the cell poisons. Vaccines to treat cancer are under study. Aberrant intracellular functions are now discovered as new added mechanisms of cancer cell growth and can be mitigated through these various newly identified mechanisms, such as tyrosine kinase inhibitors, proteasome inhibitors, angiogenesis inhibitors, and ubiquitin alteration, as represented in Figure 1.6.

In summary, what are the health care trends in the 20th century and the birth of the 21st century that have and will impact research and development of drugs and biologicals? We are observing changes in the patient pool, science, health care delivery, and its finances. Ten factors impacting R&D are listed in Figure 1.7 and discussed here [4, 10–14]. (1) The aging population creates a rapidly growing pool of older patients who also are living much longer, especially over the next 30 years as the baby boomers reach 60 years plus of age. Also, these people will retire and seek more active lifestyles and demand a better quality of life than previous generations. The older patient has multiple diseases and often more advanced disease. Prescription drug use in the elderly (65 years old and over) is 84% vs. 35% for 18–44 years old and 62% for 45–64 years old. The larger number of older and more complex patients creates an opportunity for research and new products for the industry.

(2) Sciences of molecular biology, genomics, proteomics, and pharmacogenomics, along with medicinal chemistry, are discovering new disease mechanisms and possibilities for drug intervention. More drug use is occurring in the population in general, especially multidrug use (three or more drugs; e.g., 12% of population in 1988–1994 vs. 17% in 1999–2000). Medical visits with prescriptions for five or more drugs rose from 4.1% in 1995–1996 to 6.7% of population in 2001–2002. Categories for increases in drug use were broad (e.g., drugs for pain, depression, hyperlipidemia, hypertension, asthma, allergies, and diabetes). (3) Oncology therapy has advanced with more and more patients living for years instead of months, because of novel therapies that are more effective and less toxic, along with more supportive protective products. (4) The health care providers and payers want new data for drugs regarding their overall impact on health, quality-of-life, and delivery of care, in addition to safety and efficacy. Pharmacoeconomic studies now need to done routinely by pharmaceutical companies before the health care systems will accept new drugs. (5) Medication safety causing morbidity, lost work, and mortality continues to be a major health issue, especially for adverse events and their prediction, prevention, and management. The Institute of Medicine (IOM) in 1999 raised the public awareness along with business groups, such as the Leapfrog group. The cost of adverse events and medical errors includes an estimated 40,000 to 100,000 deaths per year and a cost of $29 million for health care costs and lost productivity. Adverse drug events have increased, for example, to being responsible for 4.8 emergency room visits per 1,000 persons by 1999–2000, doubling from 1992 to 1993. More safe drugs are needed.

Health care delivery changes also significantly impact R&D for drugs and biological products. (1) The site of care of patients is moving from hospitals to outpatient environments, which changes the types of drugs needed to care for patients. Of course, oral agents are the preferred choice in an ambulatory or home setting, but as many more patients with more serious diseases are treated more often at home, injectible drugs are being used much more frequently. Many examples now exist for such injectible drugs being used for chronic conditions and at home (e.g., Aranesp® and Procrit® for anemia of cancer, beta-interferons [Avonex® and Rebif®] for multiple sclerosis, and Remicade® and Enbrel® for rheumatoid arthritis). (2) and (3) Guidelines for the diagnosis and treatment of diseases in most organ systems are now commonplace. Many groups create such guidelines; for example, government (National Institutes of Health, and Agency for Health Care Research); societies (American Heart Association, American Society of Clinical Oncology); institutions (National Cancer Center Network; and individual university hospitals). A company must keep abreast of all these sources of health care decisions, which can change therapy and drug choices while they are studying a new drug based on prior guidelines in place. (4) Payers' role in health care has grown and changed how health care is delivered and financed. Insurance companies, for example, employ a variety of mechanisms; they use prescription benefits managers to track drug use and even change a physician's drug choices, require health maintenance organizations to deliver care instead of private physicians and offer drug choices at reduced costs, demand more novel products with overall health cost data, and have referral systems and negotiated rates for physician services. Medicare now pays for oral drugs for chronic disease in the elderly population, extending access to drugs but then influencing health delivery and drug choices. (5) The rising cost for health care and especially the cost of drugs at double digit rates can become an impediment to R&D investment, if perceived by the public and government as excessive and not of sufficient value. Companies are and must look at the financial return on R&D investment for new drugs, such that, for example, antibiotics are being developed less often, related to their short-term use and restrictions in use for new, even advanced, drugs. As noted earlier, requirements for pharmacoeconomic research have grown substantially in the past 10–15 years to meet the demands for such data by health systems and to establish the overall value of a new drug.

Industry Statistics (The Pharma and Biotech Industries)

The worldwide pharmaceutical marketplace is composed of four geographic areas; the United States, Europe (European Union), Asia (Japan, China, Australia), and the rest of the world (ROW). In addition, pharmaceutical companies are generally divided into five categories; pharmaceuticals (brand drugs, also known as "ethical" drugs), biotechnology products, generic drugs, over-the-counter (OTC) drugs (nonprescription), and devices. Support companies for the industry exist in many categories as well. Seven categories are suggested as follows; research or discovery technology (e.g., high-throughput screening, genomics, antisense, monoclonal antibodies), venture capital companies (financing support especially for small companies), clinical research organizations (generally operations and management for clinical research), specialty services companies for conduct of clinical trials to supplement company staffing (e.g., statistics, patient recruitment, medical writers, regulatory), medical education and/or communication companies (symposia and educational materials developed and implemented), advertising and/or promotion agencies, market research and marketing data companies, and law offices (patent and regulatory work).

The medical university setting serves as a source of several critical functions and expert individuals, such as basic science laboratories for disease and drug research, medical experts for disease and drug advice, clinical investigators to conduct the studies, health economists to assess a product's humanistic and financial utility, health care systems to understand the product's full impact and use of the product, and access to patients in the hospitals and clinics, all of which need to be used by pharmaceutical and biotechnology companies. The measurement of success of a company involves several sets of statistics that we will review in this chapter, including sales of products, New Drug Applications and approvals, research pipeline, alliances and collaborations, and reputation. Several other factors, especially many financial statistics, are important metrics, but are outside the scope of this book, such as profitability, market capitalization, profits to earnings ratio, which will only be mentioned in context.

Product sales are reported to the investment community and general public on a quarterly basis by companies. Sales figures that are reported usually include total sales of all products and services for the company, sales data for each product, regional/global sales, growth over time (quarter to quarter and year over year), market share within a therapeutic or pharmacologic category, gross margin (sales less all expenses), and profit. For a single product, the term "blockbuster" product usually refers to a sales level of $1 billion per year. Generic drugs are copies of the original patented product that have proven equivalence primarily of ingredients and pharmacokinetic parameters, especially bioavailability.

Sales data for all pharmaceuticals are reviewed for the years of 2001 to 2004, depending on availability of the data (Fig. 1.8). Prescription drug sales were $550 billion (B) worldwide (WW) in 2004, a 7% growth over 2003 ($492

- $ 550 Billion WW (2004)
- $ 248 U.S. (9% of Health spend in 2000 & 2003)
- Blockbusters: # 94 with > $ 1B sales ($186 B)
- Market WW % 2004: US 48%, EU 30%, Jap 11%, Asia/Pac 8%, ROW 4%
- Top 200 products (2004): CV $56B-1st , CNS $59-2nd, Onc $27B-3rd, ID $22-4rd, GI $18B-4th, Resp $17B-5th, MSK $22B-6th
- Generics 50% of all Rxs; 2/3rd market share in 12 mo.s after originator patent loss
- Company growth target: Increases of > 10% /year

Fig. 1.8. Industry Statistics in Sales
Source: © 2005 Thomson Center Watch

Lipitor	11.59	Pfizer/Astel	Diovan/Co-Diovan	3.09	Novartis
Plavix/Iscover	5.39	BMS/S-A	Risperdal	3.05	J&J
Zocor	5.20	Merck	Remicade	2.90	J&J/Sch-Pl
Advair/Seretide	4.50	GSK	Cozaar/Hyzaar	2.82	Merck
Norvasc	4.46	Pfizer	Neurontin	2.72	Pfizer
Zyprexa	4.42	Lilly	Pravachol	2.64	BMS
Nexium	3.88	AstraZen	Singulair	2.62	Merck
Procrit/Eprex	3.59	J&J	Rituxan/MabThera	2.62	Roche/Gen
Novo-Insulins	3.43	NovoNord	Epogen	2.60	Amgen
Zoloft	3.36	Pfizer	Prevacid	2.59	Tap
Effexor	3.35	Wyeth	Enbrel	2.58	Amgen/Wy
Celebrex	3.30	Pfizer	Aranesp	2.47	Amgen
Fosfamax	3.16	Merck	Lovenox/Clexane	2.37	San/Aven
Source: Med Ad News May, 2005			Duragesic	2.08	J&J

FIG. 1.9. Blockbuster Products (04-94 BBs = $186 Billion)

- Pharma: Pfizer, GSK, Sanofi-Aventis, J&J, Merck, Novartis, AstraZen, Roche, BMS, Wyeth, Abbott, Lilly

- Biotech: Amgen, NovoNordisk, Genentech (Roche), Serono, Biogen-Idec, Genzyme, Chiron, Gilead, MedImm

- Industry goal = Blockbuster model 1-2 per year

- Product development costs: over $800 MM per product

- Merger mania: Pfz-WL-PD-Phca-Upj; Glaxo-Wlc-SK-Bech; Amg-Immun-Tularik-Synergen; MMD-Aven-Sanof-Synth.; Sandos-Ciba-Geigy (Novartis); Yamanouchi-Fujisawa (Astellas)

- Alliances & Licensing for technology, molecules & products (Pharma-Biotech = 383; Bio-Bio = 435 in 2003)

FIG. 1.10. Statistics – Top Co. (Sales & Research)
Source: Pharmaceutical Executive 2005; Med Ad News 2005; Ernst & Young 2004

billion). About $248 B occurred in the USA, 48% of the WW sales; Europe had 28% of WW sales, $144 B, and Japan was $58 B (11%). This sales growth has been much lower since 2000 (about 5.3%) vs. double digits 11–13% over the 1980s and 1990s. An example of a desirable annual target for growth in sales by a company is about 10%. When a product goes off patent, generic products will be substituted for 55% of the prescriptions for that product within 1 year and will be two-thirds of the sales market share (85% for Blockbuster drugs). Over 4 years (2005–2008), 17 BBs will lose their patent protection. Generic drugs accounted for over $40 billion in worldwide sales ($17.1 B in USA) and 50% of all prescriptions in 2002. Over-the-counter drug sales were $30 billion in 2002. Medical devices is yet another major health cost reaching $143.2 B in 2003 (U.S. $63.2 B), led by Johnson & Johnson with about $15 B [15–23].

Blockbuster (BB) drugs in 2004 included 94 worldwide accounting for $186 B; these 94 products were 33.8% of total sales (Fig. 1.9). The top therapeutic category in 2004 for the top 200 worldwide products was cardiovascular products ($56 B, 14 products), followed closely by central nervous system ($45 B, 17 products), oncology ($26 B, 8 products), infectious diseases ($22 B, 11 products), gastrointestinal diseases ($18 B, 8 products), and respiratory areas ($17 B, 7 products) (Fig. 1.8). The top companies (worldwide) in the marketing of BBs were three European and three U.S. companies; GlaxoSmithKline (GSK, 12 BBs), Pfizer (10 BBs), Sanofi-Aventis (S-A, 9 BBs), Johnson & Johnson (J&J, 8 BBs), Merck (6 BBs), and AstraZeneca (AZ, 6 BBs) (Fig. 1.10). On a smaller but significant scale, the top biotechnology companies were Amgen (5 BBs), NovoNordisk, (2 BBs), Genentech (1 BB), Serono (1 BB), Genzyme (1 BB), and Biogen-Idec (1 BB). Hematological and diabetes products led the biotechnology areas, for example, hematopoiesis (six products, $10.37 B), diabetes (five+, $6.57 B), inflammation (three products, $6.33 B), multiple sclerosis (four products, $5.34 B), cancer (3 products, $4.24), hepatitis/cancer (four products, $3.03), myelopoiesis (three products, $2.92 B), and growth hormones (five products, $1.92). Biological products were significant also for Lilly, Johnson & Johnson, and Roche [17, 18, 23].

The top 10 pharmaceutical companies are listed in the next table (Fig. 1.10); their sales were $240 billion in 2004, 44% of all company sales worldwide. The top 11 leading companies with over $10 billion in worldwide sales were Pfizer at $46 B, GlaxoSmithKline at $31 B, Sanofi-Aventis at $32 B, Johnson & Johnson at $22 B, Merck at $21 B, Novartis at $18

B, Roche at $17.3 B, Bristol-Myers-Squib (BMS) at $15 B, Wyeth at $14 B, Abbott at $14 B, and Lilly at $13 B. The top Japanese companies were Takeda ($6.3 B), Astellas ($6.9 B, Fujisawa-Yamanouchi), and Sankyo ($2.9 B). The most successful biotechnology company was Amgen at $11 B in sales and five BBs, followed by NovoNordisk at $4.85 B, Genentech at $4.6 B (a division of Roche), Serono at $2.5 B, Genzyme at $2.2 B, Biogen-Idec at $2.2 B, and Gilead at $1.2 B. The top biotechnology products (blockbusters = 19) yielded sales of about $34 B WW in 2004.

Collectively, pharmaceutical companies spent $74.8 billion worldwide in 2004 on research and development of products, 19.4% of gross sales (U.S. PhRMA was $38.8 B). The public biotechnology companies spent $16 B on research and development, 34.4% of revenues. The cost of product development in the industry, now at $800 to 900 million per product, has grown substantially over the past several decades; 1970s, $138; 1980s, $350; early 1990s, $500; late 1990s, $800+ millions. These figures were generated by independent research organizations, Boston Consulting Group (BCG) and the Tufts Center for the Study of Drug Development (T-CSDD). Data were real costs from pharmaceutical companies, plus widely accepted economic calculations for after tax cost of R&D and the opportunity costs of capital. Other groups have used different assumptions, including the OTA, both corroborating and challenging the above costs (e.g., average costs of $137 million in 1976, $149 & $173 million in 1987, $293 & $445 million in 1990, all in 2000 U.S. dollars). A low of $110 million for 1991 is suggested by Public Citizen, a consumer group, but their assumptions were very limited and in conflict with Office of Technology Assistance (OTA), BCG, T-CSDD, and others. The cost of postmarketing clinical trials, which are commonly required by the FDA or needed to understand the full use and safety of a product in more traditional settings, adds about another $90–100 million to product development costs [13, 17, 18, 22–25].

A key goal of the each company in the industry is to launch annually at least one new product that will be a blockbuster product within 5 years of its approval. This cost of research for a new product, estimated to be $800–$990 million, necessitates quite large R&D budgets, fosters the need to launch blockbuster products to meet the financial expectations of the investment community, and creates a drive for operational efficiency and synergies in both the research and the sales areas. These three reasons also are three primary reasons for consolidation in the pharmaceutical industry over the past 10 years. Warner-Lambert acquired Parke-Davis, while Upjohn and Pharmacia combined; Pfizer consumed all four companies in a mega merger in the industry. Other merger or acquisitions were Glaxo – BurroughsWelcome – SmithKline – Beecham (now GlaxoSmithKline); Marion – Merrell-Dow (Aventis) – Sanofi – Synthelabo (now Sanofi-Aventis); Sandoz – Ciba Geigy (now Novartis), and Fujisawa–Yamanouchi (now Astellas). Biotechnology companies also are acquiring other biotech companies to achieve the same kind of synergies (e.g., Amgen–Synergen–Tularik–Immunex, now Amgen; and Biogen–IDEC. Besides the full incorporation of one company into another, alliances between separate companies are a necessity for successful R&D as well. One company cannot have all the expertise and resources to cover all the basic science areas germaine to their therapeutic areas of interest. Such that, one company will have access to a particular added technology or product, which is shared through alliances and licensing deals. For example, monoclonal antibody expertise is found especially with Abgenix, PDL, and Immunomedix companies, who collaborate with many pharmaceutical and biotechnology companies. In 2003, over 800 such collaborations were signed for the pharma to bio and the bio to bio agreements. Also, in the 2003–2004 time frame, 14 research partnerships were created between pharma and biotechnology companies that could be worth up to $100 to $535 million each if the research and marketing milestones are met [17–22].

Besides success in gross sales and new product approvals, a host of factors are used to measure the success of pharmaceutical companies as represented by the 15 parameters in Fig. 1.11. A trade journal for the industry is *Pharmaceutical Executive*, which performs an annual assessment for the top company performers using these standard business operating parameters that are heavily focused on financial issues; that is, sales, earnings, profits, revenues, assets, and equity in various combinations. Ratios among these parameters are a key focus (e.g., earnings per share, profits to assets). Just few key nonfinancial factors are incorporated, such as contribution of new products, brand power (the value of a company's name and the product names), and enterprise value (overall company operations, productivity, profitability, reputation, and sales success). Companies are ranked for each parameter, which are then integrated. The top 2003 industry performers in order were GlaxoSmithKline, Merck, and Johnson & Johnson, based on these 15 criteria; they also were in the top five companies in prescription product sales in 2004. In 2003, two relative

1. GSK	5. Forest	9. Genentech	13. Abbott
2. Merck	6. AZ	10. Pfizer	14. Biogen-Idec
3. J&J	7. Lilly	11. BMS	15. Wyeth
4. Amgen	8. Novartis	12. Aventis	16. SchPlgh

Metrics:
- Enterprise value to sales
- R&D spend to sales
- Earnings per share
- Price to earnings
- Sales growth
- Gross margin
- Percent of revenues from new products
- Profit to sales
- Sales to as sets
- Profit to assets
- Net profits to net worth
- Domestic sales per rep
- Sales per employee
- Brand power : knowledge capital equity
- Return on shareholder equity

FIG. 1.11. Top Industry Performers (2003)

newcomers broke into the top five; that is, Amgen and Forest Labs. Also, biotech was well represented for the first time in the financial success assessments by three companies, Amgen (at #4 ranking), Genentech (#9) and Biogen-Idec (#14) [18].

Research Activities and Costs

Important general information regarding research is included in this section, especially research activity by the number of molecules and therapeutic areas and costs by stage of research, company, and changes over time. Worldwide, pharmaceutical research has reached a cost of $67.9 billion (52% spent in the USA), of which clinical research was about $47 B. To put these statistics into perspective, all government R&D spending for clinical research in the USA was $26 B vs. $31 B by the industry in one report [22].

The basic goal of the research division for a new product is to try to create a novel compound with a competitive advantage over existing products, in regard to mechanism of action, site of action, efficacy, safety, dosing schedule, formulation, administration, or convenience of use by providers and patients (Fig. 1.12). The commonly held rule of thumb for a success rate in product development from research molecules to approved products is a story of very heavy attrition and follows: 5,000 compounds in research, to 500 in preclinical study, to 50 into Investigational New Drug applications (INDs) and early clinical research, to 5 into late-stage clinical research, and then only 1 product approved. The 2003 research activity is shown on Fig. 1.12; 10,000 new projects in the laboratory, 2,100 new candidates, 100 New Drug Applications (NDAs), and about 15–30 new molecular entities (NMEs). For the about 10,000 active INDs that existed in 2001, the figure gives us an idea of future areas for potential products, being led by CNS area, oncology second, and immunology with infectious disease (ID) third, followed by metabolism and endocrinology fourth, inflammation and analgesia fifth, cardiovascular (CV)/renal ranked down to sixth, gastrointestinal (GI)/coagulation seventh, and urology/reproduction eighth. Biotechnology companies have become major drivers of research in collaboration with pharmaceutical companies; the 4,400 worldwide companies have thousands of research projects and from 300 to 600 products in clinical trials [22, 27–29].

Research (basic and clinical) costs are substantial and increasing in the pharmaceutical industry. About 18% of gross sales are invested in research and development, and it has grown by almost 100% in a short 5-year period from an industry total of $21 billion to about $40 billion in 2004 (Fig. 1.13). The number one provider of clinical research investment in the USA is the pharmaceutical industry, almost triple the National Institutes of Health (NIH) budget. Clinical spending rises as you proceed from phase 1 to 3, directly related to the size of the studies and their greater diagnostic and monitoring complexity. Also, an estimate of the growth in clinical research costs is about 10% per year. To give you another perspective on the costs for clinical research, a per patient fee is provided to an institution/investigator for conducting a clinical trial, and it was $6,716.00 per patient across the industry in 2002. On top of this university grant support, the cost of the company efforts per patient well exceeds this figure to pay for the work of its research staff, the clinical managers, study monitors, statisticians, data managers, regulators, auditors, and others. Most companies employ clinical research organizations (CROs) to perform a large percentage of this workload (in 2003, companies spent $10.4 B with CROs, about 30% of R&D budget). The total cost to develop a pharmaceutical product by the industry has been calculated to have risen to about $900 million per successful drug approval. This cost has risen by more than 200% over the past decade, related to increased regulatory hurdles, greater patient study sizes, more complex disease diagnostics and assessments, the cost of product failures, postmarketing research costs, and inflation in health care costs [4, 15, 21, 22, 30].

- **Research:**
  - WW Projects: 10,000 + in 2003 ➡ INDs: 2100 + /yr ➡ NDAs: 100 + /yr (>60% approved) ➡ NMEs: 15-30/yr

- **INDs & Therapeutic Areas :**
  - #9, 704 INDs (# active in 2001)
  - CNS #1, Oncol #2, ID/Immun #3, Metab/Endocrin #4, Inflam/Analg #5, Viral #4, Radiolog #5, CV/Renal #6, GI/Coag #7, Urol/Repro #8

- **Biotech:**
  - 4400 companies (WW) in research
  - over 300 products in clinical trials

FIG. 1.12. Research Activity

- Clinical Research budgets WW / U.S.:
  - 1998 $39B / 21B; $ 47B / 33B in 2003 (18% of sales in U.S.)

- Grants:
  - Clinical grants: $10 Bin 2003
  - All R&D: $31B (industry) vs $26B (government)

- Spend for clinical phases: ($33B total in U.S.)
  - P.I - 11%; P.II - 22%, P.III - 48%, P.IV - 19%
  - Growth: >10% per year

- Costs per patient: $6, 500 GI to $9, 800 oncology in 2003

- Drug development cost increases:
  - '92 - $259MM, ➚ '96 - $499, ➚ '99 - $635,
  - ➚ '01 - $800+, ➚ '04 – $900+

FIG. 1.13. Research Activity & Costs

The total R&D spend by the top 10 drug companies in 2004 was $49.1 billion, led by Sanofi-Aventis at $9.31 B with over $2 B for the 10th company (Fig. 1.14). For the top 50 worldwide companies, R&D spend as percentage of total sales was 19.4%, ranging from 12.2% to 31.2% excluding the generic companies who primarily perform pharmacokinetic equivalence studies and few clinical trials. These top 10 companies equal about two-thirds of the total R&D spend of the industry. Biotech spending is rising as well, with the top 10 reaching over $5 B in 2004, which is dwarfed by the top pharma companies in dollars, but represents more than 25% of biotech company revenues. Amgen leads biotechnology segment by far in R&D investment, doubling the second company, Genentech [15, 17, 19, 21, 30].

Another look at research spending is across all the stages of R&D, as shown on Figure 1.15. In 2002, R&D spending also can be broken down into three general segments; nonclinical drug work at 21.4%, animal testing at 16.2%, and clinical trial costs at 35.9%. The nonclinical work is comprised of four components: (1) the laboratory efforts in the synthesis and extraction of the drug, (2) the creation of the product formulation (and testing of its viability and practicality), that is, tablet vs. liquid vs. injection, (3) process development, which is the work on manufacturing of the drug, and (4) all the testing of the processes and interim product at various stages of its evolution. Animal testing includes primarily pharmacology and toxicology studies with some pharmacokinetics efforts. The major cost area is clinical testing as shown at 35%. It is worth noting that the regulatory process of preparing both the IND and NDA at the company appears to be only 3%, but the real dollar number is about $24 million, a significant expense for meeting the regulatory application processes [22].

A representative research pipeline at a major pharmaceutical company is provided on Figure 1.16 for Novartis in 2003. Dr. Garaud presented these statistics at a conference in February 2004 for all the projects at four stages from preclinical to their phase 3 and regulatory submission stage. The table of 125 projects includes some drugs being studied for multiple indications; please be reminded that two different indications are potentially two different NDAs. Their business including research planning (product portfolio) is divided into eight distinct business units, and also two major areas, first into primary care areas and second into specialty therapeutics. Five areas exist for primary care therapeutics for Novartis and three specialty areas. Several criteria are important in establishing a successful and robust pipeline, some of which are demonstrated on this slide: product candidates in all four stages of development; a reasonable number of product candidates at each of these four stages, especially given that most will ultimately fail and not be carried forward; a reasonable number and balance of product candidates across the eight business/research areas; and therapeutic areas being represented with important patient care and significant business opportunities (sales potential). Other major criteria for a robust pipeline, not represented on the slide, are blockbuster potential for several candidates,

Pharma	$ Billions	% Sales	Biotech	$ Millions	% Sales
Sanofi-Aven.	9.31	29.3	Amgen	2,028	19.2
Pfizer	7.52	16.3	Genentech	948	20.5
Roche	5.40	31.2	NovoNord.	664('03)	18.9
Johnson & J	5.20	23.5	Biogen-Idec	688	31.1
GlaxoSK	5.20	16.6	Serono	595	24.2
Merck	4.01	18.7	Chiron	431.1	25.0
AstraZen.	3.80	17.7	MedImmun.	403	35.3
Novartis	3.48	18.8	Millenium	403	89.9
Lilly	2.69	20.6	Genzyme	392	17.8
BristolMS	2.50	16.1	Cephalon	274	27.0

FIG. 1.14. R&D Spend –Top 10 Companies (2004)

Phase	% Spend	Phase	% Spend
Synthesis & Extraction	9.3	Bioavailability	1.5
Formulation & Stability	5.1	Regulatory IND & NDA	3.0
Process devel. & Qual. Control	7.0	Clinical Phases 1, 2, 3	25.6
Toxicology & Safety testing	4.1	Clinical Phase 4	8.8
Biolog. Screen & Pharmacol.	12.1	Other	12.6

FIG. 1.15. R&D Spend by Research Phase (2002) (Adapted with permission from Thomson CenterWatch. Boston, MA. From An Evolution in Industry 4th Ed. 2003 Lamberti MJ Ed. Graph – Distribution of U.S. R&D spending, 2000. Pg 59.)

Bus. Un.	Bus. Franchise	Precl.	P.1	P.2	P.3/Reg	Total
1* Care	Nerv. System	5	3	6	5	19
1* Care	Cardiov./Metab.	10	0	3	6	19
1* Care	RA/Bone/GI/HRT	9	3	7	6	25
1* Care	Anti-Infectives	2	1	1	1	5
1* Care	Respirat./Derm.	5	3	5	4	17
SP-Onc.	Oncology	7	5	7	4	23
SP-Tnpl.	Transplatation	6	0	0	3	9
SP-Oph.	Ophthalmic	3	0	2	3	8
	Total Projects :	47	15	31	32	125

FIG. 1.16. Pipeline (Projects) – Major Pharma Company
Source: Garaud J-J, Novartis, R&D Directions Conference, Feb. 2004

unmet medical needs being addressed (more provider, payer, and health system acceptance), early stage research projects that are sufficient in number and fit their eight research areas, novelty in mechanism of action and competitive edge for as many products as possible in the pipeline, and best in class potential for some candidates. Another approach in research pipelines at many companies is a business unit that is an exploratory area for the company. It does not really fit into their designated therapeutic areas but has some very interesting science in important therapeutic areas for patient need and market opportunity. This area later may become a major company research area, but the company is keeping their opportunities open on a small scale, and often it is done through research collaborations with small companies who are expert with the new technology or with a university laboratory [31].

R&D Productivity

The next two figures present the research and product productivity for the industry from 1995 to 2003, based mostly on FDA statistics for major regulatory milestones. Research spending by the industry according to Pharmaceutical Research and Manufacturers Association (PhRMA) statistics has grown quite dramatically over the past 5 years (almost 100%) from $21 billion to about $40 billion worldwide. However, the number of new drugs reaching the public has not kept pace with this massive growth in research investment [15, 17].

First, we address in Figure 1.17 how many molecules (product leads) are found in the research stage from 1995 to 2003. We also list the number of INDs at the FDA for the same time period. The molecules in basic research have increased steadily and fairly well from under 5,000 to more than 9,000 in this 5-year period. Even though you would expect the INDs to increase coming from the growing number of leads (molecules), the INDs submitted to the FDA however have remained almost the same, hovering around 2,000 per year [22, 32].

Second, we display the success rates for product approvals through the FDA in the USA by documenting the NDAs submitted for 1994 up to 2004 vs. the NDA approvals and also how many of the approvals were NMEs (new medical entities) (Fig. 1.18). An NDA is not only for a new product being approved and now available for use but also includes new products, new indications for approved products, and new

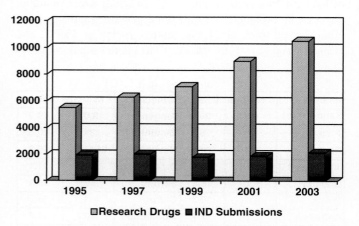

Fig. 1.17. R&D Productivity: Research Drugs & INDs (Copyright 2005, Thomson CenterWatch)

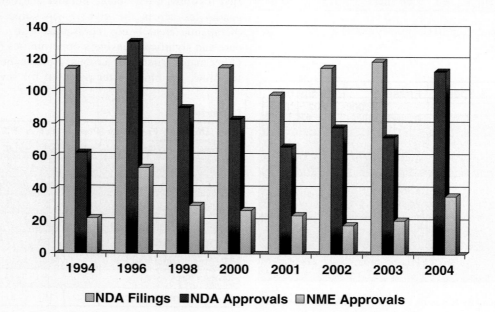

Fig. 1.18. R&D Productivity: NDAs, NDAs & NMEs (Copyright 2005, Thomson CenterWatch)
Source: © 2005 Thomson Center Watch

formulations. Generic product approvals are a separate statistic. Again, we can observe that the added research investment has yet to pay off for the public and the industry. NDA filings reached a high above 120 for two years in the late 1990s, but they were at the rate of 90–110 annually for the following 4 years. The year 2004 is an encouraging sign with an increase in NDA approvals. NDA approvals followed a similar pattern, from more than 100 down to 65–75 in any one year. NMEs similarly changed from 30–40 down to about 20 in any one year. Success rates for product approvals by the FDA were studied. For an NME (drug), only about 15–20% of them were successfully approved; the rate for individual biologicals was 30–40%, according to work at the Tufts Center for the Study of Drug Development. The therapeutic category of an NME can impact success rates, for example, infectious disease, 28.1% (highest); cancer, 15.8%; immunology, 15.4%; and CNS, 14.5%, in the estimates for products approved for 1983–1992 in one study [10, 22, 26, 32].

The research activity for individual companies is presented in the next table for the years 2000, 2002, and 2004 (Fig. 1.19). The top 12 companies had 1,105 product candidates in late research (usually indicating preclinical research) or in development, which increased by 13% to 1,250 product candidates in 2002 and also 2004. The range in the number of product candidates per company was wide, even for the top 10 companies, from 56 to 188 in 2004. GlaxoSmithKline with 188 candidates led the pack by a large margin, followed by Pfizer, Johnson & Johnson, and Aventis, which is not surprising given their high R&D budgets noted earlier in the chapter. Consolidation in the industry alters this line-up with Pfizer adding in Pharmacia and its candidates to its portfolio for 2002 and Aventis being added with Sanofi-Synthelabo for 2004 and forward in time. Usually, companies report the product candidates licensed in from other small companies, such as the biotechnology companies, along with their own discovered molecules, in such figures [22].

Figure 1.20 lists more than 1,800 molecules in clinical research pipeline, in phases 1 through 3 or under consideration for approval by regulatory authorities in 2004, for many diseases responsible for much morbidity and mortality in the world. Cancer was and is the predominant disease category for new products, well exceeding (twofold) the next categories, infectious diseases, cardiovascular/heart diseases, and collectively, central nervous system/depression/migraine/Alzheimer disease. This high number of potential cancer products should be no surprise in 2004, given the number of diseases in the cancer area, the mortal consequences of them, the excellent payer environment in oncology, the many discoveries for cancer mechanisms in the past decade, the engagement of most pharmaceutical and many biotechnology companies, as well as the National Cancer Institute, and the dramatic benefits now occurring from drug therapy, cures and longer life with better quality of life.

The time frame for product development is lengthy, often requiring 4 to 5 years for research and preclinical workup, followed by 4 to 5 more years or more for all the clinical development work, and finishing with an FDA review process that can require as little 3–4 months but can take up to 2 years or more (Fig. 1.21). The total time for the clinical phases of development is now very similar for biologicals and drugs, 6.1 years vs. 6.8 years, respectively. The average FDA review time for a standard NDA in the USA is now down to about 12 months for all drugs (average of 1.6 months) and for priority reviews at 7 months. Regulatory authorities around the world vary in the time necessary for review and approval (U.S. NME review time of 384 days); Europe and Japan with more time for their product reviews (460 and 508 days, respectively). In the USA, laws and regulations have been promulgated to strengthen the review process and shorten the

Company	2000	2002	2004
GlaxoSmithKline	194	204	188
Pfizer	93	108	154
Johnson & Johnson	90	109	134
Aventis (pre-merger with Sanofi-Synth)	142	128	123
Novartis	96	92	109
Bristol-Myers Squibb	62	108	94
Merck	72	87	93
Wyeth	91	95	84
Abbott	72	138	79
Lilly	79	64	76
Roche	86	88	60
Sanofi-Synthelabo (pre-merger w Aventis)	28	29	56

FIG. 1.19. R&D Productivity: Molecules in clinical trial (Copyright 2005, Thomson CenterWatch)
Source: © 2002 Thomson Center Watch all right reserved

Diseases	Products/Uses	Diseases	Products/Uses
Alzheimers	33	Heart disease	73
Arthritis	60	Infectious disease	234
Cancer	478	Migraine	15
Cardiovascular	122	Ophthalmology	32
Central nerv. system	153	Osteoporosis	22
Depression	30	Pain	60
Diabetes mellitus	63	Respiratory disease	78
Digestive diseases	74	Other diseases	251
Totals: Phase 1-620, Phase 2 - 687, Phase 3 - 404, NDA awaiting - 167			

Med Ad News July 2004 Supplement, 35-74.

FIG. 1.20. Product Pipeline (Adapted with permission from Engel Publishing Partners, Newtown Square, PA 2004. From Med Ad News July 2004 Suppl. 35-74.)

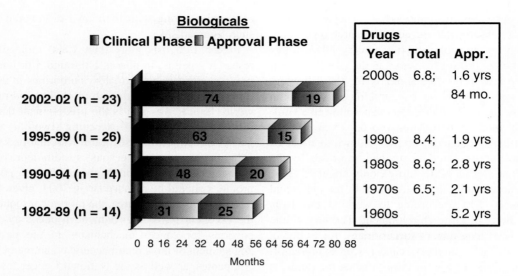

FIG. 1.21. R&D Productivity: Development Times (Outlook 2005. Boston: Tufts Center for the Study of Drug Development, 2005, http://C500.Tufts.edu/InfoServices/OutlookPDFs/OUtlook2005.pdf)
Source: Kaitin KI. Tufts Center for the Study of Drug Development, 2003 & 2004

review time. PDUFA I (1992), II (1997), and III (2002) are laws providing income to the FDA for more reviewers by requiring application fees from the pharmaceutical industry; about $150,000 would be a typical fee. The FDA is obligated to provide a 1-year turnaround time for the review time and an official response regarding approval status. The figure demonstrates for both biologicals and drugs the FDA success in reducing review times in half. Also, in response to the high mortality in the AIDS crisis and with cancer along with their acute need for novel drugs, a fast-track status was created and can be given to a product for life-threatening diseases. This status creates a 6-month or less review period for a product and has worked well for patients, the FDA, and companies. Clinical development times for fast-track drugs were about 4 years vs. 5–6 years for standard drugs, and review times were 6 months vs. over 1 year, respectively, in a Tufts CSDD report. In return for the fast-track status, companies have been required to perform more postmarketing clinical trials to further study safety especially and establish efficacy further with more data, which also adds to cost (about $90 million). The total number of postapproval trials committed to be done in FDA statistics has increased from about 130 in the 1992–1994 time period to 170 in 1995–1997 to 230 in 1998–2000.

Further FDA regulations were promulgated to allow for more official meetings between the FDA and companies at major milestones (e.g., pre-IND and end of phase 2), to clarify research and regulatory requirements. Therefore, NDAs would contain the best possible trials and information to facilitate the review and approval process. However, although review times with the FDA were reduced significantly and the new review time commitments were generally met, the overall time for clinical development rose for biological products to equal the times for drugs (now about 7 years), mostly because much more clinical work needed to be done especially for biological products (Fig. 1.21). The added clinical work for both drugs and biologicals includes more trials, with larger trials (higher number of patients), over a longer time period, being done with greater workup of patients (more procedures), and with more complex and expensive monitoring. For example, procedures per patient numbered 100 in 1992 and in 2002 were 153. The number of studies for an NDA is substantial (37 in total: 21 in phase I, 6 in phase II, and 10 in phase III/IIIB). The number of patients and study sites per trial is also substantial (phase I, 33 patients/trial at 2.4 sites; phase II, 133 patients at 14.1 sites; and phase III, 1,367 patients at 110 sites). From the 1960s to 2000, the time for clinical work increased to 6 years (50% increase for drugs and over 100% for biologicals); however, for drugs, the 2000s showed a decrease of more than 1 year for clinical work; perhaps, it is some evidence of efficiencies effected by the industry and the better working relationship with the FDA. Clinical costs are $250 to 500 million out of the $800–$900 million for drug development. Growth in this clinical work has greatly increased costs of product development (e.g., estimated clinical cost increases from $106 million in 1991 to $467 in 2000) [10–14, 22, 25, 27, 32, 33].

Speed to market is a key measure of operational and regulatory efficiency of pharmaceutical companies, and the regulatory authorities as well (Fig. 1.22). Every day extra needed for research and regulatory review time is a day lost on the patent life of a product, a day lost for improved patient care opportunity, and a day lost of sales for a company (an average of $1.3 million per day for all products, and up to $11 million per day for blockbusters). The time for exclusivity of product

- **Industry metrics:**
  - Ave. Time: 5.7 yrs ('81-'99), 6.6 yrs ('85-'01)
  - Range: 4.4 to 9.7 yrs
  - Cost of lost sales = $1.3MM/day (up to > $11MM)
- **Top 10 companies:**
  - Schering - Plough: 4.6 yrs
  - Roche, Merck, Novartis, J&J, Pharmacia: 5-5.6 yrs
  - AstraZeneca, Aventis, GSK, BMS: 6-6.6 yrs
- **Disease categories:**
  - ID: 4.8 yrs, Anal. 4.8, Resp. 6.2, CV: 6.5, Endo. 7.8, Oncol. 7.9, CNS: 8.9, GI 9.7 yrs
- **Best practices:**
  - Global PPM
  - Realistic protocols
  - Collaboration with regulatory authorities
  - Technology for project planning & communication
  - Project team operations

FIG. 1.22. R&D Productivity: Speed to Market Times
Source: Getz KA, deBruin A. Pharm Exec July 2000; Center Watch. State of the Clinical Trials Industry 5th Ed. Thomson Pub. 2005 (data 1981-1999)

manufacturing and sales by the drug's originator company is often short after product approval. Most drug patents have only about 5 years left after approval, although a patent exists for 17 years (20 years with North American Trade Agreement; NAFTA). This situation relates to patenting of a drug during the early research stage and the long time frame for R&D, which uses up the patent life before approval.

Figure 1.22 presents statistics for clinical development times for all product approvals during the 1980s, 1990s, and 2000s. The overall average time was about 6 years to perform all the clinical research studies but ranged widely from 4 to 10 years. The therapeutic category for a potential product dramatically influences the time for clinical research, with the shortest being infectious disease at 4.8 years, related to the simpler studies and the longest for gastrointestinal (GI) and central nervous system (CNS) categories at 8–9 years. Oncology products are being studied over shorter times, as the fast-track status for many of the newer products in the 2000s has become the norm. Individual companies benchmark each other with this statistic, addressing their relative efficiency. Schering-Plough ranked at the top with a 4.6 years average for their products with all the top companies at or under about 6 years, based on 1996 and 2001 data. In an earlier publication, AstraZeneca ranked first with a 3.7 years average and GlaxoSmithKline was second at 4.1 years for 1981 to 1999 data. It should be noted that the companies identify global portfolio planning management (PPM) as one of the best practices to achieve the better (shorter) time frames. PPM also directly and favorably impacts a couple of other best practices, project team operations and use of technology for planning and communications by the teams. Realistic protocols are important in speed because they especially will be easier to conduct for the sites and investigators, easier to interpret in statistical analysis, and likely require less review time for regulatory authorities [22, 34].

Organization of a Pharmaceutical Company

A pharmaceutical company that is complete in all the necessary research and business operations and divisions is called a fully integrated pharmaceutical company (FIPCO). A FIPCO contains eight operating divisions.

The eight divisions of a FIPCO are presented in the next diagram (Fig. 1.23). Research and development (R&D) and

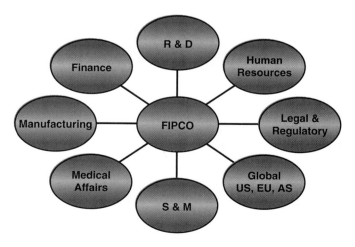

FIG. 1.23. Fully Integrated Pharmaceutical Company

sales and marketing (S&M) divisions house the largest number of staff members, and the largest budget outlays by far occur there. To the outside world, most of the "action" at a company appears to occur in the R&D group and the S&M division, because their staff interacts with so many people, that is, the public, investors, investigators, providers, payers, vendors, and regulators. However, success of a company needs major engagement of six other key functional areas: human resources (HR, personnel), legal and regulatory, finance, manufacturing, medical affairs, and global operations, each of which will be discussed in this section of the book. Each division is headed by a senior or executive vice-president, who all usually report to the chief operating officer (COO) or president. The senior leadership team is composed of the chief executive officer (CEO), COO or president, chief financial officer (CFO, who leads finance group), chief scientific officer (CSO, who often leads the research group), chief medical officer (CMO, who usually heads the development group), and the chief information officer (CIO), along with the other senior division heads in manufacturing, sales and marketing, legal, and HR. They usually constitute the operating committee or team that runs the company on a day-to-day basis. Global operations will often be led by an executive vice-president, reporting to the CEO, or U.S., Europe, and rest of world will have equivalent leaders in a hierarchy reporting to the COO or CEO. Sitting above the operating committee and the CEO, a board of directors exists for oversight of the business and operations and is reportable to the stockholders, primarily, and government agencies (e.g., Securities and Exchange Commission and the Justice Department) and the public at large.

The research division is composed of predominantly PhD scientists, as expected, running laboratories dedicated to specific basic research areas that are technology platforms, such as genomics or protein chemistry or medicinal chemistry, biology or therapeutic focused, such as cardiovascular disease, or functional oriented (e.g., high-throughput screening or x-ray crystallography) (Fig. 1.24). Research role is the discovery and characterization of potential disease targets and possible molecules as interventions for the targets. Research needs to deliver product candidates to the development division for clinical work. Vertical organization based on drug categories often is done throughout a company from research, to development, and through marketing, forming cross-functional business units to optimize communication and coordination leading to drug development and product marketing in a specific therapeutic area.

The diagram in Figure 1.24 displays a core group of representative functions in a research division. Disease biology is a starting point in discovery process to explore disease pathophysiology, especially to understand existing and new mechanisms for disease. Targets for disease intervention are identified and validated in other laboratories. Molecules are created or discovered in yet other laboratories (hits) that need validation, resulting in leads and later drug candidates. Animal testing is done (preclinical work) for pharmacology and toxicology of the drug candidates. The metabolism and pharmacokinetics group examines drug disposition and drug interactions in animals and then humans. The pharmaceutics group formulates a product into a specific dosage form (e.g., oral capsule or injectible liquid), a container system (e.g., bottles and vials), and a delivery system (e.g., a syringe), based on the disease, patient, and health care system. A variety of specialty research groups may exist to explore a technology area (e.g., anti-RNA). An Investigational New Drug application (IND) is their successful end-product. Collaborations

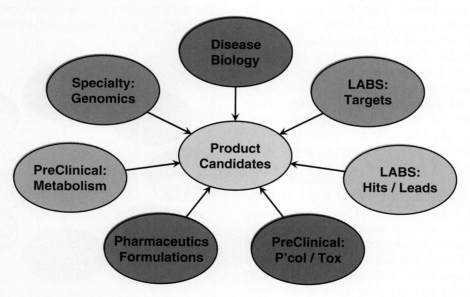

FIG. 1.24. Industry Organization: Research

with universities are very common mechanisms for a company to expand its research portfolio in related research fields and especially to tap into the wealth of basic research performed at universities. A company often allows lab heads to each create several research alliances with several universities to expand the search for novel mechanisms of disease and drug action. Basic research has collaborations with specialty companies as well to expand access to more technology platforms, delivery systems, and functional areas.

The development division creates the protocols and performs the clinical studies (phases I, II, and III) and related work, leading up to a New Drug Application (NDA in USA, Common Technical Document [CTD] in Europe) for drug products or biologics license application (BLA) for biological products. More than 10 different functions are involved and need to be integrated in the operations and planning of clinical studies for timely and targeted drug development (Fig. 1.25). The coordination role falls to the project management department, which usually chairs the product development teams composed of these development groups. Clinical management is the unit that most investigators work with, because they are responsible for protocol writing, investigator training, patient recruitment and selection, drug disposition, study monitoring visits to the sites, and final study reports, all of which the investigators are intimately involved in. The staff includes physicians, clinical pharmacists, clinical pharmacologists (PhD), and clinical research associates (CRAs).

The data management department creates the case report forms with the CRAs and performs data entry from the completed study case report forms. The biostatistics group writes the statistics section of protocols, ensures adequate study design, performs analyses of all the study data, and writes a statistical report for each study and for all studies in the NDA. Most companies will have a writing group to author the draft of the final study reports from the stat report and also help write protocols and publications. A quality assurance group exists to perform audits of case report forms, processes and procedures, and study conduct at the company and at study (investigator) sites. Their goal is to assure compliance with protocols and procedures and avoid government regulatory bodies finding data deficiencies or procedural lapses that will nullify a study's credibility and even possibly throw out an NDA. The regulatory affairs group ensures compliance with worldwide regulatory laws and regulations, organizes and files the INDs and NDAs with regulatory authorities, and is the primary interface with any regulatory authority. A metabolism group, if not housed in the research division, performs the pharmacokinetic and drug metabolism studies, including ADME (absorption, distribution, metabolism, and elimination) trials and drug interactions work. The safety department records, summarizes, monitors, and reports the safety data from clinical trials and spontaneous reports for a company's drugs before and after marketing the product.

Although economic data is not required for product approval by regulatory authorities, the health care systems in the USA and government product pricing groups in the rest of the world demand such data to more globally understand a product's health care impact. Therefore, a pharmacoeconomics group usually exists to perform these studies before and after marketing. Finally, an outsourcing group exists to select vendor companies (Clinical Research Organisations, CROs) and coordinate the work of these; the CROs commonly will perform overflow work in any of the above development work areas. PMS is postmarketing surveillance for adverse events, usually through clinical trials. Epidemiology groups perform studies to generate disease-related data to help understand diseases (frequency, presentation, and their treatment), which in turn can help in the design of future drug studies.

The marketing division often is considered the lead group at a pharmaceutical company, guiding especially corporate and product strategy, but also creating plans, objectives, and action items for the whole organization around each of the products. They are organized by therapeutic areas and customer groups. The sales division often is combined with the marketing teams into product, therapeutic, or customer groups. The figure in Figure 1.26 shows seven possible functions that

- Project management (U.S. & Global)
- Clinical management:
  - PhD, MD, PharmD, BS, MPH)
  - Managers (protocols & reports)
  - Monitors (sitevisits, CRFs, data)
- Outsourcing - CROs
- Data management (forms development & data entry)
- Biostatistics (protocols, analyses, reports)
- Medical writing
- Quality assurance (Audits)
- Regulatory affairs
  - Safety (adverse events)

Phases of Clinical Research: I – IIa – IIb – IIIa – IIIb
Other Trials: Metabolism (ADME), Economics, PMS, Epidemiology

FIG. 1.25. Industry Organization: Development

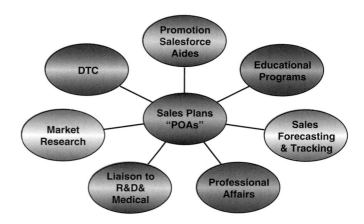

FIG. 1.26. Industry Organization: Marketing

compose a marketing group. Marketing directs the sales organization with strategies, sales plans, tools (promotional and educational), and sales targets (providers, institutions, payers). Periodic sales plans (Plans of action, POAs) or *promotion* plans, often quarterly, are created to achieve a certain sales level for all the products. Sales pieces are created to describe the features, benefits, and limits of the company products. *Professional affairs* role involves liaison work with the health care provider groups and their professional societies, for relationship building and in sponsoring education, especially about new products for new disease targets, where the medical community is not familiar. Advisory groups are used by a company and are composed of investigators and providers who are experts for a disease. They are organized early in a product's life cycle, usually by phase 2 in 88% of companies. *Educational programs* are sponsored with universities, societies, or educational vendors to discuss the company products and related diseases, starting well before product approval, even at the phase 2 research time frame. A medical education group will exist for this purpose, separate from promotional marketing. *Market research* is performed in at least a couple ways; first, *forecasting* and tracking of sales is done for marketed or soon to be marketed products (the company's and competitors) for senior management; second, various *market analyses* are done for patient and provider preferences and product usage, utility of sales aides for marketing groups, and comparative product profiles and desirable new product profiles for the R&D groups in planning for new drugs and studies. Much market research occurs during research to understand product opportunities and advise R&D; a 229% budget increase occurs just at phase 2 to the tune of $1 million for a likely blockbuster or about $500,000 for other products, and about $5.5 million overall for one product according to cutting edge company. *Direct to consumer* (DTC) advertising has become a major marketing role to reach patients and improve both disease knowledge and access by the public to products through print media, Internet, radio and television.

Finally, a key *liaison role with R&D* exists to help focus their work on unmet medical needs in the medical marketplace, optimal product characteristics, provider and patient preferences, and health care system needs. The budgeting for a marketing team increases dramatically as a product moves through research and becomes more likely to be marketed as you would expect. When a drug enters phase 3, the budget goes up by 400–500% and jumps up again by 300% at launch. The launch spend by marketing for a potential blockbuster can be $500 million from 1–2 years preapproval to 1 year postapproval.

The sales organization is composed of the field sales persons (PSRs, professional sales representatives) and their management (district and regional mangers), who are responsible for achieving the sales of the company products (Fig. 1.27). In the USA, there are about 100,000 PSRs costing the industry about $2 billion per year. One large pharmaceutical company will have several thousand PSRs. The cost for one sales person in the field has been estimated to be about $150,000–$350,000, including salary, benefits, bonuses, a vehicle, entertainment account, and an educational account. The size of a sales force is predicated on the number of sales calls, that is, the number, and frequency, and also the type of contacts to health professionals (physicians, nurses, pharmacists, and administrators), as well as anticipating turnover (estimated to be on average 10% per year). A typical sales person will make 150 sales calls, visits, to customers in a month.

The primary role of a sales call is to promote their product, but they often offer educational materials and programs and other services provided by the company, such as reimbursement support, and can help the provider with access to the company's home office research people. Sales people need constant education to keep up to date with the new sales POAs, new clinical data and publications, and company services. Compensation is often more than $100,000 per PSR, composed of a base salary ($62,000 to $100,000) plus bonuses in cash or stock options. Bonus is based on exceeding sales targets, new product sales, and special achievements.

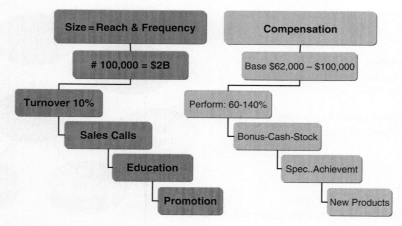

Fig. 1.27. Industry Organization: Salesforce

Besides PSRs, a sales organization will also have specialized, more senior sales people in a national or corporate accounts group to call on health system groups, for example, managed care organizations (MCOs), preferred provider organizations (PPOs), group purchase organizations (GPOs), payers, and insurers [35].

The manufacturing division creates the final product using all the necessary ingredients for a tablet or injectable or otherwise, packages it into an appropriate container system, and distributes it to the wholesalers, providers, and health care institutions. Of course, tremendous differences exist in manufacturing between drugs and biologicals and between different formulations, such as injectables and oral tablets. The core functions in manufacturing are elucidated in Figure 1.28 for all products. The process engineering group works on the manufacturing operations to improve its efficiency, reduce costs to manufacture, and improve quality of the final products. The formulations group, if not already housed in the research group, may work on later generations of a product improving the product's shelf-life or its form to increase provider acceptance. The quality control department tests the purity and stability of the product and audits all manufacturing processes to ensure integrity of the final product. Package engineering works on the container system and its labeling to maximize product integrity, information availability, and utility of the product to patients, providers, and distributors in the vials, bottle, boxes, or whatever packages needed.

The physical plant where the product is manufactured must be planned well before product approval to meet projected market needs (scale-up) and keep up with changes in the marketplace. A new plant for a new product can cost $100 million to several hundred million dollars to plan and construct. The decision to build a new plant for a new product is a risk spend needed at least 5 years before product approval for construction, staffing, validation, trial runs, and regulatory approvals. The inventory and distribution system must be able to meet the needs of the product, market, and health care system, regarding storage, shipping, the distribution centers, the distribution channels, and locations of care. Finally, the product's needs in the USA and all the world must be met, including the differences of geography, health care systems, culture, and language.

Medical affairs, also called professional services, is a group of health care professionals who serve as the primary clinical interface with health care providers, patients, public, and customers for the marketed products (Fig. 1.29). They perform the types of services outlined in this diagram: medical information (questions and answers to patients and providers), medical science liaison (education of health care professionals by field-based professionals), marketing support to home office with technical information, clinical trials on the marketed products (phase IV), and pharmacoeconomic research on the marketed products. The basic goals are to support health care customers who use the products with information, education, and clinical research, mostly within the approved indications for the products, as well as the marketing department. This group is permitted by regulatory authorities to address unsolicited questions about nonapproved uses ("off-label") from customers.

Global operations (Fig. 1.30) replicate the roles of the U.S. operations for the rest of the world, usually divided into the three other major markets: European Union (EU), Asia (Japan, China and Australia), and rest of the world (ROW). The most significant parallel groups are S&M, development, and manufacturing. A key challenge is integration of the plans and actions between the USA and all other countries in the global markets for the company and its products. The unique challenges for global operations are manifold; the unique cultural differences in business operations, regulations and laws, medical practice, health systems, general culture, and language, as well as integration with the U.S. operations. The approval processes are through separate regulatory authorities for each of the other three major markets and even individual countries (e.g., EMEA, European Medical Evaluations Agency). Outside the USA, socialized medicine predominates as the health care system in general, and most countries have

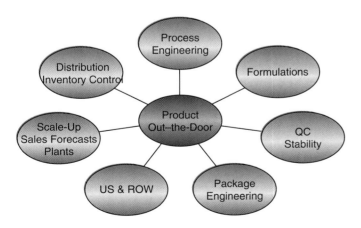

FIG. 1.28. Industry Organization: Manufacturing

FIG. 1.29. Industry Organization: Medical Affairs

Fig. 1.30. Industry Organization: Global Operations

mostly a single payer, the government. Each country will have a pricing committee or its equivalent to negotiate with the company the price that the government will pay for a product after it is first marketed and then periodically to change pricing thereafter. Manufacturing is challenged by the different languages and labeling needs, as well as differences in health care and medical practice necessitating possible different dose sizes or formulations.

R&D Challenges and Issues

Research and development in the industry has to deal with much more than just the biology of a disease, or the creation of drug category, or filing the NDA. Many external forces impact a company's goal to develop blockbuster products; although most are not under a company's control, they must be dealt with or they will become inordinate barriers to research and marketing of products. Here we will discuss also some financing issues particularly for start-up companies. Company collaborations are a necessity for access to novel science and also efficient operations as we will discuss. Company culture is a major enhancement or barrier to operations' effectiveness. Finally, resource focus on specific technologies is a best practice to be reviewed as well.

The research and development process is complex (hundreds of actions in many stages by thousands of people) and lengthy (about 10 years) as already stated, and Figure 1.31 displays the situation well with a myriad of plans, activities, and regulatory issues. Four phases at the company are presented; that is, discovery, research (preclinical animal and related work), clinical research, and postapproval activities. Two other phases occur with and at the regulatory authorities for the review and hopeful approval of applications to proceed on to the next stage of R&D: Investigational New Drug application (IND in the USA) or Clinical Trials Application (CTA in Europe), and new drug application or biologics license application or common technical document (NDA or BLA in the USA or CTD in Europe). The regulatory authorities regulate and assist companies with product development and at the same time protect public safety, ensuring efficacy, safety, and quality product production. Each phase contains a variety of actions as summarized in the slide; each phase will be discussed at length in subsequent chapters in the book. Overriding these phases and activities is the planning processes for the portfolio, products, and projects, all of which will be discussed in the governance and planning chapter. Throughout the R&D process, certain activities are done, repeated, and refined in the areas of market research and manufacturing, in order to support the molecules and products as they progress and evolve though R&D requiring new

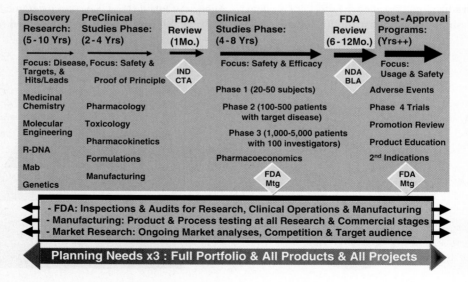

Fig. 1.31. Phase (6) – Content-Processes in R&D

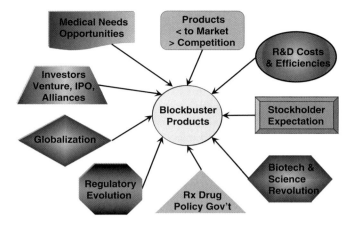

FIG. 1.32. R&D challenges in drug development
Source: Tufts Center for Study of Drug Development, Outlook 2003. phRMA, Prescription Medicines 24 Years ago and Today: Changing Trends, Enduring Needs

information and new needs. Also, throughout the R&D process, interactions with regulatory authorities (e.g., Food and Drug Administration [FDA]), are required often and desired for the filings of applications, reports, and communications for study and manufacturing issues.

In this next figure (Fig. 1.32), we have suggested nine significant external challenges involving science, business, health care, government, and the public at large that ratchet up the pressures on a company to be successful in creating the next blockbuster product for patients. (1) A company must identify what are current and especially future *medical needs* of patients and health care systems, projecting 10 years into the future, given the time necessary to develop a new product. Above, we have discussed public health statistics for deaths and disabilities, which must be tracked and examined to identify the opportunities for their molecules in their pipeline to impact diseases with sufficient advances in health for the public, which in turn offers financial opportunity for the company. (2) The success rate by the industry for *new products* has diminished over the past several years, which will be discussed later in this chapter, and competition has become more intense (multiple products in one drug category, more drug categories for one disease, and much more research investment with fewer companies). (3) The *cost of R&D* for a drug has ballooned to about $900 million for each successful product, such that good decision-making and efficiencies in clinical trials work and all operations has become even more paramount for the industry. (4) Pharmaceutical companies are public companies with *stockholders* who have high expectations for their investments and demand short-term return, often quarterly, which, for a business with a 5–10 year time frame for new products, further accents the pressure for a robust pipeline. Scientific milestones need to be met throughout the year with new products at least annually to satisfy the investment community. (5) The *advance of science* continues to accelerate at a dizzying pace, well represented by the 2001 publication of the sequencing of the human genome. But it is sobering to realize that proteomics, the protein fingerprint of the human body and disease, is even more complex. We have already shown the scientific change in oncology as one clear example of the scientific explosion of information and in turn potential product opportunities. (6) *Government policy* is an ever-changing landscape, because of new science, the need to protect the public, and the political arena that we live in. The new Medicare drug law in 2004 offers payment by government for seniors for drugs, expanding access to drugs, but interjects more government involvement in drug issues. A major drug withdrawal especially for adverse events, such as with Merck's Vioxx® for cardiac and stroke problems in 2004, raises public outcry and congressional demand to tighten the drug approval and monitoring processes. (7) *Government regulation*, especially from the FDA and the Office of Inspector General (OIG), also is an ever-changing world related to science, health care, and public pressures. Orphan drug and accelerated drug approval regulations have helped to bring life-saving products to market faster in the 1990s. However, new regulations have added onerous processes and costs to product development. For example, the new risk management guidelines from the FDA and guidelines for protection of privacy of patient information create new processes and information requirements for the research and marketing of products. The goal of more patient safety is most laudable, but the new rules carry a big price tag attached to them. (8) *Globalization* has been occurring in all business segments including the pharmaceutical industry. The bases are; better communication exists across the globe, business opportunities open around the globe vs. one nation, disease needs are universal around the world, consolidation creates operational efficiencies, and consolidation increases size and breadth to meet global business needs. New science and business models are being created to operate on this global scale. Also, the FDA now allows studies to be done with worldwide patient enrollment and even European studies being used for U.S. product approvals. (9) The *investment community* provides financial resources and can foster profitability but demands both short-term and long-term positive outcomes. They now track the industry ever more closely for their scientific advances as part of the business, using scientists and clinicians at their companies attending medical meetings to track a product's research progress. You must not only have product approvals but show achievement of interim scientific milestones, such as IND submissions, completion of phase 3 trials, and publications. In the 21st century, the challenges have only magnified in R&D. Alliances with many companies who are expert in specific fields are needed to exploit all the research opportunities of the disease targets and product candidates that a company has discovered [4, 10–16, 21, 36–48].

Although hundreds of drugs exist in the health care system to treat patients with almost all disease, opportunities for new products remain because treatments often only deal with symptom control or only partially control the disease presentation or progression. The expense of health care in the USA is $1.6 trillion dollars in 2002, of which drugs comprised about $250 million at that time. The cost for caring for patients is quite high in dollars and percentage of gross domestic product (USA, about 15%). Figure 1.33 lists 10 common maladies, their prevalence, estimated cost to the health care system, and a conservative estimate of the number of products in the pipeline. These 10 diseases involving many organ systems afflict about 177 million people in the USA and are mostly chronic conditions (some of these diseases overlap in the same patients). The costs to the health care system for these 10 disease are staggering, $1.7 trillion in direct health costs and indirect costs like lost productivity, and range from $15 billion for migraine to $879 billion for all cancers. With an aging population, the economic impact and costs of Alzheimer's ($100 billion), arthritis ($65 billion), and osteoporosis ($18 billion), to name just three cases, will be rising significantly over the next 20 years. Their control is variable; excellent disease control in some, partial in most cases, and with many patients responding poorly. Therefore, many disease opportunities exist for new products for patients to improve disease control, remove some costs in various health care management (although adding drug costs), and hopefully slow or stop health problems lowering excessive economic impacts. The industry has more than 1,800 products in various stages of clinical trials as of 2004, including almost half of the products (890) for these 10 diseases. Cancer has the most products in trials for the collection of more than 20 diseases, followed by cardiovascular disease [49–57].

The financing of the industry involves a variety of sources of money needed to pay for the expensive and lengthy R&D, wherein the promise of a return (a new product) is far down the road, 5–10 years, from the first investment (Fig. 1.34). The biotechnology segment of the pharmaceutical industry includes more than 4,000 small companies worldwide (about 1,500 in the USA) and needs to employ a full range of financing to maintain their viability. It is estimated in the annual Ernst and Young report on the biotechnology industry that usually 25% or more of companies only have 2 years of financial capital left before they have to replenish their capital or go out of business.

In Figure 1.34, the sources noted include six areas. Pharmaceutical companies can *partner* with a small company and provide up-front dollars, followed by further investment

Selected Diseases	U.S.A. Prevalence	Economic Costs (Annual)	Number of Products in Clin. Research
Alzheimer's[a]	4,000,000	100	33
Arthritis[b]	66,000,000	82	60
Cancer[c]	8,000,000	879	478
Cardiovascular[d]	70,000,000	394	122
Depression[e]	18,800,000	44	30
Diabetes mellitus[f]	18,200,000	132	63
Migraine[g]	28,000,000	15	15
Osteoporosis[h]	44,000,000	18(fx)	22
Schizophrenia[e]	1,500,000	23	14

FIG. 1.33. Diseases – Cost, Prevalence & Research
Sources: a–Alzheimer's Assoc., 2004 data, b–Arthritis Fndtn, 2005, c–Nat'l Cancer Institute, 2005,
d–Am. Heart Assoc., 2002, e–Nat'l Institute Mental Health, 2002, f–Am. Diabetes Assoc., 2002, g–Nat'l Headache Fndtn, 2004, h–Nat'l Osteoporosis Fndtn, 2004, i–R&D Directions Spl.7:2004, 2002–fx-fracture only

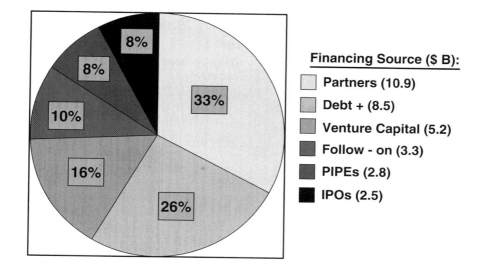

FIG. 1.34. Financing in Biotechnology Industry

as scientific milestones are achieved (e.g., a regulatory application being submitted, or the successful completion of a key phase 2 study, or an approval of a product). The small company usually gives up one of their lead products to the partner or has a co-marketing agreement. The investment here can be $25 million to $600 million spread out over 5 or more years. Usually, all these monies will run out over the 5 plus years that will occur during the basic science research and early product development stages. After a new company has progressed in their research with unique targets for disease mitigation or especially has products in a pipeline, they can *"go public"* and offer stock (IPO, initial public offering) to the investment community and public. In 2004, IPOs were $2.5 billion providing $50 million to $160 million to one company. Becoming a publically held company is a large source of income for operations, and it gives the company freedom to operate without the control or oversight of a partner. A *follow-on* is a later stock offering following an IPO. A company can certainly go to banks and investment companies, creating *debt* by offering corporate bonds or taking on a loan of money, both of which pay interest to the lender, a bank or investor. The lender expects near-term good news in product advancement and approval as their collateral for the loan. A *PIPE* is a private investment in public equity, which is special funding by outside investors outside of the typical stock purchase.

Venture capital (VC) from such financial companies is often the first area of financing for a new company, often started by a scientist from a university who has a significant scientific advance with good promise for a new drug down the road to favorably impact a disease. The size of the investment may be $5 million to $250 million, often at the smaller end of this range. The VCs become company owners with the founding scientists, holding equity or debt convertible to equity, and often sit on the board of directors. The VC is a wealthy individual (also called an "angel") or a VC company. Venture capital is provided usually in stages (up to seven) as the company advances its science and product development; starting at seed stage ($1 MM) to form the company, create a business plan, and early validation of the science, series A ($1–5 MM)/series B ($5–20 MM), series C/D ($15–50 MM) to cover through preclinical and early clinical product development, mezzanine (before an IPO or acquisition), and bridge (before an IPO or buy-out by the VC group), based on the maturity of a company and its needs. For any one company to move from a new start-up company to marketing a product over the 10 years, they usually will need to employ a variety of these funding sources at different times. Another common outcome related to funding, especially for a small biotechnology company with a unique technology or a lead product, will be an acquisition by a larger company, who needs the technology or product and values it highly. Alternatively, a merger between two small biotech companies, who perceive synergy in their technology and operations, can assist in achieving product approval and both science and business success [21, 22, 23, 30].

R&D in the industry at any one company must decide on which areas of science to focus for research on at least two dimensions, platforms of basic science (technologies) and clinical areas, that is, either therapeutic areas, or pharmacologic categories, or disease areas (Fig. 1.35). This focus is a critical decision for a company, because everyone has limits on resources, that is, financial (budget) and personnel (number and expertise), and the potential science areas number over 100 for the possible disease areas, involving any organ system and research platforms. Also, a company wants to invest

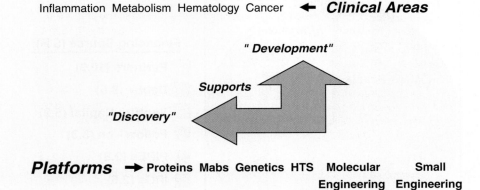

Fig. 1.35. Discovery Technology Supports Development

significantly and enough in specific areas to build the necessary depth of experience of its scientists and marketing staff to assume a leadership role in those areas. Such expertise often will foster better science-based decisions in choices for targets and product candidates and permit the funding to make the decisions potentially more successful. The investment is not only in their bright scientists but also in sufficient lab space, high enough budget for the work, and enough budget for appropriate collaborations in their focus area. Figure 1.35 displays the focus areas in the 1995–2002 time frame for the Amgen company with five platforms (proteins, monoclonal antibodies (MABs), genomics, high-throughput screening (HTS), and small molecules) and four clinical areas (hematology/renal disease, oncology, inflammation, and neurology). Flexibility in these focuses with exploratory lab operations is needed to take advantage of a unique discovery or license opportunity in a new area, which will allow for expansion into a significant new research area, as science evolves and the company evolves.

Along with an internal focus in technology represented in the above discussion, every company must look outside their own laboratories for scientific discoveries to universities, research centers, government research (e.g., NIH), and other small companies worldwide. A basic premise often mentioned is that 90% of new discoveries in your own area of expertise will occur somewhere else at these other research places. Besides this discovery phase of R&D for diseases and new molecules, the standard operations of clinical research and product marketing, which are core functions, will need supplementation to meet all the episodic work demands. Figure 1.36 lists many of the collaborations for the Amgen company during the 1990s up to about 2002. For core functions, clinical research organizations (CROs) are companies that are dedicated to performing clinical research for FIPCOs, because the work demands for a newly advanced product may exceed the work capacity of the company at a particular point in time. A marketing core function would be market research for competitive product assessments or direct to consumer advertising, wherein specific expertise is needed and found outside your company. Also, the company collaborations will involve technologies in which a company has no expertise, but it might be important in developing their products. For example, a protein company wants to expand to monoclonal antibodies and genomics or needs more high-throughput screening for a new set of targets. Access to more molecules and products to put into a company's pipeline is probably from a business perspective the most important collaboration. A company licenses in a product from a university or other company and shares the costs of R&D, costs of marketing, and later future revenues. Mergers and acquisition usually occur when a company elevates their decision for access to a product or technology area, which is principal to their R&D and business success, and it needs to be fully incorporated into their operations through acquisition and integration.

Five areas for the collaborations are listed on Figure 1.36 along with the company partners: (1) *core operational functions*, expanding opportunity to complete standard work projects on pipeline and marketed products; (2) *technology*, expanding the search for molecules and creating other types of products or formulations for existing pipeline products; (3) *product* candidates (individual products or a new family of compounds), licensed in from companies, government, or a university; (4) *research centers*, wherein labs are funded and a company has access to scientists and their discoveries; and (5) *mergers* or acquisitions [58].

A major impediment to success of mergers and even collaborations between companies can be the culture of each company, which will thwart realistic communication, collaboration, shared operations, and decision making. The business culture for small biotechnology companies with a new product often based on a new technology is discussed in Figure 1.37. This small company is more representative of a university-type environment. The companies usually were started by a university professor, who hires the early basic scientists from other universities. Scientists and research predominate in the culture, which is almost the full-time focus of more than 90% of the employees, including often the CEO and the board of directors. The science is novel, cutting edge. The primary topic at management meetings is the

➤ **General - Core Operations :**	**Company / Institution:**
○ Global Partners in Marketing & Research	Roche and J & J
○ Clinical Research Organizations	Quintiles/Radiant
➤ **General - R&D - Research centers:**	MIT
➤ **General - Merger** (Products, R&D, Manufacturing):	Immunex
➤ **Specific - Products :**	
○ Leptin & Lab output (D/C later)	Rockefeller University
○ Keritinocyte Growth Factor	National Institutes of Health
○ Calcimimetics multiproduct (Cinacalcet)	NPS Pharmaceuticals
○ Abarelix (D/C later)	Praecis
○ Neuroimmunophilin products (D/C later)	Guilford
○ Fibrolase	Hyseq
○ Interferon alfacon-1 (Out-license)	Yamanouchi
○ Epratuzumab (D/C later)	Immunomedics
➤ **Specific - Technologies:**	
○ Manufacturing & Inflammation products	Synergen (Acquired)
○ Small molecules	Kinetix (Acquired)
○ Mab technology & Panitumumab	Abgenix (Acquired)
○ Signaling drug discovery	Tularik (Acquired)

FIG. 1.36. Alliances & Collaborations by a Company*
*Amgen, 2002; Public information

- University style – academic (origin)
- Research predominates
- Scientists predominate
- CEO & Board scientists
- Dress casual
- Communications very open & challenging
- Cutting edge science
- Small companies
- Team concepts for decision making
- Best ideas predominate
- Naivete' in marketing & product needs

FIG. 1.37. Biotech vs. Pharma Companies Culture

latest scientific developments and their related product opportunities. Dress is casual as in universities, helping to foster a relaxed environment. Communication is very open and challenging among all levels of the organization. Disagreement is fostered as in universities to get to the best answers in problem solving. Independence of scientists is common in their work decisions. Processes are much less structured. A team of scientists that work on the project usually form the decision-making group. The best ideas in science predominate, which often is a quite good outcome. However, the best business assessments and plans may be missing, because of the naivete of the scientists and even their leadership. These cultural factors will inhibit collaboration with a major FIPCO, which normally has a hierarchical structure, many specific procedures for work and decisions, slower decision-making, and more management oversight.

Summary of Research and Development Approaches

A complex matrix exists in the industry for research and development that incorporates diverse groups at all stages of product evolution which can be called portfolio management (Fig. 1.38). A myriad of research technologies need to exist (e.g., genomics, medicinal chemistry, and transgenics). Also, standard product development functions in both research and development areas are required (e.g., toxicology, pharmaceutics/formulations, process engineering, and regulatory affairs). Marketing engagement is needed for strategic leadership and their product and health care data and planning. Integration, communication, collaboration, prioritization, and guidance are absolute requirements, which also demands planning, tracking, and decision making. Project management appears at the head of this matrix, in order to pull it all together and keep it progressing to the ultimate outcome of approved new products.

How can we describe success in R&D? The operational paradigm for R&D ("P to the eighth power") includes six key sets of factors that lead to product approvals and the portfolio. A recent pharmaceutical conference of industry leaders in 2004, organized by R&D Directions trade group, focused on research success factors; presentations were made by senior research leaders and senior portfolio planners from, for example, AstraZeneca, Lilly, Pfizer, Novartis, GlaxoSmithKline, and

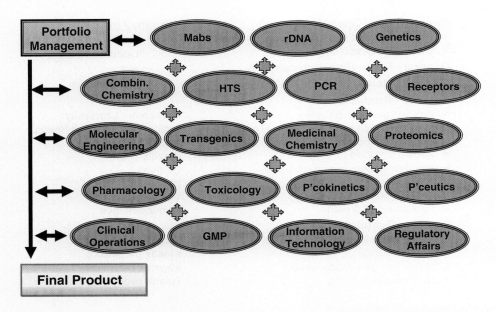

FIG. 1.38. Portfolio Planning in R&D (Copyright 2006 from *Encyclopedia of Pharmaceutical Technology, 3rd Ed* by Swarbrick. Reproduced by permission of Routledge/Taylor & Francis Group, LLC)

- **1 - People:**
  - Build expertise
  - Recruit & Retain talent
  - Defineroles & responsibilities
  - Have & Support investigators

- **2 - Process:**
  - Use life cycle management
  - Use portfolio management
  - Share knowledge & information
  - Operate globally
  - Work in product teams
  - Kill compounds early

- **3 - Pipeline:**
  - Pursue multiple indications
  - Maximize product options
  - Use novel technologies
  - Use enabling technologies
  - Create full pipeline at all research stages

- **4 - Profits:**
  - Meet unmet medical needs
  - Develop best in class
  - Focus resources
  - Support optimal product pricing & reimbursement

FIG. 1.39. R&D Success Factors – 1
Source: R&D Conference, Med Ad News, Feb 2004

Genentech. They particularly suggested the success factors for research at their company individually and for the industry as a whole as well. The common factors from more than six of these expert presentations are combined, realigned, and summarized in the next two figures in Figures 1.39 and 1.40. The realignment of the 30 factors categorizes them into the 6 organizational and operational parameters of the R&D paradigm, "P to the eighth power" (People, Process, Pipeline, Profits, Principles of the Pharmaceutical organization, and Performance) for 2 outcomes, a successful product and portfolio. Finally, one of the most important tools is communication; that is, communicate, communicate more, and communicate some more. Effective communication is vital to a successful development program across and up and down any company, especially as they grow

5 - Pharma Co. Organization:
- Recognize R & D Excellence
- R & D, True Integration
- Common Objectives X Organ
- Freedom to Operate (patents)
- Reward Speed AND Quality
- Create Research Alliances

6 - Performance:
- Maintain Accountability
- Perform Consistent Decision Making (Standards - Process- Players)
- Use Efficient Data Capture & Analysis & Reporting
- Use Innovative Technology
- Have Internal Champions

R & D 6 "Pillars" ⇒ **7th - Portfolio & 8th - Products**

FIG. 1.40. R&D Success Factors – 2
Source: R&D Conference, Med Ad News, Feb 2004

in size. Communication is a two-way process that requires sending and receiving information. Characteristics include open, honest, clear, concise, constructive, timely, and targeted [59].

References

1. Cowan C, Catlin A., Smith C, Sensenig A. National Health Expenditures, 2002. Health Care Financing Review 2004; 25(Summer, #4):143-66
2. Anonymous. Program Information on Medicare, Medicaid, SCHIP, and other Programs. Centers for Medicare & Medicaid Services. Department of Health and Human Services. June 2002, 1-18
3. Anonymous. Health, United States, 2004. National Center for Health Statistics, Centers for Disease Control and Prevention, U.S. Department of Health and Human Services. 2004
4. Vander Walde L, Choi K, Higgins J. Health Care Industry Market Update. Pharmaceuticals. Centers for Medicare and Medicaid Services. Department of Health and Human Services. January 2003, 1-52
5. Anonymous. Health Care in America. Trends in Utilization. Centers for Disease Control. Department of Health Human Services. 2002, 1-90
6. Lethbridge-Cejka M, Schiller JS, Bernard IL. Summary of Health Statistics for U.S. Adults; National Health Interview Survey for 2002. USDHHS, Centers for Disease Control and Prevention, National Center Health Statistics, July 2004
7. Anonymous. Surveillance for morbidity and mortality among older adults—U.S. (1995-96). MMWR 1999;48(S508):7-25
8. Anonymous. Chronic Disease Overview. Centers for Disease Control. National Center for Chronic Disease Prevention and Health Promotion. US DHHS. 2004, 1-6
9. Anonymous. The Value of Investment in Health Care. Pharmaceutical Research and Manufacturers Association. 2002, 1-8
10. Tufts CSDD. Outlook 2005;1-8
11. Tufts CSDD. Outlook 2004;1-8
12. Tufts CSDD. Outlook 2003;1-5
13. Tufts CSDD Outlook 2002;1-5
14. Tufts CSDD Outlook 2001;1-5
15. Anonymous. Profile pharmaceutical industry 2004, Focus on innovation. PhRMA, 2004
16. Anonymous. PhRMA. Industry profile 2003. Prescription medicines 25 years ago and today: changing trends, enduring needs. 1-81
17. Gray N. Our 6th annual report of the world's top 50 pharma companies. Pharmaceutical Executive 2005;25(5):83-94
18. Humphreys A. Mayer R. 11th annual report, World's best-selling medicines. Med Ad News 2005;24(5):1, 24-40
19. 15th annual report top 50 pharmaceutical companies. Med Ad News 2003;22(9):4-19
20. Ernst & Young. The economic contributions of the biotechnology industry to the US economy. May, 2000, Biotechnology Industry Organization
21. Ernst & Young. Resurgence: The Americas perspective. 2004
22. Lamberti MJ (Ed). State of the Clinical Trials Industry, 5th Ed. Thomson, CenterWatch. 2005
23. Lahteenmaki R, Lawrence S. Public biotechnology 2004—the numbers. Nature Biotechnology 2005;23(6):663-7
24. Tufts CSDD. Backgrounder: a methodology for counting costs for pharmaceutical R&D. Recent News. November 1, 2001
25. Tufts CSDD. Total cost to develop a new prescription drug, including cost of post-approval research, is $897 million. Recent News. May 13, 2003
26. Dickson M, Gagnon JP. Key factors in the rising cost of new drug discovery and development. Nature Reviews Drug Discovery 2004; 3(5):417-29
27. PhRMA. 2004 Survey. New medicines in development. Biotechnology. Pharmaceutical Research and Manufacturers Association, Washington, DC, 2005, 1-44
28. From pipeline to market 2004. Areas of interest. R&D Directions 2004;10(6):8-18
29. King J. Advances in medicine. 100 great investigational drugs. R&D Directions 2004;10(3):31-51
30. Ernst & Young. Progressions. Global Pharmaceutical Report. 2004

31. Garaud J-J. The Most Diversified Development Pipeline, Innovation, Focus, Productivity. R&D Directions 2004 Drug Development Summit, How to Build and Maintain a Winning Pipeline. Feb. 8-11, 2004
32. Anonymous. Report to the nation 2003. Improving public health through human drugs. Center for Drug Evaluation and Research. Food and Drug Administration. U.S. Department of Health and Human Services
33. Tufts CSDD. Postmarketing studies becoming essential to new drug development in the U.S. CSDD. Recent News. July 6, 2004
34. Getz KA, de Bruin A. Speed demons of drug development. Pharmaceutical Executive 2000;20(7):78-84
35. Goldberg M, Davenport B, Mortellito T. The big squeeze. Sales forces are still growing, says a new survey. Pharmaceutical Executive 2004;24(1):40-6
36. Guy P. Rising to the productivity challenge. A strategic framework for bipharma. Boston Consulting Group. July 2004
37. Lawyer P, Kirstein A, Yabuki H, Gjaja M, Kush D. High science: a best-practice formula for driving innovation. In Vivo The Business & Medicine Report 2004;22(4):1-12
38. Engel S, King J. Biotech innovation. Pipeline gaps. Can biotech fill them? R&D Directions 2004;10(7):42-56
39. Booth B, Zemmell R. Prospects for productivity. Nature Reviews Drug Discovery 2004;3(5):451-6
40. The pipeline problem. The pipeline solution. R&D Directions 2004;10(5):42-54
41. Rawlins MD. Cutting the cost of drug development. Nature Reviews Drug Discovery 2004;3(4):360-4
42. Cohen CM. A path for improved pharmaceutical productivity. Nature Reviews Drug Discovery 2003;2(9):751-3
43. R&D business. Mostly growth for big pharma in 2003. R&D Directions 2004;10(3):14-15
44. Preziosi P. Science, pharmacoeconomics and ethics in drug R&D: a sustainable future scenario? Nature Reviews Drug Discovery 2004;3(6):521-6
45. Balekdjian D, Russo M. Managed care update: show us the value. Pharmaceutical Executive special report. September, 2003
46. Tollman P, Guy P, Altshuler J, Flanagan A, Steiner M. A revolution in R&D. How genomics and genetics are transforming the biopharmaceutical industry. Boston Consulting Group, November, 2001
47. R&D Directions 2004 Drug Development Summit, How to Build and Maintain a Winning Pipeline. Feb. 8-11, 2004
48. Edwards MG, Murray F, Yu R. Value creation and sharing among universities, biotechnology and pharma. Nature Biotechnology 2003;21(6):618-24
49. Alzheimers Association, accessed June 2004 at www.alz.org
50. Arthritis Foundation, accessed June 2004 at www.arthritis.org
51. National Cancer Institute, accessed June 2004 at www.nci.nih.gov
52. American Heart Association, accessed June 2005 at www.americanheart.org
53. National Institutes of Mental Health, accessed June 2005 at www.nimh.nih.gov
54. American Diabetes Association, accessed June 2005 at www.diabetes.org
55. National Headache Foundation, accessed Junes 2005 at www.headache.org
56. National Osteoporosis Foundation, accessed June 2005 at www.nof.org
57. Engel S, King J. From Pipeline to market 2004. R&D Directions Suppl. 2004;10(6):35-74
58. Personal Communications & Experiences, Amgen 1989-2002
59. R&D Directions, Summit on Product Development, February 2004

2
R&D Planning and Governance

Ronald P. Evens and Joel Covinsky

Development Schemas and Leadership	34
Clinical Development Strategy	36
Leadership	38
Product Development Paradigm	39
Portfolio Planning Management (PPM) Description	43
PPM Process Components	47
PPM Analyses (Examples)	54
Project Management	60
Summary	62
References	64

A company needs the best people, advanced science, great products, the physical plant, the right operations, the financial wherewithall, the right leadership, appropriate patients, experienced investigators, a good dose of common sense, a vision of the future, and some luck to develop significant new products. Many companies may possess these attributes; however, how well companies can pull all these manifold, often disparate, and sometimes conflicting resources together in turn will differentiate themselves as leading companies. A metaphor for this situation is the pack elephant ("the product") and the five different blind passengers ("the departments") needing to reach a common destination ("approvals"). Each may touch the elephant at different areas, large flappy ears, stumpy legs, long flexible trunk, hard pointy tusks, the tail, or the broad back. Besides their different personal experiences and expectations, they each believe to be holding a different animal with different possible benefits and risks in reaching their destination. Someone or some group must be able to step back and see the whole animal, as well as its parts, to coordinate the individual players, help make the best judgments, and set the best direction. Governance and planning can be best practices in pulling it all together in product development and in differentiating the top companies from the pack. These best practices are needed for the products (the elephants), by the departments (the elephants' passengers), and by the whole company (the elephants' whole environment), as well as for all stages at which a product exists or at any stage of a company in its evolution. It is very easy to be myopic and become lost in the details and not focus on the big picture. In this chapter and book, we are focusing on the many common challenges associated with product development, always keeping in mind the big picture (the big outcome)—product approvals.

Product development over the past 20 years has and is experiencing significantly reduced output of new products, which has been discussed earlier, while costs have risen sharply. Numerous publications discuss a variety of solutions to this dilemma, often highlighting improved drug discovery, more in-licensing deals, better governance and decision-making, operational efficiencies such as use of enabling technologies, teamwork, and also portfolio planning on a global scale. We will attempt to bring these issues together and summarize them in this chapter for the new product researchers in academia and industry [1–22].

The chapter outline is presented above. We will focus on leadership, organizational effectiveness, global resource allocation, and portfolio planning, which directly impact governance and planning. Representative models are presented here for each of these three areas, not as a panacea or *sine qua non*, but to primarily provide the students and fellows reading these materials with key sample operational frameworks in the industry regarding plans, organizations, and decisions. An overview of the complete product development process first provides the content as a framework for planning and governance for our discussion. We will present the phases of research and their components and then a construct for a global product development plan. Leadership is discussed next as a primary skill set that underpins optimal decision-making, governance, and operational effectiveness. Then, the organization and its operational effectiveness will be addressed through a model, the paradigm of product development ("P to the eighth power"). The model incorporates eight major parameters to be outlined below. Then, a description of the portfolio planning management (PPM) is provided, answering the questions why do it, what can be done, and by whom is it done? PPM contains a variety of components, which are presented to address the question how is it done? Analyses are required to establish goals, assess risks, plan resources, and judge progress; some of these analyses of PPM are discussed as examples. For completeness in our planning discussion, a brief review of project management is given, because it contributes to better PPM through planning, coordination, and execution that is directed toward specific projects, such as a product or a component in the plan (e.g., an individual study or new formulation). A well-organized plan should provide a foundation for good timely decisions.

Development Schemas and Leadership

The drug development process, some key milestones, and many of its activities are summarized in this figure to help present in one picture the overall process and its elements (Fig. 2.1). This figure, its many facets and complexity, reiterates well the foremost challenge for company leadership, that is, to pull together these six phases of product development with all their possible projects and outcomes (50 plus noted in the figure), accomplished by many different people, over a long timeline, into a coherent flexible achievable plan, and with product success as the end product. The product development schema for a biotechnology or pharmaceutical company is a lengthy (about 10 years or more from disease biology to product approval) and a complex scientific *and* business endeavor, as the diagram suggests. The six typical phases are displayed in this figure: (1) discovery, (2) preclinical research, (3) government regulatory review (e.g., U.S. Food and Drug Administration [FDA] review of the IND

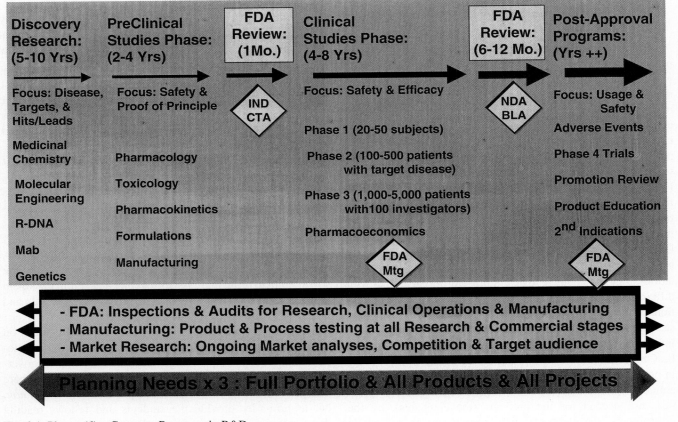

FIG. 2.1. Phases (6) – Content - Processes in R&D

[Investigational New Drug application] or CTA [Clinical Trials Application] for Europe), (4) clinical research program, (5) regulatory review of the NDA (New Drug Application), BLA (Biologics License Application), IDE (Investigational Device Exemption), or CTD (Clinical Trials Document), and (6) postapproval phase. Within each phase many activities are included, some of which are listed here as examples, but each phase will be discussed at length later in subsequent chapters. The regulatory authority (e.g., FDA) and company interactions are now several-fold throughout the product development timeline, intended to be an iterative, supportive, and collaborative process, while regulatory compliance and public protection remain as paramount functions of government. Four other activities that occur throughout the drug development process are presented. Planning for the products, overall portfolio, and individual projects is an overarching process. Government regulatory inspections of research, manufacturing, or clinical operations can occur at any time. Manufacturing is refined repeatedly to provide the best quality product, improve the output, and reduce costs, lowering the cost of goods. Quality assurance (QA) is performed continuously to assess and assure attainment of quality goals. Marketing obtains information repeatedly about the unmet need, diseases, therapies, the ideal product, marketplace, competition, thought leaders, providers, payers, patients, and sales opportunities, in order to fine tune the strategies, targets, research plans, and the marketing and sales plans [23].

Planning and governance in R&D requires corporate plans (both strategic and operational) that must be global, comprehensive, and focused on critical pathways in the product research and approval process. A representative global plan involves, for example, seven possible overarching programs or functions of the company that comprise such critical pathways for product development: global project teams, clinical (and marketing) strategy, market development, clinical operations plan (studies), regulatory milestones, safety updates, and manufacturing updates (Fig. 2.2). Their integration, the tracking of progress, and the decisions (go–no go) are the responsibility of senior management, through optimal leadership, organizational effectiveness, and portfolio product planning. The global product plan also will include three other elements, as shown in the diagram, to plan for and help assess progress on product advancement from the lab toward approval in the seven critical pathways: (1) decision points at key milestones to anchor and guide senior management and the company, (2) a timeline that integrates the critical pathways and for gauging progress, and (3) a planned evolution in the nature of the team as the product status matures. Within each of these critical pathways, major milestones are established that must be achieved and then reviewed and approved by senior management for a product to advance. Costs and budget projections must be identified annually as well.

The *global project* team creates the global plans for product advancement incorporating the major work outcomes or milestones from the other pathways, for example, target indications, desirable product profile, studies to be done, safety reports, manufacturing needs, budget needs, and sales forecasts. Membership on the team evolves over time as the product matures and the focus of work changes from a purely research focus to market launch mode; however, always a dual science and marketing approach needs to exist. The *clinical*

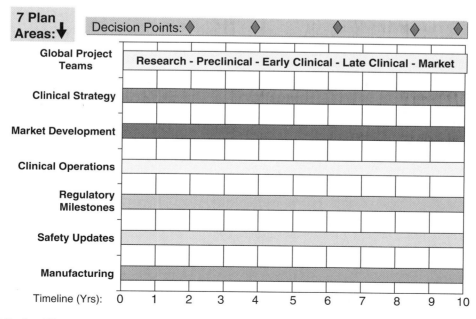

FIG. 2.2. The Global Product Plan

strategy even at an early stage incorporates the target indications, ideal product profile, key disease features, key data on the company's product as it exists at that time, marketplace factors (target prescribers, competition, health care system, market size, and payer issues with the value proposition for the product), manufacturing and cost of goods issues, patent status, and any known regulatory hurdles. The clinical strategy must be adjusted over time based on new data becoming available on the science, regulatory requirements, and market from the other pathways. This strategy guides all the other pathways.

Market development pathway includes internal and external roles. Internal company awareness of the drug and health care issues is developed by the team, the planning process, and senior management, including for example likely target audiences, optimal product profile (especially efficacy needed, safety, dosing, and formulation), competitive products, health care environment for the product, reimbursement issues, market size and segmentation, and sales forecasts. A primary question to answer is whether the projected revenue is worth the development costs for a product, which is asked throughout development as more product and related market data become available. This question can be answered "no" even at the end of phase 3 trials. The hurdle is defined differently depending on the size, structure, research interests, budgets, and philosophy of the company. For an Amgen example, they were developing a product for prostate cancer with another small company. The clinical data did not achieve the level of acceptance (efficacy and safety advantages compared with the proposed product profile and competitive products in their opinion) compared with the future research and market costs to Amgen, especially with all its other priorities. Amgen discontinued the license and gave the product back to the originator company, who subsequently continued the work and have received approval, but the drug was taken off the market for past sales. Externally, a novel product, or really any product, presents to marketing the exciting challenge of preparing the marketplace. For example, activities include educating the medical community about the new scientific advances, working with and seeking input of thought leaders for the target disease, creating public awareness, preparing the distribution channels, setting up reimbursement systems, preparing the sales force (their education of disease, product, health systems, providers, and marketplace), and creating the promotional materials.

The *clinical operations (studies plan)* includes all the different studies from across the company and around the world to be done to create the product profile, the data for product approval, and the data for product acceptance by providers, payers, and health systems in all countries. Certain countries may have specific data needs for approval and special studies to be done. The potential indication is the primary target for the studies' plan, along with the ideal package insert (PI). The proposed PI can guide the studies plan to fill in all the necessary elements (e.g., mechanism of action [MOA], efficacy, dosing, formulation, and stability). *Regulatory milestones* in the planning are probably the most significant landmarks for a product's advancement, where regulatory authorities have established requirements in content, formats, and timing of applications and reports that must be met, along with formal submissions to or meetings with them (e.g., IND, end of phase 2, NDA).

Safety of a product in the studies plan is such a major clinical question in product development plans that it usually is set apart from the rest of clinical work as a separate critical pathway. It needs special attention to ensure patient safety in the thousands of patients who will use the product after marketing. In the role of protecting the public, safety is the most significant focus for regulatory authorities, as well as the company. Safety has huge marketing implications as well in limiting the utility of the product in the mind of the provider and limiting the market. The worst case scenario needs to be anticipated as best as possible and avoided; that is, removal of a product from marketing after its approval because of serious and unexpected adverse experiences. Certain types of adverse reactions may require special attention in product development and added study and resolution, especially the ones that have led to such withdrawals (e.g., hepatotoxicity and cardiovascular events). Over the past 20 years, 21 drug products have been withdrawn from the market because of adverse events; the two aforementioned problems led to 12 of the 21 withdrawals. We need to realize that the majority of clinical studies are designed to evaluate efficacy as the primary end points with safety information always collected but derived from signals that appear during the course of studies. These signals may lead to the need for studies designed to adequately capture the safety profile of the product. *Manufacturing* issues are another critical pathway and have substantial impacts on clinical development and marketing. Can the formulation perform in the marketplace? Can it be manufactured in sufficient quantity, scale-up, for clinical trials and then meet marketing demand? Is the formulation for dose-response trials and phase 3 trials identical to the intended marketed product, as required? When does the new physical plant need to be built (a risk spend, capital outlay, initiated often 5 years before approval)?

Clinical Development Strategy

Drug development is a team sport. Corporate management determines which disciplines will provide the organizational leadership in the drug development process. Companies may drive the development process through their program management organization, their clinical organization, or their regulatory organization. Regardless of the source of development team leadership, as in any sport, team chemistry and good communication are the critical underpinnings required to be successful. The players and leadership involved in these global development teams tend to change as programs progress from the preclinical phase to the market place. Good processes alone doe not guarantee success. While this chapter will focus on the overall portfolio management process and global program management, a brief discussion of clinical strategy and clinical operation provides a view of the drug development process from the vantage point of one of the critical segments. The clinical strategy discussed below is a central component in portfolio and program management for an individual compound. This strategy may be focused on a particular claim or a variety of indications.

Clinical development requires both a strategic and an operational plan. The strategic plan includes the overall direction and the operational outline for how the compound will be developed. This information is compiled in the clinical development plan (CDP). The CDP evolves over time, generally in three stages (Fig. 2.3): preclinical phase; the early development phase; and the late development phase. Each plan will build on the former plan. A multidisciplinary team with specific expertise in those aspects of development adds their unique perspective to this evolving plan. The leadership, as well as the membership of these clinical teams that writes the CDP, also may change as the project evolves from the preclinical phase to the development phase.

The initial CDP (CDP-1) may be driven by a product champion originating in one of the preclinical disciplines. Generally, the product champion is someone who was instrumental in the compound's discovery and has an understanding of the clinical environment. In leading development companies, this person and process is considered a best practice. This individual introduces the organization to the unique features associated with the new chemical entity and helps others understand how this new compound will address an unmet medical need. The original product champion will lead the team in producing the first CDP (CDP-1) and stimulating and exploring the organization's interest in pursuing subsequent investment. The preclinical development team will prepare a development plan for R&D management to consider and approve. It will consist of a discussion of the product opportunity; any early preclinical data that supports the proposal, toxicology needs, pharmacokinetic and pharmacodynamic profile, patent potential, the studies needed to bring the product to the next stage (go–no go decision point), compound available, and the potential investment costs. During this phase, someone with expertise in clinical pharmacology will work with the preclinical scientists to determine what information (studies) is needed to advance the project to the subsequent phase. This individual may subsequently lead the next development phase. Early collaboration facilitates a smooth transition between phases. Preliminary discussions will be held with marketing to garner their interest and support. This work will be initiated approximately 1 year prior to the preparation of the Clinical Trial Application (CTA)/Investigational New Drug (IND) application. The information collected during this phase of development will be shared with the regulatory authorities to determine whether the proposal is suitable to study in man.

The early development plan (CDP-2) builds on the information collected in the preclinical phase and includes both the strategy of where and how the new compound will be first investigated in man. Strategic issues include global considerations, such as which countries have both the regulatory expertise and clinical expertise to enable study initiation in

- **Principles:**
    - Strategy – overall direction & operational plan
    - Goals - strategy, shared learning & generation of support
    - Product champion (scientist)
    - Multidisciplinary team
    - Strategy & plans evolve over time
- **Pre-Clinical Phase (CDP-1):**
    - Initiated 1 year prior to CTA/IND
    - Start of phase 1
- **Early Development Phase (CDP-2):**
    - Initiated 9-12 months prior to phase 2B
    - Lead country chosen
    - Global leader needed
- **Late Development Phase (CDP-3):**
    - Initiated at phase 2B
    - Continues through phase 3

FIG. 2.3. Clinical Strategy 3 Phases

the most efficient fashion. Study design issues will be shaped based on previous experience in a particular area. Companies with expertise in modeling and simulation may be able to optimize both their development strategy and study design by creating proposals based on their previous experience collected in their centralized databases. During this time period, key thought leaders and people with unique expertise both within and outside the company are identified to help the team. In large companies, a global clinical team leader is identified to begin working with clinical pharmacology to facilitate a smooth transition to the subsequent development phase. Collaboration among the team members will lead to the identification of biological markers and analytical tools designed and refined to assist in the evaluation of the new chemical entity as it moves through the development process. This phase begins prior to the phase I studies and ends 9–12 months before the phase IIB dose-response trial. Each CDP serves multiple purposes. These plans are designed as a strategic document to enhance the understanding regarding the evolving profile (target product profile = TPP) of the new compound; share the learning and the perspective gained; and generate support for the plan (the studies planned, understand the risks, contingency plans, and the investment) and commitment to go forward at the critical decision points. Safety and proof of concept are key targets in this plan. Feedback from outside experts and regulatory review are all incorporated in the evolving strategic plan.

The building of CDP-3 begins with patient enrollment in the phase IIB, dose-response trial. The development costs escalate substantially at this point and really increase as you enter phase IIIA. Strategic issues at this point include consideration on where to conduct the critical phase III safety and efficacy studies. Considerations include where the best clinical expertise resides, availability of patients, past performance of investigators, regulatory environment, costs for doing the studies at particular locations, internal company expertise and monitoring resources, and availability of comparator drugs that will meet the broadest regulatory requirements and management buy-in. This plan will be refined and adjusted as study results become available and feedback from outside experts and regulatory authorities accumulate. Communication with all segments of the organization is critical throughout the development process and becomes more complex as the program progresses and more people get involved in contributing to the process.

The strategic elements in CDP-3 provide the organizational framework for the detailed clinical operational plans. A detailed global clinical operational plan (G-COP) in large companies is generally managed centrally with delegation of responsibilities to individual countries. Ideally, these operational plans should be constructed based on feedback from the individual countries who have the best insight regarding local medical practices and unique elements of their regulatory environment. The global clinical leader (GCL) or a global clinical director may coordinate the CDP-3 (strategy), as well as leading the cross-functional product team. Global clinical operations may work with the GCL to coordinate the operations plan, which includes all the details associated with conducting the individual studies in the various countries. Program management coordinates the interactions with all the departments involved in the overall.

Leadership

Better governance starts with better leadership in any organization. Leadership is a tone and set of behaviors and skills necessary for any person in the head role of a group or organization to work with, assist, and stimulate success of the people in that organization. The senior management of a company is a group of leaders who need to make the best possible product development decisions and then lead their departments to achieve product success. Leadership is a core requirement for success in all company operations including R&D, and its importance necessitates more full elaboration.

A leadership model is offered here in abbreviated form, the 6 + 6 = 6 "Essentials of Leadership Excellence" (Fig. 2.4). The leadership essentials involve two distinct but complementary domains composed of six skills each that can result in six end points. The two domains are group effects, leading by "e"nabling others to succeed, and individual effects, leading by setting an "e"xample for the group. Another way these two domains are expressed is leading from within the group and leading from the front, respectively, and we suggest that both are optimally needed for ideal leadership and corporate success. The six "enabling" effects to support the group's staff to succeed are "employ" (hire the best, often called the A players), "envision" (set a forward-thinking vision and the goals that support and stimulate a group into a better future), "excite" (motivate, foster being the best, and expect to beat the competition), "equip" (train and educate to sustain cutting edge abilities and efficient operations, plus give them the tangible tools in technology to succeed), "environment" (create an organized, challenging, stimulating, and friendly workplace, where staff experience support from management and desire to perform at their best), and "encourage" the staff members (set high goals, give positive reinforcement of best practices, critique with feedback, and as needed push to sustain the cutting competitive edge).

The second indispensable half of leadership is the individual set of six leadership principles for setting an "e"xample, that is, "ethics" (personal ethical practices of leaders that stimulate a staff person to follow a leader), "edge" (a personal demonstration of competitive zeal, creative thinking out of the box, and commitment of time, energy, and smarts above the norm), "engage" (personal involvement in the planning or decision-making or feedback processes demonstrating commitment to the organization), "execute" (personal follow-through on the goals and responsibilities established by the leader in the organization, demonstrating accountability and operational excellence), "experiment" (a willingness to try

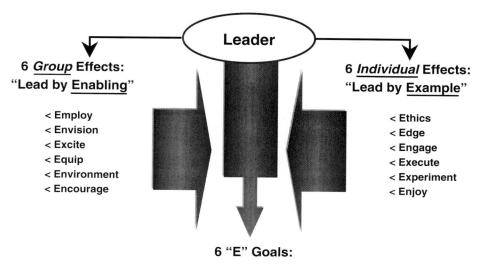

FIG. 2.4. The "E"ssentials of "E"xcellent Leadership

new ideas and approaches that fosters creativity and innovation), and finally "enjoy" (having fun in the workplace with our respective job; a happier workforce is a more productive workforce in general). The ultimate corporate goal in leadership, as stated succinctly in the diagram, again is six elements; an *enterprise* that becomes *eminent* yet continues to *evolve* through *employees* (*and executives*) who *excel* and *enjoy* what they do. This leadership concept ("Essentials of Excellent Leadership: 6 + 6 = 6") is based on [personal] observations of leaders [and leadership experiences] over 30 years at five pharmaceutical companies, Bristol-Myers, Marion, Marion-Merrell-Dow, Aventis, and Amgen, and five leading universities (Buffalo, New York; Memphis, Tennessee; Lexington, Kentucky; Kansas City, Missouri; and Gainesville, Florida); leadership training [experiences] programs (e.g., Center for Creative Leadership in Colorado Springs, Linkage Leadership Conferences); [exposure to] seminars from leadership experts (e.g., Warren Bennis, John Kotter, Madeleine Albright, and Jack Welch); boards of directors of five professional societies; and the extensive published literature in books by thought leaders on leadership [24–38].

Product Development Paradigm

The R&D division needs an operational framework (paradigm) in which to function and achieve organizational effectiveness. The following proposed paradigm for product development at a pharmaceutical company, "P to the eighth power," provides such a general working framework. This paradigm is based on a distillation of many observations and experiences over 20 years by the authors. Plus, a large mass of selected published literature addresses problem areas in drug development, environmental factors for research, diseases and industry opportunities, and ideas to maximize organizational structure, processes, and productivity [1–22, 17–21, 39, 40].

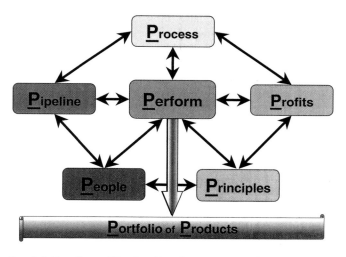

FIG. 2.5. Paradigm of Product Development (P to 8th)

This paradigm of eight "Ps," "P to the eighth power," suggests eight major parameters exist that together enhance success in product development (five components through one set of actions yielding two outcomes equals the eight parameters) (Fig. 2.5). The five parameters are (1) *processes*, (2) *profits*, (3) *principles*, (4) *people*, and (5) *pipeline*, which need to be executed and integrated well, that is, (6) excellence

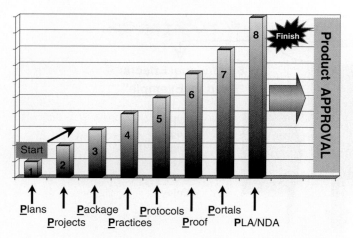

FIG. 2.6. Processes in the Paradigm (P 8th Power)

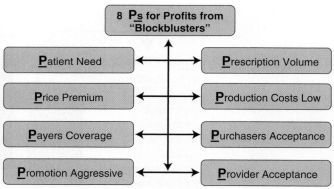

FIG. 2.7. Profits in the Paradigm (P 8th Power)

in *performance,* and collectively they result in (7) *product* approvals, and (8) a *portfolio* of products in R&D. The paradigm of eight parameters (Ps) and an overview of the process of product development in the pharmaceutical industry are displayed in the following commentary and diagrams.

Regarding *processes* (Fig. 2.6), the rule of Ps and eights continues in the operational components at a company culminating in product approval. These eight Ps for processes start with plans for research and build all the data and documents that finish with a BLA, NDA, CTD, or IDE. A *plan*, for example, includes all the projects, goals (target indications and product profile), protocols, resources (internal and external, people and budgets), and time frames. The *project* can be a single clinical trial or the collection of all trials that comprise the planned trials to be done for the NDA, BLA, or CTD. The *package* is the container system of the product, as well as the final formulation, that is, the physical vial, bottle, or boxes containing the product, and the product labeling on these packages. The *practices* are the company's operational guidelines and the standard operating procedures (SOPs), as well as their values and ethics, in conducting their work, including regulatory requirements. Compliance with these SOPs is an absolute requirement for an NDA, BLA, or CTD, and the regulatory authority performs audits to assure the compliance. *Protocols* are the study documents wherein each summarizes the intended conduct of a study, such as objectives, justification, and background for the product use in the target disease, patient selection, drug administration, monitoring parameters, study controls, and intended statistical analyses. *Proof* includes the clinical evidence from the clinical trials for safety and efficacy, as well as pharmacology, toxicology, pharmacokinetic, formulation, and manufacturing data, in the form of investigator brochures, final study reports, abstracts for medical meetings, and publications. *Portals* are the decision points by management for go–no go decisions over time for the studies and projects for a product. The PLA (BLA, NDA, or CTD) is the complete and final set of documents filed by the company with the regulatory authorities intended to achieve product approval.

The next parameter of the product development paradigm involves the eight Ps for the *profits* necessary to fund all the work (Fig. 2.7). Profitability will only occur with the development of both medically and then financially successful products. A blockbuster describes such a product that achieves a sales level of $1 billion per year. Profitability of a product and the company is predicated on eight parameters to be attained; meeting an unmet *patient* medical need; good *provider* acceptance for an innovative product based on its safety, efficacy, and convenience (sound scientific information); *purchaser* (hospitals and physician offices) acceptance to minimize barriers to access to products; *payer* acceptance with willingness to pay for products based on clinical utility and value to health care system; a competitive *price* premium that can be charged for an innovative product; *patent* protection offering opportunity for sustained sales over several years without generic competition; aggressive *promotion* to achieve the sales level (e.g., $50 million–$100 million for launch phase of marketing); high *prescription* volume from the providers; and low *production* costs in the manufacturing and distribution of the product (reasonable cost of goods [COGS], that is, less than 25% with a target of 10% of the sales price.

The company leadership, the senior management team, needs to exemplify and foster certain *principles* or behaviors that will stimulate creativity, productivity, and success in their staffs. Yet again, this area of the paradigm involves eight Ps or eight principles (Fig. 2.8). *Purpose* is needed to generate motivation and offer overall direction for the organization, which often is captured in a corporate vision and in mission statements that must be reality and not a set of words. Strategic *planning* frames the vision and mission with strong disease, product, and market opportunities, appropriate resources, and measurable targeted goals, all integrated across the organization. *Principles* about how the company operates day-to-day for patients, employees and stockholders serve powerfully to motivate, encourage, and guide the organization. A sense of satisfaction from the contribution to better

2. R&D Planning and Governance

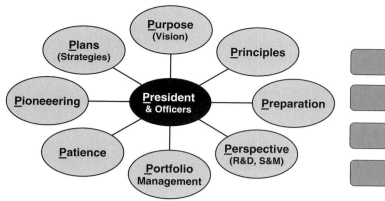

FIG. 2.8. Principles in the Paradigm (P 8th Power)

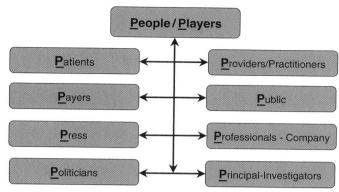

FIG. 2.9. People in the Paradigm (P 8th Power)

quality health care for their fellow man often is a an important principle and motivational driver at companies. *Preparation* of the senior team means each part of their organization is ready to optimally perform their share of the product development process with an experienced and trained staff, efficient systems, sufficient budget, and informed and organized management. The executive team needs a dual *perspective*, that is, to both realize and create the best science (R&D) and also the best sales opportunity (S&M) for the product portfolio. A *pioneering* spirit is needed to foster smart and innovative risk taking in the organization that can create and sustain scientific advances with market potential. *Patience* is also a proverbial golden rule for research success, given the time it takes for R&D (5–10 years per product) and the risk spends to be made. However, time is money in business and product development with limited patent lives of products, such that operational efficiency and optimal research planning and execution must be demanded. Finally, the corporate leadership must embrace and consistently support *portfolio* management, which takes the resources of specialized people with budget authority, includes both support and oversight over departments and their management (coordination and leadership rather than actual supervision and decision-making authority), and fosters the communication and collaboration in goals, work items, and outcomes that result in successful product development.

The *people* that influence the R&D of a product in our paradigm of Ps include eight groups of internal and many external players (Fig. 2.9). For the company players, we have discussed previously in this chapter the operating divisions and the roles of the *professionals* in general and R&D in particular. The *patients* are the study subjects participating in the clinical trials to establish safety and efficacy of the products. *Principal investigators* are the research experts in health care settings, most often at universities, that provide input on study design and perform the clinical trials. *Providers and practitioners*, especially physicians, pharmacists, and nurses, help identify the optimal product profile by providing market

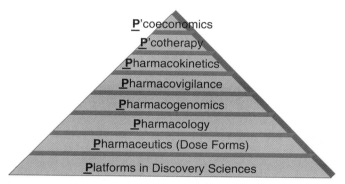

FIG. 2.10. Pipeline in the Paradigm (P 8th Power)

research information, as well as also providing a source of patients for clinical trials and ultimately write the prescriptions. *Politicians* and *policymakers* in government create and execute the regulations and laws that govern research requirements, product applications, approval of the products, and marketing practices. *Public* creates health care demands, and also influences health care systems, providers, and government by prioritizing disease care needs and choices available, as well as impacting policies and practices. *Payers*, private and government, help set health care treatment goals, product access, and product choices through payment policies. The *press* serves as a forum for the public, providers, politicians, and payers to communicate and discuss health priorities, express health care or drug needs or problems, and offer some level of education of the public and others about product development challenges, success and failures.

Pipeline component of the R&D paradigm is comprised of eight scientific disciplines, all of which comprise the work of basic research and clinical development and create the information that supports the approval and use of the product (Fig. 2.10). Ultimately, the pipeline also is commonly considered the portfolio of products, in either basic research or clinical development stages that the company is studying. These terms for the disciplines are defined in the glossary of terms appendix.

The first five Ps, parameters, in the R&D paradigm (processes, people, profits, pipeline, and principles) need to be executed, coordinated, and integrated. The sixth P in the product development paradigm is *performance* of the overall company organization and all the personnel, that is, in the departments, the consultants and service vendors, and the investigators at the universities and health care institutions (Fig. 2.11). Eight distinct and complementary practices are expressed as adjectives focused on individual performance. As a composite, they can be characterized as doing more, better, smarter, faster, now and over time. The eight practices although being distinct personal actions need all to be utilized at the appropriate times, in the appropriate setting, and integrated. A person needs to be *prepared* in education and motivation to perform the work, *precise* in their work given its scientific nature and impact on diseases and patients, *prudent* in their decisions, *partnering* with others in a department, on teams, and with outside experts to be successful, *persuasive* at times to convince others to use the best ideas and follow their lead, *persevering* through all the challenges and time required in research and working with projects and other people, *pioneering* in their idea generation without losing focus of the routine job at hand to be done, and finally *productive* to get the all the work done in the right way at the right time with the right people and following the right plan, to achieve the right result.

The first five parameters of the R&D paradigm through an effective sixth parameter, performance, result in the final two Ps, *product approvals* and a *portfolio of products*. The product approvals and the portfolio are the ultimate goals to determine organizational effectiveness of R&D, but they must incorporate the operational elements described above regarding the people, processes, profits, principles of leaders, pipeline of science, and performance behaviors of individuals, in order to be considered successful in the minds of all the stakeholders, that is, staff and leadership of the company, patients, providers, public, and stockholders. Figure 2.12 presents the eight stages of product evolution at a company (from a target, hits to leads to candidates to product approvals to new indications and label extensions, and then to new molecules and new formulations); of course, product approvals are the ultimate products of R&D. For the portfolio of products, eight different factors could be used to categorize the products and organize the business, based on the corporate strategies, science issues, health care structures, and business organizations: general medicine versus specialty practice, disease areas, therapeutic areas, pharmacology groups,

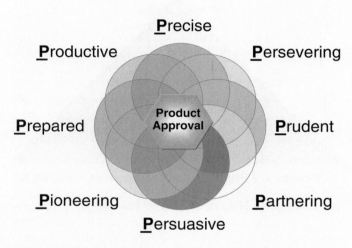

FIG. 2.11. Performance – the Paradigm (P 8th Power)

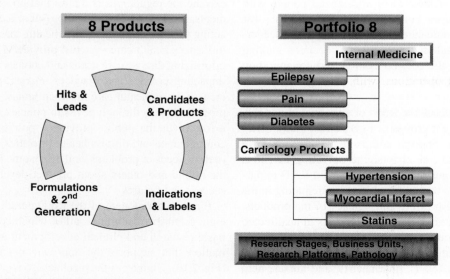

FIG. 2.12. Products & Portfolio in Paradigm (p8th)

pathologic mechanisms (e.g., inflammation), stage of research (e.g., lab–preclinical–clinical phases I to IV), research platforms (e.g., drugs vs. proteins vs. monoclonal antibodies), and/or business units of the company.

Portfolio Planning Management (PPM) Description

The overriding purpose of PPM is to improve productivity in R&D, obtaining more regulatory approvals in number, on a global basis, with better quality, at a faster rate, and at less cost, through better planning, decisions, and execution. Companies further desire to avoid failure in regulatory approvals. A formal and well-run PPM process can make for better communications, better decisions, better process indicators, more efficient work, better product candidate choices, better indications and safety with the products, better labeling, better global coordination, better relations with investigators, better transparency, and thus better relations with regulatory authorities, less cost, and better postapproval marketing, and, therefore, more effective drugs for patients and better business success. Even though the advantages of PPM are manifold and principal to success, it is quite hard to achieve because of the complexity and high costs of product development, pressures from corporate leaders and boards for rapid progress, personnel variables (biases, knowledge or experience deficits, territoriality), patient and disease variability, product performance questions, the unpredictability of research, evolving knowledge base, global company challenges, many types of markets, volume of customer and research data needed, and evolving regulatory hurdles. The industry and our PPM practices need to always realize that major breakthroughs occur when innovators challenge the current paradigm and investigate new options even when all the facts are not well established. PPM on a global scale has become a certain best practice for success in R&D to manage the above-noted problems and help create the blockbusters.

Essentially, PPM should sit in a company at the center of a crossroads where six roads come together representing preclinical research, clinical research, regulatory, manufacturing, marketing, and global operations, with oversight from the senior management team. However, effectiveness of PPM also is dependent on established cooperative and communicative relationships without controlling, intrusive (operations), or appearances of spying behavior. It is good to be reminded that PPM does not set the mission and strategy, does not lead the departments and divisions, does not make the decisions, and does not do all the work. PPM is a process and organization involving well-informed people (organizers, collaborators, communicators, and planners with vision) who assist the development and business leaders (executive management, departments heads, and team leaders) to make it all happen within a framework; that is, they know what we are doing and why, when and how it is being done, and they are getting there as intended. In short, PPM should be facilitating good quick decision-making [22, 39, 40].

A useful and practical list of pitfalls in filing of regulatory applications was published in 2003. We will list them here and will return to them after our PPM discussion to judge how well PPM, as we discuss it in this chapter, will help minimize these 15 pitfalls: (1) not communicating with the regulatory authorities, (2) avoiding the safety issue, (3) lack of planning, (4) omitting data or including unnecessary data, (5) not paying attention, (6) not documenting manufacturing process, (7) ignoring the investigators, (8) forgetting conflicts of interest, (9) hiding something, (10) being too eager to disclose to the public, (11) not thinking globally, (12) not thinking about electronic submissions, (13) rushing, (14) not choosing wisely, and (15) filing a drug that is not needed [43]. This section of the book will include why do it (PPM), the elements of it, the players to do it, manager concerns about it, and metrics to measure it. Later sections will cover the process elements of PPM and types of analyses.

What is the rationale to conduct portfolio project or product management (PPM) as a core process and skill set for senior management in a company's decisions and operations for product development? Six benefits of PPM are addressed in Figure 2.13. The executive committee of the company needs a process to assist in a structured way to implement their corporate strategy across the whole organization and to understand gaps in strategy, planning, operations, or resources. Please be reminded that the FIPCO includes at least eight divisions that all have their individual roles in executing the corporate strategy; the operational divisions of research, development, marketing and sales, manufacturing, and the support divisions of law and regulatory, information, human resources, and finance. Coordination, integration, progress tracking for execution, and goal achievement can be done with a PPM organization for product development across these eight varied organizations. Resource allocation of staff, systems, and budget needs to be done in a balanced and organized way for all the various projects for individual products and of overall portfolio levels, based on availability, need, capability, and priority. Again, PPM offers the company

PPM (Portfolio Project Management)

- Execute corporate strategy
- Allocate resources
- Determine hiring needs (resource needs)
- Assess & limit risk management
- Improve budget management
- Underpin corporate planning

FIG. 2.13. Portfolio Project Management – Why do it?

and management the process to do it. Another part of resource assessment is to determine strengths and gaps in staffing, based on the global product portfolio plans. Individual departments and divisions will perform this analysis, but PPM will guide the process to assure consistency and matching with the global plans. The end result would be that integrated needs are established for staffing for product development, and the best people with proper expertise and skill sets for work to be done can be hired at the company or employed through external consultantships.

Product development is a very risky and expensive business with only one in 5,000 products from research eventually being approved and marketed. Every decision will have pros and cons and carries a risk of success or failure, the later of which a company will try to reduce. Through PPM processes, risk assessment for plans and decisions can be done to give this added perspective to management in making more informed and better decisions; thereby risk of failure is lowered by factoring in risks in the outcomes of decisions. Does this study have a 25% or 75% chance of success? Does the manufacturing decision for a new plant lower risk of product outage postmarketing or add unnecessary cost? Budget decisions are always a primary focus of an organization. PPM offers methods to improve budget decisions across an organization, for example, through an integrated and executable strategy, more organized resource allocation, more useful progress reports, and especially better go–no go product decisions. Finally, PPM is the support (underpinning) organization and process for corporate, portfolio, and product planning and tracking of execution, as we have discussed previously. The Boston Consulting Group has done a variety of studies with medical technology companies to determine their set of best practices. In looking at the best high science companies versus all others, they found that PPM was defined by 89% of the best high science companies versus 66% of all others and actually followed by the companies in 83% of the cases versus 57%, respectively [8–10, 22, 41, 42, 44].

This description of PPM will involve seven topics; leadership, planning, teamwork, process and methods, organizational participation, portfolio data, and decision making (Fig. 2.14). The senior management group, *the leadership*, must create the vision, mission, and overall strategy for the corporation. This information is the critical framework for the whole organization, all staff, all plans, all operations, and all products. Basically, a company needs to know where it is going first for PPM to assist the company in executing on the strategy. PPM involves all *levels of planning*, starting from the top, the portfolio, through the products, and to the individual projects. All levels should not be just layered on top of each other, but integrated for a fully effective PPM. Another best practice within PPM, actually a requirement for success, is *teamwork*. PPM guides, coordinates, integrates, helps communicate, and measures across all the various operational groups who are needed for a project or product. The team of people performs all the work. The team also needs to follow through on the plans and work projects within the goals and plans, that is, operational excellence of teams is another best practice. PPM incorporates a specific set of *processes and methods* to assist the company to carry out the product and portfolio plans. Previously, we discussed a global product development plan and process that would be used by a PPM group to do portfolio planning and management. Product profiles are used to frame the targets for R&D in product development. Tracking resources is another key practice of PPM and includes what is available, what is being consumed, and what is needed. Practice-practice-practice for PPM is a final

- **Corporate strategy & leadership**
- **Planning, Planning, Planning:**
    - Project plans
    - Therapeutic area plans
    - Product plans
- **Teamwork AND Follow-through**
- **Process and methods:**
    - Global plans
    - Product profiles
- **Organizational participation**
    - Mandate from the top
    - Buy-In@all levels
    - Engagement@all levels
- **Portfolio data:**
    - Availability
    - Types
    - Analysis
- **Decision-making:**
    - Decision points/gates
    - Decision criteria
    - Timing
    - Priority setting
    - Killing projects
    - Commitment of management@all levels

FIG. 2.14. What are the Elements of PPM?

rule of thumb to share, especially as the PPM system is first being put into place in a company. Continuous improvement requires a feedback loop for lessons learned.

A critical success factor for PPM is that *organizational participation and commitment* must be from top to bottom for all the processes. Senior management must be unequivocal in their support. Often, the PPM group reports into the senior management team to demonstrate this commitment. All levels (division heads, department heads, and staff) must buy-in to PPM, the benefits to the organization and to themselves, the processes, and the reports of progress, be they success or failure. An effective PPM group engages the whole organization in the processes, the decisions, and the planning from its inception through its conduct and in all its operations. Any methodology employed by PPM would need validation for credibility to the organization and insure outcomes are reasonable. Much *data* about the products, the science, and markets must be available for PPM to work. Key internal data would include each product profile (expected and ideal), all projects (e.g., studies in labs or clinics and at all stages, formulations, manufacturing, stability), timelines of all projects, status of all projects, budget available and being used, staffing (how many, who, when), costs for projects, equipment, and staffing, and systems available. External data also is highly important to assess the marketplace, such as competition (products, companies), target audiences, treatment opportunities, regulatory requirements, and sales projections. Data sources are extensive, need to be manifold, and are both internal and external to validate the information used in planning; for example, the company's study reports, the sales force, market research department and their studies and reports, medical affairs staff expertise, Internet, library and publications, trade journals, focus groups and advisory panels of customers, customer-based companies in key markets (group purchase organizations of institutions, or GPOs; managed care organizations, or MCOs), presentations at medical meetings, competitor materials, consulting organizations, financial companies, government reports (e.g., FDA, CMS, CDC, OIG, EMEA), trade associations, and individual industry experts. Timely access to targeted information when there is a need to know is vital to a successful operation. The analyses are extensively done for PPM, especially for projecting possible outcomes and assessing risks, two core components for PPM. This chapter will present the types and a variety of these analyses later.

Decision making is a core operational focus for the organization in implementing the strategy and the plans. Decision making gives us direction and outcomes for the company. PPM helps foster and improve (with corporate approval) the decision making process, guides the decision making, and measures success of the decisions. In making *decisions*, PPM assists with what are the decision points or gates at key milestones, use of consistent and appropriate decision criteria, the appropriate timing for work to be done and for decisions to be made, and the engagement from the organization to follow whatever the particular processes may be at their company. Priorities must be established for products within a portfolio and projects (including indications) within a product plan. The criteria, which we will discuss later, must be appropriate and complete for the decision at hand and consistently applied, in order to be credible, supportable, and useful. Go–no go decisions are always difficult because you must ultimately kill some projects (the hardest decision) that either are not successful at that point or that you do not have the resources to do it, or where the project is outside of the corporate strategy and plans. Some companies may let some projects linger consuming resources that could be utilized more effectively elsewhere. The decision-making process, the criteria, the players making decisions, and the decisions themselves must receive full support (buy-in) from all levels of the organization to be successful. The aforementioned BCG study of best practices also identified three performance categories (governance, organization, and process), and 15 specific practices within the three categories that are optimal PPM-type practices. Well developed PPM-related practices, as judged by the employees, were evident for 14 out of these 15 practices at best high science companies, much more often than the corporate averages in these medical technology businesses [8–10, 22, 41, 42, 44].

Participation in PPM involves the whole company at many levels for it to work properly and achieve corporate product development goals (Fig. 2.15). Sitting at the top of the pyramid for PPM players is senior management, either the whole executive committee, or a subset called the product development senior team, often composed of the chief operating officer (COO), chief financial officer (CFO), senior vice-president (VP) for R&D, senior VP sales and marketing (S&M), usually the head of development who often is the chief medical officer (CMO), senior VP manufacturing, sometimes the senior VP quality assurance and control, and even the chief executive officer (CEO) at some companies. Their role is twofold. They support PPM through the mandate given to coordinate and assess the product development process, and they make the decisions for product advancement or killing projects in a fair, consistent, and timely manner. The decisions are made at the decision points for milestone achievement, basically addressing the question, was the milestone achieved with sufficient scientific, marketing, and organizational information to proceed forward with least possible risk, and funded to complete the prospective work plans as presented by the teams? The portfolio planner oversees [monitors] the individual product and portfolio process leading to the decisions but also possesses evaluation roles; gauges progress versus timelines in the plans, coordinates assessments of risk, and estimates likelihood of success going forward. Risk assessment is a dynamic process especially with regard to optimizing resources (people and money). These evaluations become important additional decision criteria for teams and senior management. The product teams and team members are responsible to prepare the presentation of data, accomplishments, and outstanding issues to senior management.

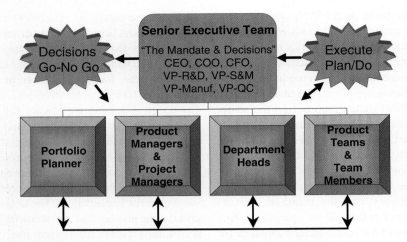

Fig. 2.15. Who are the players in PPM?

Sources – department managers & scientists, based on changes in operations and decision-making.

- Loss of independence/autonomy
- Less engagement in decision-making
- Loss of access to senior management
- Stifled creativity with more structure
- Resources lost to other projects or departments
- Greater scrutiny by management
- Less information available on products or projects
- Loss of openness in communication
- Lack of management resource commitment, or unclear priorities, yet expectation to perform work

Fig. 2.16. What are Managers Concerns with PPM

The department heads provide operational leadership to ensure all the work is done by their respective staffs for the products and plans. Individual product managers in various departments (marketing, development, formulations, manufacturing, etc.) on the teams perform the work in the plan with assistance from project managers to coordinate and communicate across departments. Therefore, execution of the plan is the shared responsibility of the individual managers, their department heads, and the teams [8, 9, 22, 41, 42]

Portfolio planning and project planning often create operational and organizational concerns for department heads, who control their staff, budgets, and processes to execute the plans of product development. Figure 2.16 lists nine potential problems scientists and department managers may perceive. An appearance of taking control by PPM and project managers (PMs) can exist in an organization, because their involvement changes how information flows through the organization and how decisions are made, even though their role is support to the teams and management, assisting in planning, process follow-through, and decision making. A common and difficult challenge to collaboration is to create sharing of information openly related to issues and options for studies, manufacturing, stability, formulations, analytical problems, and slow enrollment. PPM must realize these concerns and minimize their perception by engaging these scientists and managers in the processes (planning, analyzing, tracking, and reporting) and even seeking their input in the creation of the PPM for a company. Frequent communication of what is being done, why, how, and by whom is critical to success of PPM. However, managers at all levels must get on board with PPM, and at the end of the day if anyone does not, they need to move on to some other function. Roadblocks associated with silent consent of staff or managers dooms programs to failure because cooperation is lacking, even resistance exists through lack of cooperation, or slow contribution occurs [9, 45].

A key success criterion of PPM in support of a company is the measurements employed to assess achievement of milestones at the decision points (Fig. 2.17). PPM supports department heads, team leaders, and senior management with better execution of the plan especially by gauging progress on fulfilling the plans. The 2004 BCG study and report lists 20 different metrics that could be employed by a company and PPM related to strategy, operational improvements, and project team functions. *Progress reports* must be regular, consistent, honest, meaningful, and *relevant* for the company's processes, products, and culture. The department heads need to contribute to such reports and sign-off on them (*buy-in, essentially own them*), in order for them to carry credibility and follow-up. Tracking of *achievement of milestones* includes quality (was the question answered?), timeliness (did it get done on time?), and follow-up (what is left to be done?).

The metrics chosen need to be *quantifiable, measurable, accurate, fair, efficiently done*, and *improvement oriented* (Fig. 2.17). Efficiency in reporting and its metrics is important so that the work to collect the data is not cumbersome and actually slows work down. Improvement orientation of metrics is a necessary positive approach, not punitive, and always will be better received by staff and managers whose work is being evaluated. Metrics need to change as the product evolves during the course of a program. Use of consultants is common for metrics implementation in order to access the specialized expertise, but the company needs to ensure they have real-world experiences and employ practical metrics in data collection. Implementation of even a simple metric (e.g., a system for performance relative to quota) has been shown to improve employees' agreement toward their own efficiency by a large margin; for example, such a quota system increased employees from 50% to 75% stating their division is well developed in project initiation, management, and milestone achievement in one study. Also, monthly project review meetings, versus none or less frequent, were judged by employees that their products got to market 25% sooner [8, 22, 30, 42, 45].

PPM Process Components

The process components for PPM in supporting product development are discussed in this next section. The 10 constructs, plans, and processes to be discussed demonstrate PPM's potential breadth of support for R&D. They also show some of the details of PPM in its activities, planning, and decision making: global product plans (presented above), budgets, product profiles, project plans, enabling technologies, decision points, decision criteria, pharmacogenomics, drug delivery/formulations, and product life cycle management.

In the global product development plan, each pathway involves a budget to be planned, especially for the clinical plan, market development plan, and manufacturing plan, wherein most of the dollars are spent (Fig. 2.18). The responsibility for budgeting in R&D is usually shared between department heads and their superiors, the finance division's accountants, and PPM. This slide presents the budget by its organization on the left side (categories and cost centers) and the processes on the right. Each company will vary all the headings or various categorizations of a budget, but the slide gives us a representative example. Four *budget categories* are outlined in the slide. Other budget categories may be used (e.g., buildings, overhead). In the five *cost centers*, human resources (HR) department includes salary for all staff, bonus

- Progress reports of key functions and milestones
- Relevant to corporation
- Buy-In from all levels of management
- Quantifiable
- Measurable
- Efficiently done
- Improvement oriented
- Designed to help teams and departments
- Fair
- Accurate

FIG. 2.17. What are Characteristics of PPM Metrics

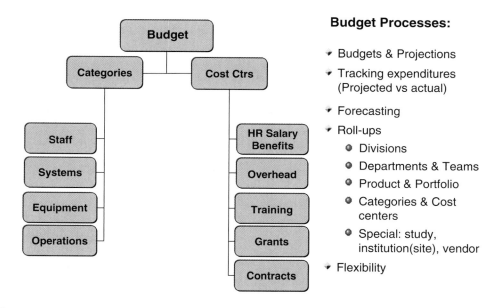

FIG. 2.18. Budgets

plans (cash and or stock options), and benefits, such as health care, vacation, sick leave, retirement program, and savings plans. Training is listed separately because of its importance to the skills, knowledge, and satisfaction of employees, which directly relate to productivity, but it may be included in HR costs. Overhead pays for the building construction and depreciation and utilities. Grants are paid to investigators and institutions for all the various studies (clinical, basic science, pharmacoeconomics, pharmacokinetics, stability, etc.), consultantships, and educational programs, including their institutional overhead, usually 25% to 50%. Contracts are payments to vendors for various services (e.g., to clinical research organizations, site management organizations, data centers, market research, and off-site manufacturers).

Processes for budgeting are many in number and variety to meet all the operational needs of different groups. Budgeting is a critical success factor in the tracking function for R&D. We need to create a *budget* for each year for each product and all the R&D departments, within some corporate set of targets, dependent on expected revenues, work objectives, and operational needs. *Expenditures* and costs during the year are tracked for planned versus actual, which is a major operational goal for the management in any for-profit company. Departments and teams are obligated to stay within their forecasted expenditures and are measured against them. The cost of goods projections need to be compared and tracked very carefully to actual costs as more work on scale-up is done over time with a product; an excessive added cost could adversely impact the viability of a product. *Forecasting of costs* is done over multiple years going forward based on the portfolio plans and compared with potential revenues, especially sales, or other income sources (stock offerings, investments, loans, or milestone payments for small new companies). *Roll-ups* of the budget for cost centers and categories need to be done for products, teams, departments, and portfolio. Also, each category and cost center has a budget roll-up, as part of the assessment of operational success. This information generally is reviewed at least quarterly. Special reports are commonly done for specific vendors, institutions, investigators, individual studies, so that managers can understand and control cost more effectively. *Flexibility* needs to be built into the process to adjust allocation of budget across the organization, as unexpected R&D outcomes or external events occur (e.g., study failure, unexpected observations [AEs], product fails at IND or NDA stage, vendors change costs, research alliance being terminated, or an important new product is licensed in from an outside company).

A core document in product development and PPM is the "product profile." *Ten factors* commonly are considered as the core information needed to describe the product in both its science and marketing potential, as listed in Figure 2.19 (e.g., efficacy, safety, and formulation). In essence, we start with an ideal product profile that comes from market research with the disease needs, patient convenience, health care system issues, medical thought-leaders' input, cost of care for the target

"First-In-Class" versus "Best-In-Class"
"Need" & "Novelty" & "Competitive Advantage"

- Unmet medical need (Stand alone vs Add-on)
- Market (# addressable patients; 5 year revenue)
- Competition (Products, Company, Tactics)
- Category & MOA (Novelty)
- Efficacy (Competitive advantage)
- Dose/Schedule (Practical)
- Formulation (Convenience)
- Safety (Less AEs)
- Pharmacokinetics (Disease fit)/Drug Interactions (Less)
- Economics/reimbursement climate (Favorable)

FIG. 2.19. The Key Factors in a Product Profile

indication, and competitive product issues. This profile is the target for research to try to achieve in formulation, indications, efficacy, dosing, and adverse effects. As studies are done and more information is known about the company's product, the profile incorporates these data, and then the product profile is for *the* product being developed including its good and bad properties.

Three goals serve as overriding considerations for all the factors in the product profile; *need, novelty, and competition*; that is, is such a product needed to treat this disease? are we a unique and innovative product? and do we have advantages over competitive products? Positive answers on these three questions will increase the scientific acceptance in the medical community and health systems and substantially expand the market and possible revenues over the life of the product. Also, a particular philosophy regarding a new family of products being developed by a company must be addressed, that is, do we want to be *first-in-class* to market or later and *best-in-class*. First-in-class historically has the prime marketing position, with more rapid and complete formulary acceptance and then loyalty of prescribers and health systems. In some cases, a breakthrough product changes the treatment paradigm, such that the medical community has to be educated to facilitate the uptake of the new approach and put the new benefits in perspective. However, a best-in-class product has the benefit of knowing the product weaknesses of the first product that can be exploited. The second or third product can have significant advantages in formulation, dosing, efficacy, or safety, such that prescribers switch to the best product. The benefit of being first-in-class has decreased also because the time for market approval for the first to the second product was reduced from 8.2 years in the 1970s to 5.5 years in 1980s to 1.8 years in 1995–1998 time period [7, 17, 22].

Project or product planning in R&D is a very detailed oriented process, certainly covering all the critical pathway studies. The project plans must be quite comprehensive incorporating the work of all departments, worldwide efforts, all staff involved, any budget issues, and timelines for all projects (Fig. 2.20). Actually, every work activity that significantly

- PreClinical (Tox - P'col - Pkin)
- Formulations / Pharmacokinetics
- Manufacturing:
  - Clinical trial & Marketed product
  - Process engineering & scale-up
  - Packaging engineering
  - Quality control (analyses, stability)
  - 5 year plan, inventory, SKUs
  - Distribution channels
- Clinical Trials:
  - Indications, dosing
  - PIs, sites, patients
  - Data management, stats, writing
  - Phase 1, 2, 3 & 4, QOL, economic
- Regulatory:
  - IND, NDA
  - FDA meetings, Audits
- Safety
- Market research:
  - Disease & Product
  - Pricing & Sales
  - Competition
- Business plans:
  - Publications
  - Thought-leaders
  - Reimbursement
  - Launch [market programs, medical, staffing/sales]
  - Partnerships
  - Life cycle
- Patent status & Trademarks
- Financial:
  - COGS, Budgets, Costs
  - P&Ls, Revenues, NPV

FIG. 2.20. Project Plans (WW, All groups, Timing, All Resources)

impacts any section of the package insert, the regulatory package, NDA/CTD, and any marketing launch program must be integrated and tracked. The apparently minor change in the diluents of an injectable product could cause an NDA to be not approved because their potential side effects or their impact on stability of the active drug was not sufficiently addressed. In marketing, launch of the product will require often extra analyses of clinical data to assist in educational or promotional material preparation, and market research will assist R&D in the types and amount of research to be done. One small study in one country anywhere in the world may surface a serious adverse experience that may require further evaluation and could hold up a regulatory submission. Figure 2.20 lists for us 10 major areas in bullet points and more than 50 work areas or items to be done and tracked by project planners for the product teams, some of which will be included in portfolio analyses. The timing of the different projects in preclinical, manufacturing, clinical, marketing, and other areas need to be integrated and especially sequenced because one project may depend on information from another in a completely different division of the company.

Research contributes the mechanism of action, assessment of adverse experiences in animal models, the preliminary pharmacology, toxicology, and pharmacokinetics in animals or *in vitro*, formulation workup, and stability studies. *Manufacturing* provides the manufacturing process and of course the product during clinical trials. Process engineering and product packaging needs to be addressed early during the clinical development period, because minor product changes and stability issues can create breakdown products (ingredients) and impact clinical and regulatory requirements. Later, manufacturing and marketing of the product with product scale-up is addressed further by process engineering and product packaging, along with needs for inventory of product and its control, an sku (shelf keeping unit), and a distribution system. *Clinical development* work is substantial as expected: all the necessary studies (investigators, sites, forms, and data) and reports for approval, labeling, and marketing (publications, education, and promotion) and safety labeling and reports. Drug performance may deviate from the original hypothesis and necessitate new plans and studies to be done. The critical role of safety data in an NDA or BLA warrants special attention through a separate medical unit for *safety* in clinical development focusing on recording, analyzing, and reporting on adverse experiences with all pipeline and marketed products. Internal audits are done by a *quality assurance group* for both all processes (procedures and operational guidelines) and outcomes of R&D for an NDA. *Regulatory* manages the applications to and interactions with regulatory authorities (meetings, letters, calls, and audits). Regulatory also performs an oversight role for compliance with government regulations.

Marketing performs all the market research analyses and generates related reports for disease, product profile, competition, pricing, and potential sales, identifies thought leaders and investigators, and produces for launch all the plans for publications, educational programs, promotion and advertising, and reimbursement. The product team is the recipient for the analyses from marketing and the provider of data for product to be marketed. The team also addresses the first and subsequent indications, integrated into their goals and timing as necessary. *Law department* is responsible for the product patent with research and manufacturing providing the patent data; they assesses competitive patents; and they ensure compliance with laws. *Finance group* tracks and examines budgets, expenditures, cost of goods, profit and loss (P&Ls), and revenues for a team. Resources of staffing are evaluated by departments and/or *human resources* to provide enough of the best possible people with training as needed. A company faces a big challenge making sure everyone is up to date on both the clinical and regulatory science to enable them to contribute to optimal performance. External resources are used most often to conduct many of the development projects by clinical research organizations, and also market research is done by outside companies with appropriate access to data and expertise. Life cycle plans are done by the team and PPM from discovery to approval and throughout marketing to maximize the franchise, which will be discussed later. In global organizations, both clinical researchers and medical affairs practitioners provide critical input and feedback about the unique features and practices in their individual countries.

PPM depends on the availability of systems and databases to gather and process the data regarding product status. "Enabling technologies" are the systems, equipment, or processes that are intended to improve operational effectiveness of department functions, such that more and better products can be developed faster and at less cost (Fig. 2.21). Many of them are needed for PPM, and actually, PPM is an enabling technology for the planning, tracking, and analysis functions. For example, e-clinical is electronic data capture for the data in clinical trials to facilitate ongoing analyses and decision making by the R&D management team. PPM is central to coordinating the collection of the information and the data for progress reports in clinical trial progress. Currently, research is going to develop spp. tools that will directly both identify and collect patient data (e.g., blood pressure readings) and send it to the company's study data center without handling by the site. This technology will facilitate decision making and save valuable time. Informatics can be structured to collect, store, and analyze data from any part of the organization. Informatics certainly can incorporate the global product plans with milestones for and progress on study plans, regulatory milestones, timelines, safety reports, costs, and future projected sales.

One of the major challenges faced by the pharmaceutical industry is how to optimally handle the massive amount of information collected by all levels of R&D. A global standard for the collection, organization, storage, and analysis opens the door for substantial enhancement in organizational performance. With global standards in place, a data warehouse can be established, which modelers can draw on to conduct simulations designed to assist in evaluating program potential and greatly improve the design efficiency of individual studies. The building of these data warehouses requires collaboration and communication across the R&D organization. Informatics in the research arena can store, manipulate, and analyze varied databases for genomics, pharmacology, receptors, ligands, proteomics, structure-activity relationships, to help identify targets and leads for further research. High-throughput screening (HTS) in lead analyses is estimated to have increased discovery rate of molecules by 6% from 1994 to 1998 and by 11% from 1998 to 2002 [20].

Other enabling technologies are considered to improve efficiency in establishing effectiveness and safety in clinical trials by using biomarkers and surrogate markers of efficacy early in development, even in the preclinical stage, to kill the poor performing product candidates sooner and advance the more likely winners faster. Conducting clinical trials by having more practical study protocols and with better use of outsourcing are enabling technologies for efficiency, too, which can reduce costs of operations for trials by getting only needed data (less) collected faster that still meets NDA/CTD requirements for a disease [8, 9, 11, 17, 19, 22, 44, 46, 47].

PPM monitors *decision points* in research and development, also called decision gates, related to milestones and timelines in a product's advance toward approval. A paramount area of governance is decision making, including decision authority, decision criteria, time frames, communication of decisions (informing and consulting), performance management, and incentives, all of which needs to be engaged in PPM. A product team presents and must show the data,

FIG. 2.21. Enabling Technologies

2. R&D Planning and Governance

defend conclusions, and offer recommendations to support a go–no go decision to be made by the senior management team. Figure 2.22 presents seven representative decision points, along with some of the new commitments being made at that point to fulfill in the next phase of product development. The decision gates involve regulatory and safety hurdles most often, along with the decision of the acceptability of the data for exceeding that hurdle, and the organization's willingness to expend available resources (dollars, people, and systems) to continue onto the next stage. Information and data come in from all the critical pathways; as progress, or lack thereof, occurs in study plans, safety, regulatory, marketing plan, and manufacturing. The information available grows and changes in quality and completeness helping make more informed decisions over time. Data gathered at decision points may lead to changes or define protocol and programs based on the new information and decision. Other areas of the company contribute as well with very important information (e.g., patent status from law department, budget status from finance, and staffing levels from human resources). If a product's performance cannot measure up to the hurdles at the decision gate, then the decision needs to be to kill the project or product, so that resources will be applied to more likely successful product candidates [10, 12, 22, 48].

At these decision gates, a set of questions needs to be addressed by the teams to senior management to permit as informed a decision as possible at that point in time (Fig. 2.23). The decision gates employ milestones that elevate the decision

When & Why move forward?

Decision Gates:
- Discovery (from Lead to Candidate) ....
- PreClinical work to be done ......
- IND/1st in man ....
- End of phase 1 ....
- End of phase 2/PrePhase 3 ....
- End of Phase 3/Filing ....
- FDA approval ....

Commitments:
- Full laboratory resources (Space & scientists)
- Do animal studies & Form a team
- Start human trials & Create profile
- Show proof of principle in disease
- Launch major efficacy trials (Finances +++ & Resources +++)
- File NDA & Prepare S&M for launch
- Launch product to market with label

Fig. 2.22. Decision Points/Gates

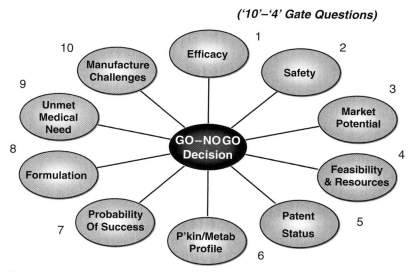

Fig. 2.23. Decision Criteria (Adapted with permission from Pharmaceutical Executive. Advanstar Communications. Cleveland, OH. From Table – Factors in R&D Decisions. From Lam MD. Knowing when to pull the plug on your experimental drug. Pharmaceut Exec 2004;24(2):56)

usually to a go–no go level for senior management involvement. The same set of questions usually will be addressed at all the gates and for all products to help set expectations around the company for information needs, which should improve decision making in its consistency, fairness, and quality of the outcomes. Of course, the information available will be different at early versus later points in the product's advancement through the plan. Special questions will be added at certain stages because it is the most appropriate time for the question, for example, if and when do we create a new plant to manufacture the product? or do we have a backup molecule in a family because the lead one failed in preclinical efficacy stage? or what product or program changes need to be made due to safety different than anticipated? or if and when do we perform a pharmacoeconomic study in managed care area? The 10 questions suggested are fairly standard, including *unmet medical need, efficacy, safety, market potential, patent status, pharmacokinetics/metabolism, formulation, manufacturing issues, resources and feasibility to be able to continue,* and *probability of success*. The 10 gate questions each need to incorporate 4 consistent questions within them (the "10-4" gate questions); does the data give us some *novelty* for the product? is the *data available* and sufficient? what is the *health and market impact* of the information or data? and do we have the *resources* for the work going forward?

Regarding these decision criteria, how often are they used by pharmaceutical companies? Each company creates their own list and uses them to varying degrees, based on personal management preferences, the experience base of the company, and the relative use of PPM. Figure 2.24 displays a table produced by Thomson-CenterWatch in 2004 for frequency of use of decision criteria at the phase 2 or 3 points in time. Consistency in their use is pretty good, 54% to 89%, but deficits are surprising. For example, 25% of companies did not use competitive activity, projected peak sales were not used by 32%, and company staffing was not used by 43%.

Termination of an R&D project is done when the decision at a milestone is go–no go, and the data allows management to determine that the product has failed the expected milestone outcome (e.g., phase 3 data indicates inadequate efficacy, or safety is unacceptable, or low revenue projections because of high production costs, or given a too low level of efficacy to warrant future high expenditures for questionable research outcomes). The Tufts Center for the Study of Drug Development studied drug termination in 2004 and found three primary reasons, safety failure (20% of the time), inadequate efficacy (almost 40%), and economics (about 35%). The time to termination during development was approximately 2 years, 3 years, and almost 4 years for these three reasons, respectively. For products terminated during development, much cost already had been incurred for research, formulation, clinical trials, manufacturing workup, and market preparation. Improving predictability of failure is a major need for companies; approaches to hopefully kill projects earlier and incur less costs are biomarkers for disease, validated surrogate end points, computer modeling techniques for disease, and maximizing FDA or EMEA interactions with the company to help design pivotal studies [17, 45, 48].

Pharmacogenomics is a relatively new discipline combining pharmacology, genetics, pharmacokinetics, and pharmacodynamics. Genetic differences among the population can greatly impact drug activity, metabolism, or pharmacokinetics. Patients' phenotype can lead to either selective advantages, unexpected serious toxicity, or lack of drug effect. In the future, drugs may be developed for smaller target populations. Higher specificity may allow for improved efficiency in clinical research but also reduce the size of the target patient population (market). Phenotypes may identify a population of patients who require long-term prophylaxis or who must avoid certain treatment options. Single nucleotide polymorphisms (SNPs) are estimated to occur in the human genome at about 1.4 million, with 60,000 in the coding exon regions. Numerous

Factors	% Companies	Factors	% Companies
Clinical trial data	86	Projected peak sales	68
Financial assessment	86	Commercial uncertainty	68
Time required	79	Novelty	57
Market attractiveness	79	Staffing at company	57
Competitive activity	75	Portfolio fit	54

FIG. 2.24. Factors in R&D Decisions (Example @ Phase 2/3) (Copyright 2006 from *Encyclopedia of Pharmaceutical Technology, 3rd Ed* by Swarbrick. Reproduced by permission of Routledge/Taylor & Francis Group, LLC) (Reprinted with permission from *Pharmaceutical Executive*, Vol 24, No. 2, 2004, page 58. *Pharmaceutical Executive* is a copyrighted publication of Advanstar Communication Inc. All rights reserved.)
Source: CenterWatch 2004

drug examples in many pharmacologic classes and subsets of patients exist where individual patients experience adversity, but finding those patients in a practical, cost-effective, clinically useful manner remains a major challenge to health care. These diagnostic tests are not routinely done today in clinical practice, which could change especially in oncology.

Figure 2.25 presents the pros and cons in incorporating pharmacogenomic approaches in product development plans. Drug developers suggest that many of the 20 drug withdrawals from the market after approval over the past 20 years was due to unexpected serious adverse experiences potentially related to pharmacogenomic variation in the population. Pharmacogenomic variation is starting to be used in drug development in cancer area. In efficacy, Herceptin® (monoclonal antibody) is only effective in breast cancer patients with her2neu oncogene present. Gleevec® is tyrosine kinase inhibitor and highly effective in chronic myelogenous leukemia patients with the c-abl oncogene fused to the bcr cluster protein region. For toxicity, the hepatic enzyme system cytochrome p-450 is involved in metabolism of many drug categories (e.g., beta-blockers, tricyclic antidepressants), with numerous different mutations common often in ethnic groups; they alter drug clearance and lead to toxic effects. However, identification of these patients is not done routinely in clinical practice, they add cost to health care system, the tests are not readily available, and education of professionals is lacking. Furthermore, FDA has only recently written (end of 2004) proposed optional guidelines for use of pharmacogenomics in drug development. This discipline could help identify the best responders or patients more likely to experience adverse effects. However, added cost to drug development without proven diagnostic capability, with high variability, unsure ethics, unsure insurance issues, and negative impact on the market, are some of the current challenges [13, 49, 50].

Administration of products to patients depends on the properties of drug or biological products, disease characteristics, human physiology, and health care system variables, which will impact the formulation options desirable for a new product (Fig. 2.26). Products must cross several biologic membranes from their site of administration and encounter endogenous enzymes that could alter their activity. Goals of formulation development include mechanical and physical stability that is compatible with how product will be used, long shelf-life, maximum patient acceptance, provider utility, bioavailability, lowest cost of goods to manufacture, competitive advantage, and less local adverse effects with use. Most products are intended for oral use as tablets, capsules, or liquids, which is preferred by patients, but the disease or drug metabolism may not permit this route of administration.

Pros (Proposed):
- More Safety in phase 1
- Identify best responders
- More efficient trials
- Identify outliers with SAE
- Explain unusual AEs
- Fewer late stage failures
- Avoid PI restrictions
- Less cost of drug development

Cons (Costs & Challenges):
- Cost of genotyping
- Patient consent x 2 needed
- SNP diagnostic tests not done
- Clinical significance?
- SNPs polygenic, utility?
- SNPs predict disease?
- Insurance Impact
- Ethics & Access to data
- FDA regulation of PIs & trials
- Market of drug reduced

FIG. 2.25. Pharmacogenomic Issues in Development

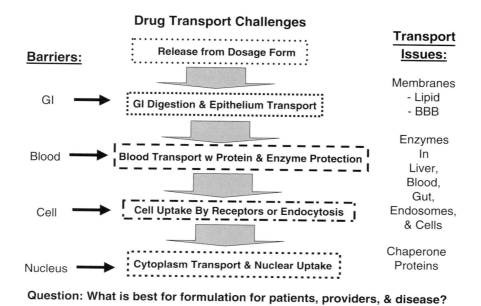

FIG. 2.26. Formulation Issues in Drug Administration

Active pharmaceutical ingredients need to meet specifications (e.g., purity, potency, physical/chemical/biological properties). For injections, we need to consider the solution's pH, ionic strength, solvents, buffers, stabilizers, and preservatives. Special formulations will have their own unique considerations, for example, inhalation (particle size), topical (skin penetration), and ophthalmic (mix with tears).

A product is not one single entity that is discovered, studied, and marketed at one point in time. A life cycle exists for all products, wherein it evolves in its uses, properties, and formulation over time (Fig. 2.27). A company attempts to get the most scientific benefits from its investment in a product for patient care improvements and, of course, maximum financial gain. However, to fully capitalize on the product, a life cycle plan (LCP) must be created as early as possible to literally map out the full potential of a product and the resources that it might take to achieve all the opportunities. The LCP is a dynamic document; as more data becomes available over time, you iterate the LCP. You can suggest more than 10 different stages in a life cycle; that is, discovery of molecule, preclinical product, product in clinical trials (first indication), approved and marketed for one indication (hopefully novel, first to market), added countries for marketing, product with competition, added indications, expanded labeling for added dosing schema, unexpected adverse experience limiting use, new formulations, follow-on molecules in the family or by in-licensing, off-patent with generic substitution, and over-the-counter product. Many of these stages will require added product development and costs regarding clinical trials to create the data and file an abbreviated NDA to obtain labeling changes. New formulation work may require clinical trials to establish efficacy, but at least new manufacturing, pharmacokinetic, and stability work will be necessary. Sustained-release formulations are often developed later in the product's life cycle once the initial efficacy is established with the simpler first-generation form. Then, enhanced convenience (e.g. once daily vs. four times a day dosing) becomes an important focus. The LCP needs early development to plan, schedule, and integrate milestones, timelines, work requirements, cost projections, market research, and revenue impacts. For example, Enbrel® was approved first for rheumatoid arthritis, with later indications established for juvenile arthritis, psoriatic arthritis, psoriasis, and ankylosing spondilytis, as well as follow-on molecules and other dosing schema being evaluated. The statin anticholesterol drugs were over-the-counter products in Europe by 2004. Pegylation was done for interferons and filgrastim to slow their clearance from the body and improve their clinical utility with less frequent dosing and also extending their patent protection [51, 52].

PPM Analyses (Examples)

As noted before, analyses are a critical function for PPM to coordinate in all parts of the organization and at many points in time, as they relate to product development. A myriad of analyses exist where whole books are dedicated to listing and describing them. In this book, we are presenting some representative examples of analytical techniques that are often employed by pharmaceutical companies to evaluate a product or related projects from both scientific and marketing perspectives. Many analyses attempt to give projections of future possible outcomes, as well as determining the status of projects. Data formats are also highly variable and often can depend on personal style (e.g., tables, graphs, histograms, bubble diagrams, flowcharts, and decision trees). The PPM planners and product team leaders must address the questions at hand, use the necessary data available, know the preferences of management, present their recommendations, and help produce a decision [22, 41, 42].

Figure 2.28 presents five categories for the PPM analyses to fit into; opportunities for the product (usually the scientific, medical, and public benefits, as well as internal product profiles and compound ratings), sales in the marketplace (short-term and long-term plus by indication, prescriber groups), risks of failure or probability of success (modifiers of the scientific and sales opportunities and their impacts on success), resources in both budget and staffing (cost projections to operate), and

Planning early & late & broadly = Product Success

- Develop a map for molecule progression
- Organize corporate resources
- Plan for multiple indications
- **Think international**
- Consider alternate drug delivery systems
- Consider labeling expansion
- Get the most out of your patent & defend it
- Plan for follow-on molecules
- Consider in-licensing or blocking strategy
- Be prepared for competitive molecules
- Be prepared for generic competition

FIG. 2.27. Product Life Cycle Management

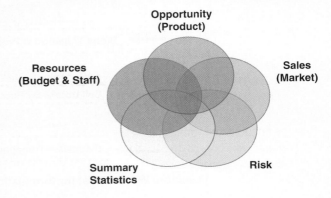

FIG. 2.28. Types of PPM Analyses

finally some overall summary statistics to be elaborated later [22, 41].

In Figure 2.29, four opportunity and three sales analyses are listed. The opportunity analyses noted are compound ratings, product profile comparisons, safety versus efficacy bubble diagrams, and NMEs versus MOAs. First, compound ratings are used to help set priorities among the product candidates in a portfolio. The ratings engage the teams to individually score their product with a standardized list of properties or parameters for new products. Often, the set of decision criteria discussed previously are used (e.g., efficacy, safety, formulation utility, market size, sales potential, competition, patent status, probability of success, and cost of goods). Each criterion is scored on a 1 to 5 or 1 to 10 scale from worst to best or lowest to highest. These scores are highly subjective, educated opinions and must be defended by the teams in front of management. An inordinately high score will diminish the credibility of the team and hurt the product's chances of moving forward. The scores are summed and all the products are then compared with each other with a minimum score required for a product to advance. The scoring is not an answer in and of itself but offers two benefits; creates some degree of standardization across a portfolio and is intended to stimulate a good discussion of products across a portfolio leading up to decisions of go–no go.

A second opportunity analysis is for product profiling, which can be combined with sales forecasts to assess a product's range of possible future revenues. The product team will create three profiles, one that fits the most desirable product profile for that disease, these type of products, and optimal research outcomes for the company's product. The most desirable (ideal) profile is compared with a second profile for the most likely product profile with the company product's performance in research. Finally, sometimes a third profile for the least acceptable profile for the product is created. Market research takes these three profiles to groups of thought leaders and customers to evaluate the merits of the product and its potential use by itself and in comparison with competitors. Marketing then has to transform the written feedback into sales possibilities based on experience in this particular marketplace of products. Sales over time for about 5 years is projected for the three products (three product profiles) and compared to guide R&D on the importance of particular profiles and even specific properties in a profile impacting future potential use of such products. Such profile and sales data is used to move a product forward to market or to kill it.

Bubble diagrams are used to identify and gauge opportunities (or lack there of) in a graph for two key parameters describing the products (science or marketing data). Each parameter on the x- and y-axes is rated from lowest to highest values. Each bubble is a molecule or product. A third parameter can be introduced by altering the size of the bubbles to indicate, for example, sales potential or likelihood of success.

Sales analyses are generated by the marketing groups who already are presenting some sales numbers on a weekly, even daily, basis to senior management. Prescriptions are generated for products used in the retail markets distributed by community pharmacies, mostly for oral products and a few other formulations to some extent. Other channels for product distribution (e.g., hospitals, clinics, physician offices) purchase products through wholesalers or directly from a company. Sales can be reported in dollars or a combination of dollars and prescription levels. Common time periods for sales reports are monthly, quarterly, and annually. Sales projections for a new product in development are based on a likely price often similar to other marketed products for the same disease. Competitive products need to be taken into consideration, either ones already or soon to be on the market. Usually a 5-year sales forecast is desired for a new product, and a goal is to produce a blockbuster, $1 billion, in this time frame. Peak sales are compared with the priority rankings previously noted and to R&D costs.

A sample bubble diagram is presented here (Fig. 2.30), wherein safety is combined with efficacy and market size in one comparison for the product portfolio of the seven products. Product no. 7 receives a very high assessment with good efficacy and good safety and pretty good market size; this product would be advanced quickly. High efficacy and safety does not ensure high sales if the product has a small niche

Opportunity analyses examples:
- Compound ratings
- Product profile: minimum & desirable vs. sales forecast
- Bubble diagrams for efficacy vs. safety with market size
- NMEs vs. MOAs

Sales analyses examples:
- Revenue projections over time (5 + years)
- Peak sales vs. priority ranking
- Peak sales vs. R&D costs

FIG. 2.29. PPM Analyses: Opportunities? & Sales?

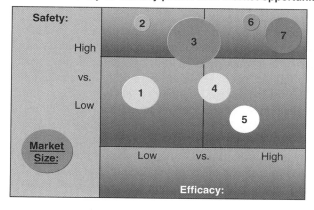

FIG. 2.30. PPM Analysis: Safety, Efficacy & Market Size

type market, basically a narrow indication and/or small market size. In general, low safety dooms a product in decisions of advancement unless the severity of disease is high, other treatments options are not available, market size is large, and treatment failures are common with existing products allowing for use of such a product. This situation is common in cancer, and antimetabolites are used for other chronic serious inflammatory conditions such as arthritis, lupus, psoriasis [41, 42].

A comparison of potential pipeline product sales can be presented over time, which is a very common calculation for companies (Fig. 2.31). The y-axis is sales in dollars per product and the x-axis is years. The slope of the line is desired to be as upwardly sharp as possible after approval and launch, indicating more rapid and broad acceptance of the product by the medical and payer communities. If the slope of the sales line is positively higher in the first and second years, it predicts high annual sales for the life of the product. If a product has lower sales early, it will have lower sales throughout the life of the product. You have only one first year of launch for a new product to get the full attention of the health care community and product acceptance; if you miss it, you never get it back, unless a major new positive event happens with your product. How well you prepare the medical community to accept a new or novel treatment influences the curves. Often work is started several years before the formal launch to facilitate adoption of such treatments. This work may require helping the medical community understand a new evolving area of science. If a product has taken many years to reach the marketplace, prelaunch preparation is critical to assuring a positive return on investment (ROI). A product falls off or levels off in sales after initial rises when a superior competitor comes to market, and especially if the product goes off patent and a generic drug enters the market [41, 42].

Risk and resource analyses are suggested on Fig. 2.32. Four risk assessments are listed; time to market versus risk of failure, probability of success by phase, risk scoring, and decision trees. Resource analyses noted are budget actual spend versus projected spend over time, a gap analysis for resources available versus project resource demands (for staff or budget), and sales projections versus resources needed for development and launch (research and marketing costs). Resource assessment also must take into account appropriate expertise of staff.

A pipeline is shown in the table in Figure 2.33 for a 9-year period and presents projections and expectations by the R&D organization for the annual number of research projects, INDs/CTAs, NDAs/CTDs, approvals, and NMEs. The goal of R&D for the pipeline is to have a reasonable number of products in each phase of research, so that there can be a steady stream of products moving through the organization to occupy the existing staff levels and especially to avoid any years down the road where products applications to regulatory authorities are lacking. The R&D management team and PPM planners then can perform a gap analysis of the pipeline. This sample pipeline has some problems in the out years after 5 years with inadequate number of NDAs/CTDs, plus a deficiency in NMEs, the novel products. The research projects fell to half in the middle years of the planning period. A solution for a company to consider, in the 5- to 10-year period prior to this gap in NDAs/CTDs, is to build up research with outside research organizations, especially with future potential NMEs (novel MOAs or product categories). Two approaches would be to license in early stage novel targets and molecules from universities, government (NIH), or small

Risk analyses examples:
- Time to market vs. Risk of failure
- Probability of success by phase
- Risk scoring
- Decision trees

Resource analyses examples:
- Budget
- Department staff & resource gap analysis (by department)
- Sales vs. resources

FIG. 2.32. PPM Analyses: Risk? & Resources?

What are Revenue forecasts per pipeline product & total over 5 years?

Revenue $ Millions	2005 Tot. $1.8	2006 $3.7	2007 $6.2	2008 $7.6	2009 $9.0

FIG. 2.31. PPM Analysis: Pipeline Revenue Forecasts

Are there any gaps over time in any stage of research to fill in?

Year → R&D Deliv	'04	'05	'06	'07	'08	'09	'10	'11	'12
Res. Proj.	150	105	88	75	70	?	?	?	?
INDs	47	56	66	41	20	22	24	27	37
NDAs	32	40	28	32	15	12	10	15	22
Approvals	8	15	15	10	6	2	3	4	6
NMEs	3	5	4	3	3	2	1	0	0

FIG. 2.33. PPM Analyses: Risk, Pipeline Gap Analysis

specialized companies, or create strategic research alliances with biotechnology companies in areas new to the company. In this example, gaps appear also in late-stage clinical phases (phase 3 trials) near regulatory submissions, or it could be that NDAs are not being approved by the regulatory authorities. A reason may be poor molecules are advancing into late-stage research that should have been killed at an earlier time frame, such that process changes in decision making need to be entertained for the company.

A risk assessment in the early discovery phase of research is a bubble graph of the novelty of the chemical entity and mechanism of action of the molecules, as shown in Fig. 2.34. The projected market size is added by altering the size of the bubbles commensurate with potential market. Although the molecules are over 5 years away from the market and without any clinical data, markets are estimated based on assumptions, which need to be plausible to both research and marketing management. Assumptions can be any existing product utilization, growth in markets over time, novelty of mechanism or chemical entity (competitive advantage), growth in number of addressable treatable patients with such a novel product proposed product profile of such a molecule, and price of such a product. Company senior management needs to decide on how much risk they desire to take.

In this sample graph, molecule no. 3 has high risk with an unproven chemical entity and speculative MOA, but the market potential is huge. Perhaps, existing drugs work poorly, many patients go untreated, and the medical community is very hungry for a novel product. In this case, it may very well be worth forging ahead with such a high-risk, high-reward molecule. A research group with this molecule especially needs to nail down the novelty of the MOA for the product to reach its full medical benefits and financial gain. Conversely, molecule no. 6 has the least risk but a small market that may not be worth spending the vast amount of resources for little financial gain. Along with this analysis, the research group needs good feedback from medical and marketing groups about the disease opportunity to make good research decisions on choosing molecules to advance out of the laboratory. The challenge is to make sure the vision is not a hallucination, vis-à-vis, balancing a product champion's view with marketing reality. The gap analysis for basic research (discovery) can look also at process issues to improve productivity, as suggested by one analyst, with a survey of 20 questions to examine discovery. Always, a process is needed to take the gap findings to management, communicate them to appropriate managers, have them identify corrective action, and implement changes [41, 42].

The probability of success for molecules to move through the stages of research and development can be estimated for a company and compared with industry standards. Figure 2.35 presents some summary statistics for the probabilities of advancing a product candidate through the clinical research stages. Twelve pharmaceutical companies were surveyed in this 2003 assessment; the median value and the low and high values were included. Phase 1 involving a safety assessment and some pharmacokinetics has a relatively high success rate, but phase 2 studying a product's activity in disease patients is much lower. Proof of pharmacologic principle in real patients is almost always a major and commonly unpredictable challenge for a new product. Animal models often are helpful and required to be done before introduction to humans, but they are vicarious predictors of drug activity in disease in people, related to species, genomic, disease, and metabolic variations, to name a few. Most products that have good activity in phase 2 will proceed to phase 3 and be fairly successful (73%), but 27% still fail to advance because of safety concerns, lack of sufficient efficacy, pharmaconomic deficits (efficacy exists but insufficient to garner enough market share). Probabilities can be calculated for preclinical success rates of molecules, but this data source did not have them. Termination in preclinical stage occurs because of tissue or organ toxicity, no product activity, excess metabolism, high drug interaction potential, and formulation difficulties compared with existing products and disease needs. A failure to advance after NDA submission relates to

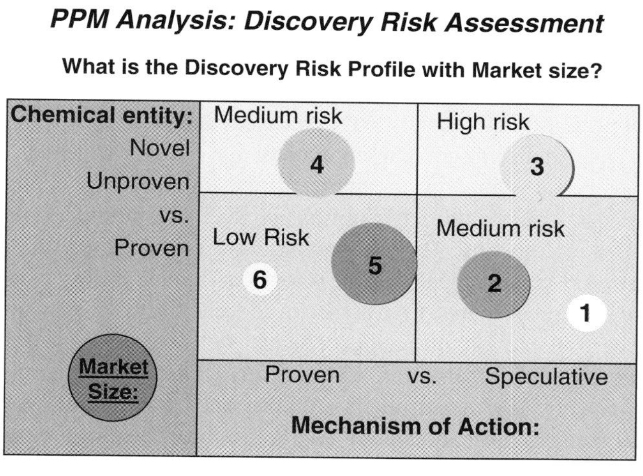

FIG. 2.34. PPM Analysis: Discovery Risk Assessment

Has company made estimates for success for all products?

Stage of Development	High %	Industry median %	Low %
Preclinical	?	?	?
Phase 1	80	70	60
Phase 2	52	46	30
Phase 3	85	73	65
Filing or Launch	?	?	?

FIG. 2.35. PPM Analysis Risks, Probabilities of Success
Source: Survey of 13 Industry reports. DIA Annual meeting, June 17, 2003, San Antonio, Tx.

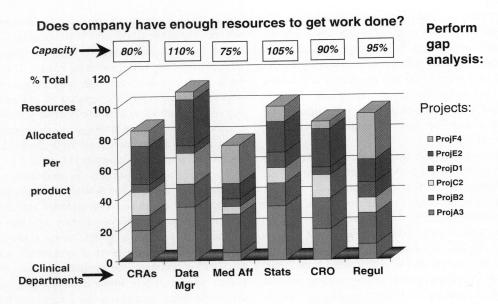

Fig. 2.36. PPM Analysis: Resource Gaps? Availability?

- Progress of activities over time vs milestone targets
- ROV of product (Return on Value)
- Balanced Scorecards
- NPV of product (Net Present Value)
- Value of Pipeline; Criteria Scoring

Fig. 2.37. PPM Summary Product Analyses

disagreement between regulatory authorities and the company with the NDA because of safety or inadequate efficacy. Another reason to kill a product after phase 3 is that the market assessment, given the full knowledge of the product profile after phase 3, suggests unmet need and related market share will be inadequate compared with sales and marketing costs including launch costs ($25 million to $100 million) and annual selling costs. A measure of a best practice company is it's products fail early versus late in the development cycle [45].

A company's R&D success will be dependent on their staffing and expertise as well. Do we have enough people with the correct expertise to perform that work needed to be done for all our molecules? The staffing for the development operation is displayed in Figure 2.36 with a histogram approach. The key departments are listed on the x-axis for CRAs, data managers, medical affairs or safety group (AEs), statistics, regulatory, and the CRO outsourcing groups. The y-axis shows the percentage of resources allocated per project. Each product project is given a resource use value by a department and entered on the graph; and a total of all product resource use per department is given. Then, a gap analysis can be performed to determine excesses capacity and underage needing shoring up with outside resources (more CROs). Another solution is to do less work and slow down the advancement of a molecule that is lower priority or kill a very low priority project altogether. This gap analysis is best done as projections going forward into future years as well, which will permit the hiring and training of new staff members or education of internal staff, which easily can take from 3 to 6 months for this orientation process [9].

The last set of analyses to be discussed covers overall assessments of pipeline status or product status, especially the concept of value, which incorporates scientific and marketing principles (Fig. 2.37). Two pipeline assessments will be addressed in subsequent discussions (e.g., the balanced scorecard and then pipeline value with criteria scoring). More general product assessment may include progress of activities over time, an example of which will be described in the later project management topic, return on value of a product, and net present value, which will be covered below.

The balanced scorecard approach for R&D can assess a very broad scope of parameters focusing on components, process, and evolution (flexibility) of R&D and not just products at a company. The assessment has a positive outlook for staff and management, in that it focuses on making organizational improvements. Five areas are addressed and scored: customer satisfaction, process excellence, sustainable innovation, learning and growth, and value in financial and commercial terms (Fig. 2.38). Customers are potentially any internal staff member or external resource. Within each of the five areas, a set of key characteristics is assessed at four levels, ongoing and okay, ongoing and needs improvement, inadequate

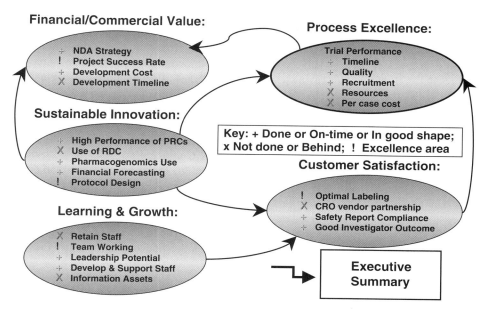

FIG. 2.38. PPM Analysis: Balanced Scorecard for R&D (Reprinted with permission from *Pharmaceutical Executive*, Vol 23, No. 10, 2004, pages 84-90. *Pharmaceutical Executive* is a copyrighted publication of Advanstar Communication Inc. All rights reserved.)

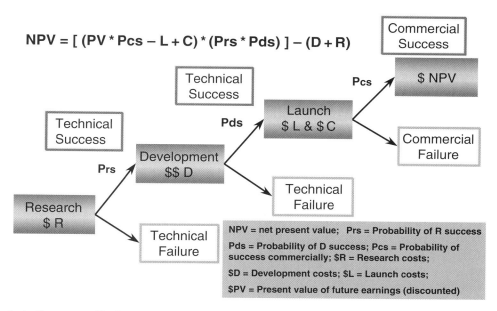

FIG. 2.39. PPM Analysis: Summary – Net Present Value (Copyright © 1998 by R.G. Cooper, S.J. Edgett and E.J. Kleinschmidt, reprinted by permission of Perseus Books PLC, a member of Perseus Books, L.L.C.)

performance or not being done. The diagram shows these three scores by the "×" for not done, "+" okay, and "!" needs improvement. An underpinning of this type of assessment is a focus on improving productivity especially with the high and rising cost of R&D. Cost savings and efficiencies can be made by examining processes or the work activities, which should result in more productivity. The final product is an executive summary identifying strengths and weaknesses of R&D operations and outcomes for improvement [53].

Net present value is a frequent calculation made to gauge the overall future value for individual pipeline products considering the whole life product cycle for R&D and S&M (Fig. 2.39). The key variables incorporated are costs of research, development, and marketing (launch and commercialization), the estimated probability at three transitions (research to development, development to launch, and launch to commercial success), and future earnings (discounted for inflation). A net present value (NPV) is calculated for each product in the particular review cycle, (e.g., annual) and is used with other data to help make the go–no go decisions for product advancement. For more innovative products, prediction of its value is more difficult with more assumptions lessening the reliability of the assessment [42].

Scoring of the pipeline to judge potential success of the product portfolio can be done using the company's own key corporate criteria and a standardized rating scale that broadly engages the whole organization. Individual products are scored and then ranked together to get the collective picture of the pipeline and its components. Figure 2.40 presents a sample pipeline scoring system, incorporating a scale from 0 (worst) to 15 (best) and seven criteria, such that a product score can theoretically range from 0 to 105. The seven criteria cover science and marketing, and operations and strategy: strategic alignment of the product to corporate overall science and marketing strategy, novelty in the science (MOA, product category, actions safety), market attractiveness (number of patients with disease, addressable patients with such a product, growth in patients, chronicity of therapy), competitive advantage (no classes to treat disease, first-in-class, best-in-class for a important product property), resources available to do the research (R&D), capability to produce the product (manufacturing capacity and skills), sell the product (marketing expertise, sales people), and the NPV. The ratings are quantifiable but still subjective, warranting caution in fair scoring between products with a minimum of bias. The senior team may score the products and not the team members. Let us use the fictional example of a FIPCO company that researches and markets cardiovascular products for hypertension, infectious disease drugs for pneumonia, and neurologic drugs for epilepsy. A product like Vytorin®, a new and more effective hyperlipidemia product, could be scored as strategy (3), science (3), market (4), competitive advantage (3), resources (3 and 4), and NPV (3), which results in a good score of "80." Another example is an antibiotic for urinary tract infections in the existing quinolone family that can be given once a day (a moderate opportunity); its ratings could be Stra-3, Sci-1, Mkt-2, Comp-2, Res-3&4, NPV-2, with an overall moderate score of 50. This company may market both products because the first one is expected to be a market leader in cardiovascular, and the second product could be approved when a gap exists in new products for the company and the sales force will have time to devote to such a product, even though it will be a smaller financial opportunity [42].

Project Management

Project management is a discipline in companies that focuses on processes and their continued improvement, as well as achieving outcomes, for specific projects. Planning, program design, process design, communication, coordination, resource allocation, and timing of people, systems, information, and their projects comprise key elements of project management. Although the common goal across the company is developing a product, a company has many distinctly different departments all contributing to product development with very different projects and outcomes, different operations, different expectations, different education, and different cultures; however, they all need to be brought together to function as a unit. Efficiency in operations of R&D has repeatedly been identified by analysts of, and senior management at, pharmaceutical companies as an important means to reduce cost and improve productivity. Of course the drivers of this efficiency requirement are the very high cost of R&D, now $800 million up to $1 billion per new product; the longer development times and slowing product approval rates, especially for NMEs; and the complexity of R&D. R&D can be very amenable to various efficiency improvements as follows. Many work items are done for an IND and NDA for a product by many separate departments by a large number of people with varied amount and levels of experience. In addition, the processes and outcomes (studies, reports, and applications) are highly structured and detailed. In the past, managers from research or clinical trial areas with technical expertise became the team leaders and coordinators of team projects. However, the role of coordination, tracking, and communication of a team of technical people requires an additional skill set and tools for these operational functions. This section of the chapter discusses the process of project management and a couple of roles that PM people can offer to an organization. A complete presentation of project management is found in book publications and is beyond the scope of this book.

Rating Scale ⇒	15	10	5	0
Strategic alignment	Yes, 4+	Supportive 3+	Neutral 2+	No, 0
Novel science	4+	3+	2+	1+
Market attractive	High 4+	Medium 3+	Low 2+	Poor 1+
Competitive advantage	Strong 4+	Moderate 3+	Some 2+	Equal 1+
Resources available	R&D Yes 4+ S&M 4+	4+ 2+	2+ 4+	Low 2+ 2+
NPV	>$ 5 Billion 4+	$1-4 Billion 3+	$ 0.5 - 1.0 Billion, 2+	< $0.5 Billion, 1+
Each product score is 0 to 105; How is each product in portfolio ranked?				

FIG. 2.40. PPM Analysis: Criteria Scores for Pipeline (Copyright © 1998 by R.G. Cooper, S.J. Edgett and E.J. Kleinschmidt, reprinted by permission of Perseus Books PLC, a member of Perseus Books, L.L.C.)

A structured and ideal planning cycle exists for project management (PM), involving eight steps or phases in chronological order: conceptualize, design, plan, allocate, execute, deliver, review, and support in the model from Kennedy (Fig. 2.41). As you can observe, a full cycle of activities is incorporated sequentially and then returned to the origins of the project (concepts and plans) to reinvigorate the process and improve them and their outcomes. Also, it should be noted that "support" is the final phase of PM, suggesting to the team members that the bottom line for project management is to help (support) the team plan and achieve its outcomes. The project planner does not do all the planning, or execute the work, or deliver the outcomes, but assists the team to make sure that all the steps are followed through by the participants to improve the outcomes. A common complaint about PM is that it may appear initially to take added work and time of individual departments and team members away from their specific duties. A PM manager needs to demonstrate the value of PM, which includes some training of all team members for PM and especially the team leaders. Corporate commitment to PM is another leadership requirement of senior management in order to achieve the efficiency opportunities [41].

Project planners can use a variety of tools to identify, track, and coordinate progress on all the different projects for a product's development in R&D. Pictorial or graphic representation of such project data assists department managers, team members, and team leaders, and senior management understand the scope of work being done, time frames for the work, deadlines for any one area and the collective process, and how one area's work may fit into the bigger development picture in work flow, sequence, and goal achievement. A typical graph is a Gantt-like chart that lists all the key projects over time, showing their start times and projected completion dates and current status (Fig. 2.42). Such project presentation allows a team to identify what all is being done by whom and its timeliness. Any delays can be noted, readily identified, and discussed at team or ad hoc meetings, with formulation of a resolution to the issues. PPM can use these Gantt-like charts as well collectively for all the products and identify areas of the company where consistent problems may occur in getting work done on time. Then PPM can help that department manager identify the specific problems and add resources or

FIG. 2.41. Project Management, The Planning Cycle

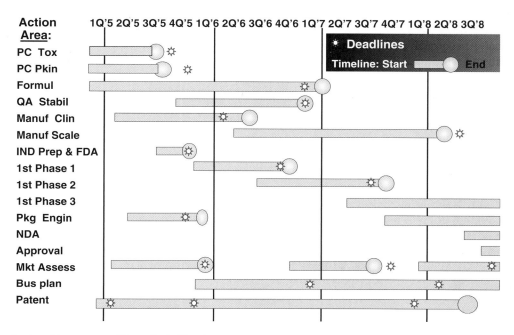

FIG. 2.42. Product Project Management (Gantt Style Chart)

Process Steps [6]:

- Identify all risks (take the time)......
- Analyze both probabilities & impacts of risks.........................
- Prioritize risks on drug development, esp. timeline..........
- Plan responses to risks (avoid, mitigate, accept)...............
- Monitor for risk occurrence & Implement responses..................
- Repeat cycle as a risk occur.........

Benefits [6]:

- Prioritize project work
- Supports decision making & planning
- Avoids crisis management
- Encourages management of opportunities
- Creates project credibility for milestones and time lines
- Provide more realistic control of project outcomes

FIG. 2.43. Project Risk Management

change processes to rectify the delays. Also, areas of the company consistently exceeding deadlines can be identified for awards of excellence [41, 42].

Risk management (RM) for projects is a permutation of project management wherein the PM manager and team members try to anticipate future problems and mitigate them in advance (Fig. 2.43). The axiom, "an ounce of prevention is worth a pound of work or problems," is an excellent justification for RM. Although extra team member time is taken upfront delaying some work, the total time for the projects is lessened along with less anxiety, more timely execution, better team morale, and product success. Some areas of the company may be known for delays; which is where RM should be done. For high-priority products that are on a fast track, you want no down time or delays, because of the high corporate profile and the greatest need for optimal follow-through. Here especially is where RM will serve an organization well. Crisis management is needed in all companies but creates major delays with low morale and higher costs when one project has to be redone, upsetting sequence of work in other areas as well. Avoidance actually will take less time, benefit the company by avoiding angst, and offer more productivity in R&D and more commercial success as well. Also, RM helps an organization learn from its mistakes or problems, which is a strong learning tool for process improvement and efficiency. A full discussion of RM is not possible here, but a summary is provided in Figure 2.43 to offer at least an overview of this important process.

One way to enhance performance is to do a postmortem examination of a program to determine what went well and where there were problems. If the data from this sort of evaluation can be collected in a systematic way and reviewed honestly and openly with senior management, problem areas can be identified. Solutions found in one project may be utilized in others to improve the development process. PPM can collect this data and is in a good position to share the learnings with new teams early as a preventative and educational measure to avoid pitfalls.

Summary

PPM is considered a very important process for companies and has potentially quite significant impact on overall corporate performance. An assessment of the importance of portfolio management was published in 1998 in *Research-Technology Management* journal, as abstracted in Figure 2.44. Importance of PPM was rated on a 1 to 5 scale from not to critically important. Five levels of management were surveyed in three specific areas (technology, production, and sales and marketing) and two higher levels, senior and corporate levels. Also, the top 20% of performers were compared with all businesses and the bottom 20% of performers. In every management category, top performing companies assessed PPM significantly much more highly (one whole level in the scale) than poor performers and even more than all businesses. Also, senior and executive management assessed PPM quite high. Technology management felt PPM was the most important among all managers, up to 4.6 (between critically and very important). The production and operations managers felt PPM was the least important, from somewhat to quite important [42].

The summary for planning, governance, and execution for products in R&D is a picture of the key elements that comprise R&D; Figure 2.45 displays them in an interactive integrated matrix. They need to be done to fulfill the strategy, on target (specifications), on time, in sequence, adjusted as necessary as the product matures and evolves, and performed by a team of staff in many distinct areas of the company. Twenty plus unique and different departments or project types need to be accomplished to result in product approvals and a product portfolio, as shown here. The acronym "PICTRS" fits well the summary of optimal planning, governance and execution through PPM, leadership, and organizational operations (P to the eighth power): P, plan; I, involve and implement; C, coordinate, and coerce if necessary; T, track; R, review and re-energize; and S, succeed and satisfy (team members, research

2. R&D Planning and Governance

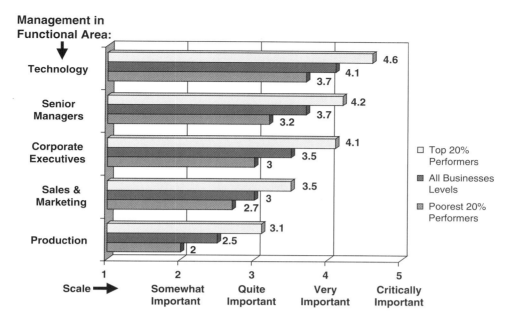

FIG. 2.44. Importance of Portfolio Management (Copyright © 1998 by R.G. Cooper, S.J. Edgett and E.J. Kleinschmidt, reprinted by permission of Perseus Books PLC, a member of Perseus Books, L.L.C.)

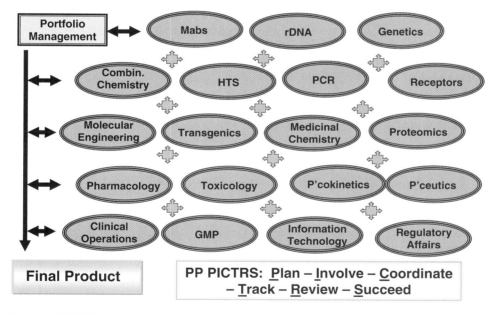

FIG. 2.45. Portfolio Planning (PP) PICTRS in R&D (Adapted with permission from Routledge/Taylor & Francis Books LLC, NY, NY. Freom Figure in Evens RP. Biotechnology and Biological Products. Figure 1 – Technologies of biotechnology. In Swarbrick S. ed. Encyclopedia of Pharmaceutical Technology, 3rd Ed., 2006.)

department managers, marketing and sales management, and senior management).

Now, as a summary to the planning and governance chapter and our discussion, let us revisit the list of the 15 potential pitfalls in the regulatory process for IND/CTA filing and NDA/CTD filing (Fig. 2.46). We will make a subjective assessment of the impact on avoiding the pitfalls by the three major planning and organizational effectiveness tools discussed in this chapter; portfolio project management (PPM), the paradigm of product development (P to the eighth power), and leadership (6 + 6 Essentials of Leadership Excellence). We will use a Likert-type scale of agreeability for the impact of each tool on

Pitfalls with INDs Scores:	PPM	P-8th	Ldr	All 3
• Not communicating with FDA	1	1	0	2
• Avoiding safety issue	2	2	0	3
• Lack of planning	3	2	2	3
• Omitting data or unnecessary data used	1	2	1	3
• Not paying attention	3	2	3	3
• Not documenting manufacturing process	2	2	0	3
• Ignoring investigators	1	2	2	3
• Forgetting conflicts of interest	1	1	1	2
• Hiding something	3	2	3	3
• Being too eager to disclose to the public	1	2	1	2
• Not thinking globally	3	2	2	3
• Not thinking about electronic submissions	1	2	0	2
• Rushing	1	1	2	3
• Not Choosing wisely	3	2	3	3
• Filing a drug that is not needed	2	2	2	3
Total Scores	28	27	22	+41

Impact Scale: –2 Very poor, –1 Poor, 0 Neutral, +1 Fair, +2 Good, +3 Excellent

FIG. 2.46. IND Benefits: PPM, P-8th paradigm & Leaders

avoiding each pitfall from strongly disagree (−2), disagree (−1), neutral (0), fair (1), good (2), up to excellent agreement (3). A total score can be as low as −30 to as high as +45, and we would suggest that a high score at or above 20–30 indicates significant impact (overall at least good) on NDA/CTD process and approvals. These scores will be generated by the authors, and you as the reader can determine your own scoring. Pitfalls with the scores are found in the table in Fig. 2.46.

PPM should have certain strengths as can be observed by the five high scores for "planning," "paying attention," "not hiding something," "global thinking," and "choosing wisely." Subtotal score for impact of PPM on pitfalls is 28, a good score. Pitfalls scores for the organizational effectiveness model (P to the eighth power) are suggested to be fairly consistent across most of the pitfalls in avoiding them, without any particular single strength. The model has broad potential beneficial impact on all the various operations of R&D product development because the paradigm basically is intended to impact all the elements of product development to some extent. The subtotal score for use of the organizational model, P to the eighth power, is 27, also a good score by itself [43].

Leadership and specifically the model of leadership offered in this treatise can be scored on the 15 pitfalls in filing INDs/NDAs. The strengths that leadership offers for the pitfalls is suggested to involve particularly "paying attention," "hiding something," and "choosing wisely," plus favorably impacting six other pitfalls. Its subtotal score is 22, a reasonably good score. Then, when you combine all three operational and planning models (leadership, organizational effectiveness, and PPM), the total score for avoidance of failure with INDs/NDAs is suggested to be quite high, possibly 41 out of 45. The caveat is that the three models are fully implemented, operational, and running reasonably smoothly, as well as reassessed periodically to refresh and upgrade them.

References

1. Guy P. Rising to the productivity challenge. A strategic framework for biopharma. Boston Consulting Group. July 2004
2. Engel S, King J. Biotech innovation. Pipeline gaps. Can biotech fill them? R&D Directions 2004;10(7):42-56
3. Booth B, Zemmell R. Prospects for productivity. Nature Reviews Drug Discovery 2004;3(5):451-6
4. Rawlins MD. Cutting the cost of drug development. Nature Reviews Drug Discovery 2004;3(4):360-4
5. Cohen CM. A path for improved pharmaceutical productivity. Nature Reviews Drug Discovery 2003;2(9):751-3
6. R&D business. Mostly growth for big pharma in 2003. R&D Directions 2004;10(3):14-15
7. Booth B, Zemmel R. Quest for the best. Nature Reviews Drug Discovery 2003;2(10):838-41
8. Lawyer P, Kirstein A, Yabuki H, Gjaja M, Kush D. High science: a best-practice formula for driving innovation. In Vivo The Business & Medicine Report 2004;22(4):1-12
9. Malek J. The path to smart R&D. Pharmaceutical Executive 2003;23(11):70-80
10. Engel S. How to manage a winning pipeline. R&D Directions 2002;8(2):36-41
11. The pipeline problem. The pipeline solution. R&D Directions 2004;10(5):42-54
12. Tollman P, Goodall S, Ringel M. Good governance gives good advice. BCG Focus. Boston Consulting Group. 2004

13. Tollman P, Guy P, Altshuler J, Flanagan A, Steiner M. A revolution in R&D. How genomics and genetics are transforming the biopharmaceutical industry. Boston Consulting Group, November 2001
14. Preziosi P. Science, pharmacoeconomics and ethics in drug R&D: a sustainable future scenario? Nature Reviews Drug Discovery 2004;3(6):521-6
15. Balekdjian D, Russo M. Managed care update: show us the value. Pharmaceutical Executive special report. September 2003
16. Kaitin KI (ed.). Decline in drug approvals is driving industry, FDA to enhance efficiency. Impact Report, Tufts Center for the Study of Drug Development 2005;7(1):1-6
17. Tufts CSDD. Outlook 2005;1-8
18. Tufts CSDD. Outlook 2004;1-8
19. Tufts CSDD. Outlook 2003;1-5
20. Tufts CSDD. Outlook 2002;1-5
21. Tufts CSDD. Outlook 2001;1-5
22. R&D Directions 2004 Drug Development Summit, How to Build and Maintain a Winning Pipeline. Feb. 8-11, 2004
23. Anonymous. Report to the nation 2003. Improving public health through human drugs. Center for Drug Evaluation and Research. Food and Drug Administration. U.S. Department of Health and Human Services
24. Kotter J. What Leaders Do. Harvard Business Review Book. Cambridge, MA. 1999
25. Walton M. The Deming Management Method. Putnam Books, New York. 1986
26. Covey SR. Seven Habits of Highly Successful People. Simon & Schuster, New York. 1989
27. Heider J. Leadership of Tao. Humanics New Age, Atlanta, GA. 1985
28. Pitino R. Success is a Choice. Ten Steps to Overachieving in Business and Life. Broadway Books, Bantam Doubleday Dell Publishing Group, New York. 1997
29. Maxwell JC. The 21 Irrefutable Laws of Leadership. Maxwell Motivation Inc, Thomas Nelson, Nashville, TN. 1998
30. Bynum WC, Cox J. Zapp! The Lightening of Empowerment. Harmony Books, New York. 1988
31. Peters T, Waterman Jr RH. In Search of Excellence. Lessons from America's Best-run Companies. Warner Books. Harper & Row Publishers, New York. 1982
32. Dupree M. Leadership Jazz. Dell Publishing, New York. 1992
33. Kouzes JM, Posner BZ. The Leadership Challenge. How to Keep Getting Extraordinary Things Done in Organizations. Josey-Bass Publishers, San Francisco, CA. 1997
34. Frankel LP. Overcoming Your Strengths. Harmony Books, New York. 1997
35. Welch J. Leadership and Mangement. The 2002 Linkage Excellence in Management & Leadership Series. Linkage, Inc. Lexington, MA. 2002
36. Bennis W. The Essence of Leadership. The Linkage Leadership and Strategy Series and PBS The Business Channel. Lexington, MA. 1999
37. Shelton K (ed.). A New Paradigm of Leadership. Visions of Excellence for 21st Century Organizations. Executive Leadership Publishing, Provo, UT. 1997
38. Albright M. Global Leadership. 2001 Linkage Excellence in Management & Leadership Series. Linkage Inc. 2001
39. Vander Walde L, Choi K, Higgins J. Health Care Industry Market Update. Pharmaceuticals. Centers for Medicare and Medicaid Services. U.S. Department of Health and Human Services. January 2003, 1-52
40. Anonymous. Profile pharmaceutical Industry 2004, Focus on innovation. PhRMA. 2004
41. Kennedy T (ed). Pharmaceutical Project Management. Drugs and the Pharmaceutical Sciences, Vol. 86. Marcel Dekker, New York. 1998
42. Cooper RG, Edgett SJ, Kleinschmidt EJ. Portfolio Management for New Products. Perseus Publishing, Cambridge, MA. 2001
43. Boesig C. Why NDAs fail. 15 pitfalls and how to avoid them. R&D Directions 2003;9(2):26-42
44. Razavi H. Managing and valuing the development pipeline in a major pharmaceutical company. DIA Annual Meeting, June 17, 2003, Chicago, IL
45. Lamberti MJ (ed). An Industry in Evolution, 4th ed. Thomson, CenterWatch. 2003
46. King J. Today's drug discovery unlocking greater potential. R&D Directions 2004;10(2):28-39
47. Engel S. Drug-discovery companies aim for commercialization. Good leads good medicine. R&D Directions 2002;8(7):34-42
48. Lam MD. Knowing when to pull the plug on your experimental drug. Pharmaceutical Executive 2004;24(2):55-60
49. Bernard S. The 5 myths of pharmacogenomics. Pharmaceutical Executive 2003;23(10):70-8
50. Shah J. Economic and regulatory considerations in pharmacogenomics for drug licensing and healthcare. Nature Biotechnology 2003;21(7):747-753
51. King J. Can a drug live forever? R&D Directions 2003; 9(4):40-52
52. Ho JH. Extending the product lifeline. Pharmaceutical Executive 2003;23(7):70-76
53. Li G, Dalton D. Balanced scorecard for R&D. Pharmaceutical Executive 2003;23(10):84-90

3
R&D Outcomes

Ronald P. Evens

Public/Patient Outcomes	66
Product Outcomes	70
Research Outcomes	73
Data/Information Outcomes	76
Company/Business Outcomes	79
References	82

When outcomes from R&D are addressed, new products are certainly the first outputs that are thought of as being produced by pharmaceutical and biotechnology companies. New products often are prescribed preferentially by physicians for the mitigation of disease and improvement of patient care because of their novel features over existing treatments. These new products will be the primary driver of innovations in health care, research advances, profitability, and business success for a company. However, many other important outcomes are needed routinely to be delivered by the R&D division and need support from all the rest of the company in order for the company to achieve four goals: demonstrate their scientific and medical prowess and productivity, meet the needs of the public and health care community for the best products used optimally, meet the needs of the shareholders, and sustain the company's research edge against the competition.

Several organizations evaluate the industry and create periodic reports for use by the industry, the medical and research communities, and public consumption as well. They address various industry outcomes, challenges, and improvement opportunities, involving diseases, technologies, products, processes, and business issues. The organizations include the following seven categories, along with examples and references provided: private consulting companies (e.g., Boston Consulting Group [1, 2], Ernst & Young [3, 4], and CenterWatch [5]); government (e.g., Food and Drug Administration [6] and Center for Medicare and Medicaid Services [7, 8]); publications (e.g., *Nature Drug Discovery* [10, 11] and *Nature Biotechnology* [9]); trade publications (e.g., *Medical Advertising News* [12], *Pharmaceutical Executive* [13, 14], and *R&D Directions* [15-17]); trade organizations (e.g., Pharmaceutical Research and Manufacturers Association [18, 19] and Biotechnology Industry Organization [20]); university research centers (e.g., Tufts Center for the Study of Drug Development [21–25]); and clinical research organizations (e.g., Parexel [26]). This chapter summarizes many of the key outcomes provided by these analyses and more.

More than 20 different outcomes related to R&D are expected by various external stakeholders listed above. In this chapter, these outcomes of R&D will be discussed subsequently in five categories of outcomes; that is, public's/patients' expectations, the various products, research outcomes, data/information generation, and several company-related deals.

Public/Patient Outcomes

The benefit to patients for R&D outcomes is first and foremost the new products that will improve their disease status with the least amount of side effects, but it is not just any product (Fig. 3.1). The 2004 Health Report for the United States from the Centers for Disease Control and Prevention documents the major reductions in death rates for heart diseases, cerebrovascular disease, and cancer and states that new drugs and more use of existing drugs were primary contributors to this improvement in health in America [27]. The optimal product should meet an unmet medical need of patients,

Fig. 3.1. Patient Benefits

Fig. 3.2. Relationships

significantly advancing the care of patients (e.g., statins in hypercholesterolemia in the 1990s). More *untreatable diseases* are finding amelioration or improvement through product innovation over the past 20 years (e.g., HIV infections with new classes of antiviral drugs, anemia of kidney disease and cancer with epoietin alfa, and enzyme deficiency diseases such as Gaucher disease with enzyme replacement). A *novel product choice* has been created because of its unique mechanism of action different from existing products, altering a key newly identified pathophysiologic process for a disease (e.g., aromatase inhibitor [Arimidex®] or oncogene inhibitor [Herceptin®] for breast cancer), or a better side effect profile has been achieved (e.g., Nonsteroidal anti-inflammatory drugs for arthritis versus aspirin). The new product achieves *patient care improvement* with higher efficacy over prior therapy, becoming a clinically superior or even best-in-class product (e.g., Crestor as a statin for high cholesterol vs. Pravachol; or Gliadel wafers [BCNU drug] for glioblastoma with local cranial placement after brain surgery and higher tumor resolution with less systemic adverse effects). *Convenience* for the patient and/or health care provider and/or health care system is created (e.g., pegylation of proteins substantially extends the half-life of biological products, such that injectible interferon for hepatitis C can be given weekly, Peg-Intron® or Pegasys®, instead of thrice weekly as Intron®; or filgrastim can be given weekly as Neulasta® for neutropenia correction with cancer chemotherapy, instead of daily as Neupogen®; or, insulin can be given by inhalation vs. injection, which is in clinical trials in 2005; or oral migraine products such as Imitrex® vs. its prior injectable form; or long-acting oral forms with daily vs. multiday dosing, Cardizem LA®).

A second outcome for R&D is relationships of the company with various constituencies that a company needs to be working with or serving. Favorable relationships with these varied audiences will assist not just the reputation of the company, but also good working relationships will enhance the process of product development (Fig. 3.2). The management and research staff at the company will work with their *investigators* at study sites (institutions), who at some point in time down the road after product approval and marketing will be both product experts and customers as well. Adherence to protocols, flexibility with study changes, and later product usage all may be improved with such good working relationships. *Patients and the public* favor companies with good reputations as well as good products, for example, in their product choices, in a willingness to be study subjects, by giving them the benefit of the doubt when serious adverse effects arise, and even in stock investments, which is part of the profitability and cash a company needs for its research. Good working and ethical relationships with *regulatory authorities* help in their receptivity in negotiations with the company for product approvals (even speed of approvals), audits by government, possibly labeling changes, and advertising approvals. Such good working relationships are based on good regulatory practices; for example, following FDA guidances in all operations, not having warning letters regarding adverse effects or advertising, experiencing audits without compliance problems, and having complete NDA or BLA applications in data or manufacturing. *Payer* relationships foster favorable payment policies and procedures, when a company tries and meets payer needs as much as possible, such as indigent care programs (free drug), or payment assistance programs for providers, or applicable cost-benefit and related data being provided. Providers are the prescribers and gatekeepers for product usage. With most products, multiple choices exist to treat a disease. A *provider or their institution* will work with a company and prescribe their product that of course is safe and effective, but also meets their needs for information, or education, or reimbursement assistance. These providers and the investigators who are experts for a disease are considered thought leaders, who advise the company about the diseases and products. Most companies (about 88%) have interactive programs in place with thought leaders early in research during at least phase 2 research. The *legislative and executive branches* of the government write the laws governing research investments, product approvals, and access to products (Medicare, Medicaid, VA, DOD), for example. A company with good relationships will have the opportunity to be consulted and influence outcomes in these and other areas, which assist in product development. Finally, the *investment community* needs as much information as possible about a company to make the best decisions for recommendations for stock purchases by their clients. A robust product pipeline of a company and progress of it over time are closely followed by investors as measures of corporate success. Furthermore, companies that are cooperative without compromising confidentiality will receive the benefit of the doubt if problems arise down the road in their R&D.

Relationships between a company and the university are core to the successful function of an R&D operation and carry benefits and risks for both sides (Fig. 3.3). The company receives direct R&D benefits and potential future benefits as well from the university in their collaboration. For their clinical trials, access to investigators to conduct the study and access to patients as subjects for the trials are obvious absolute needs for a company. Besides these clinicians having both therapeutic/disease expertise, their research expertise can help get the work done more effectively. Even before the trial starts, the university experts, often called thought leaders, provide consultation to R&D for the optimal product profile in a particular disease and study designs. Ultimately, the NDA completion requires this collaboration to create the data, reports, and possibly even FDA testimony. While a study is being conducted, the institutional staff of health care professionals (HCPs) is being given an education about this new product in a well-controlled situation. The pharmacy often group as the investigational distributor and quality control person in the institution. This working relationship can give the pharmacy, a gatekeeper for product usage, knowledge about and experience with the product, which may help future discussions and deliberations toward formulary review and approvals after the product's marketing. Data can be shared between the university and the company, including the data required for the study and possibly other research data, product usage, or patient care information. Data sharing may assist the investigator in furthering their research and the company with better knowledge of their product's fit in the university setting. A pivotal study in phase 2 or 3 for an NDA usually involves a highly selective group of patients, based on a strict study design, which may not be fully representative of typical patients. The university can additionally provide health outcome information beyond such clinical data, that is, practical health care information to be used in pharmacoeconomic assessments about the care of these patients (e.g., charges for tests and procedures, or quality of life).

On the other side of the ledger, risks or added demands exist for a company. Research costs in pivotal studies include grants to the university and investigator for patient accrual at costs of $5,000 to $15,000 per patient, plus institutional overhead of 25% to 75%. The study's patient accrual may go too slowly or there just may be an inadequate number of patients at one institution, straining their relationship. Study data is owned by the company as is the norm in study contracts, especially for investigational products. With major new findings for a product, especially unexpected outcomes beyond the study design, universities desire to patent their discoveries and share the downstream revenues, creating legal battles and even lost revenue for a company. Confidentiality of the product and study data is important for the NDA, patent issues, and competition issues with other companies. Timing of release and placement of new data are major issues with the study data for companies, which needs to be negotiated with investigators. Publications and their reception by the medical community can have significant impact on the success of an NDA, future prescription potential, patenting especially of new indications, and even stock investment impact. Publication rights (independence in content, publishability, and placement) are concerns of the research faculty, but most companies are flexible. They may require company preparation of the draft

Company
▸ **Benefits:**
- Access to patients
- Access to investigators
- Access to thought leaders
- NDA trial completion
- Data sharing from university
- Health outcome data
- Work with pharmacy
- Educate practitioners
- Formulary opportunity

▸ **Risks:**
- Cost
- Slow patient accrual
- Patents
- Confidentiality of NDA data

University/Medical Center
▸ **Benefits:**
- Clinical research trials
- Investigator initiated trials
- Access to novel products
- Data sharing from company (lab, animal, & clinical)
- Presentations
- Publications
- Education (products, research)
- Funding

▸ **Risks:**
- Conflicts of interest
- Ethics
- Independence
- Targeted research

FIG. 3.3. Company-University Relationships

manuscript for major publications from a study, especially because data analysis is performed by the company for study regulatory reports and publications as well, but the investigators contribute principally to the construct of the publication and edit all content. Companies will require the opportunity to review and comment about any other publication generated by investigators who are faculty from study data.

The university receives a host of benefits in their research collaborations with the industry (Fig. 3.3). Access to novel products to improve disease mitigation is an obvious prominent benefit. Access to the pivotal clinical trials for novel products is a related benefit as a research opportunity, to their stature in the medical community, and for future publications. A university often will have access to company data beyond the study that they are participating in because most pivotal trials are multicenter, including laboratory data, animal studies, and data from other institutions also participating in the studies. Certainly, the research grant funding is a major benefit to university faculty, who are measured by their ability to bring in grants to their university for promotion in rank. One grant, for example, for a pivotal study in which their institution contributes only 50 patients can be as much as $250,000 to $750,000, with a generous overhead that goes to the general university and department coffers. The study also offers a publication opportunity with novel science or new products for patient care, yet another criterion for faculty advancement. Educational benefits are obtained even somewhat passively by working on and learning about novel products in a research setting. At a major study site, thought leaders at the university become product experts who will conduct CME programs for regional clinicians. The university thought leader will also benefit from consultantships to companies, serving on expert panels for R&D and also marketing, assisting in identifying best products opportunities and their profiles, providing or critiquing research ideas and designs, presenting to FDA and regulatory authorities, and even market research on the company products. The consultantships provide several benefits, such as fees and honoraria for the work and also prestige in the research community for this recognized expertise. Another research benefit for the university faculty is follow-up investigator-initiated research projects (after an NDA approval), which become more likely to be funded, based on their experience with the product and their existing relationship with the company. However, some risks must be dealt with by the university clinician, related to independence already mentioned above, access to clinical data, publication opportunity, presentation (of data) opportunity, and possibly patent opportunity. Conflict of interest, or ethical conduct, is such a major issue that some institutions limit such research collaborations with companies. Another question with drug company–sponsored research is that the research work is targeted by the company, related to a specific disease and product, and the company writes the draft of the protocol, which are significant limits to independence for faculty [28].

The reputation of a company is influenced by the R&D operations. The next figure (Fig. 3.4) summarizes the assessments (2003 and 2004) by *Pharmaceutical Executive* trade magazine of the reputation of 19 pharmaceutical companies [29, 30]. Interviews were conducted (about 400) with industry executives and industry analysts, who were asked to rank the companies from highest, no.1, to lowest, no.10, on nine (2004) and ten (2004) parameters, which are defined in the two publications, and an overall assessment. Reputation strength scores were calculated based on their model (proprietary to authors) and the company executives' assessments with the ten criteria listed in Figure 3.4, which are prioritized in the model. Each criterion is scored separately, the companies are rated on individual criteria, and a composite score is created. Six issues are the expected criteria for business effectiveness (ranking of importance noted in parentheses): workforce (2), financial

FIG. 3.4. Pharma & Biotech Company Reputations
Source: Pharm. Exec. Vol. 24, 2004, Vol. 25, 2005

stability (3), leadership (4), marketing effectiveness (6), strategy (8), and global effectiveness (9). Four other criteria involve issues well beyond the sales and marketing–related operations and business issues, looking at the societal issues of social responsibility to the public and community: ethical behavior (1), third-party relations (5), community outreach (7), and charitable support (10). R&D has a direct impact on at least seven factors: overall company status, ethical behavior, strategy, financial stability, third-party relations, global effectiveness, and leadership. The top companies changed significantly from 2003 to 2004: Merck, Pfizer, Johnson & Johnson, Amgen, and GSK to Lilly, Genentech, Amgen, Johnson & Johnson, and Novartis, respectively. Biotechnology companies reached the top five in 2003 with Amgen for the first time and for Genentech and Amgen in 2004. The criteria were changed year after year to add strategy, global effectiveness, and third-party relations; two areas were deleted, manufacturing effectiveness and competitiveness. These additions are very compatible with major business and social commitments that comprise successful pharmaceutical companies. Employee retention was expanded to address workforce, a more global analysis of personnel issues.

The stock price of a pharmaceutical or biotechnology company is impacted by a host of factors well beyond just the sales and profitability statistics. Ten factors are listed in the next list (Fig. 3.5). As expected, the R&D *track record* for product approvals by regulatory authorities would be the biggest factor to move the stock price, and approvals in all three major markets worldwide is then a global goal. Also, product failures commonly occur and are assessed closely as well; less than 1 product in the 5–10 products entering clinical research will be approved. The breadth and depth of the product portfolio in research and in development change every year as products move along their timelines with good and bad results. A robust pipeline must be sustained to keep the stock price up. Even interim steps (milestones) in the product timelines are being watched closely for success of R&D, such as IND filings and phase 3 completions. The investment community seeks *data and information* on pipeline status, which emanates from R&D on a regular basis through a variety of avenues, such as presentations at medical society meetings, press releases from a company, and the quarterly investor meetings. *People* issues in R&D influence stock price as well, including hiring and retention of quality scientists and staff (their expertise and track record in science and industry experience). The research collaborations or alliances with external entities suggest the desirability of a company as a partner in research and their product development success. Sustaining your reputation and a positive history (*growing and leading*) will contribute to the idea of successful operations and indirectly lessening the volatility of the stock price as good and bad R&D news occurs.

Product Outcomes

The criterion of success associated with R&D most sought after by the company and the most monitored by the press and investment community is the product approvals, especially products with blockbuster potential ($1 billion in sales within 5 years of marketing approval). Both the number of blockbusters on the market by a company and a steady stream of them going forward are currently thought of as the holy grail of pharmaceutical success. A company also can interject into their track record of product approvals products with more modest sales potential of $250,000 to $500,000 per year, demonstrating a sustained research productivity, as long as blockbusters come along periodically.

The drivers to achieve the blockbuster level of product sales are mostly patient, health care, and product issues, as outlined on this next list (Fig. 3.6). Disease and patient criteria included

---

Why can stock be so volatile?

- **Track Record :**
    - $ - Product approvals
    - $ - Product failures
    - $ - Product portfolio
    - $ - Research milestones achieved
- **Data & Information:**
    - $ - Presentations at society meetings
    - $ - Investment community meetings
- **People & Processes:**
    - $ - Hiring - technical & leadership expertise
    - $ - Alliances with companies, universities, gov't
    - $ - Reputation & history
- **Growing & Leading**

Fig. 3.5. Stock Price

Peak Sales of at least $1Billion & $ 5 B by year 5

- Large patient population
- Unmet medical need
- Chronic disease treated
- Competitive superiority in efficacy
- Competitive superiority in toxicity
- Premium pricing
- Global approvals (US, EU, Japan)
- High marketing spend @ launch (market penetration)
- Patent life is long (5+ years) & protected
- Label extensions planned

Fig. 3.6. "Blockbuster" Products – The 10 Drivers

Product	Sales	Company
Lipitor	11.59	Pfizer/Astel
Plavix/Iscover	5.39	BMS/S-A
Zocor	5.20	Merck
Advair/Seretide	4.50	GSK
Norvasc	4.46	Pfizer
Zyprexa	4.42	Lilly
Nexium	3.88	AstraZen
Procrit/Eprex	3.59	J & J
Novo-Insulins	3.43	NovoNord
Zoloft	3.36	Pfizer
Effexor	3.35	Wyeth
Celebrex	3.30	Pfizer
Fosfamax	3.16	Merck
(Ref.: Med Ad News May, 2005)		

Product	Sales	Company
Diovan/Co-Diovan	3.09	Novartis
Risperdal	3.05	J & J
Remicade	2.90	J&J/S-P
Cozaar/Hyzaar	2.82	Merck
Neurontin	2.72	Pfizer
Pravachol	2.64	BMS
Singulair	2.62	Merck
Rituxan/MabThera	2.62	Roche/Gen
Epogen	2.60	Amgen
Prevacid	2.59	Tap
Enbrel	2.58	Amgen/Wy
Aranesp	2.47	Amgen
Lovenox/Clexane	2.37	San/Aven
Durasegic	2.08	J & J

FIG. 3.7. Blockbuster Products (94 in 2004 = $186 Billion)

Blockbuster products with sales over $1 billion numbered 94 in 2004 worldwide (Fig. 3.7). They accounted for $186 billion in sales out of total worldwide sales of $550 billion (34%). Cardiovascular products with 14 (e.g., Lipitor®, Zocor®, Plavix®, Norvasc®) and central nervous system products with 17 (e.g., Zyprexa®, Effexor®, Zoloft®, Neurontin®) were the leading blockbuster categories. The top single product was Lipitor® for hyperlipidemia at $12 billion in 2004, the first time a product exceeded $10 billion in worldwide sales. Two gastrointestinal products and two respiratory products hit the top products in 2004 ($2 billion plus), Prevacid® and Nexium®, and Advair® and Singulair®, respectively. For the first time, oncology products moved strongly into the top used products with eight blockbusters (e.g., Rituxan®, Taxotere®, Gemzar®, Cozaar/Hyzaar®, and Gleevec®). Biotechnology products as blockbusters expanded significantly to 20 in 2004, led by erythropoiesis products around the world (all forms of epoietin alfa, Procrit®, Epogen®, Eprex®, Neorecormon®, and Epogin®, and Aranesp®), the insulins (Humulin/Humalog®, Novolins/Novolog®, and Lantus®), the Neupogen® and Neulasta® franchise in oncology supportive care, oncology therapy products Rituxan®/MabThera® and Herceptin®, the inflammation products Enbrel® and Remicade®, and the multiple sclerosis products Avonex® and Rebif®. Collectively, the erythropoietin products became the first $10.3 billion product franchise in 2004. The top companies with blockbuster products were GlaxoSmithKline with 12, Pfizer with 10, Sanofi-Aventis with 9, Johnson & Johnson companies with 8, Merck with 6, Astra-Zeneca with 6, Amgen with 5, and Novartis with 5 [3, 9, 12, 13, 18].

In addition to new product approvals, a variety of other product-related outcomes can be accomplished by R&D (Fig. 3.8). New products ideally need to be *new molecular*

- NMEs – New Medical Entities (Novel) (e.g., kinase inhibition in cancer)
- Second generation molecule (e.g., Neulasta® vs Neupogen®)
- 1st-in-Class versus 2nd-in-Class products (e.g., Prevacid® vs Crestor®)
- Molecular manipulation (e.g., TNKase® vs Activase®)
- Route of use additions (Oral vs injectible)
- Formulation improvements (XL)
- Product delivery systems (Insulin pen)
- Manufacturing process improvements

FIG. 3.8. Molecule & Product Opportunities

a *large patient population* with the disease that are addressable patients for such a new product, an *unmet medical need* being addressed wherein the product offers novel therapy for a disease not controlled significantly enough, and a *chronic disease* that is treated with the product for months to years. The product profile for a blockbuster usually requires sufficiently strong data demonstrating *superiority* in efficacy or toxicity. With the above-noted sufficiently positive societal and product attributes, a company can charge a *premium in the price* that will be paid by the health care systems, adding to its profitability. Process issues for company operations that contribute to blockbuster achievement are suggested to be threefold now: *global approvals* in a timely fashion across the world, a sufficiently *high marketing budget* at launch to reach the providers, the information and education to warrant a high prescription volume (good market penetration), and execution of a plan to perform further research *expanding the approved labeling* with new indications, or formulations, or new doses. Finally, the protected *patent life* needs to be as long as possible after marketing, ideally at least 5 years, in order to recoup all R&D and operational costs, as well as pay for the product failures [31–33].

entities, which the regulatory authorities consider as major advances in treatment of a previously untreated or already treated disease. A product candidate may be a *second-generation molecule* with some patient care advantage being established, for example, pegylation of proteins to stretch out the injectible dosing of these products from daily to weekly with Neulasta® or from thrice weekly to weekly for interferons (Pegasys®). Some products may be approved for use as the second or third or later product within an existing therapeutic category with already approved products, but their success is based on the R&D organization doing the research to show *best-in-class* properties of the product (e.g., better cholesterol reduction with Crestor® vs. Pravachol®). *Manipulation of molecules* can be done to change their properties and possibly improve their efficacy or toxicity or utility. For example, Genentech created Activase® first for clot lysis in acute myocardial infarction (AMI), and then several years later created a follow-up molecule, TNKase®. Protein changes allowed intravenous bolus versus slow infusion, a major advantage in the acute setting of AMI. Other product changes (improvements) are goals and outcomes for an R&D operation, such as a new *route of administration* for treatment flexibility, *formulation improvements* for extended release and less frequent daily doses, more stability for a better shelf-life, or a new *product delivery system* with convenience for health care delivery, such as self-injector pens for insulins given by injection once or more per day. A new formulation is a patentable new product, which continues a company's dominance in patient care and the related product sales. For example, the calcium channel blocker Diltiazem® for hypertension originally from Marion Laboratories was a major therapeutic advance for hypertension (new mechanism of action) at its time of marketing 20 years ago but was given multiple times a day. The extended release form with once-daily dosing offered much better patient convenience and compliance especially for a silent killer like hypertension and gave them continued market exclusivity for several years further. R&D or the manufacturing division has a unit called process engineering that does research on *improved manufacturing processes* to improve the yield, remove contaminants, or reduce cost of operations, using less manpower, fewer steps in the process, faster process, more automation, or less ingredient costs [17, 33–36].

A very undesirable outcome of the R & D organization is product failures at various stages in a product's life cycle (Fig. 3.9). A company wants to kill a product at the earliest possible stage if it eventually believes that it will become a failure, thus not wasting research dollars and better utilizing resources for faster product approvals with more likely better products. The kill decision is by far one of the most difficult ones for a company. A significant number of scientists devoted time, energy, creativity, and emotions into their work, which is very hard to turn off and redirect sometimes into whole new therapeutic or disease areas, where the scientists may not be as comfortable or as capable. Too often the company will try to do one more study to tease out some benefit, but it may be only marginal. On the other side of coin, you do not want to abandon a molecule completely if it may have other indications. One of the best examples is etanercept (Enbrel®), which was first studied for sepsis, based on the major inflammatory problems in sepsis and the significant role tumor necrosis factor (TNF) plays in sepsis. However, it was a complete failure in phase 3 trials with marginal benefit being observed at best. The Immunex company continued to look for other applications for its molecule, other inflammatory conditions where TNF is a major mediator, because they knew their molecule favorably lessened TNF effects in several disease models. Rheumatoid arthritis was studied, and about 5 years later it was not only approved for use, but etanercept is a major advance to control arthritis and slow

Biotech drugs in 1980s & 1990s, Success Rates = about 30-35% (P.1 to Approval)

Products	(Company)	Indication	Products	(Company)	Indication
Ad5FGF4	(Schering)	Angina	GDNF	(Amgen)	Parkinsonism
Astenose	(Glycomed)	Restenosis	Hirulog	(Biogen)	Angioplasty
Antril	(Synergen)	Sepsis	Lamin gel	(ProCyte)	Diabetic ulcers
Arasine	(Gensia)	AMI (bypass)	NS2330	(Neurosearch)	Alzheimers
Argidene	(Telios)	Diabetic ulcers	Pactimibe	(Daiichi)	Atherosclerosis
Glitazone	(NovoNord)	Diabetes	Repinostan	(Bayer)	Ischemic stroke
Betakine	(Celtrix)	Macular holes	REN-850	(Renovis)	Multiple sclerosis
CB001	(Viacell)	Cell Transplant	Sumanirole	(Pfizer)	Restless legs
CEP-1347	(Cephalon)	Parkinson's	T-88	(Chiron)	Sepsis
CD-5	(Xoma)	Transplant	Tezosentan	(Actelion)	Heart failure
Centoxin	(Centocor)	Sepsis	Therafectin	(Greenwich)	Arthritis
CNTF	(Regeneron)	ALS	Thymosin	(SciClone)	Hepatitis
CS917	(Metabasis)	Diabetes	Vitaxin	(MedImmune)	Arthritis

Fig. 3.9. Product Failures in R&D (Biotechnology)

Drug Name	Company	Use	Safety Problem
Alosetron (Lotronex)	GSK	Crohns	Ischemic colitis
Astemizole (Hismanal)	J/J, Janssen	Antihistamine	Irregular heart rhythm
Benoxaprofen (Oraflex)	Lilly	NSAID	Fatal cholestatic jaundice
Bromfenac (Duract)	Wyeth	Pain	Hepatotoxicity
Cerivastin (Baycol)	Bayer	Cholesterolemia	Rhabdomyolysis
Cisapride (Propulsid)	Janssen	PUD	Arrythmias
Dexfenfluramine	Various	Diet	Aortic & mitral regurgitation
Encainide (Enkaid)	BMS	Arrythmias	Deaths after AMI
Fenfluramine	Various	Diet	Aortic & mitral regurgitation
Flosequinan (Manoplex)	Boots	Heart failure	Heart failure benefit < short
Grepafloxacin (Raxar)	GlaxoWelcome	Infection	CV events
Levomethadyl (Orlaam)	Roxanne	Narcotic addict	CV events
Mibefradil (Posicor)	Roche	Hypertension	Interactions, CYP450
Nomifensine (Merital)	Hoechst	Depression	Hemolytic anemia
Phenylpropanolamine	Various	Diet	Hemorrhagic stroke
Rapacuronium (Raplon)	Organon	Muscle relax	Bronchospasm, deaths
Temafloxacin (Omniflox)	Abbott	Infection	Renal failure, Anemias plus
Terfenadine (Seldane)	Hoechst M-R	Antihistamine	Cardiac arrythmias
Ticrynafen (Selacryn)	SKF	Hypertension	Hepatotoxicity
Troglitazone (Rezulin)	PD / WL	Diabetes	Hepatotoxicity
Zomepirac (Zomax)	McNeil	NSAID	Anaphylactoid reactions

FIG. 3.10. Drug Failures, Post Approval (1982–2002)

progression of the disease. Etanercept now has four approved indications in inflammatory conditions, rheumatoid arthritis, juvenile rheumatoid arthritis, psoriatic arthritis, and psoriasis. Figure 3.9 shows product failures over a ten year period (1983 to 1994) for biological products.

Product failures can continue to be a problem after product approval and even years after its marketing (Fig. 3.10). At this late date, a failure with a product recall or even removal from the market can have disastrous effects on an organization, its profitability of course, but also morale, stock price, and staffing (downsizing). The most common reason for product withdrawal is a serious and unexpected adverse product experience in patients identified by practicing physicians in their routine use of the product. These serious adverse effects most often are very infrequent. In the few hundred to a few thousand patients in all clinical trials work, only a very few cases occurred, and they could not be solely associated with the new drug under study. However, now after marketing thousands or even millions of patients have been exposed to the product, and the adverse drug effect has occurred in a few hundred patients. We now can describe the drug-induced problem more fully in its onset, time course, and signs and symptoms, examine temporal relationships, compare it with mechanistic data on the drug, and exclude other causes. This situation is a medical, public relations, and financial disaster for a company and a proverbial black eye for the R&D organization that did not identify the problem during R&D before marketing. The company spent hundreds of millions of dollars for the R&D and then millions dollars more to market the product. The marketed product is built into the profit picture of the company for the next 5 years or more of its patent life as well. Follow-up studies may be required to examine the problem adding huge costs. The company's relationship with the regulatory authorities may be tarnished, negatively influencing the reviews of future drug applications. The table of product withdrawals lists 21 products removed form the market over a 20-year period, 1982 to 2002, all related to serious adverse experiences that involve various organ systems, but liver (4) and heart (9) problems predominate [37].

Research Outcomes

In addition to various product approvals, the R&D organization has a variety of further substantial outcomes for the organization and public as well, based on their research. In this section, we will present science leadership, overview of study types, pipeline, research techniques, and investigators.

In science, that is, the discovery phase, the scientists have major interim scientific goals and accomplishments that eventually lead to product candidates, such as four discoveries noted in the next figure (Fig. 3.11). New *disease biology* or mechanisms responsible for disease pathogenesis are uncovered (e.g., in the late 1990s and early 2000, the proteasome pathway and the impact in cancer). *Novel targets* that are associated with a disease and favorably influenced by drug therapy are discovered (e.g., TNF and ulcerative colitis with Remicade®, or protein kinases in cancer and Iressa®). Chemicals and biologicals are screened and created to influence the *target (hits)*, which are confirmed and validated to become drug candidates. Whole new *drug categories* can be discovered that address the new targets and mitigate a new disease pathway (e.g., statins in high cholesterol, or oxicams for infections). Finally, *novel mechanisms of action* may be discovered (e.g., receptor antagonists for various diseases with identifiable cell receptors involved in a disease [e.g., oncogene her2neu in breast cancer mitigated by Herceptin® monoclonal antibody]). Process improvements such as in biological manufacturing or product analysis can be scientific advances as well [35, 36].

In product discovery over time, whole new techniques are created to identify product leads that could develop into new product categories for one or several diseases. Medicinal chemistry with structural modification of a drug, along with structure-activity relationships and product screening, have been and remain hallmarks of product development in the industry to develop new drug categories or follow-on molecules with improved properties to treat a disease. Over the past 20 years, the approaches to product discovery have grown immensely especially in the biological arena, as represented on Fig. 3.12. Process improvements in research include, for example, transgenic animals to create more reliable and predictable disease models, high-throughput screening to accelerate and increase amount of work per unit of time 10- to 1000-fold, pharmacogenomics to identify the best or worst responders or more susceptibility to adverse drug effects, and bioinformatics to store and manipulate the vast volume of data available. New product categories for drug discovery over this 20-year period include the following examples each for a different technology: (1) recombinant DNA technology to reproduce proteins as therapeutics (e.g., Kepivance® for mucositis in cancer patients), (2) monoclonal antibodies for 20 different diseases (e.g., Rativa® for psoriasis), (3) molecular engineering to have improved second generation protein molecules (e.g., Pegasys® for hepatitis C), (4) nucleotide therapeutics with, for example, antisense antiRNA (e.g., Vitravene® for CMV retinitis), (5) tissue engineering (e.g., Fortaflex™ for rotator cuff repair), (6) protein kinase receptor interference (e.g., Gleevec® for acute myelogenous leukemia), (7) peptides (e.g., Fuzeon® for HIV infections). Proteomics, ribozymes, combinatorial chemistry, and more are being studied in laboratories to find yet new generations of products.

Studies are yet another set of outcomes from the R&D organization, which will be discussed at length in Chapters 4 and 5 and are listed in the next figure (Fig. 3.13). These studies often are done at a university through research grants. The seven study types encompass the full cycle of research at a company from early work in the basic sciences for disease pathology and mechanism of action of products; through preclinical work in animals for pharmacology, toxicology, and pharmacokinetics; human trials for metabolism and pharmacokinetics, early small clinical trials (phases 1 and 2) to demonstrate proof of principle; to full large pivotal trials for the marketing application to establish safety and efficacy (phase 3) and also postmarketing (phase 4 and postmarketing surveillance); economic trials to establish cost-effectiveness, quality of life improvements, and the value of a product to the health care system; and other studies as needed for disease epidemiology, drug interactions, or product stability in various patient care situations.

The outcome for R&D that measures overall productivity is the pipeline, usually organized by business areas for a company, and/or therapeutic categories focused on by the company, and/or the stages of research. In Figure 3.14, the data covers the pipeline for Novartis company by the end of 2003, which was presented in February 2004 by a lead scientist at an pharmaceutical conference. A robust pipeline possesses several characteristics, which are represented in this table for Novartis. A reasonably large number of molecules is needed in general. A sufficient number of molecules in each cell is the next prerequisite for a robust pipeline. All the therapeutic areas and businesses are covered. Each stage of research is covered, such that, as the pipeline evolves and products advance to approval, no gap will exist in a year with no approvals, and the businesses will have a steady stream of new products. In addition, the following characteristics are important for defining a robust pipeline: unmet medical needs being met, diseases with high patient populations, chronic diseases being treated, high sales

FIG. 3.11. Science Leadership

- Disease biology
- Compound screening
- Recombinant DNA technology
- Monoclonal antibodies
- Polymerase chain reaction
- Nucleotide blockade
- Gene therapy
- Combinatorial chemistry
- Molecular engineering
- Medicinal chemistry
- Transgenic animals
- Structure/activity relationship
- High-throughput screening
- Micro-Array assays
- Genomics
- Proteomics
- Pharmacogenomics
- Ribozymes
- Protein kinases
- Receptorology
- Cell therapy
- Tissue engineering
- Bioinformatics

FIG. 3.12. Techniques in Product Discovery

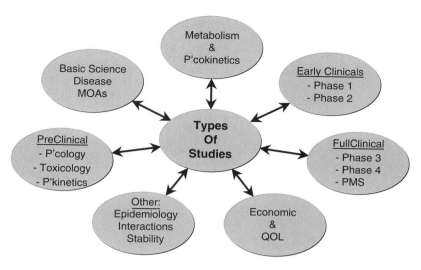

FIG. 3.13. Types of Studies

Bus. Un.	Bus. Franchise	Precl.	P. 1	P. 2	P.3/Reg	Total
1* Care	Nerv. System	5	3	6	5	19
1* Care	Cardiov./Metab.	10	0	3	6	19
1* Care	RA/Bone/GI/HRT	9	3	7	6	25
1* Care	Anti-Infectives	2	1	1	1	5
1* Care	Respirat./Derm.	5	3	5	4	17
SP-Onc.	Oncology	7	5	7	4	23
SP-Tnpl.	Transplatation	6	0	0	3	9
SP-Oph.	Ophthalmic	3	0	2	3	8
	Total Projects :	47	15	31	32	125

FIG. 3.14. Pipeline (Projects): Major Pharma Company*
Source: Garaud J-J, Novartis, R & D Directions Conference, Feb. 2004

potential for products, good fit of products for the company's therapeutic areas in research and sales, and in-license and proprietary products in pipeline [38, 39].

Another pipeline for an R&D organization is displayed in Figure 3.15 for a biotechnology company. The company shown here, Amgen, is the largest by far among biotechnology companies with over 14,000 employees in 2005, over $8 billion in sales, about $1.5 billion in R&D expenses worldwide, and more than 20 alliances or partnerships. However, the pipeline size is about 20 molecules and projects in clinical trials in four focused therapeutic areas, which is about one-tenth the size of the major pharmaceutical companies. Compared with a drug company, a biotechnology company has less financial resources (R&D budget), fewer research alliances, smaller R&D staff, usually fewer research areas of expertise, all of which creates a leader organization with fewer possible outcomes. However, this size can be an advantage in several ways: flexibility to move into new research areas, more focus in research with less internal competition for resources, being a preferred partner with small biotechnology companies, and hopefully more streamlined or less complex decisions with fewer research molecules [40, 41].

Research creates an important outcome through external research collaborations with universities and health care institutions, specifically with the scientists and clinicians at those institutions, that is, the investigators for a company. Certainly, a company needs to work with experienced clinical researchers to get the work done to the quality necessary for product

Molecule:	Area:	Usage:	Status:
AMG 108	I,B&M	Arthritis	Early - P. 1/2
AMG 162 (OPG)	O	Metastatic Bone	Early - P. 1/2
AMG 162 (OPG)	I,B&M	Osteoporosis	Early - P. 1/2
AMG 531	I,B&M	I. Thrombocytopenic P.	Early - P. 1/2
AMG 714	I,B&M	Rheumatoid Arthritis	Early - P. 1/2
Alfimeprase	I,B&M	Peripheral Artery Occlusion	Late - P. 3
Aranesp®	H & N	Anemia in Kidney Disease	Approved
Aranesp®	O	Anemia in Cancer	Late P. 3
Enbrel®	I,B&M	Arthritides, Psoriasis	Approved
Epogen®	H & N	Anemia in Kidney Disease	Approved
GDNF	N	Parkinson's Disease	Early - P. 1/2
Kepivance® (KGF)	O	Mucositis in Transplants	Late P. 3
Kineret®	I,B&M	Rheumatoid Arthritis	Approved
Neulasta®	O	Febrile Neutropenia in Ca	Approved
Neupogen®	O	Febrile Neutropenia in Ca	Approved
NGF Antagonist	N	TBD	Early - P. 1/2
P38 Antagonist	I,B&M	TBD	Early - P. 1/2
Panitumumab	O	Colorectal Cancer	Late - P. 3
Sensipar™	H & N	HPT in Kidney Disease	Late - P. 3
11B-HSD1	I,B&M	TBD	Early - P. 1/2

FIG. 3.15. Biotechnology Product Pipeline*
*Amgen, 2004; Public Information

- U.S. 52,000 with at least one trial
- Education & Training
  - Degrees: MD 96%, DO 3%, Other 1%
  - Certification: IM 50%, N/Psy 9%, Ped 8%, FP 5%
  - Age: 45-50 yrs, 45%; 51-60 yrs, 32%
- Trial experience: < 1yr 63%, 2-4 12%, > 5 25%
- Rationale for involvement
  - Scientific reward
  - Financial reward
  - Better patient care
- Setting: Group practice vs University

FIG. 3.16. Clinical Investigators (A Profile)

applications to regulatory authorities and with reasonable alacrity. However, new products in research may pose educational challenges for all investigators, even experts, related to new mechanisms and new protocols. Training of investigators is a major undertaking for pharmaceutical companies, related to each and every specific protocol for a product, especially related to patient recruitment and eligibility, product administration, monitoring and case report forms, all of which are very specifically spelled out in study protocols. Figure 3.16 documents that the industry used about 52,000 investigators in the USA alone, which can be characterized as mostly M.D. physicians with relatively limited research experience (63% less than 1 year), older practitioners by age and time in practice, and practicing in both clinical settings of the university or clinical office practice. The university expert usually leads the study effort for the company, but often many practitioners are needed to recruit sufficient patients that qualify for a study. University-based patients are often tertiary care, complex patients that may be too sick or have too many complications to participate. Also, the practitioners have access to many more patients in number. The rationale for a clinician to participate in a trial includes three major motivations; first, the science, engagement in novel products to advance science; second, financial reward from the compensation for participating in a trial (research grants), especially for the university where grants are a major part of advancement criteria; and third, an opportunity to help create better patient care through improved therapeutic products [5].

Data/Information Outcomes

A variety of types of information are produced by the R&D organization as outcomes, including labeling for the product (package insert), extensions to the labeling often for expanded indications, regulatory applications for product approvals or expanded product information, presentations at scientific meetings, publications of the studies conducted, and educational materials, to be discussed below.

First, as part of the new product application to regulatory authorities, the company writes the draft of the official package insert (PI). The PI is tightly controlled by regulators in its organization and content (12 standard sections), as described in this next figure (Fig. 3.17). The PI must use these terms for subheads and follow this order of information. The label is reviewed in detail and approved by the regulators, but much negotiation between the company and the regulatory authorities occurs because the company wants accuracy and completeness but as much flexibility as possible in wording. The PI gives the clinician background information and usage information to guide its prescribing and monitoring for efficacy and side effects. The words in the PI for the business are the limits of any advertisement or sales person activity, and a competitive edge versus other products is highly desirable if the clinical data supports the statements, which the regulators demand of course. The importance of the PI content is critical for patient care and business opportunity.

A second labeling outcome is an expansion of the labeling by the company and/or required by the regulatory authorities (Fig. 3.18). The company will perform a great deal more research after the product is approved and marketed to obtain approval for a new indication, or expanding usage and marketing activity, or offer more safety information. For existing indications, new clinical research may demonstrate additional dosing approaches, new administration techniques, special subpopulations that may respond better or worse to the product, and perhaps added quality of life benefits for the patients and health care systems. New studies may require new safety precautions as we learn more about the product. With the much broader use of a product postmarketing, more side effects or more severe manifestations of listed side effects may arise, all requiring labeling changes to better guide the clinicians in using this product more appropriately. The majority of pharmacoeconomic studies is done after a product is marketed, which may yield information for labeling, such as quality of life improvements (QOL) [44].

Regulatory applications will discussed in full in Chapter 7, and the many types are listed here (Fig. 3.19) as another informational and data outcome from a company. The NDA, BLA and PLA are an exceptionally complete set of documents to establish safety and efficacy of the product, along with manufacturing information and labeling. They are thousands of pages in many volumes for any one product application. A regulatory application will be filed for most products in all the three worldwide markets, USA (FDA), EU (EMEA), and Japan. Prior to human studies in the USA, an Investigational New Drug (IND) application is required to be filed with regulatory authorities. Labeling changes in the USA require a supplemental NDA document. Generic drug applications in the USA require an abbreviated NDA demonstrating pharmacokinetic equivalence (bioavailability), assurance of the same ingredients, and manufacturing processes. Many European countries have dual sequential and separate approval processes; documents are required for the regulatory authority for safety and efficacy and then the pricing committee for approval of reimbursement for the product [6].

A pharmaceutical company performs many studies that are internal standard documents for company use, incorporated later as key parts of a regulatory submission, and are presented at scientific meetings, as shown in Figure 3.20 Each study will result in a statistical report, tables of data, and their statistical interpretations, and the final study clinical report, adding to the stat report all the clinical background, clinical interpretations, and conclusions. External presentations of the data are done by company clinicians and especially and more often by their university collaborators in two major settings; investigator meetings and medical society meetings. The presentations include written abstracts for posters at meetings,

Name: Trade, Generic, Chemical
- Description
- Pharmacology
- Clinical Data
- Indication & Usage
- Dose
- Warnings
- Precautions
- Adverse Reactions
- Dosage & Administration
- Overdosage
- How Supplied

FIG. 3.17. The "Label" (Package Insert)

Regulatory Bodies: FDA, EMEA, Countries, Japan

~IND (Investigational New Drug App)
~NDA (New Drug App)
~PLA (Product License App)
~sNDA (Supplemental New Drug App)
~ANDA (Abbreviated NDA)
~IDE (Investigational Device)
~Pricing Committees (each country)

FIG. 3.19. Regulatory Applications

The Product Label (Package Insert)
- Second Indications
- Safety information
- Dosing choices
- Administration (Route)
- Special populations
- QOL

FIG. 3.18. Labeling Expansion

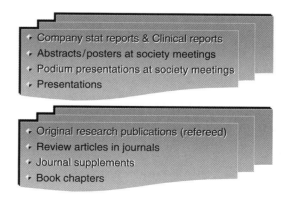

- Company stat reports & Clinical reports
- Abstracts/posters at society meetings
- Podium presentations at society meetings
- Presentations

- Original research publications (refereed)
- Review articles in journals
- Journal supplements
- Book chapters

FIG. 3.20. Data Publications & Presentations

podium verbal presentations, and then later full publications. All such presentations are refereed by other independent experts, assessing the design of the studies (at least, appropriateness, quality, and novelty), the data and observations, and the conclusions. The publications will take several forms: the original research paper for the study to be published in a medical journal, review articles summarizing the product's use in a disease published alone or in journal supplements for the disease or drug category and later in book chapters about the disease or drugs [32].

Educational materials are developed by a company during the clinical research phase for a new product, and, of course, especially after a product is marketed (Fig. 3.21). The most significant educational document produced by the development (clinical research) group is the investigator's brochure (IB), which is prepared by all companies for a new product while their clinical trials program is being done. The IB is a summary of all primary data from the animal studies, metabolism, clinical trial summaries, and formulation data, in order to educate the investigators and their institutional review boards for patient safety. Certainly, safety is the most significant focus followed by product activity and efficacy, up to that point in time. Any other indications or studies in process for other indications need to be discussed in the IB also. This document must be updated regularly as new data on the product comes to light through the clinical trials or other research. Prior to marketing, investigator meetings will be held to educate them about the various properties, safety, and uses of the new product, as well as training about the new protocol being initiated. The principal investigators, co-investigators, and study coordinators are all trained with the new protocol and its requirements, especially for patient enrollment, study conduct, and monitoring. If the product is novel and a major advance in science and patient care, symposia also are conducted by groups of scientists to advance the scientific dialogue, share information in the medical community, and receive input about the potential role of the product in patient care for the target disease(s). The programs must be independent of the company control for legal, regulatory, and ethical reasons. The company may create slide materials about the product and protocols for educational use by the investigators. Monographs are produced about the product and the disease related to the new discoveries to educate the medical community. Administration guides might be required for products with novel or more complex methods of administration. As part of the approval process, the R&D organization needs to prepare a patient package insert to explain how to use the product to obtain optimal benefit, including not just administration

- Investigators brochure
- Investigator protocol training
- Symposia publications
- Slide programs
- Disease monographs (PreLaunch)
- Product monographs (Launch)
- Administration guides (Launch)
- Patient package insert (Launch)
- Product inquiries

FIG. 3.21. Educational Materials

- **General - Core Operations:**
  - Global Partners in Marketing & Research
  - Clinical Research Organizations
- **General - R&D Research centers:**
- **General - Merger** (Products, R&D, Manufacturing):
- **Specific Products:**
  - Leptin & Lab output (D/C later)
  - Keritinocyte Growth Factor
  - Calcimimetics multiproduct (Cinacalcet)
  - Abarelix (D/C later)
  - Neuroimmunophilin products (D/C later)
  - Fibrolase
  - Interferon alfacon-1 (Out-license)
  - Epratuzumab (D/C later)
- **Specific Technologies:**
  - Manufacturing & Inflammation products
  - Small molecules
  - Mab technology & Panitumumab
  - Signaling drug discovery

Company/Institution:
Roche and J & J
Quintiles/Radiant
Toronto Research
Immunex

Rockefeller University
National Institutes of Health
NPS Pharmaceuticals
Praecis
Guilford
Hyseq
Yamanouchi
Immunomedics

Synergen (Acquired)
Kinetix (Acquired)
Abgenix (Acquired)
Tularik (Acquired)

FIG. 3.22. Alliances & Collaborations by a Company*
*Amgen, 2002; Public Information

guides but also potential benefits and risks. Before product approval, questions will be received about the new product from providers, patients, and the press, which need to be anticipated, and responses to product inquiries prepared [32].

Company/Business Outcomes

In order to perform all the research and market preparation for a new product, a company will need to create various collaborations and alliances to complete all the work. In the industry overview chapter, a figure (Fig. 1.33) for research and business collaborations was provided. In this outcomes discussion, we need to briefly reiterate a few points about collaborations as outcomes for R&D. Clinical research organizations are research companies focused on performing any or all of the clinical trials, on behalf of a pharmaceutical companies, picking up the overload of clinical trials work that almost always will occur. They will conduct the whole study or any segment (e.g., patient recruitment or statistical analysis). A full discussion of CROs will appear in the clinical operations chapter in this book. In the basic science area, collaborations are very common with universities especially or small companies with very specialized expertise, in order to expand the opportunities for discoveries in disease biology, target identification, lead identification, or new molecules with different mechanisms of action (e.g., monoclonal antibodies, small-molecule drugs, and antisense molecules for the same disease mechanism). In clinical research, the university is the site where the clinical trials usually are conducted. A small company or other company collaboration may provide access to key technology, such as throughput screening or x-ray crystallography. A small company may need a larger pharmaceutical company collaboration to perform the clinical research in the expanded late phase 2 and especially phase 3 clinical trials work.

Product licensing is yet another outcome for the R&D organization, and it is a major process to obtain molecules for clinical research and expand the pipeline of a company in complementary areas from outside the company (Fig. 3.23). Some market analysts consider in-licensing a key success factor in product development. In order to license in molecules, a company needs to be a desirable partner for a smaller company. The risk of failure is high in most collaborations, but the benefit is high also, if the product works and makes it through the development process to the market. Such licensing is done by all companies, in addition to the internal discovery activity of a company, which is demonstrated in this next diagram for the Amgen company from 1980 to 2002.

The sources for research outside a company include universities, the government (NIH), and especially other, often smaller, companies. The acquiring company will completely take over all the research and development of a molecule or share the research work. The acquirer will receive the vast bulk of sales revenue after approval, usually over 90%. The out-licensing partner will receive usually payments in a variety of installments, for example, up-front cash payment or stock purchase, milestone cash payments as research is done successfully (e.g., phase 2 vs. phase 3 vs. NDA filing), and likely royalties on future sales of the product after it is marketed. The size of the payments from the acquirer company to the discovery partner is based on the novelty of the molecule, any competitive advantage or being first in a class to market, size of the future market, stage of research, and the risk of failure (later stage molecules have successfully passed

Internal Discovery		
Yr	Trade name	Generic name
'83	Infergen®	Interferon a-con
'84	Epogen®	Epoetin alfa
'85	Neupogen®	Filgrastim
'88	--	GDNF
'89	Stemgen®	Stem cell factor
'92	Aranesp®	Darbepoetin
'98	Neulasta®	Pegfilgrastim
'01	--	Osteoprotogerin

Licensing & Acquisition		
Yr	Product Name	Company
'90	BDNF & NT3	Regeneron
'92	Kepivance®	NIH
'94	Kineret®	Synergen
'95	Leptin	Rockefeller
'96	Sensipar®	NPS
'96	Leptin receptor	Progenitor
'96	Neuroimmunophil.	Guilford
'99	Abarelix	Praecis
'00	Epratuzumab	Immunomed
'02	Enbrel®	Immunex

FIG. 3.23. Discovery and Licensing for Products*
*Amgen, 2002, Public disclosures; Dates approximate

research milestones and have less risk of failure), one molecule versus a family of related molecules, and the amount, extent, and timelines of research to be done. The discovery partner will obtain a higher percentage of future royalties for late-stage molecules.

In this example, Amgen obtained neurogenic molecules (BDNF and NT3) from Regeneron company in early 1990s, but both molecules have failed. Keratinocyte growth factor (KGF) was obtained from the NIH in 1992 for all indications involving the epitheilium and epidermis. The good news is that the product research was successful with approval in 2004 for a mucositis indication (Kepivance™), but it took 12 years and the indication is fairly narrow, mucositis in stem cell transplants in hematologenous cancers receiving high-dose chemotherapy. In the inflammation area, one of Amgen's research focuses, two molecules were obtained from Synergen through acquisition of the whole company. One molecule has been approved for rheumatoid arthritis in 2002, IL-1ra (Kineret®), and the other remains in clinical trials and formulation development, TNF binding protein over a 10-year period. A calcimimetic family of compounds was licensed in from NPS company, a small biotechnology company in Salt Lake City, Utah, in 1996 with an initial focus on hyperparathyroidism. The first molecule from the company was already in phase 2 trials but failed to be continued in its development due to excessive drug interactions found with further clinical work. Fortunately, another molecule in the family was available for clinical trials, which were done successfully culminating in product approval as Cincalcet® in 2003 for secondary hyperparathyroidism.

A potentially major acquisition was made in 1995, which hit the front pages of the science literature and public press, regarding the molecule leptin for obesity. The competition for the molecule was intense between Amgen, Pfizer, and other major players, because of the exceptionally huge market in the many billions of dollars (obesity) and the molecule's specificity for obesity mechanisms. Rockefeller University and the scientist received a $20 million up-front payment and funding of their laboratory. Amgen pursued the leptin mechanism for obesity further with licensing of the leptin receptor from a biotechnology company. This molecule failed to produce sufficient weight loss in most patients.

Licensing activity around a new disease mechanism or target is an optimal approach to ideally protect an acquisition from future competition; a company will acquire the target molecule, related molecules, receptors for the molecule, and related mechanisms of actions and their targets. Another neurogenic molecule acquisition was done by Amgen with Guilford for a family of molecules, called neuroimmunophilins, but in the ensuing years, they proved to possess insufficient activity for Amgen to continue the research. The collaboration was terminated. In the late 1990s, Amgen licensed in a late-stage product, Abarelix, from Praecis company for prostate cancer as an alternative to existing treatment with a new mechanism of action. Phase 3 work needed to be done, and the work was a collaboration between the two companies. This area was a whole new market for Amgen in urology, requiring building a clinical research team and a marketing team for the molecule. The product worked well in its phase 3 work but was judged by Amgen to be not sufficiently greater in activity versus already marketed products, and the product agreement was terminated, and the molecule was returned to Praecis. The cost of further research and marketing (and sales) build-up was too much for Amgen, given their other pipeline and marketed products and needs versus the return (sales) on the investment for abarelix. However, Praecis continued the research and NDA filing resulting in a successful product approval. The benefit versus expense profile was favorable for a young new company needing their first product approval. In the oncology area, Amgen licensed in a monoclonal antibody (Mab) from Immunomedics company for lymphoma with a novel cell target, CD22 antigen. The research area was compatible for Amgen, that is, oncology, although they had no specific expertise in Mabs, but they were moving favorably into another mechanism for cancer therapy. The product did not perform as well as expected and did not move forward as hoped. In 2002, Amgen made its biggest product acquisition in the future blockbuster, Enbrel, from Immunex, by acquiring the whole company. Besides the product revenue being brought to Amgen, this company acquisition had many potential benefits for Amgen, because the lead research and marketing areas of Immunex were highly symbiotic to Amgen's in inflammation and oncology, adding more pipeline molecules, and adding substantial scientific expertise and marketing savvy in the personnel acquisition [40, 41, 46–47].

A patent is a critical success outcome for any company to protect future revenue from any molecule. The intellectual property through patents must be protected by a company to be successful in sales for the longest time possible and minimize competitive products from becoming available. In biotechnology, given the complexity of the discovery research, the newness of these types of molecules and processes, and the complexity of manufacturing, patents are just being adjudicated now and over the past 10 years, even though thousands of patents are issued each year by patent offices around the world. What is the next major advance in science for a product area by competing companies in their research? This will remain the key question for the courts to decide. Figure 3.24 lists the five main questions to address in establishing a new patent in the USA; subject matter, utility, novelty, obviousness, and disclosure. Products can be patented if they are not a naturally occurring compound. In the drug world, relatively minor chemical modification creates a new patentable compound. In biotechnology, the process to create the molecule, and possibly the molecule as well, are major patentable outcomes. Amgen patent for epoietin alfa is the process to create the molecule, recombinant DNA process, as well as other patents. Amgen won a patent suit against a potential competitor who had isolated erythropoietin from urine, based on novelty and utility for its recombinant process [48–50].

Mergers and acquisitions are commonplace in the pharmaceutical and biotechnology industries over the past 20 years. Figure 3.25 gives six sets of mergers. In the Pfizer example, Upjohn and Pharmacia merged separately as did Warner/Lambert and Parke/Davis. Pfizer acquired first W/L-P/D and then acquired U/P. In these consolidations in the industry, they are intended to create the critical mass of expert scientists, number of quality pipeline products, and research dollars for development of blockbuster products, especially in the face of exceptionally high and rising costs of R&D and the high risks of failure in the industry. A company looks for compatibility in their product lines, along with the scientific and marketing expertise of the two staffs, either to complement an existing business and clinical focus or move into a whole new therapeutic area with the acquisition. Efficiencies in operations are an expected outcome with less costs to operate (e.g., one sales force can handle the combined products with small additions or minor reorganization). Also, the human resources, law, and finance divisions are often downsized in the combined company. A merger is very rarely a combination of two equals; one company predominates or is the acquirer and makes the key organizational, staffing, and operational decisions. One easy way to tell the predominant company is to look at the chairman of the board and CEO positions and which company filled them.

Challenges to the combined operations are substantial and manifold, related to the staff's worries about job loses with loss of productivity, integration of two different staffs with same responsibilities, co-mingling of two business cultures that impacts operations and even communications, and integration of two different sets of operating procedures. Office and research buildings in disparate locations may be a bane or boon. Restraint of trade needs to be addressed with the U.S. Securities and Exchange Commission and European equivalent, such that a monopoly of products in one therapeutic area is avoided. The merged company often must out-license one of their products to avoid this problem. Merger costs at the beginning are huge (e.g., severance packages to managers, golden parachutes to senior management being pushed out, closing of some offices, labs, or operations, and moving people to new locations). Several years often are required for the combined company to assimilate the costs of merging and return to a level of profitability above the added profits from each company. The slide also notes the number of alliances in the industry, intended to produce a symbiosis between two separate companies, usually a smaller one and a large FIPCO to better perform the research and marketing of a product [51–55].

The last outcome, but certainly not least in importance, to discuss involves the staff members in R&D, that is, the scientists and clinicians. What outcomes do these staff and managers receive? Six areas include financial, the research work, scientific advancement, public good (health care), education, and philanthropy (Fig. 3.26). Most of these researchers have come out of academia to the pharmaceutical industry, often harboring questions about scientific integrity, independence, loss of collegiality with their university brethren, and an inordinate focus on products. These concerns will be

Issues to preserve intellectual property of inventions for Material (drug), Manufacturing, & Formulations

1. Subject matter	Identify area in which the invention fits.
2. Utility	Establish invention has real value to humans, and show how it is to be used.
3. Obviousness	Demonstrate the inventor took more than "the next obvious step".
4. Novelty	Document invention is different and not known or previously published in the "prior art".
5. Disclosure	Publish enough information to allow one skilled in the art to repeat invention.

FIG. 3.24. Product & Process Patents

- **Merger mania (1980-2005):**
  - Pfizer-Warner/Lambert-ParkeDavis-Pharmacia-Upjohn = Pfizer
  - Glaxo-BurroughsWelcome-SmithKline-Beachum = GSK
  - Amgen-Immunex-Tularik-Synergen-Kinetix-Abgenix = Amgen
  - Marion-Merrill Dow-Aventis-Sanofi-Synthelabo = Sanofi-Aventis
  - Sandoz-Ciba-Geigy-Chiron = Novartis
  - Yamanouchi-Fujisawa = Astellas
- **Alliances & Licensing (2003):**
  [Focuses: Technology, Molecules, & Products]
  [Areas: Research, Clinical Development, & Marketing/Sales]
  - Pharma-Biotech = 383
  - Biotech-Biotech = 435

FIG. 3.25. Mergers & Acquisitions

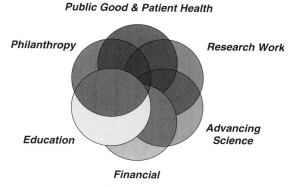

FIG. 3.26. Staff Members

addressed with the many benefits received by industry researchers. Financial reward is usually the first outcome for industry staff that is recognized. Salaries basically need to be competitive with the private sector, and they are. In addition, scientists can receive a bonus in the forms of cash and stock options, based on exceeding their objectives, which is product advancements, publications, and product approvals. Two significant professional benefits, not related to any compensation of any kind, are twofold; helping patients by developing new products to improve their care, thus advancing public good; advancing science by their novel product-related work that is shared with all scientists through presentations at scientific meetings and publications. Another intangible reward to research scientists in the employ of a company is the philanthropy that companies provide to their community, institutions, and patient groups through donations, in the millions of dollars per year per company. Their research work is yet another personal reward; the opportunity to perform research on novel products during their career, present and publish their discoveries and findings, and work with thought leaders from around the world. Some academician may look with a jaundiced eye at industry-based research, both basic and clinical, but the work stands on its own merit based on its scientific quality and innovation. Educational opportunities abound for industry scientists who can learn and then use the latest technologies, attend educational conferences, and for those interested and capable take on management and leadership development.

In this chapter, we have discussed many outcomes of research for a company beyond the ultimate outcome, new products. Company-based research is measured by a company's ability to be successful with all these interim outcomes, all of which essentially are steps, the building blocks, that lead up to product approvals and a successful company.

References

1. Guy P. Rising to the productivity challenge. A strategic framework for biopharma. Boston Consulting Group. July 2004
2. Lawyer P, Kirstein A, Yabuki H, Gjaja M, Kush D. High science: a best-practice formula for driving innovation. In Vivo The Business & Medicine Report 2004;22(4):1-12 (BCG authors)
3. Ernst & Young. Resurgence: The Americas perspective. 2004
4. Ernst & Young. Progressions. Global Pharmaceutical Report 2004
5. Lamberti MJ (Ed). An Industry in Evolution, 4th ed. Thomson, CenterWatch. 2003
6. Anonymous. Report to the Nation 2003, Improving Public Health through Human Drugs. Center for Drug Evaluation and Research, Food and Drug Administration, U.S. Department of Health and Human Services.
7. Vander Walde L, Choi K, Higgins J. Health Care Industry Market Update. Pharmaceuticals. Centers for Medicare and Medicaid Services. Department of Health and Human Services. January 2003, 1-52
8. Anonymous. Health Care in America. Trends in Utilization. Centers for Disease Control. Department of Health Human Services. 2002, 1-90
9. Lahteenmaki R, Baker M. Public biotechnology 2003—the numbers. Nature Biotechnology 2004;22(6):665-670
10. Booth B, Zemmell R. Prospects for productivity. Nature Reviews Drug Discovery 2004;3(5):451-456
11. Rawlins MD. Cutting the cost of drug development. Nature Reviews Drug Discovery 2004;3(4):360-364
12. 15th annual report top 50 pharmaceutical companies. Med Ad News 2003;22(9):4-19
13. Sellers LJ (ed). Our 5th annual report of the world's top 50 pharma companies. Pharmaceutical Executive 2004;24(5):60-70
14. Trombetta B. Industry audit & "fab four' companies of the year. Pharmaceutical Executive 2003;23(9):38-64
15. From pipeline to market 2004. Areas of interest. R&D Directions 2004;10(6):8-18
16. King J. Advances in medicine.100 great investigational drugs. R&D Directions 2004;10(3):31-51
17. Engel S. How to manage a winning pipeline. R&D Directions 2002;(2):36-41
18. Anonymous. Profile pharmaceutical industry 2004, Focus on innovation. PhRMA, 2004.
19. Anonymous. PhRMA. Industry profile 2003. Prescription medicines 25 years ago and today: changing trends, enduring needs. 1-81
20. Ernst & Young. The economic contributions of the biotechnology industry to the US economy. May 2000, Biotechnology Industry Organization
21. Tufts Center for Study of Drug Development. Outlook 2005;1-5
22. Tufts CSDD. Outlook 2004;1-8
23. Tufts CSDD. Outlook 2003;1-8
24. Tufts CSDD. Outlook 2002;1-5
25. Tufts CSDD. Outlook 2001;1-5
26. Matthieu M (ed.). Parexel's Pharmaceutical R&D Statistical Sourcebook 2005-2006. Parexel Publishing, Boston, MA. 2005
27. Anonymous. Health, United States, 2004. National Center for Health Statistics. Centers for Disease Control and Prevention, U.S. Department of Health and Human Services.
28. Edwards MG, Murray F. Yu R. Value creation and sharing among universities, biotechnology and pharma. Nature Biotechnology 2003;21(6):618-624
29. Resnick JT. Reputation the inside story. Pharmaceutical Executive 2003;23(6):40-48
30. Gasorek DW, Resnick JT. The rise and fall of pharma reputations. Pharmaceutical Executive 2005;25(2):76-83
31. Maggon K. The ten billion dollar molecule. Pharmaceutical Executive 2003;23(11):60-68
32. Bogan C, Wang D. Launching a blockbuster. Pharmaceutical Executive 2000;20(8):96-104
33. Booth B, Zemmel R. Quest for the best. Nature Reviews Drug Discovery 2003;2(10):838-841
34. King J. Today's drug discovery unlocking greater potential. R&D Directions 2004;10(2):28-39
35. Lindsay MA. Target discovery. Nature Reviews Drug Discovery 2003;2(10):831-837
36. Lombardino JG, Lowe III JA. The role of the medicinal chemist in drug discovery—then and now. Nature Reviews Drug Discovery 2004;3(10):853-862

3. R&D Outcomes

37. Product failures
38. Garaud J-J. The most diversified development pipeline, innovation, focus, productivity. R&D Directions 2004 Drug Development Summit, How to Build and Maintain a Winning Pipeline. Feb. 8-11, 2004.
39. Bellucci NM. The top 10 pipelines. Novartis: most productive and strongest primary care pipeline. Stroing and steady. R&D Directions 2005;11(1):50-53
40. Amgen Web site (amgen.com), December 2004.
41. Bellucci NM. Top 10 pipelines. Amgen: most advanced pipeline/best biotechnology pipeline. R&D Directions 2005;11(1):32-35
42. Bellucci NM. Top 10 pipelines. R&D Directions 2005;11(1): 32-66.
43. Ashburn TT, Thor KB. Drug repositioning: identifying and developing new uses for existing drugs. Nature Reviews Drug Discovery 2004;3(8):673-683
44. Frei P, Leleux B. Valuation—what you need to know. Nature Biotechnology 2004;22(8):1049-1051
45. Tollman P, Goodall S, Ringel M. The gentle art of licensing. Rising to the productivity challenge in biopharma R&D. Boston Consulting Group. July 2004
46. Arnold K, Coia A, Saywell S, Smith T, Minick S, Loffler A. Value drivers in licensing deals. Nature Biotechnology 2002;20(11):1085-1089
47. Featherstone J, Renfrey S. From the analyst's couch. The licensing gamble: raising the stakes. Nature Reviews Drug Discovery 2004;3(2):107-108
48. Webber PM. Protecting your inventions: the patent system. Nature Reviews Drug Discovery 2003;2(10):823-830
49. Rubinger B, Davis H. Protecting IP throughout the product lifecycle. Pharmaceutical Executive 2003;23(8):40-48
50. Steffe EK, Shea Jr TJ. Protecting innovation in biotechnology startups. Nature Biotechnology Supplement 21:BE51-BE53, 2003
51. Lam LD. Biotech+Pharma dangerous liaisons. Pharmaceutical Executive 2004;24(5):72-80
52. Engel S. Drug-discovery companies aim for commercialization. Good leads good medicine. R&D Directions 2002;8(7):34-42
53. Lam MD. Biotech+pharma. Why alliances fail. Pharmaceutical Executive 2004;24(6):56-66
54. Ansell J. The billion dollar pyramid. Megamergers' greatest challenge. Pharmaceutical Executive 2000;(8):64-72
55. Bogan C, Symmers K. Marriages made in heaven. Pharmaceutical Executive 2001;(1):52-60

4
Discovery and Nonclinical Development

Stephen F. Carroll

The Discovery Process	85
Targets	93
Products	95
Nonclinical Development and Testing	98
IND-Enabling Studies	102
Added Discovery Work	103
References	106

The discovery of new products for patient use takes place in laboratories at universities, in the government, or in pharmaceutical companies; actually, it starts in the minds of scientists with a scientific innovation or idea for creating a new therapeutic molecule that may be a biological or drug. This research is performed through carefully done studies, either with the physiology of humans or other species, disease models, or some core structure of a molecule, through a host of different scientific technologies. Sometimes, a drug discovery is an accidental finding related to an unexpected action of a drug being studied for other uses, such as Viagra® for impotence. Each molecule may have an impact on a general physiologic process such as inflammation and thus have the potential to be used in many organ systems and diseases, or it may impact a specific receptor on a cell, such as a tyrosine kinase, and be used only when the receptor system goes awry. Knowledge of the discovery and early development process creates a basis for understanding how potential new therapeutics advance from the research laboratory to the clinic and some of the issues involved.

Molecules designed for therapeutic use come in many sizes and shapes. Although most of the therapeutics on the market today are "small molecules," also called drugs, such as aspirin (13 atoms, excluding hydrogen atoms), Viagra® (33 atoms), and Taxol® (62 atoms), increasing efforts are being placed on the development of larger "biological" molecules. Included among the biologics are molecules like insulin (408 atoms), erythropoietin (2,634 atoms), and antibodies (10,402 atoms). Most small molecules are produced by excretions from microbial fermentation or produced by chemical synthesis, often in combination with structural modifications produced by techniques of medicinal chemistry. Biologics are manufactured by complex living systems (e.g., recombinant techniques), where transfection of the appropriate gene into a suitable cell line (either microbial or mammalian) allows production of the therapeutic protein. For some products like monoclonal antibodies, hybridomas that produce the antibody initially are created from murine and lymphoid myeloma cells for antibody production. These hybridomas then can be used to produce the monoclonal antibody or, more commonly, the antibody genes are transferred to another host cell, such as Chinese hamster ovary (CHO) cells for clinical production. Each class of compounds, small molecules and biologics, has unique advantages and disadvantages that can be utilized and tailored to address specific therapeutic needs.

In this chapter, we will discuss the steps involved in the discovery and early development of new therapeutics, leading up to an Investigational New Drug (IND) application. We will also try to discuss some of the issues encountered along the way. These topics will be covered in seven sections (discovery process, targets, products, nonclinical development and testing, IND-enabling studies, added discovery work, and a summary) and 42 figures.

4. Discovery and Nonclinical Development

The Discovery Process

Before we begin our discussion, let's first review terms particularly relevant to discovery and early development. These include the six terms shown in Figure 4.1

Target: A protein, enzyme, receptor, signaling or other molecule that may play a role in a particular disease process. It is the target molecule or process upon which the discovery and therapeutic strategy will be focused.

Hit: A test protein, peptide, or compound that appears to act on the target. Depending upon the target and the biological or chemical system, thousands of hits may be evaluated, looking for the most active compounds to test further.

Lead: Among numerous hits or variants, the protein, peptide, or compound showing the highest degree of activity. It is the lead compounds that will be further examined in greater detail.

Candidate: A protein, peptide, or compound that has most or all of the properties of the desired therapeutic (a development candidate). Incorporated into the thinking here is not only the level of activity that a lead has, but also how easy is it to formulate and manufacture, how safe is it, and does it meet the *in vitro* and *in vivo* requirements and medical needs. Candidates usually enter into clinical trials to establish safety and then efficacy.

IND: Investigational New Drug application, filed for the initial testing of each new drug in humans. This is the actual document filed with the FDA or other regulatory body requesting their approval to begin clinical testing in humans. It contains a summary of the compound to be tested, especially all the animal pharmacology and toxicology data, the rationale for testing in a particular indication in humans, a detailed description of the clinical protocol itself, as well as the methods used to manufacture and test the compound.

Product: A marketed therapeutic drug or biological, approved for use by regulatory bodies.

With these terms in mind, let's take a look how new drug candidates are identified and moved toward clinical testing. In its simplest form, drug development can be viewed as a stepwise process involving a series of sequential discovery and development decisions that are based on the target and the potential product (12 such steps in two phases are shown in Figure 4.2). This process is commonly separated into two sections, discovery and early development (often referred to as nonclinical or preclinical development, as it relates to studies needed prior to clinical testing), because they involve different approaches, skills, and facilities. A definition for both terms is provided on figure 4.2.

The first part of the drug development process is called discovery. Discovery is driven by unmet medical needs and financial opportunity and focuses on understanding the disease process and the identification of disease targets and potential therapeutic compounds. This stage of the process is perhaps the most variable and least successful of all aspects of drug development. These difficulties are due in part to the fact that discovery research is highly dependent upon a detailed knowledge of the disease in question and because it involves the isolation, production, and testing of compounds that may not have existed previously. Thus, if the disease biology is only poorly understood, it is difficult to know what an appropriate target for intervention might be. Similarly, even if the disease biology is quite clear and a suitable target can be readily identified, it is not uncommon that many thousands of compounds may have to be synthesized, purified, and screened in an effort to find initial "hits" that can be further developed.

Once a target has been identified, methods to influence that target are then considered. Typically, this involves the design or identification of compounds that either stimulate or inhibit the actions of the target, initially *in vitro* (in the test tube, a "hit") and then *in vivo* (in animals, a "lead"). Preliminary studies are also conducted to evaluate the *in vivo* properties of the leads as possible therapeutic candidates (pharmacokinetics, pharmacodynamics, efficacy, toxicity, etc.), and those compounds that have acceptable activity and safety profiles may be passed along to development as "candidates".

At the early development stage, a new group of scientists gets involved who have expertise in translating what has been accomplished at the laboratory scale into methods and systems that will ensure the reliable and reproducible manufacture, also called process engineering, and testing of the product. Thus, it is at this stage that robust methods for purifying, formulating, manufacturing, and testing (e.g., analysis, stability) the product candidate will be developed. As an example, groups studying antibodies will initially work with material derived from tissue culture systems or small fermentors (<10 L), but during process development systems may be scaled up to 30 L, 150 L or 500 L fermentors, depending upon the initial development and clinical needs. Subsequently, for large-scale manufacturing and product sales, multiple 15,000 L fermentors may be utilized. A more detailed description of the steps involved in development is presented elsewhere in this volume. Here we will focus only on those tasks that have relevance to the filing of an IND.

- **Target** – A protein, enzyme, receptor, signaling or other molecule that may play a role in particular disease proces
- **Hit** – A test protein, peptide or compound that appears to act on targets
- **Lead** – Among numerous hits or variants, the protein, peptide or compound showing highest degree of activity
- **Candidate** – A protein, peptide or compound that has most or all of properties of desired therapeutic (a development candidate)
- **IND** – Investigational New Drug application, filed for initial testing of each new drug in humans
- **Product** – A marketed therapeutic

FIG. 4.1. Important Terms – General

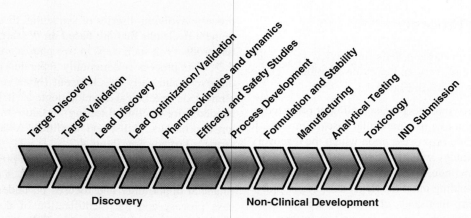

- Discovery - Identification of disease targets and potential therapeutic compounds:
  - The most variable and least successful aspect of drug development
- Non-Clinical Development - Translates discovery science into therapeutic candidates:
  - Involves modifications, scale-up, purification, test methods and production

FIG. 4.2. Drug Discovery & Non-Clinical Development

- Target identification – Process by which potential targets are investigated, screened and prioritized:
  - Involves a detailed knowledge of the disease process, such as up-regulation of certain proteins in cancer cells
- Target validation – Process by which role a target plays in a disease is characterized and established:
  - Involves a combination of invitro and in vivo functional studies
  - Common tools are cellular-based assays, antisense, RNAi and knockout mice
- Lead identification – Process by which potential therapeutics are screened and prioritized:
  - Utilizes knowledge of the specific target to identify/design an appropriate agonist/antagonist
- Lead optimization and validation – Process by which actions of products on diseases are characterized and confirmed:
  - Compound is tested in animal models of target disease
  - Improvements ar edesigned and evaluated

Not all aspects must be completed to move candidates into clinical trials

FIG. 4.3. Important Steps in Discovery

Focusing more closely on discovery, we see that there are essentially five main steps; target identification and validation, and lead identification, optimization, and validation. Figure 4.3 provides a description of these five steps. To illustrate these steps, we'll consider two different examples, a small molecule to treat AIDS and an antibody to treat psoriasis. The disease AIDS (acquired immunodeficiency syndrome) is caused by infection with HIV (the human immunodeficiency virus).

Once HIV infects cells, it produces several enzymes that are required for the replication and propagation of the virus, one of which is reverse transcriptase. This enzyme uses the viral RNA as a template and makes DNA copies of the viral genome, which then enter the nucleus where host cell enzymes are used to many more copies. Thus, because reverse transcriptase is a viral-specific enzyme, inhibiting the activity of this enzyme could reduce the spread of AIDS. This makes reverse transcriptase a potential "target" for therapeutic intervention [1].

In the early 1980s, a number of nucleotide analogues were being studied as potential anticancer therapeutics and, because these structures mimic the building blocks of DNA and RNA, many were subsequently screened for their ability to inhibit reverse transcriptase ("hits" and "leads"). One of these analogues was AZT (azidothymidine), which was found to be an effective inhibitor of reverse transcriptase (a "lead") and, when tested in patients, inhibited replication of HIV. AZT was therefore a "candidate" that became a "product."

Psoriasis is an autoimmune disease characterized by activated immune cells. Normally, the immune system acts as an internal security system, protecting the body from infection and injury. With psoriasis, however, T cells become overactive. This activity sets off a series of events that eventually make skin cells multiply so fast, they begin to pile up on the surface of the skin, forming characteristic plaques (red, scaly patches on the surface of the skin). Thus, agents that interfere with the function of T cells could reduce the signs and symptoms of psoriasis (indeed, topical steroids are used extensively), and as such are hits, leads, and candidates, depending on their stage of evaluation.

Clearly, many systemic immunosuppressive agents have been identified (cyclosporin A, methotrexate, etc.), and most provide benefit to patients with psoriasis. However, many of these agents are also quite toxic, making prolonged use difficult. As a consequence, alternative ways to interfere with the activation of T cells have been explored, and several T-cell surface structures were believed to play critical roles in this activation process. Among these structures, one (LFA-1, or lymphocyte function–associated antigen 1) appeared to be involved in T-cell activation, function, and trafficking to sites of inflammation, a new "target." Antibodies were therefore raised against human CD-11a (a subunit unique to LFA-1) and tested *in vitro* and in animals. These antibodies were hits. Of the antibodies that were generated, several effectively inhibited a number of T cell–mediated functions *in vitro* and also showed efficacy in animal models of autoimmune disease, hence "leads" [2]. Based on these data, one antibody (MHM24) was optimized and became a candidate for human use by "humanizing" it [3], a process that strives to reduce the chances of generating an immune response by converting a mouse antibody sequence into a sequence commonly found in humans. The resulting antibody, termed Raptiva®, has been shown to be safe and effective in treating patients with moderate to severe psoriasis [4], resulting in its approval as a product.

Note that, although it is desirable to have all these elements completed prior to filing an IND, they all may not be required to do so. Some of the factors that influence how much information is needed to file an IND include (i) the clinical indication, (ii) the nature of the compound (small molecules vs. biologics), (iii) the specificity of the compound, (iv) the availability of appropriate animal models, and (v) the seriousness of the disease. Thus, small molecules and biologics for use in cancer (or other life-threatening diseases) may require less nonclinical information to file an IND than therapeutics designed for chronic or more benign diseases. Similarly, small molecules often require a more detailed safety package than biologics, in part because the later agents are often human proteins that have fairly predictable actions and degradation and clearance properties. The requirements for filing an IND are also influenced by whether the agent only interacts with a human target (and thus animal studies may be less predictive) and whether suitable animals exist for appropriately testing the new therapeutic.

The term *validation* has shown up several times now and is worth additional discussion as it is frequently misunderstood (Fig. 4.4). Most commonly, the term *validation* is used to demonstrate that a particular assay or process is well controlled and reproducible. Thus, for a company manufacturing a recombinant therapeutic protein, they must demonstrate that the fermentation process, purification process, and assays used to test the activity of the product produce similar results each time they are performed (i.e., that they are reliable). The steps involved are therefore called "process validation" and "assay validation." Such validation typically involves the preparation of standard operating procedures (SOPs) that describe in detail precisely how the process or assay is to be conducted, as well as having one person repeat the assay several times and then several people repeating the assay. Only when the results of all these assays are reproducible will that assay be considered "validated."

In contrast, validation is also now being used to support the potential validity of new targets or products. For example, an investigator might say they have identified 100 "validated" targets, by which they mean to imply that a clear linkage has been demonstrated between the presence or absence of this target and the disease in question. Whereas there can be value in these data, there is as yet no clear definition of what "validated" means when applied to new targets and potential products—some have used the term to indicate that a particular target is always absent on normal tissues but is always present in every

- "Validation" has been most commonly used in biotechnology and pharmaceutical industries to reflect level of control and reproducibility for an assay or process:
  - There are FDA guidelines on process and assay validation
- More recently, it has been used to "suggest" that certain therapeutic targets or products are more likely to be successful than others:
  - In reality, some "validated" targets or products may be weakly supported by limited *in vitro* data, or they may be strongly supported by knockout and disease models
- Only "validated" targets are those for which clinically successful therapeutic products have been generated
- Only "validated" products are those with several hundred million dollars in sales

FIG. 4.4. Validation – Frequently Misunderstood

diseased tissue (a good idea), but others have used the term to indicate that certain targets are simply upregulated in a few diseased tissues (not so good). As a consequence, many people feel that the only true "validated" targets are those for which clinically successful therapeutic products have been generated (3-hydroxy-3-methylglutaryl-coenzyme A reductase, COX-2, erythropoietin receptor, CD-20, etc.), and that the only true "validated" products are those with several hundred million dollars in sales (Lipitor®, Epogen®, Rituxan®, etc.). Though validation is an important component of the product development process, it is critical to keep these distinctions in mind when listening to claims for new targets!

These key questions for discovery help guide early choices during the development process for targets (five questions) and for products at the lead stage (six questions). The target questions focus on relationships of the target with the disease and how changes in the target impact the disease (Fig. 4.5).

- **Target validation:**
  - What does target do?
  - What role does target play in the disease?
  - How specific is target for the disease?
  - If I inhibit target, is there an impact on the disease?
  - If I inhibit target, what other effects are there (toxicity, etc.)?

- **Product validation:**
  - How well does product work *in vitro* and *in vivo*?
  - How selective is product for the target?
  - How stable is product (does it break down)?
  - How long is product available in vivo?
  - Where does product go after administration?
  - How toxic is product?

FIG. 4.5. Key "Validation" Questions

The lead questions relate to an early profile of the potential product prior to human use and hopefully suggestive of human activities for the lead. Product characteristics include activity, stability, distribution, persistence *in vivo* pharmacokinetics, and toxicity.

Traditionally, drug development has been viewed of as a linear, stepwise process involving a series of sequential decisions that are based on the disease, the target and the desired product properties, such as the five steps noted in the Figure 4.6 [5]. However, as is evident from our earlier examples, this can be a long (6–12 years) and expensive process (millions of dollars per lead) that does not follow a sequential path and yields many more failures than successes.

As an illustration, let's consider the case of Lipitor, a cholesterol-lowering product that had $10.3 billion in worldwide sales for 2003 [6–9]. In the early 1980s, clinical data were accumulating that suggested a linkage between high serum levels of cholesterol and increased risk of heart attacks and stroke. Beginning in 1982, scientists at Parke-Davis (now part of Pfizer) began looking at a class of compounds called statins, which are fungal products that block cholesterol synthesis at a key step (3-hydroxy-3-methylglutaryl-coenzyme A reductase, or HMG-CoA reductase). At the time, it was unknown whether lowering plasma cholesterol levels would be beneficial and, if so, whether it could it be done safely. Thus, a clinical need appeared to exist for therapeutics that could lower serum cholesterol levels, the biosynthetic enzyme HMG-CaA reductase was a reasonable target, and statins represented an initial class of lead compounds. The particular challenges here, however, were to develop a compound that had statin activity, was safe, had potential patient benefits, and could be easily manufactured.

In 1985, a compound was developed (CI-981) that appeared to meet most of the requirements. It still had limitations, however

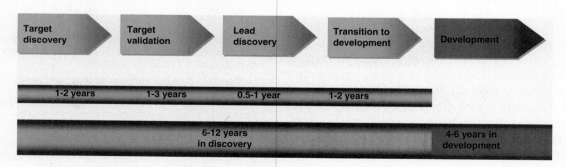

FIG. 4.6. Traditional View of Drug Discovery(Reprinted with permission from Nature Publishing Group, London, England. From Graph in Myers S, Baker A. Nature Biotechnology 2001;19(8):727. Drug Discovery – an operating model for a new era.)

(i.e., it still needed to be optimized). For one, it was a racemic mixture of two stereochemical isomers, left (L) and right (R) handed versions of the same compound, but only the L isomer was an effective inhibitor of HMG-CoA reductase. Using specialized manufacturing techniques (running certain reactions at temperatures below −80°C), a procedure was developed during the transition to development for synthesizing only the L isomer in large scale, a process that took 3 weeks from raw materials to final product. Also during this time, studies were conducted in animals to demonstrate that lowering cholesterol level was beneficial (target validation) and that CI-981 had clinically desirable properties (lead validation). By 1989, the compound was ready for clinical testing and, in 1997, Lipitor® was approved by the FDA. From start to finish, Lipitor's discovery and development took about 15 years.

As is evident from the Lipitor® example, discovery and development are not as sequentially oriented as the earlier slides suggest. Instead, the process has evolved into a more integrated and overlapping approach that seeks to streamline the identification of new targets and therapeutics (Fig. 4.7) [5]. It is thus more common (and more beneficial) that target validation occurs in parallel with lead discovery and lead optimization, with one function helping to confirm (validate) the other. Similarly, the different disciplines (biology, chemistry, and pharmacology) typically operate in a more integrated fashion, facilitating the exchange of information and conducting earlier studies in animals, thereby shortening the discovery and development timelines (3–5 years vs. 6–12 years) and possibly saving research costs. The real financial savings occur because accelerated research has potentially consumed less patent life before approval and extended it after approval, yielding higher total sales revenue before generic substitution would occur.

Such an approach also allows for the early evaluation of new biomarkers, biochemical or biological surrogates that may be used as early indicators of efficacy or toxicity. The availability of such biomarkers is extremely important, as they can greatly accelerate clinical development by providing alternative and less time consuming and less costly end points for further development decisions. One such marker is prostate-specific antigen (PSA), a tumor-specific marker currently being explored in many clinical trials in patients with prostate cancer as a possible surrogate efficacy end point [10]. PSA levels are known to be elevated in patients with prostate cancer, but if it can be demonstrated that low or declining levels correlate with drug therapy and clinical benefit, the testing of new anticancer agents would be greatly facilitated. The biomarker must be validated for its disease association, and it must change under the influence of the product to be approved. Furthermore, the regulatory bodies must also agree for the biomarker to be used in INDs and NDAs.

And why is it important to rapidly and efficiently screen and develop new drugs? Because the process itself takes a long time, it costs a lot of money, and most drug development efforts ultimately fail. These concepts are perhaps best illustrated by reviewing the efficiency with which new drugs get through clinical trials to approval (Fig. 4.8). For every small molecule that reaches the market, more than 5,000 compounds are synthesized, about 500 of these make it to preclinical studies, 10 make it to development, and 5 enter clinical trials. Similarly, although biologics give the appearance of being more efficient than small molecules (1 therapeutic approved

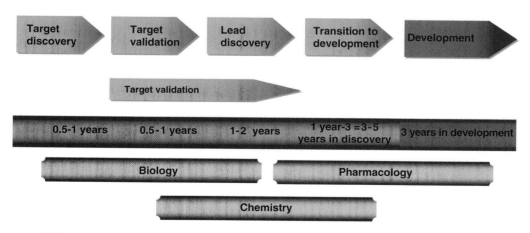

FIG. 4.7. Integrated View of Drug Discovery (Adapted with permission from Nature Publishing Group, London, England. From Graph in Myers S, Baker A. Nature Biotechnology 2001;19(8):727. Drug Discovery – an operating model for a new era.)

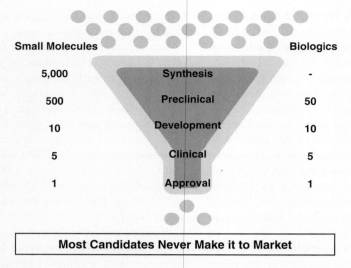

FIG. 4.8. The Problem with Drug Development

- Identify new targets for disease intervention
- Create new reagents that address important disease targets
- Develop essential assays and models
- Explore possible surrogate endpoints and biomarkers for use in monitoring clinical trials
- Conduct research needed to advance or kill each project promptly
- Determine product pharmacokinetics, pharmacodynamics and safety
- Coordinate new research to expand indications for existing products
- Generate intellectual property to create/enhance product protection

Use knowledge and experience to effectively drive and direct discovery

FIG. 4.9. Goals of Discovery Research

for every 50 entering preclinical development), even starting with a recombinant human protein with known activity is no guarantee of clinical and market success. For example, at Amgen, GDNF (glial-derived neurotrophic factor) has been demonstrated *in vitro* cell cultures to arrest death of or heal the brain cells associated with Parkinson disease and even dramatically improved the signs of parkinsonism in primate animal models. However, GDNF was a failure in human trials without significant improvement in the clinical signs and symptoms of the disease. Alternatively, an unexpected adverse effect from such a protein, which may very closely resemble the natural protein, can occur to stop its development. For example, a thrombopoietic factor for platelet disorders was found and was quite active but for some unknown reason produced antibodies against the not only the protein but also against the naturally occurring thrombopoietin, which was a life-threatening complication.

Some of the issues that complicate the drug development process include the following eleven examples:

- disease biology is incompletely understood
- *in vitro* assays may not accurately mimic disease process
- *in vivo* models may not accurately mimic disease process
- acute onset disease animal models may not accurately mimic chronic diseases in humans
- actions of the compound are inherently different in humans than in animals
- human population is very heterogeneous (laboratory animals are not)
- some toxicity issues only show up in humans
- target and compound selection are not in sync with the complexity of disease physiology
- the pharmacokinetics and clearance of molecules may differ between humans and animal model
- proteins (recombinant, antibodies, or peptides) may lead to neutralizing antibodies reducing or preventing activity
- biologic molecules may be so large or complex in structure that formulations become impossible challenges to get the product to the site of action

These issues have helped revive the concept of systems biology in drug discovery, which seeks to understand physiology and disease processes at the levels of molecular pathways, regulatory networks, cells tissues, organs, and whole organisms [11]. With such an understanding, it is hoped that drug discovery targets and their therapeutic drug candidates can be more effectively and rigorously identified and prioritized.

Here then is a summary of eight goals that discovery research is trying to accomplish (Fig. 4.9). Several topics deserve further comment:

Conduct critical studies early: It is imperative that experiments be designed to evaluate the actual validity of the target and the value of the lead compound, and that these studies be conducted as early as possible. Quite frequently, such studies are often delayed for fear that the project might be killed, but it is far better to stop an unpromising project early than to spend more time and money simply postponing the decision. Besides, terminating one program often allows more time to pursue (or create) new, more promising ones. Biomarkers as noted above are important tools to achieve these goals.

Expansion of indications: Because drug development is so costly and time consuming, one approach that takes further advantage of development dollars already spent is to explore additional indications for approved therapeutics. Such studies can involve entirely new indications, or alterations to the therapeutic (formulation, delivery route, delivery devices, etc.) for existing indication (more later). Although new indications and uses often require additional time for development and testing, they avoid additional discovery costs and effectively build on existing data.

Intellectual property: Patents are critical components of any development program, as they are a form of "property" that can be sold or traded. In essence, patents are legal documents that entitle the owner to prevent others from making, using, or selling the invention for a limited period of time. If that invention is a new therapeutic, then the owner is the only one who has the right to manufacture and market that therapeutic. Similarly, if the invention covers a specific process (such as the production of recombinant proteins in mammalian cells), then other companies interested in selling their different recombinant proteins (produced by the same method, that is, invention) may need a license to that patent in order to market their products. Importantly, although all aspects of the development process can generate useful intellectual property (including development and clinical trials), it is often the discovery phase that has the earliest opportunities to identify and protect new areas. Whether it is new therapeutic targets, new experimental therapeutics, or new indications, much of the earliest data and results that are patentable are identified during discovery. Thus, much of a product's real value comes from the intellectual property that surrounds it, and much of this intellectual property begins with discovery research. Patents usually occur early in the life of a molecule that becomes a product, but new patents are constantly being pursued throughout the product's life cycle to improve the manufacturing efficiency, protect related molecules, or find new useful formulations.

In order to accomplish these goals, what does discovery need to do be successful (Fig. 4.10)? As before, some of these six areas deserve further comment:

New targets, compounds and disease pathology: where do they come from? Historically, drug development companies relied upon internal research organizations and groups for the identification of new targets and therapeutics. Recently, however, more and more development programs are the result of strategic partnerships between drug development companies and other companies or academic laboratories

- Access to appropriate sources of new targets and compounds
- Detailed knowledge of the basic biology for the targeted disease process
- Ability to rapidly analyze targets and potential therapeutics
- Understanding of the regulatory issues and requirements
- Recognition that discovery and development decisions involve multiple groups (research, development, legal, marketing, management, etc.)
- An understanding that terminating unsuccessful projects is critical, crucial and beneficial

FIG. 4.10. Needs of Discovery Research

(more later). Biotechnology companies are a major source for new disease knowledge, targets, and compounds; about 3,300 companies existed in United States and Europe in 2003.

Interacting with multiple groups. Decisions in any organization can be a complex process, and those involving drug discovery and development are no exception. For those involved in the discovery process, it is important to recognize that decisions are multifaceted and involve numerous groups. Thus, in addition to input from the research groups, also involved are legal (is there "freedom to operate" or license issues?), technical development (can it be purified and formulated?), manufacturing (can we make it?), regulatory (is there an approval path?), marketing (can we sell it?) and management (is it good for business?). Along with such varied groups playing key roles, the processes of teams, planning, and decision making require much more emphasis even at early stages such that the right people are engaged at the right time with the right information for the best possible decisions to be made by product teams and management. Portfolio and project planning management (PPM) have become key roles at the research stage as well.

Terminating unsuccessful projects. Terminating a project is often quite difficult, as they tend to gain a life of their own. From the scientist who thought of the idea to the marketing person who really likes the idea to upper management who really wants the idea to work, everyone hopes that each project will succeed. That said, it should be clear from the foregoing discussion that in fact most projects do not. And, for this reason, it is critical to terminate unsuccessful projects as early as possible, so effort and money can be spent on potentially more promising projects.

Despite the fact that the highest drug sales are for products that treat gastrointestinal (antiulcerants) and cardiovascular (cholesterol and triglyceride reducers) diseases, both biotech and pharma companies are focusing most of their development and clinical efforts on cancer, infectious diseases, and central nervous system disorders (Fig. 4.11) [12]. One rationale for this paradox is the medical need of patients, the advancing science, and the opportunity for sales. Neurologic disorders, especially neuromuscular and Alzheimer's, are quite prevalent without good treatments, representing high

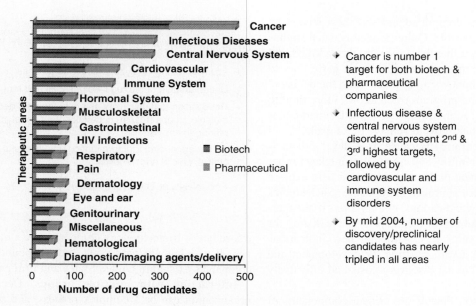

FIG. 4.11. Current Discovery Focus – 2003/2004 (Adapted with permission from Figure in Lawrence S. Acumen J Sciences 2003;1(1): 22–23. Drug development by indication.)
Source: Discovery and Preclinical Pipelines – 2003; Lawrence. Acumen J. of Sciences 2003;1:23; Biopharm Insight

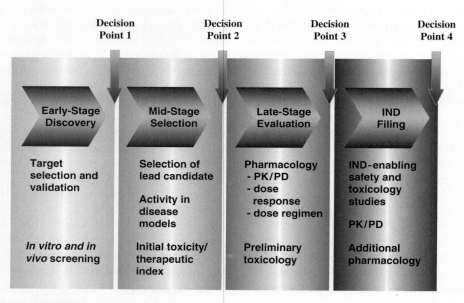

FIG. 4.12. Discovery & Development Decision Points

need and high opportunity (medical and financial) markets. The scientific advances in understanding cell growth, both normal and abnormal as in cancer, have been legion over the last 10 plus years, along with the need for less toxic and more effective treatments. Patients living longer (chronic type disease) and the fatal nature of these diseases combine to drive companies to invest in cancer research. Infectious disease area is a constantly changing arena with new product needs, based on continued evolution of resistant organisms.

Given the cost and complexity of drug development, it is essential that processes be in place that allow for ongoing review, discussion, and then decisions (go–no go or more work is needed). An example of a sequential decision matrix is shown in Figure 4.12, where discovery and early development is broken into the essential studies and information needed to move forward. The studies are further divided to address four decision points for progression of the compound, focusing on, first, targets, second, lead candidate, third, animal pharmacology, and fourth, IND enabling, especially toxicology, studies. Thus, if selection fails to identify an appropriate target at stage one, the project may be terminated or alternatives to the existing process must be investigated.

Integral to the success of this process is the establishment of criteria needed to allow a determination of "success" at each step. Thus, it generally is not sufficient to have a lead candidate that simply has activity. Rather, in order to move a lead on to the next phase, its *in vitro* and *in vivo* activity must be above a predefined threshold that, with some degree of certainty, has a high probability of being efficacious in humans. As an aid to this process, many organizations create a brief, one-page document (viz., a product profile or specification sheet) that outlines many of the biological, clinical, and practical criteria that are deemed important for product success. As development progresses, the properties of the candidates are then compared to the sheet, which becomes a benchmark that allows researchers and management to gauge progress along the development path. If a compound fails to meet the desired specifications, either new compounds need to be identified or the criteria used to develop the sheet should be reevaluated. The criteria may have set hurdles too high given the science available. Many companies use external expert groups from academia and practice to create such spec sheets, avoiding internal group-think favoring internal compounds and achieving a better profile of acceptable properties and improved care of patients.

Targets

Lets focus now on the target (Fig. 4.13). In order to have an activity, a new therapeutic must exert an action against a biological or biochemical process. As a consequence, therapeutic targets are typically enzymes, ligands, receptors, signaling molecules, or surface antigens that play a role in the biology of the disease. After an evaluation of the need for new therapeutics in a given indication, as well as the market and the potential competition, such targets are usually identified by a detailed consideration of the disease process. Importantly, throughout this process considerable attention is also placed on the intellectual property (i.e., patents) associated with the target and the therapeutic. The importance of patents and their associated know-how cannot be over stressed, as this information often forms the basis for both getting into and then surviving in the marketplace. Similarly, there may be critical intellectual property owned by others that may be essential to your product or indication. As "property," patents and know-how can be traded, bought or sold, all of which are common practice during the drug development process. Sources of new targets are identified on the slide in the four bullets.

Biotechnology particularly has focused on disease pathophysiology to uncover the secrets of human physiology and disease. Technological advances, such as in analysis of molecules and intracellular mechanisms, and whole new technologies in research methods are advancing these discoveries as well. Venture capital has been available to fund these biotech companies for these biological advances. Some examples of how biotechnology has impacted target discovery and product development are shown here (Fig. 4.14). As we have discussed, targets frequently are:

- Enzymes (aurora kinase)
- Receptors (tyrosine kinase receptors, EGFr)
- Signaling molecules (VEGF inhibitors and traps, TNF-a)
- Cell surface antigens (CD-20, CD-11a)

In addition to many of the existing tools, new approaches to target discovery have been identified over the last few years (Fig. 4.15) [13]. Some of these new tools are described below.

Genomics is the study of all of the nucleotide sequences, including structural genes, regulatory sequences, and noncoding DNA segments, in the chromosomes of an organism. When applied to target identification, genomics attempts to identify novel disease targets by comparing gene expression in normal and diseased tissues.

Proteomics is an effort to establish the identities, quantities, structures, and biochemical and cellular functions of all proteins in an organism. Said another way, proteomics attempts to understand cellular function through the measurement

---

- Most therapeutic targets are enzymes, receptors, signaling molecules, signaling cascades or surface antigens
- Initial focus is usually placed on medical need, the market, the competition, and the disease process
- Considerable attention is also placed on intellectual property:
  - If I'm successful, can I sell my product (freedom to operate)?
  - If I'm successful, can I protect my product (exclusionary rights)?
- Most common sources for targets include:
  - In-licensing from other companies, academia or NIH
  - Collaboration with other companies, academia or NIH
  - Internal research programs
  - Mining published literature

FIG. 4.13. Target Identification and Selection

Targets:	Company:
IgE antibodies	Tanox / Genentech (Xolair)
Lymphocyte CD-20	Idec-Biogen (Rituxan)
Tyrosine kinase receptors	Astra-Zeneca (Iressa)
Lymphocyte CD-11a	XOMA / Genentech (Raptiva)
Proteosome inhibitors	Millennium (Velcade)
HIV binding and cell entry	Trimeris / Roche (Fuzeon)
EGFR inhibitors	Imclone (Erbitux), OSI (Tarceva)
VEGF inhibitors	Genentech (Avastin)
VEGF receptor analogs	Regeneron (VEGF trap)
Aurora kinase inhibitors	Vertex (VX-680)
E2F decoy	Corgentech (Edifoligide)
Triple serotonin MOA	NeuroSearch (NS2359)
Reverse lipid transport (HDL)	Esperion (ETC-216 & -588)

FIG. 4.14. Biotechnology Impact
Source: Company Websites

- Genomics
- Proteomics
- Knockout and transgenic animals
- Gene silencing (antisense, siRNA)
- Pharmacogenomics and single nucleotide polymorphisms
- Microarrays (genes and proteins on chips)
- High throughput screening
- Bioinformatics
- Phage (and other) display systems
- New biology (e.g., protein kinases, proteosomes, apoptotic signals)

Although these new tools have increased number of potential targets, ability to generate successful therapeutics from these new targets has not (yet) significantly increased (target rich, product poor)

FIG. 4.15. New Tools of Discovery Research

of protein expression, activity, and interaction with other biological macromolecules [14].

Knockout and transgenic animals are animals in which specific genes have been deleted (knockout) or inserted (transgenic), allowing a determination of the consequences (phenotypes) of either the absence or presence of the specific gene, respectively. Such information can be extremely useful confirming the value of a particular target and in designing a desired therapeutic. In fact, in a recent review [15], the phenotypes of knockouts for targets of the 100 bestselling drugs showed good correlation with the known drug efficacy.

Gene silencing is an alternative to the creation of knockout animals, whereby double-stranded RNA (dsRNA) is able to inhibit the function of complementary single-stranded RNAs such as messenger RNA. This process, known as RNA interference (RNAi), is being widely used as a target validation tool in discovery research [16]. In addition, RNAi technologies are being explored as a means of generating new therapeutics useful against gene targets that may not be amenable to conventional therapeutics [17, 18].

Pharmacogenomics seeks to develop medicines on a personal level. It is the study of how an individual's genetic inheritance affects their response to drugs, and holds the promise that therapies might one day be selected for (or adapted to) each person's own genetic makeup. Variables such as environment, diet, age, lifestyle, and state of health all can influence each person's response to a drug. Thus, understanding an individual's genetic makeup may allow the creation of personalized drugs with greater efficacy and safety [19].

All of these tools are supported by a series of technologies (*microarrays, high-throughput screening [HTS], bioinformatics, phage display, etc.*) that can greatly facilitate the performance, evaluation, or interpretation of study results. For example, microarrays are now being used to study gene expression [20] and protein function [21], as well as to study compound toxicology [22]. The identification of *new biological targets* (viz., protein kinases), processes (viz., apoptosis), or structures (viz., proteosomes) has also helped focus and accelerate discovery research.

It should be emphasized, however, that all of these new tools, though they have expanded the number of potential therapeutic targets, have not yet led to the identification of successful new therapeutics. Thus, it has been said that we are currently "target rich, product poor." Because many years are required for the successful development of a new therapeutic, and because these new discovery tools have little impact on the process of drug development (purification, formulation, scale-up, manufacturing, etc.), it seems likely that the potential rapid progress touted by some for these new tools "has been greatly exaggerated." Clearly, these tools have opened up important new approaches to target and drug discovery, approaches that will most certainly have value over time.

To touch on just a few examples, here is an illustration from *Nature Reviews Drug Discovery* on how genomics and proteomics are being used in target identification (Fig. 4.16) [13]. For genomics, an RNA sample is amplified and labeled using the polymerase chain reaction (PCR), then used to probe gene microarray chips that contain a multitude of genes. Importantly, with the completion of the human genome project it has become possible to probe the expression of 32,000 human genes in a single experiment.

For proteomics, studies typically involve two-dimensional gel electrophoresis to separate the proteins in a sample, followed by excision from the gel and identification, frequently by mass spectroscopy. As in genomics, protein microarray techniques are also being utilized in proteomics research [23].

Globally, there are two broad approaches by which discovery tools are used to understand, identify, and then validate new targets, presented in this *Nature Reviews Drug Discovery* article (Fig. 4.17) [13]. The first, a molecular approach, attempts to identify new targets through an understanding of the cellular mechanisms underlying the disease. This approach is the most recent and utilizes genomic and proteomic techniques extensively.

The second approach, which has been called a systems approach, seeks to identify new targets through the study of disease in whole organisms. Throughout history, it has been the systems approach to drug development that has been the most commonly used and is particularly relevant for diseases where the observable effects can only be detected in live animals.

Importantly, there are differences in the nature of the targets identified by these approaches, as well in the types of clinical indications they can address. In terms of targets, the molecular approach is more likely to identify intracellular molecules (regulatory, structural or metabolic proteins, etc.) and has been extensively used in the investigation of oncology. Alternatively, the systems approach has been used with a broad range of indications, including obesity, atherosclerosis, heart failure, and stroke, and has identified both intracellular and extracellular targets. Even now, however, application of the molecular approach to these and other disease indications is expected to yield new therapeutic approaches.

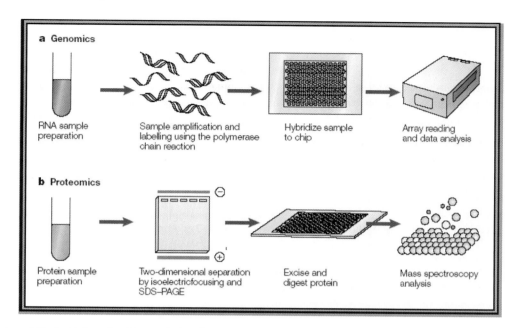

FIG. 4.16. Examples of Tools (Adapted with permission from Nature Publishing Group. London, England. From Figure in Lindsay MA. Nature Reviews Drug Discovery 2003; 2(10):831. Figure – Target discovery)

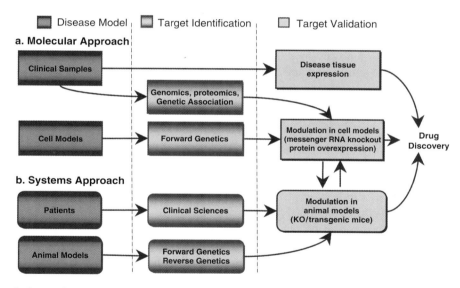

FIG. 4.17. How these tools fit together (Adapted with permission from Nature Publishing Group. London, England. From Figure in Lindsay MA. Nature Reviews Drug Discovery 2003; 2(10):831. Figure – Target discovery)

Products

Once a target has been chosen, that choice often dictates which type of product may be most useful (Fig. 4.18). Drugs, which are typically small, hydrophobic compounds, are often able to penetrate cellular membranes and thereby gain access to intracellular targets (enzymes, kinases, regulatory proteins, etc.). Similarly, because such compounds may well be stable to the digestive conditions of the stomach and intestine, they may also be suitable for oral administration. In contrast, the nature, character, and size of biologics make them best suited for addressing extracellular targets (surface structures, antigens, receptors, soluble ligands, etc.) after parenteral administration.

From the therapeutic standpoint, there are several ways in which to effect the disease process:

Inhibit the function of an enzyme or protein (ligand, receptor, signaling)
- Small-molecule inhibitor to block function
- Antibody to neutralize or remove a specific protein or structure
- Antisense RNA to prevent protein expression

Replace missing or defective proteins
- Administration of replacement protein or peptide
- Gene therapy
- Specific message induction

This point regarding regulatory considerations is quite important and emphasizes the value of frequent communication with regulatory bodies, such as the FDA (Fig. 4.19). The focuses of all regulatory agencies around the world are safety, efficacy, and manufacturing (reproducibility). Drug and biologic products usually have been regulated differently because of differences in the nature of the products, especially in the early research and development stages, as well as organizational differences. CDER, the Center for Drug Evaluation and Research, and CBER, the Center for Biologics Evaluation and Research, are the FDA groups responsible for regulation and approval of drugs and biological products, respectively. Certain types of studies cannot be readily done for biologics, such as some ADME (absorption, distribution, metabolism, and excretion). However, the core information and studies, especially the later clinical stages, have been and are pretty much consistent between drugs and biologics. Periodic changes in management and structure of the agency can certainly alter the requirements for submitting regulatory documents and initiating clinical trials, such as drugs and biological drugs being all placed under the auspices of CDER. The only biologics remaining with CBER are vaccines, cellular products, blood products, and antitoxins.

An important component of early product development is the utilization of available information to design what the product must do (Fig. 4.20). These properties then lead to the development of a document (viz., a product profile or specification sheet), which defines the properties of a successful product. Such specification sheets need only be a one-page document. Commonly included items are the unmet clinical need, a description of the product, the target indication, *in vitro* and *in vivo* potency, formulation, cost of goods, toxicity, preferred route and frequency of administration, and competition. It is also common in such documents to define "optimal" and "minimal" specifications. In this way, all research groups on the

- Federal regulations exist to ensure that new therapeutics meet 3 criteria:
  - Safe, effective, and manufactured reproducibly
- Historically, drugs have been regulated by CDER (Center for Drug Evaluation and Research), while biologics have been regulated by CBER (Center for Biologics Evaluation and Research):
  - FDA was established in 1931
  - CBER was established during 1980's to address issues specific to biologics
  - In 2003, review of all therapeutic products was transferred to CDER
- For biologics, current process often involves a combined review by both agencies, depending upon the product and use:
  - CDER – Therapeutic monoclonal antibodies, cytokines, growth factors, enzymes, and other novel proteins
  - CBER – Cellular products, blood products, vaccines, antitoxins
- Regulatory pathway to clinic for biologics has often been shorter than for drugs:
  - Certain requirements (multiple species tox, ADME, safety pharmacology, etc.) have not been relevant for most biologics
- Importantly – Regulatory process continues to evolve!

FIG. 4.19. Drugs & Biologics – Regulatory Considerations

- **Drugs** – Small molecule organic compounds obtained by screening large libraries of natural or synthetic compounds:
  - Molecular weight typically < 500 Daltons
  - Produced by chemical or semi-synthetic synthesis
  - Effective against intracellular and extracellular targets
  - Examples include most antibiotics and existing pharmaceuticals
- **Biologics** – Protein-based therapeutics obtained from humans, animals and plants:
  - Molecular weight typically > 5,000 Daltons (5 kDa)
  - Purified from natural sources or, more commonly, produced by recombinant methods or monoclonal antibodies
  - Primarily effective against extracellular targets
  - Generally administered by injection
  - Examples include antibodies, hormones, enzymes, cytokines and vaccines

FIG. 4.18. Product Choices – Drugs vs. Biologics

- Once the target has been identified, the necessary properties of a potential therapeutic can be developed (what must the product do?):
  - Inhibit target function (Lipitor, anti-TNF, anti-CD11a)
  - Stimulate target function (insulin, growth hormones)
  - Perform an enzymatic function (TPA)
  - Kill specific cells (anti-Her2, antibiotics, most anti-cancer drugs)
- Important to develop a list of product specifications (when do I have a product candidate?):
  - Provides a clear stopping point for screening
  - Must be realistic and based on current information
  - Minimizes endless discovery ("better is the enemy of good")
- Most products fail as a result of unacceptable toxicity, inadequate therapeutic index, or low potency

FIG. 4.20. Product Design and Selection

4. Discovery and Nonclinical Development

- Label claims describe FDA-approved use of the drug and are found in Package Insert:
  - Disease indication – What the drug is intended to treat
  - Target population – Ages or groups who need the drug
  - Route of delivery – IV injection, oral, nasal, etc.
  - Observed benefit – The improvement seen in clinical trials
  - Safety issues – Any toxicities noted in animals and humans
- Potential label claims influence many aspects of drug development:
  - Discovery and non-clinical expectations and plans
  - Development of product specifications (what must it do?)
  - Identification of go/no-go decision points
  - Clinical testing plan
  - Marketing plan

FIG. 4.21. Consider Potential Product label Early

- **Small molecules:**
  - QSAR and pharmacophore development
  - Natural and artificial compound libraries
  - Medicinal chemistry
- **Biologics:**
  - Phage, ribosome and bacterial display libraries
  - Protein manipulation (truncation, glycosylation, pegylation, fusion)
  - Humanization/De-Immunization
  - Chimeric proteins
- **Common to both approaches:**
  - High throughput screening
  - Informatics
  - Early toxicity studies

FIG. 4.22. Selection and Improvement Tools

project know ahead of time what the target is (the optimal specifications), as well as what would be acceptable (minimal specifications). Firm stopping rules that are realistic, practical, and set up-front avoid endless searching in discovery so a team can move on to more productive projects.

Many groups have found it useful to work backwards through the development and approval process from proposed optimal and minimal package inserts, in an effort to better define the studies that may be needed to support early clinical trials and possible regulatory questions (Fig. 4.21). The major label claims involve five areas noted in the slide and drive at least the five noted aspects of development. As an example, consider the development of a new drug for a cancer indication. What cancers could be treated? How will the drug be used clinically (stand-alone or adjunctive therapy, first-line treatment or salvage treatment, etc.)? Are there patient subsets that may respond differently? Are there certain toxicities of existing drugs to avoid or not exacerbate? How will the drug be administered (oral, intravenous, subcutaneous, etc.), and for how long (a few minutes, days, months, years)? How large do phase III trials need to be and what is the approvable end point? Answers to these questions help define the phase II program, which in turn defines the phase I, toxicology and preclinical needs. Example package inserts are recommended to be reviewed for Lipitor® [24] and Epogen® [25].

In any development program, be it a small molecule or biologic, there is often a need to improve or optimize the activity of the lead compound (Fig. 4.22). For a small-molecule drug, what if the lead compound is active in animals but too toxic? For a biologic such as an antibody, what if the antibody binds the correct target and has function, but the affinity for that target is too low for therapeutic use (i.e., too much drug would be required)? In each case, tools are available to help further refine the properties of the compounds. Some examples for small molecules follow below:

Quantitative structure-activity relationships (QSAR), a process by which the functions of all structural elements of the compound are studied, quantified, and used to direct further modifications of the compound. Such studies are typically focused on attempting to identify the "pharmacophore," the most desirable chemical structure needed to safely achieve the desired efficacy.

Natural and artificial compound libraries, which can be screened in an effort to identify more preferred compounds. QSAR data may also be used to direct the preparation of new libraries, allowing iterative screening and selection.

Medicinal chemistry, involves chemical approaches to altering the safety, efficacy, and oral availability of compounds. Because biologics are larger molecules typically produced by recombinant techniques or monoclonal antibody products, they are less amenable to modifications that are commonly used for the small molecules. Instead, alternative techniques have been devised, and include a few as follows:

Display technologies, such as phage display [26], ribosomal display [27], or bacterial display [28], allow for the production and screening of large numbers of protein variants or analogues with increased affinity or altered characteristics.

Protein manipulation techniques, such as truncation, glycosylation (more or less), peptide alteration, pegylation, or fusions, which alter the size, shape, or character of the protein, resulting in molecules that have very different physical properties (pharmacokinetics, toxicity, activity, etc.).

Humanization/de-immunization techniques, which seek to reduce the potential for generating an immune response in humans, lessens toxicity and increases activity. Most commonly, these techniques have been used to convert antibodies derived in mice into "humanized" antibodies that have the characteristics of human proteins [29].

Chimeric proteins are fusion proteins created by combining the genetic information for one protein with another. Like humanization techniques, chimeric proteins have been generated in an effort to reduce the potential for an immune response in humans, as well as to create molecules with altered pharmacokinetics or biological characteristics. Importantly, chimeric proteins can also combine the biological functions of two (or more) proteins into a new, novel recombinant form.

Although the process of research and development typically follows a logical process, it has often been through serendipity that major new drugs have been developed (Fig. 4.23). This figure

- Many drugs are result of chance observations:
  - Alexander Fleming and Penicillin:
    - Searching for agents that could kill Staphylococci
    - Observed that bacteria on culture plates were lysed around contaminating airborne molds
    - Concluded that something in the Penicillium mold was killing bacteria
- "Rational" drug design often follows a tortuous path:
  - Viagra (PDE-5 inhibitor - 2003 US sales of $1.0 billion):
    - Initially developed as anti-hypertensive drug, but specificity was low
    - Development was changed to focus on angina, but potency was low
    - During clinical trials, patients commented on decreased erectile dysfunction
- Once used to describe knowledge-based development decisions, "rational drug design" now focuses on underlying biology of disease, as an aid to appropriate target selection

FIG. 4.23. Serendipity vs. Rational Design
Source: Fleming's photo of bacteria and mold (http://www.pbs.org/wghb/aso/databank/entries/dm28pe.html)

shows Fleming's experiment with molds and penicillin production in a petri dish [30]. This serendipity follows from the fact that:

1. All the factors influencing the biology and the disease are not known.
2. Drugs often have unexpected *in vitro* or *in vivo* consequences.

Sometimes, these consequences can be too much toxicity (causing reevaluation or termination of the project), whereas other times they can be highly beneficial. Moreover, serendipitous consequences can occur anytime during the discovery and development process, as illustrated by the two examples here for penicillin and Viagra®. Viagra sales info is found in the Ref. 31.

Nonclinical Development and Testing

Having moved beyond target and product discovery, the next steps in the development process involve obtaining a more complete picture of the activities and properties of the lead compound (Fig. 4.24). Most commonly, this involves a more thorough investigation of the compound in *in vitro* assays as well as more extensive evaluations in animals. Clearly, it is the goal of these "preclinical" studies to provide the information necessary to initiate clinical trials. As such, they are heavily driven by the clinical indication and seek to define what happens when the drug enters the body:

- How long is it in the body?
- Where does it go in general and any special tissue sites?
- What happens to it, especially the elimination steps in the liver, kidney, or elsewhere?
- Does it interact with compounds in the body or ones likely to be used in patients with this disease?
- What does it do in target sites and all other tissues?

- Activity and efficacy studies:
  - *In vitro* testing:
    - Affinity, potency, minimum active concentration, physical characteristics, stability, mechanism of action
  - *In vivo* testing:
    - Potency, dose-response, drug effects
    - Models are chosen to best reflect the therapeutic indication
- Pharmacokinetics:
  - Effect the body has on the drug:
    - Clearance, distribution, degradation
- Pharmacodynamics:
  - Effect the drug has on the body:
    - Impact on disease or disease markers, PK requirements

Non-clinical development is also referred to as "preclinical" development, since it takes place prior to initiating clinical trials.

FIG. 4.24. Non-clinical Development & Testing

Answers to these questions help guide toxicity studies and how the drug will be initially used in the clinic.

Clearly, the basic goal for the *in vitro* and animal tests is to help predict the actions of a lead compound in humans (Fig. 4.25). It should be emphasized that good activity in the test tube or even in animals is no guarantee of success in humans. This result is due to various aspects of *in vitro* and *in vivo* assays, neither of which may accurately reflect the disease process in humans. This slide presents four reasons for *in vitro* tests and three representative reasons for animal tests for the lack of their predictive ability in humans.

Although pharmacokinetics and pharmacodynamics are also monitored extensively in human clinical trials, they have a special significance for preclinical development. This follows from the fact that it is the early work in animals that is used to develop the toxicology studies, which in turn are used to select the dose and dose regimens that will be initially used in humans. The animal work also will help design the type of pharmacokinetic trials needed to be done in humans, especially any special studies related to unexpected movement of a product in tissues, or special tissue effects, or special route of administration issues. Animal studies are usually quite predictive of human trials for pharmacokinetics, but surprises can occur as well. For example, protein binding can differ between species impacting pharmacokinetic parameters.

And what effects does the drug have on the body, the pharmacodynamic effects, and how do they correlate with pharmacokinetics? Again, animal research is intended to help predict human activities, and four key questions are listed on Figures 4.26 and 4.27. Ideally, there is a positive biological effect on the disease (though toxic effects must also be monitored), and it is the preclinical work that must identify the desired drug concentrations and dosing frequencies necessary to produce that effect. Similarly, it is helpful if, in addition to having readily measurable effects on the disease, other markers of therapeutic efficacy can also be identified. Such "surrogate" markers of clinical efficacy can be extremely important in monitoring the effects of a new drug in humans, sometimes even early indicators of beneficial or untoward effects before full actions of the drug and the resulting change in disease pathogenesis occurs (remember the discussion of PSA with Fig. 7 on pg. 91).

As an example, let's consider the pharmacokinetics and pharmacodynamics of a new antibody therapeutic, Raptiva®, which was approved in 2004 for the treatment of moderate to severe psoriasis (Fig. 4.28). Raptiva® (here identified as hu1124) is a humanized monoclonal IgG1 antibody that binds to the CD-11a component of human LFA-1 (lymphocyte function–associated antigen 1), a surface structure on lymphocytes that participates in T-cell trafficking and activation. As a consequence of this binding, CD-11a is downregulated and its function inhibited.

Shown in the graph is the effect of a single 8 mg/kg intravenous injection in chimpanzees [32]. Immediately after the IV injection (time 0), the concentration of Raptiva in the blood increased to over 100 µg/mL and then decreased over the next 2 months. At the same time, the expression of lymphocyte CD11a immediately decreased and stayed suppressed for the same 2-month period. Note that when the level of circulating Raptiva® fell below 3 µg/mL, the clearance of Raptiva® was accelerated and CD11a levels began to return to normal. Based on these data, mathematical models were developed (solid and dashed lines) that described the dose-dependent effects of hu1124 on antibody clearance and CD11a expression in chimps, and these models were then used to predict probable hu1124/CD11a profiles in humans. Such data also helps select initial dosing schemes for human trials.

Of all the parameters studied prior to initiating human clinical trials, product safety remains one of the most important (Fig. 4.29). This focus is because the primary decision made by the company, the FDA, and the clinicians regarding the initiation of human trials is whether the product poses a safety risk. As a consequence, toxicity studies should be incorporated not only into the final product evaluation but also into the initial product evaluation and selection process. For compounds

- *In vitro tests may not accuratelyr eflect disease process:*
  - NO access or penetration issues
  - NO clearance
  - NO metabolism
  - NO toxicity

- Animal models also may not accurately reflect disease process:
  - Animal physiology and metabolism differ from that of humans
  - Most models are acute, where as many human diseases are chronic
  - Clearance, distribution and metabolism differs between species

Generally, data from multiple *in vitro* and in *vivo* models are desirable, but are no guarantee of success in humans

FIG. 4.25. Activity & Efficacy Studies: Some Issues

- Seeks to understand and predict product levels as function of dose and route of administration:
  - Intravenous administration results in high initial concentrations in blood and then decreasing concentrations:
    - Appropriate for products requiring high peak concentrations, rapid onset and/or short exposure
  - Subcutaneous administration results in lower & delayed peak concentrations in blood:
    - Product concentrations often remain elevated for longer periods of time
    - Appropriate for products where extended coverage is desirable or high plasma concentrations may cause safety concerns
  - Oral administration provides low peak concentrations and sustained product levels:
    - Appropriate for drugs with good oral availability
    - Inappropriate for most proteins, due to reduced stability at low pH and poor oral absorption

- Important goal: Match route of administration to indication, drug, and patient population

FIG. 4.26. Pharmacokinetics (PK), Preclinical Work

- Seeks to clarify and predict relationship between product concentration and biological effect:
  - What blood concentration is required to achieve benefit?
  - How long is the effect be maintained?
  - How frequently must I dose to maintain this concentration/effect?
  - How are peak and trough levels effected by route of administration?

- Helps identify surrogate markers of disease or therapeutic efficacy

FIG. 4.27. Pharmacodynamics (PD), Preclinical Work

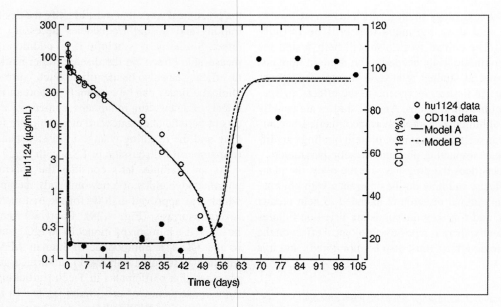

FIG. 4.28. PK & PD Example (Reprinted with permission from Springer. Heidelberg, Germany. From Graph – Population pharmacokinetics and pharmacodynamics of the anti-CD11a antibody hu1124 in human subjects with psoriasis. Bauer RJ et al. J. Pharmacokin Biopharm 1999;27(4):397)

- Important *in vivo* tool in selection of lead candidates:
  - *In vitro* tools are being developed, but are not yet reliable
- Can be monitored initially in (or prior to) efficacy studies
- Single dose and multi-dose, depending upon indication
- Dosing at multiples of expected human exposure and determination of maximum tolerated dose (MTD)
- Helps to define therapeutic index (range between effective and toxic doses of the drug)

FIG. 4.29. Toxicity Assessments, Preclinical Work

such as the small-molecule drugs, initial screening in animals can help reduce the number of candidates that require further characterization. Animal toxicology work often can be predictive of many, but not all, the major side effects to be seen in humans. To help ensure this predictive capacity, two species, one non-rodent, are used often in the animal studies. Drug doses are given singly at multiples of the human dose, acutely (daily over a few days), subacutely (daily for a few weeks), and possibly chronically (daily for months), if the drug will be used in that fashion in humans. The formulation of the product in the animal study needs to be as similar as possible to the human forms, because often drug delivery and absorption depends on the formulation. For the biologics, where it is less typical to screen large numbers of potential product candidates, initial toxicity studies can help guide the doses used in animal efficacy studies and can identify areas of potential concern. Animal studies for toxicology pose complications for most protein biologics because they are foreign to the animal and will cause an immune reaction. In all cases, compounds with high maximum tolerated doses (MTDs) and wide therapeutic indexes (TI) are more easily moved along the development path. However, in some indications (for example, certain life-threatening diseases such as cancer) a less favorable safety profile may still be acceptable.

Toxicology work comprises the majority of studies and costs in the preclinical phase of research, as shown in figure 4.30 [33]. Most of these 10 different tox studies (Fig. 4.30) are prescribed in regulatory guidelines for IND submissions; note the varied lengths of treatment with study product and the species. Three special toxicology studies are required as well for mutagenicity, carcinogenicity (a very expensive research requirement and possibly time consuming before human trails are permitted), and reproductive performance on mother and fetus. Pharmacogenomic studies are a new, either toxicology or efficacy, parameter to document if genetics plays a significant role in adverse events or patient responsiveness to the product. Guidelines are currently voluntary until the value and role of pharmacogenomics is established for diseases and therapy. This long list of studies is not a surprise given the need to find significant toxicity as early as possible and kill poorly performing molecules, thereby avoiding the expensive late termination of a product in clinical trial or after marketing, as well as the mission of the regulatory bodies to protect the public. The costs are $2.5 million to $6 million for this work, as noted in

4. Discovery and Nonclinical Development

Test	Cost Range (Euros)
Acute Toxicity (rodents)	3,900 -4,600
Subacute toxicity (4 weeks in rats-dogs)	143,000-183,000
Subacute toxicity (4 weeks in monkeys)	125,000
Subacute toxicity (13 weeks in rats-dogs)	213,000-305,000
Subacute toxicity (13 weeks in monkeys)	190,000
Chronic toxicity (26 weeks in rats & 39 weeks in dogs)	366,000-488,000
Chronic toxicity (39 weeks in monkeys)	290,000
Mutagenic potential (3 basic tests)	25,000-69,000
Carcinogenic potential (mice or rats)	1,124,000-2,287,000
Effect on reproductive performance	313,000-458,000
Complete Toxicology Budget	2,590,000-6,000,000
PLUS 1. Pharmacology (activity & efficacy, >>doses)	500,000-1,000,000
2. Pharmacokinetics (ADME) 3. Pharmacogenetics??	??

FIG. 4.30. Requirements & Costs for IND Studies (Adapted with permission from Nature Publishing Group, London, England. From Table in Preziosi P. Science, pharmacoeconomics and ethics in drug R&D: a sustainable future scenario. Nature Reviews Drug Discovery 2004;3(5):521–6. Table 2)

- Multiple assays are needed to ensure product consistency and potency
- Common consistency assays include:
  - Appearance, concentration, pH, ionic strength, sterility, endotoxin, purity
- Potency assays focus on specific properties of the product:
  - Antibodies – Binding, ADCC, CMC, functional inhibition, etc
  - Enzymes – Catalytic activity
  - Recombinant proteins – Functional activity or inhibition
  - Drugs – Functional activity or inhibition
- Current trends include greater emphasis on biological or cell-based assays versus simple binding assays
- All assays must be well controlled and reproducible:
  - Matrix effects - Blood, urine, sputum, mice vs. humans, etc.
  - Assay validation is not usually required until later in review process

FIG. 4.31. Analytical Assays
Source: Image from Amersham Biosciences (http://www.bloprocess.amershambiosscience.com)

the figure(4.30). Pharmacology work is listed here, as well the pharmacokinetic work (ADME).

As a product transitions from the research phase into process development and manufacturing, the requirements for standardized and reproducible assays increase (Fig. 4.31). Not surprisingly, many of the assays developed for the discovery work (purity, potency, concentration, function, etc.) are often further refined and characterized during the technical development phase, so as to ensure their reliability. This slide lists seven common characteristics used to judge consistency of a product. Certain types of products will have specific properties that will require special assays (e.g., antibodies need assays for binding, antibody dependent cell cytotoxicity [ADCC]). Assays basically need to be doable by anyone trained in the field, involve a well-controlled process, and be reproducible. Ultimately, as a product works its way through clinical trials toward a marketing application, the necessary assays are "validated" to further ensure their uniformity and reproducibility.

IND-Enabling Studies

The discovery and early development process culminates when a decision is made to advance the product candidate into human clinical trials. To do so within a corporate setting, an Investigational New Drug (IND) application is filed with the FDA prior to initiating trials (Fig. 4.32). The IND summarizes many aspects of the discovery and development of the product candidate, as well as how the product is manufactured and controlled and how it will be used in the clinic.

From the preclinical standpoint, a number of IND-enabling studies are typically needed that each can take a year or more to complete. Such studies typically include animal efficacy studies related to the clinical indication and detailed toxicology studies conducted under GLP (good laboratory practice). To be considered GLP, a study must follow the guidelines outlined in Part 58 of the Title 21 Code of Federal Regulations (21 CFR Part 58) that are designed to assure the quality and integrity of safety data used in support of the application [34–36]. Much of the process for filing an IND and conducting clinical trial is also covered by Title 21 CFR documents (See Figure 4.33), and guidelines similar to GLP cover good manufacturing practice (GMP) and good clinical practice (GCP) [37].

For most small-molecule drugs, an extensive package of nonclinical information is needed to file an IND [37]. This includes five following topics:

- additional *in vivo* efficacy information in animal models relevant to the clinical indication
- toxicology studies in two species (commonly rodent and non-rodent)
- safety pharmacology (most commonly effects on the central nervous, cardiovascular and respiratory systems)
- Carcinogenicity/mutagenicity (potential to induce cancer)
- ADME (absorption, distribution, metabolism, and elimination)

Both the choice of study animals and the duration of the study are influenced by the product's characteristics and intended use in humans. Note that many drugs fail due to problems with ADME (the compound isn't absorbed, goes to the wrong place, is degraded too quickly [or into toxic components], or is eliminated too quickly) [37].

Because biologics are commonly recombinant proteins of human origin, they often require less safety information prior to filing an IND (Fig. 4.34). For example, most proteins are degraded by endogenous pathways into peptides and amino acids that are then reutilized by the host. As a result, some of the studies typically needed for small-molecule drugs (safety, pharmacology, mutagenicity, true ADME studies, etc.) can be less relevant for biologics. However, if the mechanism of action of the biologic is known to affect critical systems, then additional studies may be needed. Also note that for products like antibodies that are designed to bind specific molecular targets, cross-reactivity studies with a panel of human tissues are conducted in an effort to identify other tissues that may also be reactive. Such tissues may react with the antibody due to the expression of the specific target or because the antibody cross-reacts with epitopes on an entirely different antigen, either of which can reduce drug levels or result in unexpected toxicities. As noted previously, animal studies with proteins pose the complication of immune reactions by the animal to

Guidance for Industry

Contentand Format of Investigational New Drug Applications (INDs) for Phase 1 Studies of Drugs, Including Well-characterized, Therapeutic, Biotechnology-derived Products

Center for Drug Evaluation and Research (CDER)
Center for Biologics Evaluation And Research (CBER)

November 1995

- Typical IND-enabling studies include supportive animal efficacy studies and all GLP (Good Laboratory Practices) toxicology studies.
- Toxicology studies seek to identify safe initial starting doses for human trials, potential target organs in humans, and ways to monitor these toxicities.
- For both efficacy and toxicology studies, animal species tested, doses used, and duration of the studies are chosen based on indication, product, and intended use.
- All IND-enabling work should involve regular review of theappropriate guidelines and frequent discussion with FDA.

FIG. 4.32. IND-Enabling studies

4. Discovery and Nonclinical Development

the human protein, confusing the adverse effects and safety assessments.

As was stated previously, an IND summarizes many aspects of the discovery and development process, as noted on this slide in three areas: preclinical (toxicity, pharmacoloy, ADME), manufacturing, and clinical plans (the first human protocol) Fig. 4.35. Guidelines for the IND preparation are available from the FDA in guidance documents, which can be found at

- Additional/confirmatory animal efficacy studies
- Toxicology (typically two species)
- Safety pharmacology
- Carcinogenicity/mutagenicity
- ADME:
  - Absorption – How much of administered drug is really available?
  - Distribution – To what tissues does drug localize?
  - Metabolism – How is drug broken down in body?
  - Elimination – How is drug cleared from body?
- Most drugs in development fail to reach market due to problems associated with ADME

FIG. 4.33. Typical IND-Enabling Studies for Drugs

- Additional/confirmatory animal efficacy studies
- Toxicology (one or two species)
- Tissue cross-reactivity (typically only for antibodies)

FIG. 4.34. IND-Enabling studies for Biologics

- Document summarizes the drug and the intended studies:
  - General investigational plan
  - Trial protocol
  - Manufacture and testing of the drug (CMC section)
  - *In vitro* & *in vivo* support for drug & indication (Preclinical section)
  - Essential toxicity data
- Guidelines for INDs are published in Code of Federal Regulations:
  - 21 CFR 312.22 and 312.23
- FDA has prepared guidance documents for IND submissions:
  - Guidance for Industry – Content and Format of Investigational New Drug Applications (INDs) for Phase I Studies of Drugs, Including Well-Characterized, Therapeutic, Biotechnology-derived Products.
- FDA website maintains guidance documents on many subjects

FIG. 4.35. IND Preparation

their Web site [38]. Once submitted, FDA reviews the document and responds with questions or comments as appropriate.

Added Discovery Work

Even before clinical trials have been initiated, it's important to consider other ways in which the product could be used Fig. 4.36. Such consideration not only involves alternate indications for how the existing therapeutic could be used but also includes new formulations or constructs that may allow product expansion into new areas. The advantages of these product extensions are noted in figure 4.36, along with the example of alpha-interferon. Added preclinical work in pharmacology, ADME, or toxicology may be required if the new indication or product form involves new diseases or substantially changes how the product will be used in humans.

An important component to the discovery and development process is a discussion of corporate partnerships (Fig. 4.37). Indeed, development partnerships today are more a matter of "when" than "if" [39–41]. This is in contrast with some of the concepts from the 1980s and early 1990s, when many organizations were hoping to become fully integrated pharmaceutical companies (FIPCos) capable of controlling all aspects of the drug development process (discovery through marketing). Today, even the largest pharma companies rely on in-licensing new targets or products as a means to expand their pipelines, 400 in 2003 and involving an important and growing amount of the research budget (see some examples on figure 4.14). Similarly, many companies have formed that specialize in various specific aspects of the drug discovery and development process ranging from target identification and validation to high-throughput screening (HTS), informatics, and databases. This diversification has led to a record number of partnerships, not only between pharma and biotech companies but also between multiple biotech companies [40, 41]. Some of these collaborations will result in one of the partners being acquired

- Additional formulations and new indications can expand market for a therapeutic:
  - Cost-effective way to generate new sales without requiring new drug discovery
  - Can create competitive advantages in marketplace
  - Can provide patent life extension
- Each new indication requires new IND:
  - Additional preclinical and toxicology studies may be required
- Classic biologics example is alpha-interferon:
  - Initially approved for hairy cell leukemia in 1986
  - Has received at least 6 additional approvals
  - Product life has been extended by creation of PEG-Intron A®:
    - Pegylated form provides a longer half-life and less frequent dosing
    - Other analogs also being developed

FIG. 4.36. Exploration of Additional Uses

by the other, because the technology and/or products are deemed principal to the operation and success of the company.

Alliances and partnerships between pharma and biotech companies are now the norm for research operations, as can be observed by the list of the top six pharma company partners in 2003 (Fig. 4.38). The top 10 deals involve cancer, viral disease, the cardiovascular (CV) system and central nervous system (CNS). We already discussed the value to pharma, being targets, products, and technologies. The value to biotech is first the infusion of revenue to continue research and operations; up to $4 billion was promised to biotech from pharma in top 10 deals in the forms of, for example, up-front payments, milestone achievement payments, and royalties. Biotech companies are very lean operations focused in research, such that the pharma can provide a staffing benefit in areas that are not yet ramped up for the biotech company (e.g., clinical research [clinical development] staff and operations), as well as marketing support to assist with the planning of product research, market research, and early launch preparation. The decision to form an alliance is based on a variety of factors beyond good science and product opportunity, so that it will be successful for both companies, large and small, pharma and biotech, established (structured and perhaps stodgy) and new (free-wheeling and chaotic at times). The two organizations have to fit together in some planning, financial, and operational framework. The seven issues in figure 4.38 addressing "how it (alliance) will work" must be dealt with effectively in order for both parties to benefit and products and sales to be the outcomes.

The university is a major source of new discoveries for both disease pathology and potential product opportunities across the world. Research in the basic sciences is one of the three cornerstone missions of most universities, along with education and public service. The disciplines of research applicable to product discovery area of the industry are quite broad and are listed in Figure 4.39. Research laboratories at pharmaceutical and biotechnology companies need to stay abreast of the findings emanating from the university setting, which is accomplished in three ways: (1) scientific publications, (2) scientific presentations by university and company scientists at the major science meetings in all the basic research areas, and (3) research collaborations. At most companies, research groups have a budget that often includes grants to universities who have critical research underway that may advance the company's work. The company receives access to discoveries for disease pathogenesis, product leads

- Driven by needs of large Pharma companies (expand targets, hits & unmet patient needs) and rapid expansion of new technologies
- Many drug discovery companies now exist (more than 500):
  - Tools – analysis, HTS, product systems, software, informatics
  - Databases – genome, proteome, combinatorial chemistry, biology
  - Examples – Millennium, Pharmacopeia, Albany Molecular
- Partnerships for Pharma and Biotech:
  - Pharma-Biotech - nearly 400 in 2003
  - Biotech-Biotech - more than 400 in 2003
- $3.6B in 2003 - 15% growth - 5-10 % of all research $ spent
- Acquisitions to bring in technology (M&A) - 128 in 2003

FIG. 4.37. Alliances in Discovery & Development
Source: King J. R&D Directions 2004; 10(2):28-39; Ernst & Young. Resurgence: Global Biotechnology Report 2004

- **WHO?**
  - Pharma: Astra-Zeneca, Aventis, BMS, GSK, J&J, Merck
  - Biotech: Millennium, Regeneron, TheraVax, Vertex
- **WHAT deals?**
  - $ 4.02 B (potential) in '03/'04
  - Targets: Cancer, viral, CNS, CV
- **VALUE to Pharma?**
  - New products
  - New targets
- **Value to Biotech?**
  - Revenues: Milestones, research costs, royalties, sales
  - Development expertise
  - Access to sales and marketing

- **HOW to assess value?**
  - Product potential (Mkt, Ptnt)
  - Product fit
  - Organizational fit
  - Staff – PhD & MBAs
  - Shared risk and costs
  - Bio income
  - Bio opportunity
- **HOW to make it work?**
  - "Not invented here" syndrome
  - Scientific integration
  - Culture coordination
  - Division of work
  - Decision-making process
  - Progress/Follow-thru
  - Management of alliance

FIG. 4.38. Pharma-Biotech Alliances-Value & Process

4. Discovery and Nonclinical Development

- University research (basic) scientists (many disciplines):
  - Medicinal chemists & Pharmaceutists (formulations)
  - Disease processes (physiologists, biologists, geneticists)
  - Protein chemists & Molecular biologists
  - Pharmacokineticists & Pharmacologists
- Access for company:
  - Research network (expanded brain power)
  - Technologies
  - Disease targets
  - Product leads & candidates
- Access for universities:
  - Grants
  - Patents
  - Publications

FIG. 4.39. Alliances in Discovery with Universities

- Have well developed plan for target discovery & lead identification
- Establish success criteria for making timely development decisions
- Consider anticipated product label claims early, and design nonclinical and clinical studies to address these claims
- Conduct critical go/no-go experiments quickly
- Maintain focus on target indication, but stay alert to other possibilities as well
- Discovery and development decisions should be coordinated team effort, involving many departments within organization
- Review benefits (and drawbacks) of collaboration regularly
- Activity in humans is the goal – get there as quickly AND SAFELY as possible!

Remember - "In the field of observation, chance favors only the prepared mind" **(Louis Pasteur)**

FIG. 4.40. Key Success Factors in Discovery

and candidates, disease targets, and new technologies that are created by university scientists. The university gains access to scientific and research expertise, financial support (that is, grants), patent opportunities (shared in some form with the company), product leads, possible postdoctoral training opportunities, and added publications from the collaboration. The collaboration between university and company researchers often starts with a discussion of their respective work, based on a recent public presentation or publication of a new study. If further details are needed, a confidentiality agreement is put in place to help protect both parties. If collaboration or funding appears mutually beneficial, a formal contract is created between the company and university that addresses the potential for new discoveries and patents, as well as the deliverables expected by the company from the university collaboration.

In summary, eight factors will be keys to success in discovery and early development, as shown on Figure 4.40. Excellence in governance and planning, which we discussed in earlier chapters, is also a necessity in the early research phases; five of the key eight factors, albeit in the research context, involve planning, decisions (criteria), focus (indications), reviews of work (regular), and team effort (decisions), in addition to the technical requirements of the experiments being done well and fast.

For your further education about discovery research and early development, you will find these 10 publications useful for more in-depth study (Fig. 4.41). *Nature Reviews Drug Discovery* is a particularly good source of review articles for discovery and product development.

- Bauer RJ, et al. 1999. Population pharmacokinetics and pharmacodynamics of the anti-CD11a antibody hu1124 in human subjects with psoriasis. Journal of Pharmacokinetics and Biopharmaceuticals 27:397-420.
- Dickson M and Gagnon JP. 2004. Key factors in the rising cost of drug discovery and development. Nature Reviews Drug Discovery 3:417-429.
- Hodgson J. 2001. ADMET – turning chemicals into drugs. Nature Biotechnology 19:722-726.
- Lindsay MA. 2003. Target discovery. Nature Reviews Drug Discovery 2:831-838.
- Myers S and Baker A. 2001. Drug discovery – an operating model for a new era. Nature Biotechnology 19:727-730.
- Ng R. Drugs, From Discovery to Approval. Wiley-Liss, 2004.
- Reichert JM. 2003. Trends in development and approval times for new therapeutics in the United States. Nature Reviews Drug Discovery 2:695-702.
- Reichert J and Pavlou A. 2004. Monoclonal antibodies market. Nature Reviews Drug Discovery 3:383-384.
- Zambrowicz BP and Sands AT. 2002. Knockouts model the 100 best-selling drugs – will they model the next 100? Nature Reviews Drug Discovery 2:38-51.
- Special Issue – Drug Discovery. 2004. Science 303:1795-1822.

FIG. 4.41. Further Reading

References

1. Available at www.chemistry.wustl.edu/~edudev/LabTutorials/HIV/DrugStrategies.html
2. Gordon EJ, Myers KJ, Dougherty JP, Rosen H, Ron Y. Both anti-CD11a (LFA-1) and anti-CD11b (MAC-1) therapy delay the onset and diminish the severity of experimental autoimmune encephalomyelitis. J Neuroimmunol 1995;62(2):153-160.
3. Werther WA, Gonzalez TN, O'Connor SJ, McCabe S, Chan B, Hotaling T, Champe M, Fox JA, Jardieu PM, Berman PW, Presta LG. Humanization of an anti-lymphocyte function-associated antigen (LFA)-1 monoclonal antibody and reengineering of the humanized antibody for binding to rhesus LFA-1. J Immunol 1996;157(11):4986-4995
4. Available at www.raptiva.com
5. Myers S, Baker A. Drug discovery—an operating model for a new era. Nature Biotechnology 2001;19:727-730
6. Available at open.imshealth.com/webshop2/IMSinclude/i_article_20040317.asp
7. Roth BD. The discovery and development of atorvastatin, a potent novel hypolipidemic agent. Prog Med Chem 2002;40:1-22
8. Available at www.scienceblog.com/community/older/2003/C/2003363.html
9. Winslow R. The birth of a blockbuster: lipitor's route out of the lab. New York Times. January 24, 2000
10. Lieberman R. Evidence-based medical perspectives: the evolving role of PSA for early detection, monitoring of treatment response, and as a surrogate end point of efficacy for interventions in men with different clinical risk states for the prevention and progression of prostate cancer. Am J Ther 2004;11(6):501-506
11. Butcher EC, Berg EL, Kunkel EJ. Systems biology in drug discovery. Nat Biotechnol 2004 Oct;22(10):1253-1259
12. Lawrence CE. Acumen. J Sciences 2003;1:23
13. Lindsay MA. Target discovery. Nat Rev Drug Discovery 2003;2:831-838
14. Phizicky E, Bastiaens PI, Zhu H, Snyder M, Fields S. Protein analysis on a proteomic scale. Nature 2003;422(6928):208-215
15. Zambrowicz BP, Sands AT. Knockouts model the 100 best-selling drugs—will they model the next 100? Nat Rev Drug Discov 2003;2(1):38-51
16. Sachse C, Krausz E, Kronke A, Hannus M, Walsh A, Grabner A, Ovcharenko D, Dorris D, Trudel C, Sonnichsen B, Echeverri CJ. High-throughput RNA interference strategies for target discovery and validation by using synthetic short interfering RNAs: Functional genomics investigations of biological pathways. Methods Enzymol 2005;392:242-277
17. Soutschek J, Akinc A, Bramlage B, Charisse K, Constien R, Donoghue M, Elbashir S, Geick A, Hadwiger P, Harborth J, John M, Kesavan V, Lavine G, Pandey RK, Racie T, Rajeev KG, Rohl I, Toudjarska I, Wang G, Wuschko S, Bumcrot D, Koteliansky V, Limmer S, Manoharan M, Vornlocher HP. Therapeutic silencing of an endogenous gene by systemic administration of modified siRNAs. Nature 2004;432(7014):173-178
18. Fougerolles A, Manoharan M, Meyers R, Vornlocher HP. RNA interference in vivo: Toward synthetic small inhibitory RNA-based therapeutics. Methods Enzymol 2005;392:278-296
19. Thomas FJ, McLeod HL, Watters JW. Pharmacogenomics: The influence of genomic variation on drug response. Curr Top Med Chem 2004;4(13):1399-1409
20. Harrington CA, Rosenow C, Retief J. Monitoring gene expression using DNA microarrays. Curr Opin Microbiol 2000;3(3): 285-291
21. MacBeath G, Schreiber SL. Printing proteins as microarrays for high-throughput function determination. Science 2000;289(5485): 1760-1763
22. Yang Y, Blomme EA, Waring JF. Toxicogenomics in drug discovery: from preclinical studies to clinical trials. Chem Biol Interact 2004;150(1):71-85
23. Templin MF, Stoll D, Schwenk JM, Potz O, Kramer S, Joos TO. Protein microarrays: promising tools for proteomic research. Proteomics 2003;3(11):2155-2166
24. Available at www.lipitor.com/Images/Lipitor/Lipitor_PI.pdf
25. Available at www.epogen.com/pdf/epogen_pi.pdf
26. Griffiths AD. Production of human antibodies using bacteriophage. Curr Opin Immunol 1993;5(2):263-267
27. He M, Taussig MJ. Ribosome display: cell-free protein display technology. Brief Funct Genomic Proteomic 2002;1(2):204-212.
28. Wernerus H, Stahl S. Biotechnological applications for surface-engineered bacteria. Biotechnol Appl Biochem 2004;40(Pt 3): 209-228
29. Verhoeyen M, Milstein C, Winter G. Reshaping human antibodies: grafting an antilysozyme activity. Science 1988;239(4847): 1534-1536
30. Available at www.pbs.org/wgbh/aso/databank/entries/dm28pe.html (Fleming's dish of penecillin production)
31. Available at www.drugs.com/top200sales.html
32. Bauer RJ, Dedrick RL, White ML, Murray MJ, Garovoy MR. Population pharmacokinetics and pharmacodynamics of the anti-CD11a antibody hu1124 in human subjects with psoriasis. J Pharmacokinet Biopharm 1999;27(4):397-420.
33. Preziosi P. Science, pharmacoeconomics and ethics in drug R&D: a sustainable future scenario? Nat Rev Drug Discovery 2004;3(6):521-526
34. Available at www.access.gpo.gov/nara/cfr/waisidx_02/21cfr58_02.html
35. Available at www.access.gpo.gov/cgi-bin/cfrassemble.cgi?title=200221
36. Available at www.fda.gov/cber/gdlns/ichs7a071201.htm
37. Hodgson J. ADMET—turning chemicals into drugs. Nature Biotech 2001;19:722-726.
38. Available at www.fda.gov/cber/gdlns/ind1.pdf
39. Carroll SF. Development partnerships—accelerating product development through collaborative programs. BioProc Intl 2003:48-53.
40. King J. Today's drug discovery: unlocking greater potential. R&D Directions 2004, 10(2):28-39
41. Ernst & Young. Resurgence: Global Biotechnology Report 2004.

5
Types of Clinical Studies

Lewis J. Smith

Introduction 107
Phase 1 Studies 107
Phase 2 Studies 108
Phase 3 Studies 109
Phase 4 Studies 111
Special Studies 112
Summary 120
References 121

Introduction

To obtain approval from the Food and Drug Administration (FDA) or an equivalent agency outside the United States to market and sell a new drug or biologic product for use in humans, a series of clinical studies must be performed. These clinical studies exist in four phases. Each phase has specific and differing requirements for patient types, goals, inclusion/exclusion criteria, design features, and expected outcomes. Combined, they build the patient care database for safety and efficacy that hopefully will lead to product approval.

The time frame for these clinical studies often is called "clinical development" and requires about 5 years. In the past, there were clear boundaries between the four fairly standardized phases of clinical drug development. However, the phases have become less well defined as questions previously addressed in one phase are being addressed in both earlier and later phases. In part, this new approach is designed to accelerate the acquisition of information needed for approval and successful marketing of a new drug and for collection of full and sufficient safety information as early as possible. The information that follows is designed to provide an overview of the types of studies used during clinical drug development.

Under usual circumstances, the studies progress from those designed to evaluate single and multiple dose toxicity by using a small number of normal subjects (phase 1), to define dose-response relationships and additional toxicity using a larger number of subjects with disease (phase 2), and to determine efficacy and safety with the dose(s) of interest using several thousand subjects with disease (phase 3). This chapter will address all the types of clinical studies listed in the outline above. The phase 3 studies need to provide sufficient information for a successful New Drug Application (NDA) or approval of a new biologic, that is, sufficient proof of efficacy and safety for the targeted indication.

Before receiving approval, plans are made to initiate additional studies, which are to be done during and after marketing the product. These include large, simple, clinical trials (phase 4 studies) designed to more closely resemble what occurs outside the rigid double-blind, randomized, placebo-controlled clinical trial, such as pharmacoeconomic and pharmacogenetic studies, drug-interaction studies, comparator studies, studies in special populations, and studies that primarily examine quality of life. These studies display well the need for more much information through clinical research after product approval and expand our understanding of how products will be used and perform in patients.

Phase 1 Studies

Phase 1 studies are the first studies performed in human subjects after an Investigational New Drug (IND) application has been submitted and approved (Fig. 5.1). Companies must wait

- First studies in humans after obtaining IND
- Single-dose followed by short-term multiple dose studies – follow-up for days to weeks
- Determine initial safety profile, maximally tolerated dose (MTD), and pharmacokinetic profile including ADME
- Usually 20-100 healthy volunteers
- For studies of "toxic" therapies (e.g.,Oncology, AIDS studies), "volunteers" have disease
- Duration ~1.5 years

FIG. 5.1. Phase 1 Studies

- Determine effectiveness in condition or disease of interest
- Define appropriate dose (dose-ranging studies)
- Begin to identify side effects/toxicity (over 4-6 weeks)
- Typically 100-300 patient volunteers
- Highly controlled study design
- Duration (overall) ~2 years

FIG. 5.2. Phase 2 Studies

30 days from filing the IND with regulatory authorities before starting any trials and will usually wait for their comments about their first human protocol in phase 1. Sufficient preclinical information including animal toxicology data are available to suggest that the new chemical entity or biologic may be effective and safe in humans for the proposed indication. The primary goal of phase 1 studies is to define the initial safety profile and the maximum tolerated dose (MTD).

The initial phase 1 studies use only a single-dose (often a likely no or minimal effect dose based on animal studies) and volunteer subjects. Follow-up is for days to weeks depending on the predicted pharmacokinetics. Subsequent phase 1 studies use multiple doses for several days to a week or two, with follow-up for days to weeks. The simplest dose-up schema in phase 1 studies is a doubling of the dose from the no effect dose until the MTD is achieved. These studies typically require frequent blood sampling for pharmacokinetic data. Monitoring is usually extensive in phase 1 work for various organ systems (e.g., full physical exams, blood pressure, and biological specimens), and is sometimes performed in a clinical research center in which the normal subjects are housed even overnight. Blood, urine, and other specimens are obtained to identify hematologic, hepatic, renal, and other adverse effects. If animal studies suggested any special tissue effects, extra specimens are collected and assessed to ensure subject safety. Additionally, pharmacokinetic studies are done along with the toxicology assessment, or in separate trials, to document the absorption, distribution, metabolism, and elimination (ADME profile) of the product. Pharmacokinetic differences between products in the same family of drugs can be major advantages (e.g., longer half-lives permitting less frequent dosing, nonrenal elimination permitting use in renal failure, or distribution of the drug to a site where the disease is localized).

Phase 1 studies usually require 20 to 100 healthy volunteers. However, there are circumstances in which healthy volunteers are not used. Typically, this occurs when more "toxic" therapies are being tested, such as cancer chemotherapy and antivirals for the treatment of the human immunodeficiency virus (HIV). Under such circumstances, patients with the disease are the first individuals to test the new therapy. In these populations, there is a tendency to merge phase 1 studies with early phase 2 studies to minimize the total number of subjects exposed to a potentially toxic treatment. In addition, the minimum pharmacokinetic work is done in patients to also reduce patient exposure to toxic products. It takes approximately 1.5 years to complete the phase 1 studies.

Phase 2 Studies

Phase 2 studies are performed to determine the initial effectiveness of an investigational drug or biologic in patients with the condition or disease of interest (Fig. 5.2). This phase of testing also helps determine the common short-term side effects and risks associated with the drug. Phase 2 studies are typically well controlled and closely monitored. A major focus is to find the appropriate dose(s) for the larger studies required in phase 3. The primary phase 2 studies are dose-ranging studies and are designed to provide *proof of principle*. One designs the studies such that the range of doses, typically four, includes an ineffective dose at the lower end and, at the upper end, a dose that does not add to the effect of the next highest dose. On average, 100 to 300 patient volunteers participate in the primary phase 2 studies. It takes approximately 2 years to complete these studies.

During these studies, one also begins to identify side effects and toxicity at doses to be used later in phase 3 studies and likely after marketing. Phase 2 studies of drugs that are being evaluated for chronic use are typically at least 4 to 6 weeks in duration and sometimes up to 6 months, which allows for observation of any later occurring side effects, the full action of the product to be produced, and possibly the assessment of tolerance or waning of a product's beneficial or toxic activities. Although there is a tendency to try to do as much as possible in these early phase studies, rather than focus on the essential questions, one should have a simple study design in order to maximize the probability of determining if the drug is effective and has acceptable toxicity. Many products fail at this stage and are killed, which is desirable as necessary before embarking on the very expensive and labor intensive phase 3 study program.

A feature of phase 2 studies is the use of relatively homogeneous patient populations with very tight inclusion and exclusion criteria. This is done to increase the likelihood of identifying a positive effect and minimizing confounding variables. On the other hand, the results obtained may not

- Obtain additional information about drug for publications
- Identify potential new uses
  (e.g., areas of unmet medical need)
- Examine follow-up clinical issues
  - dosing regimens
  - routes of administration
  - role of concomitant disease/drugs
  - effects on special populations

FIG. 5.3. Phase 2 Studies: Post Approval

accurately reflect the effectiveness of the drug in the more typical heterogeneous patient population.

In addition to the phase 2 studies that must be performed as part of the drug approval process, other studies may be undertaken before the phase 2 studies have been completed and the data analyzed (Fig. 5.3). Studies may be performed for several reasons: to obtain additional information for publications, to identify potential other uses (e.g., new indications, areas of unmet medical need), to examine different dosing regimens, to explore new routes of administration, to define the role of concomitant drugs, and to determine the effects in special populations (e.g., the elderly, those with renal or hepatic disease, common comorbid conditions, pharmacogenetic variables). Caution should be exercised by management at companies in doing these types of trials before approval, as they may confound or slow the approval process by creating unexpected side effect data or just more data that the regulatory authorities need to review. As a result, these exploratory phase 2 trials often will be done postapproval.

Phase 3 Studies

Phase 3 studies are expanded in controlled and uncontrolled trials (Fig. 5.4). They are intended to gather the additional information about effectiveness and safety that is needed to define the overall benefit-risk relationship of the drug in the target population. Phase 3 studies should provide an adequate basis for extrapolating to the general population and transmitting that information in the product labeling for health care providers. [3]

- Confirm effectiveness in larger studies ("pivotal studies")
- Typically need 2 positive, well-designed studies
- Monitor adverse events
  - Over a longer period (12-24 weeks)
  - More patients for more accurate frequencies and presentation
- Design of study = package insert for marketing
- Usually 1,000-3,000 patient volunteers at many sites with many investigators
- Overall Duration ~2.5 – 5 years

FIG. 5.4. Phase 3 Studies

- Scale-up in manufacturing
- Heterogeneous patient sample
- Competitive patient enrollment
- Role/use of central IRB–advantages, limitations
- Investigator & site training for product, protocol, & processes

FIG. 5.5. Phase 3 Studies: Issues

The decision to move ahead with phase 3 studies is a major one because the costs are considerably higher than for the two earlier phases combined. A drug identified as effective and safe in phase 2 studies may not enter phase 3 clinical trials for a number of reasons including insufficient efficacy when compared with its competitors, the expense and difficulty of drug formulation, especially when scaling up production, and side effects that exceed the risk profile needed to proceed.

Submission of a NDA requires at least two well-designed phase 3 studies that demonstrate both efficacy and safety in a large number of patients with the target disease, typically 1,000 to 3,000 patient volunteers. This large patient sample necessitates using many sites and investigators with their staffs, often 100 to 200 or more. The treatment period in phase 3 studies is longer than in phase 2 studies for drugs used chronically (12 to 24 weeks vs. 4 to 6 weeks). This provides an opportunity to monitor and detect adverse events and tolerance over a longer time.

A third phase 3 study may be incorporated into the initial drug development plan for at least two reasons. The risk exists that one of the two phase 3 studies may not be sufficiently positive for a new drug to be approved by the FDA. Also, procedural problems could occur in, for example, patient monitoring consistency or data collection, which results in the regulatory authority, after auditing sites and finding such a serious procedural problem, discarding an entire study from the NDA package. Disadvantages of this approach include the substantial increased costs and the longer time needed to complete the phase 3 studies. The advantage is the equivalent of an insurance policy for a more timely NDA submission, instead of waiting several years to conduct an added follow-up phase 3 study after the standard 3 studies were done. The usual time required to complete the phase 3 studies is about 2.5 years but could last 5 years. Then, data analysis and study reports are done, reviewed, and the NDA is filed, which can take 6 to 12 months or more at the company.

Phase 3 studies have several unique characteristics relative to the earlier phase studies (Fig. 5.5). They include the need to scale-up manufacturing to guarantee an adequate supply of drug for the duration of the studies, the more heterogeneous patient population than that studied in phase 2, the competitive nature of patient enrollment, the use of private as well as academic sites, and data reliability. Scaling up in manufacturing is not a trivial matter for some small molecules and even more so for biologics. For example, scaling up monoclonal antibody production may inadvertently change the characteristics of the antibody such that it is no longer identical to the antibody used in the early phase studies. Further, at this phase the formulation must be the same as what will be marketed and sold.

The more heterogeneous patient population than that studied in phase 2 creates a desirable, more representative patient sample but increases variability, the range of patient responses, provides more opportunity for side effects, and requires more patients to demonstrate a statistically significant benefit. The more heterogeneous patient population can reduce the signal to noise ratio such that a drug that was effective in phase 2 studies is no longer as effective in the pivotal phase 3 studies. Subject enrollment has increasingly become a rate-limiting step in the drug development process. Substantial efforts and funds must be expended to recruit appropriate research subjects. The need to recruit patients from a large number of sites makes it more efficient to use a central IRB, if possible. A variety of systems are used to assist in this recruitment, such as health care networks of hospitals and clinics, advertisements on the radio and in local media, Internet ads, and recruitment companies.

Academic sites are usually adds to use a central IRB, because of institutional policy and ethical concerns; however, they constitute a small percentage of the sites in phase 3 studies, which contrasts with their larger representation in phase 2 studies. Central IRBs are used by companies to expedite study approval at many sites, which may not have routine IRB access as in private physician offices. Cautions with such IRBs are their independence, sufficient expertise and appropriate representation, sufficient oversight of protocols, and appropriate oversight of the many investigative sights over wide geographic areas. Data reliability becomes a major issue in phase 3 with so many sites and people involved, which requires a significant investment in training of the site staff as well as investigators about the drug, protocol, and procedures, especially patient inclusion and exclusion, drug administration, monitoring requirements, and data collection requirements.

The typical study design used to demonstrate efficacy is randomized and placebo-controlled, but an active comparator control group can be considered (Fig. 5.6). In recent years, especially in Europe, there has been increasing concern about the safety and ethics of performing placebo-controlled trials. The most recent International Council on Harmonization/ Good Clinical Practice (ICH/GCP) guidelines recommend against doing placebo-controlled trials except under specific circumstances [4, 5]. This concern and guidelines have led to greater use of active comparator trials in which an approved, generally accepted therapy is the control arm and compared with the new therapy. When the goal of the study is to demonstrate superiority of the new therapy, the issues are the same as when the comparator therapy is placebo. However, if the goal is to show statistically equivalent benefit, it is called an equivalence or noninferiority trial (Fig. 5.6).

The major problem with noninferiority trials is the assumption that the active control treatment is effective in the trial (e.g., the trial has an assay sensitivity). Unfortunately, this situation is not always true for effective drugs and is not directly testable from the data collected, because there is no placebo group [6]. There are ways to maximize the value of noninferiority trials, such as determining from historical trials that the active control group reliably has an effect of at least a certain size, planning the trial design to be similar to that of prior trials (e.g., stage of disease, concomitant therapy, and end points), setting a noninferiority margin to be smaller than the total active control effect, and ensuring appropriate trial conduct (e.g., concomitant medications, study drug compliance). Nonetheless, because one cannot formally establish a minimal effect size, noninferiority cannot be per se taken as evidence of efficacy, and the interpretation of the trial must be based on the totality of the data, including additional analyses.

A number of additional studies may be performed during the time (2–3 years) from completion of the phase 3 studies to drug approval, also known as phase 3b studies (Fig. 5.7). This time is required to analyze and prepare the large amount of data for submission to the FDA, as well as the actual FDA review. The studies performed during this time serve to expand the adverse event database and dosing and efficacy data before approval, provide marketing support, increase physician participation (e.g., those in practice-based settings), institution familiarity with the drug prior to its approval and release, and increase the number of publications. Although the phase 2 and 3 programs may include 100 sites and 1,000 patients (smaller numbers for accelerated approvals), only 10 or 20 major universities may have participated, leaving many specialists without direct experience with novel products under study. Phase 3b allows for expansion of experts at more universities as well.

- Also called "equivalence" trial
- Uses "active" comparator
- Increasingly used due to concerns regarding placebo-controlled trials
- Non-inferiority cannot be taken as evidence of efficacy!!

FIG. 5.6. Non-Inferiority Trials

- Studies performed during time (2-3 years) from completion of phase 3 studies to approval
- Goals:
  - Expanded adverse event database and dosing and efficacy data before approval
  - Marketing support
  - More physicians (e.g., those in practice based settings) and institutions become familiar with drug
  - Additional publications
- Large size – may have sub-studies in certain populations
- Requires FDA approval (file under IND)

FIG. 5.7. Phase 3b Studies

5. Types of Clinical Studies

- Total time from beginning phase 1 studies to product launch is usually 8-10 years
- Many questions unanswered at end of phase 3 studies (e.g., longer term toxicity, use inspecial populations, role of genetic factors)
- Role of regulatory affairs department
- Requires plans to do these studies integrated phase 3b and 4 studies

FIG. 5.8. Phase 1,2 and 3

- Post-approval studies
- Multiple purposes
- Multiple types
- Unique challenges in design & conduct
- Often very large studies
- Simple design (IRB approvals)
- Failure to fulfill FDA requirements can result in withdrawal of approval

FIG. 5.9. Phase 4 Studies Overview

Many of these studies are quite large and may have substudies that utilize specific populations. All studies performed during this time require FDA approval and are filed under the original IND application. As noted earlier, some smaller studies may be performed at the time of the phase 2 studies.

In an summary of phase 1 to 3 studies (Fig. 5.8), the time from beginning phase 1 studies to product launch averages about 7 years (a range of 5–10 years). Even at the end of this lengthy period, many questions will remain to be answered such as longer term toxicity (withdrawal of Vioxx is one such example), use in special populations, and the role of genetic factors. Studies to answer these and other questions must be planned before the phase 3 studies are completed and conducted during phase 3b and 4 trials. Often, the FDA and company negotiate which of such studies need be done further as a contingency for approval. Many of such studies will enhance product use in patients, create good publications, and even improve sales.

All throughout the phases of clinical drug development, the regulatory affairs department should maintain an open, ongoing dialogue with the FDA. Reasons for doing this include (1) approval from the FDA is needed for each study (goals and designs) performed under the IND; (2) the FDA is privy to data from clinical trials of related drugs or of unrelated drugs in the same disease that may influence study design; and (3) any surprises with the FDA are avoided, which can slow the approval process. For example, the FDA may be aware of a possible toxic effect, not have anticipated from the preclinical pharmacology and toxicology. Suggestions from the FDA to incorporate additional measurements or modify other aspects of the study design should be considered very carefully.

Phase 4 Studies

Phase 4 studies, by definition, are those studies that are performed after a new drug has been approved for marketing (Fig. 5.9). Phase 4 studies serve multiple purposes and comprise many different types. They also pose some unique challenges, which will be discussed below. The FDA often requests commitments from the NDA applicants to conduct

- Compare to competitor drugs
- Use to define mechanisms of disease (often investigator-initiated)
- Identify variables in existing indications
- Identify new dosing schema
- Explore "real-world" effectiveness (e.g., in the office)
- Define effects in special populations (elderly, children)
- Examine impact of concurrent diseases
- Post-marketing surveillance
- Drug interactions

Note: Phase 4 will overlap with 3b studies.

FIG. 5.10. Phase 4 Studies: Possible Objectives

postapproval studies [7]. In general, characteristics of phase 4 studies are that they can be very large and have a more simple study design. Although these studies were not deemed essential for initial approval, they provide additional data that could change the prescribing information or the use of the drug. A major challenge to consider with phase 4 studies is that failure to fulfill FDA requirements for more data, especially for adverse events and further efficacy information, can result in withdrawal of an already approved drug.

Objectives of phase 4 studies are manifold (Fig. 5.10); determining efficacy and safety compared with competitor drugs, defining mechanisms of disease that often are performed as investigator-initiated studies (see below), exploring "real-world" effectiveness (e.g., in the office), defining effects in special populations (e.g., the elderly, children, concurrent disease), providing postmarketing surveillance for unsuspected or low-frequency adverse events, further defining potential drug interactions, and possibly pharmacogenetic assessments (to be discussed below). Some objectives done during phase 4 period may require FDA agreement and may be classified technically as phase 3b or even phase 2 studies, such as new administration schema, significantly different doses, and identifying new and expanded indications.

Postmarketing surveillance studies are meant to substantiate safety in a larger, more heterogeneous patient population than is possible during the pivotal studies performed during phase 3 with their strict, often randomized placebo-controlled double-blind designs (Fig. 5.11). The patients often have

- Safety (post-marketing surveillance) – substantiate safety in larger, broader patient populations
- Efficacy and safety – in settings of widespread subpopulations and dosing schemas
- Usage studies in varied and normal practice environments
- Substantiate product quality and consistency

FIG. 5.11. Phase 4 Studies: Types

- Food/Drug/Disease Interactions
- Investigator-initiated
- Pharmacoeconomic – may be part of phase 3 plan as well
- Pharmacogenetic – increasingly important to identify responders/non-responders and susceptibility to toxicity
- Quality of life
- Epidemiology

FIG. 5.12. Special Studies

coexisting illnesses, greater or less severity of disease, longer or shorter duration of disease, more varied signs and symptoms of disease, or are taking medications that would have excluded them from the phase 2 and 3 studies. The studies may be open-label and relatively uncontrolled, but the protocols are IRB approved usually with patient consent obtained. Sometimes these studies are required by the FDA, and then they will approve the design as well. The protocols still will contain specific dosing, inclusion criteria, monitoring, and data requirements and then they must be of sufficient quality to be publishable, so that the medical community will accept the information. Their primary focus is on serious adverse events whose frequency may be too low to identify in phase 3 studies. Postmarketing surveillance studies may also provide additional evidence of efficacy and safety in the setting of widespread related but off-label use.

There is considerable overlap between phase 4 studies and studies done as part of phases 2b and 3b in objectives, potential investigators and sites, and many design features. The major differences are fourfold: the time at which the studies are performed in development; the size of the studies differ, that is, smaller studies predominating during phase 2b and larger studies during phase 4; whether they are within versus outside of labeling (package insert); and FDA approval is required for the design in phase 3b and exploratory phase 2.

Special Studies

A wide variety of specialized studies are conducted during the drug development process (Fig. 5.12). Interaction studies include the impact of either food, other concomitant drugs, or disease on the new product when the new drug is used in these situations. Investigator-initiated studies are protocols written by a possible principal investigator and submitted to the company for approval and funding. Pharmacoeconomic and quality of life are done within health care systems or universities that have the special expertise and access to the added data on costs of care or nonclinical assessments for such trials. Pharmacogenetic studies are becoming a new requirement as we learn about ethnic and genetic differences in the population related to the actions of products, both safety and efficacy.

- Effect of meals on absorption (if drug taken by mouth)
- Interactions between new drug and other medications (e.g., Coumadin; drugs that alter activity of cytochrome P450 [CYP] isoenzymes, if drug is metabolized by these enzymes)
- Effects of disease (e.g., liver, kidney, heart failure)

FIG. 5.13. Food/Drug/Disease Interaction Studies

Phase 3b or 4 studies may examine the effect of a meal on absorption (if the drug is taken by mouth), including increases and decreases in blood levels, or its ability to reduce abdominal reactions like nausea (Fig. 5.13). Interactions between the new drug and other medications that would likely be used concomitantly are studied in phase 3b and phase 4, for example, warfarin; drugs that alter the activity of cytochrome P450 (CYP) isoenzymes, if the drug is metabolized by these enzymes. Many drug categories have impact on the CYP family of degradative liver enzymes and increase or decrease blood levels of concurrent drugs, such as antidepressants, beta-blockers, calcium channel blockers, narcotic analgesics, antipsychotics, estrogens, and nonsteroidal anti-inflammatory drugs. The effects of age may influence sensitivity to side effects, alter metabolism, or change patients' responsiveness. Diseases, especially liver, kidney, and heart, may change the pharmacokinetics of the new product, especially elimination, and side effects.

Investigator-initiated studies can be a valuable complement to the studies required by the FDA during the pre- and postapproval periods (Fig. 5.14). Investigator-initiated studies explore different uses, doses, or patient subsets, as well as basic physiologic and disease mechanisms. At the same time, the pool of investigators and thought leaders familiar with the new drug is expanded. Investigators planning to do a study using either an unapproved drug or an approved drug for an unapproved use must obtain an investigator IND. Permission to do the study must be obtained first from the holder of the original IND. Once that has been accomplished, the process for obtaining an investigator IND from the FDA is relatively simple because the FDA can reference the original IND file. A formal submission, which includes the proposed study protocol and the investigator's qualifications, is made to the FDA.

- **Goals:**
  - Explore different uses, doses or patient subsets
  - Explore basic physiologic and pathophysiologic mechanisms
  - Expand investigator pool, including thought-leaders
- **Specific investigator and manufacturer requirements:**
  - Investigator IND (reference company file)
  - Reports to FDA and company
  - Company may assist with study design and adverse event reporting, may provide drug, and may provide funding

FIG. 5.14. Investigator Initiated Trials

- Definition: Study of net economic impact of drug selection and use on total cost of delivering health care
- Health care utilization (efficiency) vs. efficacy and safety
- "Value" – a benefit for money spent
- Perceptions of value (multivariate concept) based on:
  - Perspective: patient, provider, payer, or society
  - Type of therapy, eg, new/unique vs. add-on vs. me too
  - Costs and consequences, plus economic, clinical, humanistic outcome dimensions
- Multiple study designs:
  - cost-effectiveness, cost-minimization, cost-utility, cost-benefit
- Role of FDA varies

FIG. 5.15. Pharmacoeconomics

After receiving the investigator IND, the investigator makes yearly reports to the FDA and submits any proposed changes to the protocol and any new protocols to both the FDA and the institutional review board (IRB) of record. Reports are also provided to the holder of the original IND, which are usually requested and often required by them.

The holder of the original IND (e.g., a pharmaceutical or biotech company) may assist with study design and adverse event reporting and provide study drug, especially before approval. The company grants permission for use of their product with the usual stipulations of review and even approval of the protocol, some agreement on the investigator(s)' input on any publications, and maintenance of patent rights at the company (a controversial subject). The company often also will provide some grant funding. However, the holder of the investigational IND is ultimately responsible for all activities that occur related to the study.

Pharmacoeconomic (PE) studies have become increasingly important in health care decisions for product usage in health systems and therefore for the drug development process (Fig. 5.15). They are now often incorporated into phase 3 study plans, in parallel studies, or as part of the pivotal phase 3 study. Because there are many different PE studies, they are time-consuming and expensive, they use different data than typical phase 3 trials, and they are a key component of phase 4 studies. One definition of pharmacoeconomics is the study of the net economic impact of pharmaceutical selection and use on the total cost of delivering health care [8]. A key concept in economic analyses is "value." Value can be defined as a desirable outcome or benefit for a given therapy at a certain cost. The type of product also effects value, such that a novel therapy versus a good alternative has a greater value than a "me-too" product. The assessment of value is based on one's measurement criteria and the person making that assessment. Patient, physician, health care provider, payor, or health care system have different perspectives and information needs within the realm of pharmacoeconomics. Pharmacoeconomic (PE) studies utilize several different study designs including cost-effectiveness, cost-minimization, cost-utility, and cost-benefit. Examples of each are provided below.

The regulatory role of government agencies, such as the FDA, in PE studies is variable because such studies are not requirements at all for approval of the product for marketing. Regulators may lack the expertise in assessing PE studies, and guidelines are not available for all product types and study types. However in Europe and other parts of the world, an additional government agency often exists, such as a pricing or health care payment agency, that will approve government payments for drugs and devices. Because most countries have significant government payment for medical care, the responsible agency needs to be favorably influenced by PE data, additionally motivating a company to perform these studies in their phase 3 and 4 plans [8–12].

These four types of PE studies are outlined in Figs. 5.16 and 5.17 with five elements; the name of method, a brief definition, typical outcome measures used, a description of types of results expected, and then advantages and disadvantages of the designs or utility of the information. *Cost-effectiveness analysis* is the pharmacoeconomic study most frequently used. In this type of analysis, the cost and consequences of two alternative treatments are compared and quantified. The additional cost that an alternative treatment imposes over another treatment is compared to the additional effectiveness (in terms of outcomes) the treatment provides. The main objective of cost-effectiveness analysis is to evaluate the ratio between the cost surplus associated with the new treatment (e.g., the higher cost of the new pharmacological treatment) and the efficacy/effectiveness surplus derived from it.

$$\text{Cost effectiveness ratio} = \frac{(Cost\ of\ treatment\ A) - (Clinical\ success\ treatment\ A)}{(Cost\ of\ treatment\ B) - (Clinical\ success\ treatment\ B)}$$

One example of a cost-effectiveness analysis is the treatment of diabetes. Patients who are overweight often require weight reduction to improve their diabetes. A cost effectiveness analysis could examine the cost of adding to standard

Type	Definition	Outcomes
1. Cost Minimization Analysis	Comparison of costs of alternative therapies (a ratio)	• Assumed clinical outcomes to be equal • Drug costs • Show lowest cost among those of equal benefit
	Advantages	**Disadvantages**
	• Simple cost analysis	• Outcomes of alternative treatments must be shown to be equivalent
Type	**Definition**	**Outcomes**
2. Cost Effectiveness Analysis	Ratio of costs & benefits from alternative therapies (Δ cost /Δ outcome)	• Costs expressed as monetary units • Benefits are measured as specific outcomes • Years of life saved can be given cost per specific unit
	Advantages	**Disadvantages**
	• Applicable to wide range of possible clinical outcomes • Used to compare drugs with same outcomes	• Comparisons among studies or diseases need to be same outcomes

Fig. 5.16. Pharmacoeconomic (PE) Studies 1 & 2

Type	Definition	Outcomes
3. Cost Utility Analysis	Ratio of costs & benefits from alternative therapies	• Outcome includes patient preferences (QALYs) • Cost = monetary units & benefits = preferences • Gives cost per QALY or similar measure including patient preferences
	Advantages	**Disadvantages**
	• Enables varied outcomes combined in a summary • Considers patient preferences for results	• Difficult translating Quality of Life into utility scores • Utility measure (Quality Adjusted Life Year) for preferences may be confusing tool to providers & payers
Type	**Definition**	**Outcomes**
4. Cost Benefit Analysis	Ratio of costs & benefits from two therapies expressed monetarily	• $ Dollars • Both economic inputs & outputs evaluated in monetary terms
	Advantages	**Disadvantages**
	• Compares programs with different outcomes	• Difficulty defining monetary value of health consequences • Provider & payer may lack understanding of tool

Fig. 5.17. Pharmacoeconomic (PE) Studies 3 & 4

diabetes treatment an antiobesity drug such as Xenical® (orlistat) and its effect on lowering the hemoglobin A1C with the cost and effectiveness of a different antiobesity drug or a structured dietary program.

Another example is provided by the biotechnology product Cerezyme® (imiglucerase), which is used for the treatment of Gaucher disease. Gaucher disease is characterized by a deficiency of beta-glucocerebrosidase activity resulting in the accumulation of glucocerebrosidase in tissue macrophages, which become engorged and are typically found in the liver, spleen, and bone marrow and occasionally in the lung, kidney, and intestine. The clinical consequences include severe anemia, thrombocytopenia, progressive hepatosplenomegaly, and skeletal complications such as osteonecrosis and osteopenia, with resultant pathological fractures. Cerezyme® improves anemia and thrombocytopenia, reduces spleen and liver size, and decreases cachexia to a degree similar to that observed with alglucerase. For many patients, enzyme replacement therapy has been effective, returning the liver, spleen, and bone marrow back to an effective degree of function. Cost-effectiveness studies of enzyme replacement therapy for Gaucher disease have consistently shown that the treatment is effective, safe, and associated with improved quality of life. On the other hand, it is expensive. Estimated cost of the enzyme alone ranges from $70,000 to $550,000 per year for a typical adult with Gaucher disease, depending on the dose. A cost-effectiveness analysis would determine if the additional cost associated with Cerezyme® treatment is matched or exceeded by its benefits compared with an alternate treatment.

A second type of pharmacoeconomic analysis is the *cost-minimization analysis*. It is used to define the most economical treatment among different alternatives with equal efficacy/effectiveness and safety profiles, assumed but not directly assessed in the calculations. An example of a cost-minimization analysis is the comparison of a brand name and equivalent generic drug. A generic drug is identical, or bioequivalent, to a brand name drug in dosage form, safety, strength, route of administration, quality, performance characteristics, and intended use. Although generic drugs are chemically identical to their branded counterparts, they are typically sold at substantial discounts from the branded price. The generic drug will always show advantages by cost-minimization analysis. Two antihypertensive nongeneric products with different clinical profiles could be evaluated with this method also, but the different contribution of the side effects or administration requirements will not be incorporated even though they may be important in their use.

Five growth hormone (somatotropin) products were available in the U.S. market in 2003: Nutropin AQ® (Genetech, 5 mg $441.00); Genotropin® Injection (Pharmacia, 5.8 mg $210.00, 13.8 mg $504); Humatrope® (Lilly, 5 mg $220.50, 6 mg $264.60, 12 mg $529.20, 24 mg $ 1058.40); Saizen® (Serono, 5 mg $210.00, 8.8 mg $336.00); and Norditropin® (Novo Nordisk, 4 mg $170.40, 8 mg $352.80). Based on the cost of each drug, a cost-minimization analysis is performed to identify which of these similar products has the lowest cost while providing the same benefit as the others.

A third type of pharmacoeconomic analysis is the *cost-utility analysis*. This type of analysis is based on a sophisticated methodology in which benefits are calculated using parameters that take into account the quality of life of the patient. These analyses are an extension of the lifetime cost effectiveness analysis, because they estimate both quality of life and its duration. The most utilized indicator for quality of life is the quality adjusted life year (QALY), which corresponds with a year of life adjusted for its quality.

Cost-utility analysis has been used for the drug Epogen® (erythropoietin). Assume that a patient, who has renal disease and the anemia associated with it, is treated with Epogen® and has good control. That patient is assigned a utility value of 0.9 on a scale of 0 to 1. Also assume that an untreated patient with poorly controlled disease has an average utility value of 0.5. Therefore, 10 years of life of the first patient corresponds with 9 QALYs (i.e., 10×0.9), whereas 10 years of life for the second patient corresponds with 5 QALYs (i.e., 10×0.5). The QALYs are incorporated into a lifetime cost-effectiveness analysis to determine the cost utility of each therapy.

The fourth type of pharmacoeconomic analysis is the *cost-benefit analysis*. When both costs and benefits of a treatment are measured in monetary values, cost-benefit analysis is a useful tool. Future costs and benefits are discounted to their current value and take into account the "time value of money." Because of inflation, a dollar today is not equivalent to a dollar in the future. However, the application of cost-benefit analysis in pharmacoeconomics is limited, due to the difficulties in assigning a monetary value to health outcomes and a patient's life. For example, when evaluating the cost of "statin" drugs versus the consequences of not treating patients with hypercholesterolemia, the cost associated with developing cardiac disease, a stroke, or death must be given a value benefit in dollars. Further, the statins may have additional benefits unrelated to reducing cholesterol levels, which are not measured.

Another example is provided by the drug Enbrel® (etanercept). The cost of a new biotechnology agent, such as Enbrel® for the treatment of rheumatoid arthritis, may be higher than other available agents due to its innovative mechanism of action. Not only does Enbrel® stop and relieve the pain associated with this form of arthritis, but unlike most other drugs used in this setting, such as nonsteroidal anti-inflammatory agents, it also stops joint erosion, improves mobility, and improves quality of life. As exemplified by Enbrel®, new therapies developed through biotechnology can be of great value as long as the benefits exceed the costs. Consequently, it is critically important for new biotechnology products to identify and quantify all the benefits they offer over current treatment options. Benefits may include improved outcome or efficacy including stopping and/or reversing disease progression, reduced side-effects or complications, reduced hospitalizations or bed-days, improved quality of life, improved morbidity and mortality, and reduced total health care costs. Well-designed pharmacoeconomic analyses can be instrumental in defining the overall benefits of these new therapies [8–12].

Quality of life was briefly discussed in the section on cost-utility analyses. In contrast with efficacy, safety, and cost-effectiveness studies, which are viewed by providers, investigators, and researchers as important in the decision-making process associated with drug development, quality of life (QOL) studies are often viewed as supplemental (Fig. 5.18). An exception is the role in the development of biologicals, most likely due to their higher costs [8, 11,18].

- Efficacy, safety, and cost-effectiveness studies are all viewed as important in decision making
- In contrast, QOL studies are often viewed as supplemental (However, they are key parameters with biologicals)
- QOL studies-disease specific (FACT in cancer) vs. generic (SF-36 health survey)
- What role does QOL play in drug trials and formulary decisions?
- Who benefits from QOL data?
- Who should pay for better QOL information?

FIG. 5.18. Quality of Life Studies

Although many studies provide good QOL data demonstrating additional significant benefits for patients, and their impact on the decision-making process has not been well studied. Some findings are that the role of QOL information in influencing managed care decision-making is not well understood, because research on the subject is relatively new and/or has been minimal, designs are less well understood and accepted, gold standards are not as well recognized, and applicability to specific health care settings may be missing [8, 11, 18].

For QOL studies, one can define health as "not merely the absence of disease, but complete physical, psychological and social well-being." To measure QOL, multiple tools have been developed and validated. There are generic instruments such as the SF-36 [13] and disease-specific instruments, such as St. George's Respiratory Questionnaire for COPD (chronic obstructive pulmonary disease) [14] and the Functional Assessment of Cancer Therapy (FACT) questionnaire for cancer [15]. Such generic measures have questionable applicability to certain diseases or to a drug's impact on the disease or sensitivity to pick up specific disease changes in QOL, leading to a need to develop such disease-specific instruments. However, the disease-specific tools must be repeatedly used and validated before acceptance by the medical community and health care systems.

Generic QOL instruments are used for a wide range of diseases to determine how treatment influences day-to-day activities, well-being, and social functioning. Generic instruments can be used to compare the impacts of different diseases. The SF-36 is a health survey with 36 items constructed to identify a patient's health status. It was designed for use in clinical practice and research, health policy evaluations, and general population surveys. The SF-36 includes one multi-item scale that assesses eight health concepts: (1) limitations in physical activities because of health problems; (2) limitations in social activities because of physical or emotional problems; (3) limitations in usual role activities because of physical health problems; (4) bodily pain; (5) general mental health (psychological distress and well-being); (6) limitations in usual role activities because of emotional problems; (7) vitality (energy and fatigue); and (8) general health perceptions. The survey was constructed for self-administration by persons 14 years of age and older or for administration by a trained interviewer in person or by telephone.

Disease-specific QOL instruments are usually more responsive to changes in QOL than generic and utility measures. Disease-specific QOL tools are more specific for disease but less applicable for drug formulary decision-making. Disease-specific QOL instruments require validation and applicability to the disease and disease treatment in routine clinical practice. They also must include practical measures that can generate reproducible results.

The Asthma Quality of Life Questionnaire (AQLQ) is a 32-item questionnaire that has been developed to measure the functional impairments that are most important for adults (17–70 years) with asthma [16]. A pediatric version is also available [17]. The items are in four domains (symptoms, emotions, exposure to environmental stimuli, and activity limitation). The instrument is in both interviewer- and self-administered formats and takes approximately 10 minutes to complete at the first visit and 5 minutes at follow-up. Several independent studies have demonstrated the strong evaluative and discriminative measurement properties and validity of the Asthma Quality of Life Questionnaire. It has been used successfully in a large number of clinical trials and in clinical practice around the world.

Another example of a disease-specific QOL is the Functional Assessment of Cancer Therapy (FACT) [15]. An assessment of fatigue may consider broader concerns, such as global quality of life and symptom distress. Some of the fatigue scales, such as the unidimensional three-item scale of the EORTC QLQ-C30 and the multidimensional fatigue subscale of the Functional Assessment of Cancer Therapy (FACT), are themselves modules of well-validated quality of life instruments. The larger scale may be included if additional evaluation of quality of life is valuable. For the other fatigue scales, a separate quality of life questionnaire will be needed to accomplish the same goal. Most patients with cancer or AIDS have multiple symptoms. Fatigue, pain, and psychological distress are the most prevalent in most populations. Given the likelihood of multiple symptoms, it may be informative to add a measure of symptom prevalence and distress to the fatigue-assessment strategy. This approach also can clarify the extent to which fatigue associates with other symptoms.

Although many studies provide good quality of life data demonstrating additional significant benefits for patients, the impact of the data on drug approval and on formulary decision-making is uncertain.

Who benefits from QOL studies? Pharmacoeconomics and QOL information is increasingly discussed now in formulary and drug use decisions. However, researchers have been unable to identify the extent of influence that pharmacoeconomics and QOL information has on formulary decision making. Pharmacy and medical directors in health care systems historically focused on standard clinical parameters of safety and efficacy or cost (cost-effectiveness or cost of treatment) in their decision-making process. This is largely due in part to the nature of managed care's focus on reducing cost. The concept of health care insurance or coverage was based on providing services for medical necessity. How does QOL fit into the puzzle of medical need? Should health care be responsible for providing care, services, or products that will improve the overall well-being of patient?

Who should pay for QOL? Patients reap the benefits of services or products that improve their QOL. If patients are reaping almost all the benefits, then should they be accountable to pay for these services or products? Consumers will readily pay for items that provide convenience or improve their quality of life (i.e., dishwashers and washing machines, housekeepers or gardeners). Some consumers are willing to

- Understanding impact of a product on health care system and incorporating that information into study design:
  - Clinical efficacy and safety vs. effectiveness
  - Improved outcomes (morbidity and mortality) and QOL
- Health care utilization (efficiency) vs. efficacy and safety:
  - Study Differences: setting, design, population, patient entry, intervention, outcomes, generalizability, confounding factors
- Separate from clinical trials in NDA:
  - Potential negative impact on NDA/BLA
  - Many healthcare settings and perspectives
  - Inaccessibility of data and high cost of data acquisition
  - Not needed for approval by regulatory authority
  - Needed for health system use of product

FIG. 5.19. PE Drug Development Challenges

Example: Cancer Patient (ALL)
- Drug Issue - Liver Metabolism: (+ or - activity)
  - e.g., CYP2D6, TPMT
- Drug Issue - Patient's Receptor Sensitivity: (> or < action)
  - e.g., β1AR, PXR, GR, VDR
- Cancer Genotypes: (> or < disease; > or < response)
  - Disease subtypes, e.g., Her2Neu, Bcr/Abl, P53, BEX, VEGF
  - Drug resistance factor, e.g., MDR
- Patient's Infection Defense:
  - Immune system, e.g., IL1, IL6, TNF, IL2, MHC/HLA
- Results:
  - > or < Response? and/ or Toxicity?
  - Outcomes - composite of all the pharmacogenetic changes

FIG. 5.21. Gene Variability in Pharmacogenetics

pay $6.00 a pill to improve their quality of life but will not pay $2.00 a pill to prevent them from dying of a heart attack. Are there ways or means to quantify and translate these benefits into the health care system? Other QOL studies, as in anemia in renal disease and Epogen® (epoietin alpha), used QOL as the primary end points in product approval, and patients had (have) substantial and exceptionally dramatic benefits in daily living activities such that Medicare decided to pay for the product. In order to make QOL more valuable in the decision-making process, future studies need to define more fully the economic value of QOL in the health care system [8–12].

In addition to the benefits they may provide during the drug development process in terms of added significant study end points and patient care benefits, pharmacoeconomic studies also offer a number of challenges in their conduct (Fig. 5.19). For example, it is essential to consider the potential impact of a product on the entire health care system and incorporate this additional information into the study design. It is also important to separate, when possible, pharmacoeconomic studies from those clinical trials required as part of the NDA/BLA submission because they may have a negative impact on the submission. Phase 3 studies do not use the typical patients that you find in health care systems, such as managed care organizations (MCO), and you want to use in PE studies. Furthermore, study design differences for PE versus clinical studies is quite different (e.g., setting [MCO vs. university hospital], patient entry [all comers in a health system for PE, inclusive vs. exclusive], intervention [specific drug at specific doses vs. standard of care at these institutions], and outcomes). Health care settings best used for phase 3 studies may not have the type of patients or data needed for PE trials. Training of investigators and patient monitors is a huge challenge in time, costs, and reliability of the QOL information. As noted earlier, clinical trials in a development plan are needed for approval and PE or QOL are not [8–12].

Although pharmacoeconomic (PE) and quality of life (QOL) information is being increasingly discussed in formulary and drug use decisions, it has been difficult to identify the extent to which this information influences formulary decision making (Fig. 5.20). Challenges to drug development regarding PE and QOL studies include which ones are required, the designs, their conduct, and their relationship to clinical studies. The industry must challenge itself to perform those studies that are as relevant as possible to the appropriate health settings, use easy to understand methodologies, and publish the information that is most important to health care providers, health systems, and payors. Also, gold standards in study design and application of the data do not generally exist for PE studies, especially given the many different types of studies and varied settings for drug use. When PE and QOL studies are done, the company needs to assist providers and payors in these settings to understand how this information, which may originate from a different setting, fits their institutions and systems. Because the clinical trials usually lack the PE or QOL data, the applicability and integration of both the clinical studies and the PE or QOL studies for a new product need to be addressed to also assist the payors and providers [9–12, 18].

Pharmacogenetics is a relatively new and complex discipline based on heritable or acquired genetic differences between groups of people that can change a drug's actions in the body (Fig. 5.21). About 60,000 single nucleotide polymorphisms exist on the coding regions of the human genome, and about 1.5 million exist in the full genome, creating a plethora of

- Lack of understanding of applications of PE studies
- Sub-optimal use of health outcome and PE data by health care systems
- Studies or analyses needed:
  - Easier to understand
  - Relevant to health systems
- Clinical studies lack PE content
- Unique settings impacting PE assessments
- Lack of all necessary data available
- Lack of integrated health care data

FIG. 5.20. PE Application Challenges

- Goals:
  - Identify therapy that will have high likelihood of success in groups of patients
  - Achieve improved individual responses
  - Reduce use of ineffective treatments
  - Reduce adverse events
  - Reduce cost of drug development with more efficient trials
  "Individualized Therapy"

FIG. 5.22. Pharmacogenetic Studies

- Potential disadvantages:
  - Smaller target population with reduced sales
  - Cost of genotyping
  - Additional patient consent
  - Unclear clinical significance
  - Diagnostic and assay dilemma
  - Ethics including impact on insurance coverage

FIG. 5.23. Pharmacogenetic studies

potential differences between patients' biology. This figure suggests the scope and some of the complexity of genetic variations in a cancer patient with acute lymphocytic leukemia (ALL). The cancer genotype, especially related to surface antigens, will vary in patients with the same disease and alter patient response to therapy, which is now well documented for aggressive breast cancers and Herceptin® (trastuzumab), acute myelogenous leukemia and Mylotarg® (gemtuzumab), and colorectal cancer and Erbitux® (cetuximab). Host susceptibility has genetic variation, as well as infection defense mechanisms. Drug metabolism is particularly effected by genetic variation in liver enzymes and drug clearance. In the pediatric cancer, ALL, the appropriate use (dose) of thiopurine is very dramatically changed downward tenfold by genetic variation, potentially leading to possibly fatal toxicities [19–22].

Increasing emphasis is being placed on "personalized medicine." The major goals of pharmacogenetic studies in drug development are to identify therapies that will have a high likelihood of success in individual patients and/or reduced toxicity (Fig. 5.22). An improved drug responsiveness has been demonstrated in a subpopulation of breast cancer patients with particularly aggressive cancer; that is, Herceptin® therapy significantly increases cure rates in patients with her2neu oncogene in about 25% of breast cancer patients. Another goal is to reduce use of treatments that would be ineffective in a subgroup of patients that we would know would not respond to the treatment. The current alternative is using a drug in a 100 patients, in which the response rate is 50%, but we do not know which 50 patients will be responders. Better dose selection would be possible with either less toxicity or better efficacy through genetics (e.g., thiopurines in cancer and narcotic analgesics, respectively). Reduction in the cost of drug development could be an outcome with more efficient trials; that is, products are only used in smaller groups of patients with higher likelihood of response rates to even higher degrees [19–22].

There are several potential disadvantages in incorporating pharmacogentics studies into drug development process (Fig. 5.23) [18–21]. The diagnostic use of genetics is not yet commonplace, related to, for example, the lack of knowledge of impact of genetics in many diseases, cost of tests, availability and reliability of tests, and unknown reimbursement by payors.

- Term "Compassionate" is not in IND regulations. Emergency Use and Treatment INDs
- Emergency Use - use of investigational drug or biological product in life-threatening situation when no standard acceptable treatment is available, and there is insufficient time to obtain IRB approval.
- Allows for one emergency use with out prospective IRB review. Any subsequent use of investigational product at institution requires prospective IRB review and approval.
- Company provides product after phase 2 or likely phase 3 with regulatory consent for this procedure to be done

FIG. 5.24. Compassionate Use

Other disadvantages include a smaller target population for only the genetically likely responders with reduced sales, the cost of genotyping, a need for additional and frequently separate patient consent, unclear clinical significance of pharmacogenetics to disease pathogenesis and product pharmacology, and possible ethical issues including an impact on insurability. Furthermore, large epidemiologic studies examining the associations of pharmacogenetics to diseases and with drugs are needed, which is a huge expense. This deficit is starting to be addressed in NIH funding.

The FDA has published a voluntary guidance for companies regarding the use of pharmacogenetic studies in the drug development process, their role in the approval process, and how to submit the data for its review [23].

The term "compassionate" is not in the IND regulations. "Compassionate use" studies are either "emergency use" protocols or "Treatment INDs" (Fig. 5.24). The emergency use provision governs the use of an investigational drug or biological product in a life-threatening situation when no standard acceptable treatment is available and in which there is insufficient time to obtain institutional review board (IRB) approval before treatment must be started. This provision allows for one emergency use without prospective IRB review. The use must be reported to the IRB according to federal and local requirements. Any additional use of the investigational product requires prospective IRB review and approval. This emergency use is normally only done after phase 2 is

5. Types of Clinical Studies

- Make new drugs available to desperately ill patients early in drug development process.
- Need preliminary evidence of drug efficacy & safety, documentation drug is intended to treat a serious or life-threatening disease, and no alternative therapy available to treat that stage of disease in intended patient population.
- Patient is not eligible to be in definitive clinical trials, which usually must be well underway (e.g., during phase 3), if not almost finished.
- Enables FDA to obtain additional data on drug's safety and effectiveness.

FIG. 5.25. Treatment IND

- **Definition:**
  - Observational studies of larger size
  - Prospective or retrospective
  - Longitudinal over specified time
  - Hard end point, well-defined
  - Sites more in community settings
  - Comparator control groups
  - Information sources: databases, patients, registries, charts/medical histories
  - More representative population in real world
- **Goals:**
  - Study target disease
    e.g., risk groups, disease descriptors, practice patterns, market/population size
  - Estimate rates of background events, e.g., adverse events
  - Design large simple post-marketing safety trials

FIG. 5.26. Epidemiology Studies – Definition and Goals

complete and ideally phase 3 is done or almost complete, so that a reasonable idea of both safety and efficacy exist.

For a company with such a life-saving product, emergency use is not often able to be accomplished, because of the time required to document the patient's need and diagnosis, investigator's/practitioner's credentials, and the distribution requirements for the product. These issues are not vicarious requirements from a company but minimum regulatory and especially safety issues. Usually when it is done, a protocol is created in advance to cover this usage including approval by regulatory authorities. The necessary inclusion and exclusion criteria created for this protocol can present a barrier to such open-ended use, because the individual patient and family may have an expectation of availability of the product, but the patient may not qualify. This situation can become a possible public relations boon or fiasco, which is a practical challenge to control expectations.

The Treatment IND provision makes new drugs available to desperately ill patients early in the drug development process (Fig. 5.25). Approval of a treatment IND requires preliminary evidence of drug efficacy, documentation the drug is intended to treat a serious or life-threatening disease, there is no alternative therapy available to treat that stage of the disease in the intended patient population, and the patient or patient population is not eligible to be in the definitive clinical trials. The clinical trials program usually must be well underway, (e.g., during phase 3, if not almost finished). A Treatment IND also enables the FDA to obtain additional data on safety and effectiveness. A protocol must be written by the company and approved by the regulatory authorities for this usage. Also, regulatory provisions allow a company to charge the health care system for this usage, but the company needs to share costs of production and research costs with the regulatory authority, which is proprietary information.

Epidemiology studies use observational study designs in large populations (e.g., hundreds to thousands of patients) to improve our understanding of diseases and therapies (Fig. 5.26). The sources of information about the patients, diseases, treatments, and events include large databases, such as Medicaid claims data, patient interviews (in person, telephone, mail, or Internet), patient registries, and medical record reviews. Each data source has its limitations, which will qualify the results and conclusions. For example, databases can be influenced by restrictions in formulary status, treatment guidelines in place, or age of the population exposed. Interviews are susceptible to patient memory lapses and their reliability as historians. The end points are definitive (e.g., hospitalization, death, heart failure, or gastrointestinal bleed). The data can be gathered retrospectively (e.g., chart reviews) or prospectively. The patients for epidemiology studies are found most often in community practice settings, and as a result they are more representative of the "real world." Control groups, used for comparison, are usually drawn from the same population as the patients exposed to the disease and or treatment. These characteristics should result in a representative population sample, studying typical patients receiving typical treatments in typical health care settings.

The goals of these observational studies are threefold: (1) to study the target disease, which can provide information on patients at risk for an exposure or reaction, disease or patient descriptions to be used for inclusion or exclusion criteria in other studies, practice patterns in diagnosis and therapy, and population (market) sizes; (2) to estimate rates of background events, especially adverse events, helping to identify reactions in a population and the influence of disease, risk factors, or treatment; and (3) to design large simple post-marketing surveillance trials for safety assessments.

Two study designs are predominant in epidemiology research: cohort and case-control (Fig. 5.27). A cohort is a group of patients with similar characteristics, also described as patients with an exposure to a specific product in drug studies. Two cohorts, with and without exposure (often a drug), are followed over a specific time period and compared for adverse events or practice patterns or to estimate a specific reaction, which may be rare and difficult to quantify in smaller randomized trials. Case control design involves a group of cases as defined by an exposure and set of characteristics and

- **Cohort design:**
  - Group of patients with exposure to specific products
- **Case control design:**
  - Cases of patients with a specific exposure or reaction and matched controls from same population
- **Biases to prevent; limits to observational studies:**
  - Sampling vs Selection vs Measurement vs Confounding
  - Information: subject not remember information
  - Reverse causality: exposure related to outcome
  - Detection: preferential diagnosis or selection of exposed subject
  - Healthy patient: health status influences outcomes
  - Channeling: disease severity masks drug-disease association
  - Confounding: Variable must be associated with exposure and outcome, and can not be an effect of exposure

FIG. 5.27. Epidemiology Studies – Design Issues

- 10,000-30,000 substances identified in basic research
- 100-200 reach chemical synthesis and screening
- 5-10 undergo pre-clinical testing
- 2-5 enter clinical trials
- 1 is approved and marketed

FIG. 5.28. Failure Rates of New Chemical Entities

- 33% enter phase 2
- 27% enter phase 3
- 20% undergo FDA review
- Not all that undergo review are approved

FIG. 5.29. Failure Rates of INDs

a group of matched control subjects, both selected from the same population.

Observational studies have potential biases that must be considered and either dealt with in the design and/or used as qualifications to the results and conclusions. *Information bias* involves missing information because patients do not remember events or data are missing from charts. *Reverse causality bias* is when the exposure (drug) is unknowingly used to treat an adverse event related to the outcome, such that epidemiologists will define exposures where timing does not coincide with the outcome. *Detection bias* occurs when an outcome is preferentially diagnosed in subjects who are exposed to a drug associated with the outcome. Matching well the patient characteristics, diagnoses, and other nonstudy exposures will help minimize this problem. *Healthy patient bias* is seen when a patient's health status (e.g., exercise or diet) influences the outcome and biases the result. One compensates for this problem through study design and observation or statistical analysis with stratification. *Channeling bias* occurs when the severity of a disease either masks or enhances the association between a drug exposure and the disease. Mitigation of this bias requires knowledge of the disease and modification of the study design. *Confounding bias* occurs when an external variable is mixed with the exposure and influences the outcome under study. Epidemiologists deal with such potential bias by stratification by the confounding variable or statistical analysis using multivariate analysis.

Summary

A wide variety of clinical trials intended to demonstrate safety and efficacy, which have complementary and at times overlapping goals, are required to obtain approval from the FDA or comparable government agency to use a new drug or biological product. This is an expensive and lengthy process—the largest percentage of the cost of drug development is for clinical studies, about 50% of the total cost. As we have discussed, the studies have specific stages with specific requirements; have many special design features to be used; must be acceptable to not just clinicians and investigators but also to the regulatory authorities regarding medical benefits and scientific rigor; were conducted following good clinical practice guidelines for the patients, the sites, the investigators, and the company; and demonstrate real clinical differences to reasonably meet unmet clinical needs and for competitive advantage. Further, the clinical studies must meet the needs of the marketing teams to generate data and information to help convince providers to use the company's product and payors to pay for it. As stated in a recent "white paper" from the FDA [26], novel approaches to shortening the phases of clinical testing and reducing the cost are essential if the discoveries being made in the laboratory are to be translated into improvements in preventing, diagnosing, treating, and curing disease.

Failure rates of new chemical entities (NCEs) are actually series of failures at the various stages of research at a company that are to be expected and even can be a desirable outcome (Fig. 5.28). A company does not have all the resources to advance all compounds and must be selective to advance the best compounds in activity at each stage of development. In basic research, 10,000–30,000 new substances are identified, which have increased with genome screening. Then, about 100–200 molecules reach chemical synthesis and screening. At the next step, about 5–10 undergo preclinical testing in animals. Within a family of compounds (product candidates), only 2–5 enter clinical trials. Finally, 1 is approved and marketed.

Failure rates of INDs occur commonly during the clinical phases of product development (Fig. 5.29). Another way to look at this situation is to consider that of all drugs that enter phase 1 testing in humans, 1 in 3 enters phase 2 testing, 1 in

- Inadequate characterization of dose-response profiles (peak response and time-course of response during dosing interval)
- Flaws in study design or drug development plan - inappropriate studies, difficult to interpret studies, studies based on unfounded assumptions
- Inadequate characterization of the benefit-risk profile
- Inadequate proof of improved quality of life or pharmacoeconomic benefit
- Audits of study conduct or sites find major flaws

FIG. 5.30. Why are drugs not approved?

4 enters phase 3 testing, and only 1 in 5 undergoes FDA review. The reasons for a company terminating the IND include safety issues (20% of the time), lack of sufficient efficacy (38%), economic reasons, that is, the product is too expensive to manufacture, or the potential sales are too low to justify the high expense and risks of further development, (34%), and others (9%). In addition, even at the terminal end of clinical research phase with reporting of all studies and filing the NDA, not all that undergo FDA review are approved.

An important question is why are drugs not approved, especially because over the past decade the regulatory authorities, especially the FDA, have worked more closely with companies at various stages in the drug development process, providing feedback on study design and data generated (Fig. 5.30). Most products that enter clinical trials are not approved because they fail to demonstrate sufficient efficacy or have substantial toxicity. Specifically, one finds inadequate characterization of dose-response profiles (peak response and time course of response during dosing interval), flaws in study design or drug development plan (e.g., inappropriate studies, difficult to interpret studies, studies based on unfounded assumptions, inappropriate dosing for the drug or disease), flaws in the conduct of the study and data collection (e.g., study sites not following exclusion criteria, too much missing data), inadequate characterization of the benefit-risk profile, inappropriate statistical analyses (e.g., insufficient statistical power with too few patients enrolled for the desired extent of change in end points), adverse events of a new product exceeding existing therapies beyond any added efficacy, and inadequate proof of quality of life or pharmacoeconomic benefit.

References

1. Phase 1 Clinical Trials. FDA Handbook. Accessed at http://www.fda.gov/cder/handbook/phase1.htm on February 12, 2005
2. Phase 2 Clinical Trials. FDA Handbook. Accessed at http://www.fda.gov/cder/handbook/phase2.htm on February 12, 2005
3. Phase 3 Clinical Trials. FDA Handbook. Accessed at http://www.fda.gov/cder/handbook/phase3.htm on February 12, 2005
4. E6 Good Clinical Practice: Consolidated Guidance. ICH. April 1996. Accessed at http://www.fda.gov/cder/guidance/959fnl.pdf on February 12, 2005
5. E10 Choice of Control Groups and Related Issues in Clinical Trials. ICH. May 2001. Accessed at http://www.fda.gov/cder/guidance/ 4155fnl.pdf on February 12, 2005
6. Temple RJ. Active control non-inferiority studies: theory, assay sensitivity, choice of margin. February 10, 2002. Accessed at http://www.fda.gov/ohrms/dockets/ac/02/slides/3837s1_02_Temple/sld001.htm on February 12, 2005
7. Center for Drug Evaluation and Research. Procedures for tracking and reviewing phase 4 commitment. Accessed at http://www.fda.gov/c der/mapp/6010-2.pdf on February 12, 2005
8. Cramer JA, Spilker B. Quality of Life and Pharmacoeconomics: An Introduction. Philadelphia: Lippincott-Raven. 1997
9. Balekdjian D, Russo M. Managed care mandate: show us the value. Pharmaceutical Executive 2003;23 (Special report)
10. Adapted from Venturini F, Johnson K. Introduction to Pharmacoeconomic Principles and Application in Pharmacy Practice. Can J Hosp Pharm 2002;55(2):
11. Louie SG. How to conduct a pharmacoeconomics study. Educational monograph. USC School of Pharmacy. 2002
12. ACCP – Pharmacoeconomics and Outcomes, Applications for Patient Care (Educational program, Module 1, 2, 3) American College of Clinical Pharmacy Kansas City, MO
13. Ware JE, Sherbourne CD. The MOS 36-item short-form health survey (SF-36). I. Conceptual framework and item selection. Medical Care 1992;30:473-483
14. Jones PW, Quirk FH, Baveystock CM. The St George's Respiratory Questionnaire. Respir Med 1991;85(Suppl B):25-31
15. Cella DF, Tulsky DS, Gray G, Sarafian B, Linn E, Bonomi A, Silberman M, Yellen SB, Winicour P, Brannon J. The functional assessment of cancer therapy scale: development and validation of the general measure. J Clin Oncol 1993;11:570-579
16. Juniper EF, Guyatt GH, Epstein RS, Ferrie PJ, Jaeschke R, Hiller TK. Evaluation of impairment of health related quality of life in asthma: development of a questionnaire for use in clinical trials. Thorax 1992;47:76-83
17. Juniper EF, Guyatt GH, Feeny DH, Ferrie PJ, Griffith LE, Townsend M. Measuring quality of life in children with asthma. Qual Life Res 1996;5:35-46
18. The value of providing quality-of-life information to managed care decision makers. Drug Benefit Trends 2001;13(7): 45-52
19. Shah J. Economic and regulatory considerations in pharmacogenomics for drug licensing and healthcare. Nature Biotechnology 2003;21(7):747-753
20. Bernard S. The 5 myths of pharmacogenomics. Pharmaceutical Executive 2003;23(10):70-78
21. Weinshilboum R, Wang L. Pharmacogenomics: bench to bedside. Nature Reviews Drug Discovery 2004;3(9):739-748
22. Lesko LJ, Woodcock J. Translation of pharmacogenomics and pharmacogenetics: a regulatory perspective. Nature Reviews Drug Discovery 2004;3(9):763-769
23. Center for Drug Evaluation and Research. Pharmacogenomic data submission. November 2003. Accessed at http://www.fda.gov/cder/guidance/5900dft.doc on February 12, 2005

24. Faich G, Stemhagen A. Epidemiology and drug development—a primer. DIA Forum 2004;40(3):32-34
25. Etminan M, Samii A. Pharmacoepidemiology I: a review of pharmacoepidemiologic study designs. Pharmacotherapy 2004;24(8):964-969
26. Food and Drug Administration. Challenge and opportunity on the critical path to new medical products. Accessed at http://www.fda.gov/oc/initiatives/criticalpath/whitepaper.html on February 12, 2005

6
Metabolism and Pharmacokinetics

Jun Shi, Shashank Rohatagi, and Vijay O. Bhargava

Introduction 123
Value of Pharmacokinetics (PK) and Pharmacodynamics (PD) 124
PK/PD Concepts 126
Drug Development Value Chain 132
References 147

Introduction

The role of metabolism and pharmacokinetics, under an industrial context, is to address the question of which compound should be selected for development among multiple candidates and how the compound should be dosed. As a discipline, pharmacokinetics (PK) is the study of what the body does to the drug, that is, the absorption, distribution, metabolism, and excretion (ADME) of the drug, whereas pharmacodynamics (PD) seeks to define what the drug does to the body, that is, the exposure and the response relationship. The integration of PK and PD in drug development from early to late stage can guide the decision-making process on lead generation, optimization, and product realization.

This chapter as outlined above first provides an overview about the value of PK/PD in drug development; second, it discusses the key PK/PD concepts; third, it presents the key PK/PD and metabolism studies in each developmental stage along with case studies; finally, it summarizes the regulatory expectations on PK/PD in drug development.

Pharmacokinetics is a discipline that characterizes the relationship between dose and concentration, whereas pharmacodynamics characterizes the relationship between the drug concentrations in either plasma or biophase and drug responses, including both beneficial and adverse effects.

Drug development is a sequential process involving iterative learn and confirm cycles. The strategy of the developmental value chain from discovery to preclinical through phase I to phase III and beyond is to develop and utilize new technologies, *in vitro* or animal models, that are less expensive and predictive of human pharmacokinetics *in vivo* and to maximize the information gained in humans to support the drug label. Figure 6.1 provides an overview of the phases of drug development and some of the key *outcomes for metabolism and pharmacokinetics* (MPK) in the four areas of target and compound selection, safety margin, proof of concept and dose ranging, and confirmation of safety and efficacy. The major responsibility of a drug metabolism and pharmacokinetics function within a pharmaceutical industry is to manage the exposure data generated along the developmental value chain in the four stages of research and development outlined on figure 6.1, each of which will be elucidated further in this chapter. The integration of PD information using biomarkers, surrogate markers, and clinical end points from early stage to late stage represents a more efficient and effective drug development paradigm (i.e., model-based drug development). PK/PD bridging becomes a common approach implemented in many stages or areas of drug development (e.g., bridging preclinical to clinical [allometric scaling], bridging old formulation to new formulation [*in vitro–in vivo* correlation, or IVIVC], bridging old region to new region, bridging old population [adults] to new population [pediatrics]).

Several abbreviations are presented in this figure and are described below:

C_p: Plasma concentration of a drug
C_e: Drug concentration in the effect compartment
EC_{50}: Drug concentration that produces 50% of the maximal effect
IC_{50}: Inhibitory drug concentration that produces 50% of the maximal effect
MIC: Minimum inhibitory concentration
NOAEL: No observed adverse effect level

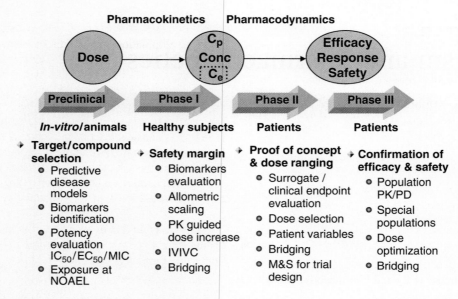

Fig. 6.1. Introduction of PK/PD-Timeline/Roles

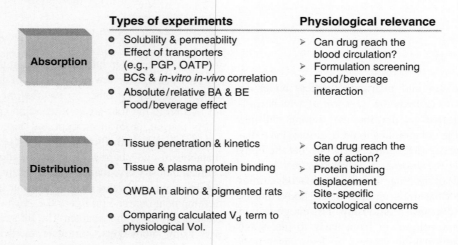

Fig. 6.2. Absorption, Distribution, Metabolism & Elimination (ADME) -1

Value of PK/PD

In drug development, the *ADME processes* of a compound are experimentally determined (Figs. 6.2 and 6.3). Absorption studies are conducted to answer the basic question whether the drug can reach the systemic circulation from the site of administration. There are several factors that can influence the absorption of the drug after oral administration. The drug needs to have a high solubility and high permeability in order to be absorbed adequately. A Biopharmaceutics Classification System (BCS) was proposed to classify drug molecules into one of four classes based on their solubility and permeability through the intestinal cell layer. The combination of BCS and *in vitro–in vivo* correlation (IVIVC) improves the efficiency of the drug development and review process—a class of immediate release (IR) solid oral dosage forms for which bioequivalence (BE) may be assessed based solely on *in vitro* dissolution results (biowaiver). *In vitro* transport studies can be conducted to evaluate the involvement of efflux pumps (P-glycoprotein) or certain molecules that serve as ligands for membrane pumps (OATP, OAT, etc.) to transport drugs across the gastrointestinal tract. Except transporters, first-pass effects including intestinal and liver metabolism and certain forms of bile excretion can affect the amount of drug eventually reaching the blood circulation. In addition, food can also affect the drug absorption. Required by the regulatory agencies, food-effect studies become standard PK trials in the industrial development of orally administered drugs.

Using radiolabelled material, quantitative whole-body autoradiography (QWBA) provides a rapid, cost-effective, and accurate assessment of the tissue distribution of radioactivity.

6. Metabolism and Pharmacokinetics

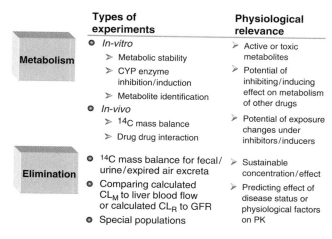

FIG. 6.3. Absorption, Distribution, Metabolism, & Elimination (ADME) -2

The study demonstrates if drug-related material reaches a target organ (e.g., CNS) and identifies sites of accumulation or unusual persistence. The measurement of protein binding is also important as it is postulated that only the unbound component of the drug is pharmacologically active and can be removed from the body. In *in vivo* studies, the apparent volume of distribution determined for a compound is a direct measure of extent of distribution, and it should be compared against the physiological volumes of plasma, extracellular space, and total body water in the corresponding species.

Metabolism studies (Fig. 6.3) identify the potential metabolites that may be active or even toxic, identify enzymes involved in the metabolism of new chemical entities (NCEs), determine the rates of these enzyme reactions, and demonstrate the inhibition or induction potential of NCEs on the enzymes. The importance of drug metabolism is twofold: (a) drugs can be extensively metabolized by a specific enzyme or by several enzymes; (b) drugs can also affect the activities of the enzymes by either decreasing their intrinsic activity (inhibition) or increasing the amount of available enzyme (induction). The consequences of the drug metabolism can lead to significant drug-drug interactions resulting in either loss of efficacy or toxicity. Sometimes, the rate and extent of drug metabolism is directly related to the efficacy of the drug, such as in the case of prodrug or pharmacologically active metabolites (e.g., terfenadine to fexofenadine, loratadine to desloratadine, leflunomide to teriflunomide). In some cases, the absence of certain drug metabolizing enzymes or significantly reduced capacity in certain subjects can have profound effects on elimination of drugs that are primarily metabolized by the enzymes leading to toxicity. In other cases, biotransformation of drugs can also lead to formation of reactive intermediates or metabolites that interact with endogenous macromolecules, such as proteins and nucleic acids. It is, therefore, important not only to study the parent compound but also to study the active/toxic metabolites during drug development.

Drugs are primarily eliminated by the feces or via the urine, and this can be determined by mass balance studies. The other routes of elimination such as via the lungs or biliary excretion can also be determined. The drug's clearance (CL) in PK studies can be compared with blood flows through the liver and kidney and glomerular filtration rate (GFR). Function of the excretory organs, especially the kidney, can have dramatic impact on organ physiology and a drug's elimination and its half-life, leading to persistent effects and lower dose needs or toxicity. PK work in this situation may be important element of product development for patient safety. Predicating elimination of a drug based on various excretory studies is a key part of MPK's contribution to product development, product dosing, and safety.

"Exposure" to a drug as defined by either plasma concentration or a surrogate of concentration, such as AUC (area under the concentration-time curve) or C_{max} (maximum drug concentration), can be correlated to a pharmacodynamic response (Fig. 6.4), either efficacy or safety data, using the following PK/PD models according to the data types (continuous or categorical), the time course of response relative to concentration, and the shape of the curve when plotting response against concentration.

- Linear or log linear: The model assumes that the effect will continuously increase with increasing concentrations.
- E_{max} or sigmoidal E_{max}. The model describes the interaction between small molecules such as drugs and large molecules such as receptors or enzymes including the shape of the response, the baseline effect, or the maximal possible effect.
- Indirect link or indirect response: Indirect link uses a hypothetical effect compartment model to accommodate the drug distribution to the biophase. Indirect response model is used if the rate-limiting step is a postreceptor event. For indirect link models, time for maximal effect ($T_{max,e}$) is independent of dose whereas for indirect response models, $T_{max,e}$ increases with increasing dose.
- Logistic: These models can correlate frequency of a categorical response to the drug concentration or dose.

It is important that there is sufficient characterization of the following parameters:

- Baseline effect: A physiological parameter is evaluated and quantified without drug dosing. Baseline can change due to circadian rhythm (e.g., circadian rhythm of cortisol or melatonin levels), food, or disease.
- Biomarker: It is a quantifiable physiological or biochemical marker that is sensitive to intervention (drug treatment). Biomarker might or might not be relevant for monitoring clinical outcome, usually used in early drug development. Validation of its relevance to disease outcomes is needed.
- Surrogate marker: If a biomarker has been shown to reflect clinical outcome, it can be called surrogate marker; for example, HIV load in AIDS patients, blood sugar in diabetes patients, FEV1 in asthma patients, and urine NTx or

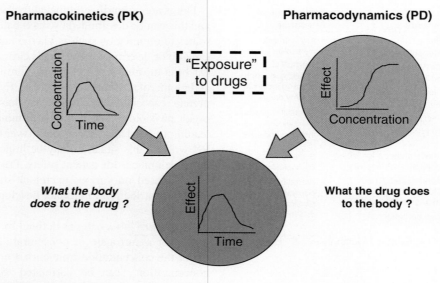

FIG. 6.4. Pharmacokinetic/Pharmacodynamic

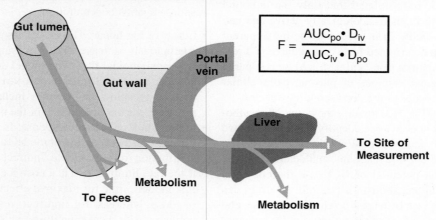

FIG. 6.5. Bioavailability and Bioequivalence
Source: Adapted from Rowland M and Tower N, Clinical Pharmacokinetics 3rd Ed., 1995. Lippincott Williams and Wilkins.

CTx (N- or C- telopeptide cross-links), and bone mineral density for osteoporosis.
- Clinical end point: A characteristic or variable that measures how a patient feels, functions, or survives and directly relates to disease outcome. However, assessment is often difficult to perform requiring a large number of patients and/or longer time frame for significant change and/or consensus of its relevance to meaningful disease change [1].

PK/PD Concepts

The oral *bioavailability* is the function of the fraction of absorption, the fraction undergoing first-pass metabolism, the fraction of loss due to efflux, and the fraction of degradation (Fig. 6.5) [2]. The amount of the drug reaching the site of measurement is the bioavailable portion of a dose administered.

It is important to distinguish between the terms *bioavailability* (BA) and *bioequivalence* (BE). Bioavailability determines the amount of drug that is absorbed in the bloodstream as compared with a standard (i.e., after intravenous administration). Absolute oral bioavailability is usually calculated as the ratio of the exposure as determined by area under the concentration-time curve (AUC) compared with the same parameter after intravenous administration that is assumed to be 100%. The relative bioavailabilty is the AUC ratio of a test formulation to a reference formulation. The comparison could be of a tablet versus a capsule or a solution and so on. If the 90% confidence interval of the ratio (ratio of the least square means using log transformed data) is contained within the limits of 0.8–1.25, then the formulations are

6. Metabolism and Pharmacokinetics

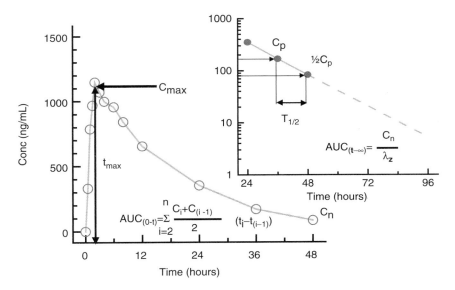

FIG. 6.6. Single Dose Pharmacokinetics

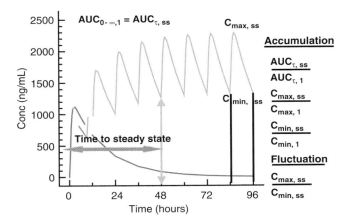

FIG. 6.7. Multiple Dose Pharmacokinetics

deemed to be bioequivalent. The bioequivalence study is the primary clinical study that a generic company has to conduct in order to get an approval when the drug patent expires. The above criteria and the two one-sided test procedure are the important means to prove that the generic formulation is bioequivalent to the innovator drug. Similar approach is also applied for drug–drug and drug–food effect studies.

Besides oral bioavailability, bioavailability can also be determined after administration via other extravascular routes, such as inhalation, transdermal, or subcutaneous injection.

Following *single dose* administration via extravascular route, the maximum concentration is defined as C_{max}, and the time to reach maximal concentration is defined as T_{max} (Fig. 6.6). The concentrations decrease in a first-order fashion, which implies that the decrease in concentration over time is dependent on the previous concentration. This type of reduction in drug concentration brings up the concept of half-life, which is defined as the time taken for the concentration to fall by one-half. The half-life can be determined as $\ln 2/\lambda z$, where λz is the slope of the terminal disposition phase on log-transformed concentration data. The area under the plasma concentration versus time curve from 0 time up to the last measurable concentration is determined by the trapezoidal rule. The AUC extrapolated to infinity is calculated as $C_{last}/\lambda z$, where C_{last} is the last observed concentration.

Following *multiple dose* administrations, it takes about four half-lives to reach clinical steady-state (i.e., 93.75% of the true steady-state) (Fig. 6.7). If the drug behaves in linear kinetics, the area under the plasma concentration-time curve up to infinity after a single dose is equal to the area under the curve for a dosing interval at steady state ($AUC_{0-\infty,1} = AUC_{\tau,ss}$). Accumulation after multiple drug dosing can be defined as the ratio of either AUC, C_{max}, or C_{min} at steady state for a dosing interval to the corresponding AUC, C_{max}, or C_{min} after single dose for the same time interval. The fluctuation is the ratio of the maximum concentration and minimum concentration at steady state.

Generally speaking, an ideal dose regimen should give both low fluctuation and low accumulation for the drug. Also, it is important to note that the time to steady state depends solely on the half-life of the drug, while the average steady-state concentration depends on the clearance of the drug and the dosing rate.

If the concentration of an NCE at any given time is proportional to the dose of the drug administered, then the PK of that drug is dose proportional (Fig. 6.8). *Dose proportionality* is necessary for linear kinetics, which implies that any concentration-time profile normalized for time and dose is superimposable. Nonlinear kinetics implies that concentration-time profiles are not superimposable due to either dose or time dependencies. The common mechanism for nonlinear kinetics is saturation in one or multiple ADME processes as described below:

1. Nonlinear absorption:
   - Saturable active GI transport (e.g., riboflavin, levodopa, β-lactam antibiotics)
   - Poor aqueous solubility or slow release (e.g., griseofulvin, phenytoin)
   - Saturable presystemic metabolism (e.g., propranolol, telithromycin)
2. Nonlinear distribution:
   - Saturable protein binding (e.g., prednisolone/prednisolone)
   - Saturable red blood cell binding
   - Saturable tissue binding (e.g., paclitaxel)
3. Nonlinear elimination
   - Saturable elimination (e.g., phenytoin, theophylline)
   - Saturable renal elimination
   - Cofactor depletion (e.g., glutathione depletion after acetaminophen overdose)
   - Mechanism-based inhibition (e.g., clarithromycin due to the formation of a stable metabolite-intermediate complex)
4. Autoinduction (e.g., rifampicin, many antiepileptics)

Lack of dose proportionality does not imply a failed compound but it has important implications with regard to safety or efficacy, depending on the mechanism involved, when adjusting dose is needed clinically. For a drug that processes a saturable absorption, efficacy can become a concern. For a drug that shows a saturable elimination, safety is a concern, especially when a drug has a narrow therapeutic window. Nonlinear kinetics usually implies larger inter-subject variability in pharmacokinetics and less predictable drug activity for a given dose across patients or in the same patient at different doses. Ideally, the drugs are easier to manage clinically if their PK are linear at their therapeutic dose range.

Although *pharmacokinetic drug-drug interactions* (DDI) can occur at any process of ADME, metabolism has been the primary site or mechanism for many clinically important drug interactions (Fig. 6.9). The emphasis of drug-drug interaction studies is on NCEs with a narrow therapeutic index and primarily metabolized via one metabolic pathway and also on potent enzyme inhibitors or inducers. Drug metabolism is primarily mediated by phase I CYP family of isoenzymes (cytochrome P450 enzymes) in the liver, which includes 1A1/2, 2D6, 3A4, 2C8/9/19, and to a small extent, 2B6. The relative amount of the isoforms of the CYP 450 enzymes is listed in this figure [3]. Phase II metabolism is also common by N-acetyl-transferase liver enzymes (NAT 1/2). Genetic polymorphism of these isozymes and DDI are common sources of variability. Drugs may be metabolized by more than one enzyme. For example, tricyclic antidepressants are metabolized by CYP2D6, CYP3A4, and CYP1A2. Also, (S)-warfarin is metabolized by CYP2C9 and (R)-warfarin metabolized by CYP3A4 and CYP1A2. Genetic absence of one isoenzyme can lead to compensation through the secondary isoenzyme pathway.

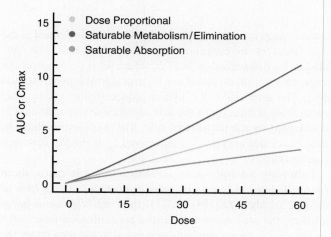

FIG. 6.8. Dose Proportionality: Linear vs. Nonlinear

- **Gastrointestinal absorption**
  - Binding/chelation of drugs
  - GI emptying/motility
- **Plasma & tissue protein binding**
- **Drug transporters**
  - Pgp, OCT, OAT, & OATP
- **Metabolism**
  - Primarily mediated by CYP isozymes: 1A1/2, 2D6, 3A4, 2C8/9/19
  - Genetic polymorphism CYPs & DDI common sources of variability
  - Drugs may be metabolized by more than one enzyme
  - Drug may induce/inhibit 1 isoenzyme but may not be its substrate
  - Inhibition or induction of an interacting drug may or may not result in clinically significant interaction
- **Excretion**

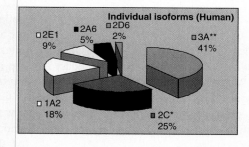

FIG. 6.9. Sites of Pharmacokinetic Drug Interactions
Source: Li AP. Advances in Pharmacology: Drug-Drug Interactions. Scientific and Regulatory Principles. 1997:43:189. Academic Press, San Diego, CA

6. Metabolism and Pharmacokinetics

CYP 450 Isozyme	Substrates	Inducers	Inhibitors
CYP 1A2	Caffeine	Omeprazole	Alfa-naphthoflavone
CYP 2A6	Coumarin	Barbiturates	Tranylcypromine
CYP 2C9	Tolbutamide	Rifampin	Sulphaphenazole
CYP 2C19	(S)-mephentoin	Rifampin	Tranylcypromine
CYP 2D6	Dextromethorphan	None	Quinidine
CYP 2E1	Chlorzoxazone	Isoniazid	Disulfuram
CYP 3A4	Testosterone	Carbamazepine	Ketoconazole
	Midazolam	Dexamethasone	Itraconazole
		Phenobarbital	
		Phenytoin	
		Rifampin	

FIG. 6.10. CYP450 Drug-Drug Interaction Probes

Inhibition or induction of an interacting drug may or may not result in a clinically significant interaction. Drug(s) may induce/inhibit one isoenzyme but may not be a substrate of it (quinidine is an inhibitor of CYP2D6 but a substrate of CYP3A4). Drug-drug interactions (DDI) can be of following types: inhibition or induction. Inhibition is the more common form of DDI.

- Enzyme inhibition: Decreased enzyme activity due to direct interaction with the drug or its metabolite(s)
  - Competitive inhibition: Inhibitor and the substrate compete for the same binding site on an enzyme. Inhibitor may be the substrate itself.
  - Noncompetitive inhibition: Inhibitor binds at a site on the enzyme distinct from the substrate.
  - Uncompetitive inhibition: Inhibitor binds only to the enzyme substrate complex.
  - Mechanism-based (or suicide) inhibition: Substrate (inhibitor) gets transformed by the enzyme to intermediate(s), which can react with the active site of the enzyme or inactivate the enzyme.
- Enzyme induction: Interaction may affect efficacy of one or more medications. Enzyme induction involves protein synthesis, therefore, requires multiple dose administration to realize.

Many phase I and phase II enzymes are inducible (e.g., CPY3A4, UDP-glucuronosyltransferases), but some are not (CYP2D6). The major enzyme that is known to be induced is CYP3A, and examples of drugs known to induce CYP3A include carbamazepine, phenytoin, rifampin, and phenobarbital. Enzyme induction potential in human is difficult to assess preclinically due to lacking of predictability of animal data.

Absorption, interaction with transportor (e.g., P-gp), elimination, and protein binding based drug-drug interactions are also possible, but they are more infrequent or less well studied.

Drug metabolism and interaction studies are usually realized using appropriate probes [4, 5]. These *probes*, which are drugs with known actions, are classified as substrates, inducers, and inhibitors of various drug metabolizing enzymes or transporters (Fig. 6.10). For *in vivo* studies, selectivity, sensitivity, safety, and availability of the probe compounds are the major factors to be considered. For *in vitro* tests, the choices are a little more broad: if recombinant enzymes (isoenzyme specific) are used, the probes can be less selective. Human liver microsomes are preferred in the *in vitro* tests. Via the probe approach, an NCE's metabolic pathways and its interaction potential can be assessed *in vitro* initially and ultimately confirmed clinically via human DDI studies. The results gathered are critical in the decision-making in drug discovery and development. An NCE that is subjected to drug interactions as a strong inducer or as substrate that is primarily metabolized by a single enzyme (e.g., CYP2D6 or 3A) may be screened out early (development stopped) because drug interactions likely will be common and significant, as long there are other similar leads with more diverse metabolic profiles (i.e., metabolized by multiple metabolic and other elimination pathways). Other types of drug-drug interactions involving non-CYP enzymes (e.g., flavin monooxygenases; FMO), drug-transporters, protein binding, or absorption and elimination related are also possible. MDR1 (P-glycoprotein) is an efflux transporter that can actively extrude or pump drugs back into the intestinal lumen, thus affecting the oral bioavailability of drugs such as paclitaxel, digoxin, and protease inhibitors [6].

When elimination occurs via a single metabolic pathway, individual differences in metabolic rates based on *pharmacogenomics* can lead to large differences in drug and metabolite concentrations in the blood and tissue. Figure 6.11 presents several CYP450 isozymes responsible for metabolism of drugs along with the proportion of drugs metabolized by particular CYP isozyme, the allele variants, and the clinical impact [7]. In some instances, differences exhibit a bimodal distribution indicative of a genetic polymorphism for the metabolic enzyme (e.g., CYP450 2D6, CYP450 2C19, *N*-acetyl transferase). When a genetic polymorphism affects an important metabolic route of elimination, large dosing adjustments between patients may be necessary to

Enzyme	% of Drug Metabolism	Allele Variants[b]	Clinical Effects
CYP1A2	5	CYP1A2*1K	Less enzyme expression & inducibility
CYP2A6	2	CYP2A6*4, CYP2A6*9	Altered nicotine metabolism
CYP2B6	2-4	-	Metabolism of cancer drugs
CYP2C8	1	CYP2C8*3	Altered Taxol metabolism
CYP2C9	10	CYP2C9*2, CYP2C9*3	Drug dosage
CYP2C19	5	CYP2C9*2, CYP2C9*3	Drug dosage, Drug efficacy
CYP2D6	20-30	CYP2D6*4,*10,*17, *41	No response, Drug efficacy
CYP2E1	2-4	-	No conclusive studies
CYP3A4	40-45	Rare	No conclusive studies
CYP3A5	<1	CYP3A5*3	No conclusive studies

FIG. 6.11. Pharmacogenetics & Pharmacogenomics
Source: Influences on Pharmacologic Responses. http://medicine.iupui.edu/flockhardt/

achieve the safe and effective use of the drug. Pharmacogenetics and pharmacogenomics are the sciences of understanding the correlation between an individual patient's genetic makeup (genotype) and their response to drug treatment. They already have influenced therapeutics. For a drug that is primarily metabolized by CYP2D6, approximately 7% of Caucasians will not be able to metabolize the drug, but the percentage for other racial populations is generally far lower. Similar information is known for other pathways, prominently, CYP2C19 and N-acetyl transferase. For example, codeine is metabolized to its active molecule, and about 10% of the population are rapid metabolizers and only need a much smaller dose for the same pharmacodynamic outcome. Omeprazole, used to treat peptic ulcers, is poorly metabolized related to SNPs in the CYP2C19 liver enzyme in 2.5–6% of Caucasians and 15–23% of Asians. For thiopurine, an antimetabolite used in cancer chemotherapy, the dose is 1/10 for the poor metabolizers, which constitute about 10% of patients related to SNPs in the N-acetyl transferase (phase II) liver enzyme [8, 9].

Genetic polymorphism is almost predominantly associated with drug metabolism and transporters; renal excretion of drugs does not appear to show genetic polymorphism. Drugs that are predominantly excreted unchanged tend to show much less inter-individual variability in disposition kinetics than extensively metabolized ones. Drug targets (receptors, enzymes, and signal transduction proteins) can have genetic variations and different drug sensitivities (e.g., ACE [angiotensin converting enzyme inhibitors], dopamine 1, 2, and 3, glycoprotein IIIa, and beta adrenergic receptors [BAR]). For BAR and the adrenergic bronchodilators, a fivefold difference in forced expiratory volume is possible because of SNPs [8, 9]. Some drugs work well in some patient populations and not as well in others. Studying the genetic basis of patient response to therapeutics allows drug developers to more effectively design therapeutic treatments. Characterization of genetic polymorphism can (1) improve candidate drug selection, (2) aid in developing new sets of biomarkers to eventually minimize animal studies, (3) help in predicting responders to a drug for enhancing desired effects and minimizing undesired serious side effects, (4) help to rationalize drug dosing, (5) improve patient selection process in studies, (6) reduce variability in drug responses in a study by excluding outliers in drug metabolism, and (7) reduce the number of subjects needed for establishing efficacy helping accelerate drug approval. These features will move from current empirical process to hypothesis-driven mechanism-based process, and thus lower cost and speed up the drug development process. However, the routine use of PG is not yet current medical practice, costs of genotyping adds to health care costs, diagnostic labs need to be better set up for this testing, PG tests need clinical validation, and legal ramifications of genetic information, its availability and use, remain a dilemma. In the field of oncology, genetic testing for responders has been encouraged as in the case of using HER-2 protein overexpression for identifying Herceptin responders.

Several definitions warrant attention on this subject.

- Pharmacogenetics: Study of hereditary variations in drug response.
- Genotype: The fundamental assortment of genes of an individual, the blueprint. Gene typing is a relatively new technique that involves the identification of genes whose expression results in a particular phenotype, such as rapid metabolites and poor metabolizers.
- Phenotype: Outward characteristic expression of an individual. Phenotyping is the expression of a genotype and usually involves ingestion of a test compound followed by serial blood or urine analysis.
- Genetic polymorphism: Defines monogenic traits that exist in the normal population in at least two phenotypes, neither of which is rare (less than 1%).
- Allele: One of two or more different genes containing a specific inheritable characteristic that occupy corresponding positions (loci) on paired chromosome. Dominant allele is expressed and recessive is not expressed.

Population-based PK/PD modeling is conducted to pool several studies with different sampling schemes (rich or sparse) and dose regimens and to describe the typical PK/PD behavior or central tendency of a population of interest. In Fig. 6.12, separate pieces of information (rich PK data, sparse PK data, efficacy data, and safety data, as well as covariate data) are combined into a pooled "mixed data set" for a population. Population-based modeling can produce unbiased estimates of PK or PD parameters, inter-individual variability, inter-occasional variability, as well as random residual variability, and can evaluate the effects of patient demographics, disease conditions, and concomitant medications on the PK/PD of the drug. Population approach allows sample numbers per subject and sample times varying from patient to patient, which fits better to the routine clinical practice or large phase III clinical trials and therefore makes it easier to obtain PK/PD information in the target patient population. Mixed effect modeling is the most commonly applied population-based approach, and it is well established. It is a fundamental tool to characterize the exposure-response relationship and help select the dosage regimen in phase III trials and labeling.

In order to determine an appropriate dose, it is necessary to establish a range of concentrations from minimally to maximally efficacious with tolerable toxicity (minimal effective concentration, MEC, and maximal safe concentration, MSC, or maximal tolerable concentration, MTC, respectively). This range of concentrations, or *therapeutic window,* usually is determined from a concentration-time curve and a dose-response curve generated from a population of patients who have been examined closely for therapeutic and toxic effects (Fig. 6.13). The graphs also may be used to determine the *therapeutic index* (TI), comparing the response versus plasma concentration curves for efficacy and toxicity on the same graph at a 50% response rate (EC_{50}). This useful measure of drug toxicity is calculated by dividing the 50% value from the toxicity curve by the 50% value of the efficacy curve. For example, in this slide, the TI is 6,500 ng/mL divided by 1,000 ng/mL, respectively, or 6.5, which is quite good for a TI. Because these curves are generated from population data, the values may not be applicable for all individuals.

One of the most important goals of PK/PD studies in drug development is to guide the determination of therapeutic dosage regimens in the clinical trials and for labeling of an NCE. To realize this goal, a number of clinical studies are required to be conducted systematically from maximal tolerated dose study (MTD) in phase I, dose ranging study in phase II, and large-scale efficacy and safety studies in phase III. In addition, PK/PD studies are often conducted in special populations for deriving dosage regimen adjustments for these patients. Drug-drug interaction or other interaction studies are also commonly conducted to guide the dosage for special conditions. An appropriate therapeutic dosage regimen is basically derived from the kinds of information

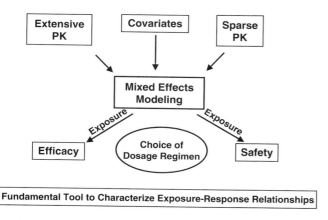

FIG. 6.12. Population PK/PD Modeling

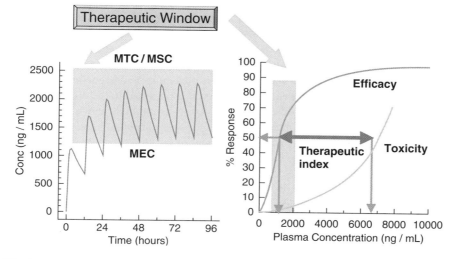

FIG. 6.13. Therapeutic Windows vs. Therapeutic Index

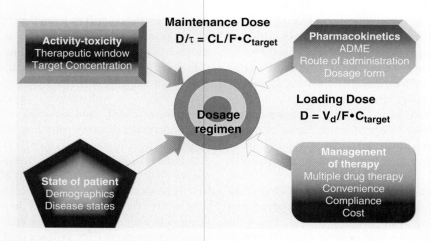

Fig. 6.14. Determinants of a Dosage Regimen

- Discovery
  - Non GLP
  - Several compounds-high volume
- Development
  - GLP
    - Toxicology species
    - Pharmacology (non-GLP) species
    - Human
- Methods
  - Separation
    - GC, HPLC
  - Detection
    - MS, ELISA, fluorescence, UV
- Sensitivity and specificity
- Assay validation

Fig. 6.15. Bioanlytical Method Development

shown in Figure 6.14, that is, the *determinants of a dosage regimen*. One consideration includes the therapeutic window and target concentration that relate both efficacy and safety of the NCE (i.e., its pharmacological response and toxicity to concentrations). Another consideration is how the body acts on the drug and its dosage form, the essence of PK, which helps to derive both loading and maintenance dose regimens. A third consideration is that of the demographics and clinical state of the patient. A fourth consideration includes all other factors such as patient total therapeutic regimen and multiple drug management including drug-drug interactions, convenience, compliance and cost, and so forth. All of these determinants are interrelated and interdependent. The loading doses or maintenance doses can be calculated based on the formulae in this figure using clearance (CL), dose (D), drug distribution (V), and fraction absorbed (F) values.

Drug Development Value Chain

Guidance for industry for bioanalysis exists from the U.S. FDA. *Bioanalytical method* validation states that "Selective and sensitive analytical methods for the quantitative evaluation of drugs and their metabolites (analytes) are critical for the successful conduct of preclinical and/or biopharmaceutics and clinical pharmacology studies" [10]. Figure 6.15 lists five issues for bioanalysis; discovery, development, methods, sensitivity and specificity, and assay validation. Development of bioanaytical methods starts in discovery stage in an early more rudimentary form and evolves through drug development in overall quality and detail of the procedures. Bioanalytical method validation includes all of the procedures that demonstrate that a particular method used for quantitative measurement of analytes in a given biological matrix, such as blood, plasma, serum, or urine, is reliable and reproducible for the intended use. The fundamental parameters for this validation include (1) accuracy, (2) precision, (3) selectivity, (4) sensitivity, (5) reproducibility, and (6) stability. Validation involves documenting, through the use of specific laboratory investigations, that the performance characteristics of the method are suitable and reliable for the intended analytical applications. The acceptability of analytical data corresponds directly with the criteria used to validate the method. These analyses must be designed to be able to be conducted by any technician trained in the discipline.

Also, the methods above apply to bioanalytical procedures such as gas chromatography (GC), high-pressure liquid chromatography (LC), and combined GC and LC mass spectrometric (MS) procedures such as LC-MS, LC-MS-MS, GC-MS, and GC-MS-MS performed for the quantitative determination of drugs and/or metabolites in biological matrices such as blood, serum, plasma, or urine.

Biological products pose additional challenges in bioanalytical development because of their nature; that is, mostly proteins, which have quite complex structures, are processed differently

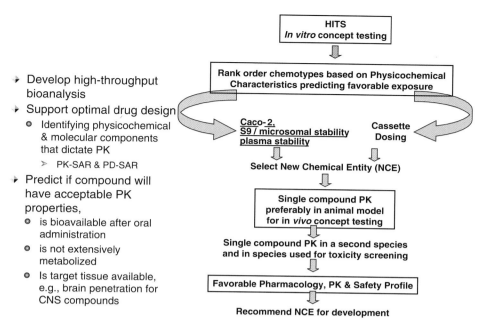

FIG. 6.16. Role of Discovery MPK

than drugs in the human body, may be duplicates of naturally occurring substances, and are susceptible to many degradative processes. Proteins are quite large molecules with specificity of their amino acid sequence, disulfide bridges, tertiary structures (carbohydrates), three-dimensional conformation, isoforms of the same molecule, and other properties. These many structural features necessitate more testing in number, variety, and sophistication to ensure the integrity of the molecule especially in the manufacturing process and complicate measurement in the MPK studies.

In the past, drug discovery focused on finding the most potent lead compounds at a particular target. However, many compounds failed in development due to poor ADME properties. At the *discovery* stage nowadays, *MPK* is used via high-throughput screening to find lead candidates that have "drugability" properties or are "drug-like" to increase the chance of success in the development (Fig. 6.16) [5]. The studies are usually non-GLP compliant. The importance of identifying the physicochemical and the molecular components that dictate pharmacokinetics has been emphasized. The early understanding of the pharmacokinetic-chemical structure-activity relationship (PK-SAR), along with the pharmacological-chemical structure-activity relationship (PD-SAR), will increase the chance of success in finding a good drug candidate. The role of MPK at discovery is to predict if the drug will have acceptable pharmacokinetic properties in man; for example, it is bioavailable after oral administration, it is not extensively metabolized, the target tissue can be reached, pharmacological activity is achievable with blood concentrations that are attainable with reasonable doses. The figure at the right panel of Fig. 6.16 describes the sequence that is generally followed in the selection of lead candidates and characterization of a lead candidate's PK properties. A target is first identified in the disease process, followed by the screening and identification of analogues that modulate the target, which are called "hits." Once a considerable number of hits have been identified, two to four representative hits that show a promising pharmacological profile are selected as lead candidates. At this point, rank ordering takes place, and usually the one lead candidate that shows the most promise is chosen for optimization and assigned as a new chemical entity (NCE), with the others reserved as backups should the lead candidate fail. After this stage, the lead candidate goes through extensive profiling for ADME in parallel with drug safety studies and reconfirmation of pharmacological proof of concept including *in vivo* efficacy in animal disease model(s).

If a compound fails during the drug development process, it is vital that the reasons for failure are clearly understood as early as possible. Understanding the reasons would help in optimizing the appropriate PK/PD or metabolism properties that would enable the next series of compounds to be successful. The schema in Fig. 6.17 shows the sequential algorithm to analyze the various reasons of *failures of drugs* due to pharmacokinetic reasons, such as low bioavailability, short half-life, high or variable metabolism, high first-pass effect, excessive drug interaction potential, or poor tissues penetration. All these can lead to low and transient exposure of the drug, and thus the failure or lack of efficacy could be due to PK reasons. However, if the drug has favorable PK properties and still does not produce the desired effect, then the reason of failure is due to the lack of appropriate pharmacodynamics, such as poor affinity to target receptors or inappropriate target.

The aim of preclinical MPK studies depends on the stage of drug development the compound is at. Eleven possible

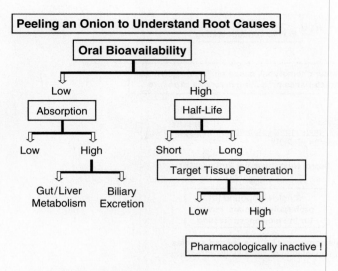

FIG. 6.17. Failure of NCEs

- Mass balance in the toxicology species
  - Metabolite identification and primary routes of elimination
- *In vitro* and *in vivo* metabolic profiling
  - Identification of major active or toxic metabolites or intermediates
  - Explanation of toxicological differences between species
- Identification of major metabolizing enzymes
  - Identification of isozymes involved in formation of metabolites and potential for drug-drug interaction
- Enzyme inhibition and induction
  - Potential effect of drug candidate on other drugs
- Formulation support studies
  - Optimizing drug substance
- PK/PD correlation
  - Rational drug development

FIG. 6.19. Preclinical DMPK to Support Development -2

- GLP bioanalytical methods development
  - Analysis of drug and metabolites
- PK profiling in two species
  - Usually toxicology species - rodent and non rodent
  - Safety margin assessment
- Assessment of PK linearity
  - Prediction of dose related adverse events or lack of efficacy
- Protein binding and erythrocyte-plasma distribution
  - Drug interaction potential
  - Determining the matrix for bioanalysis
- QWBA
  - Target organ concentrations to assess distribution and toxicity

FIG. 6.18. Preclinical DMPK to Support Development-1

types of preclinical studies or research questions are enumerated on these two figures (Figs. 6.18 and 6.19). As much information as possible should be obtained in animals and the laboratory to design the optimal human studies, to screen for the best drug candidates to move forward into humans, and especially to discontinue a molecule as early as possible to create more efficient and cost-effective product development. During development, traditional preclinical pharmacokinetic studies in animals, including toxicokinetics, will be carried out to support filing of an Investigational New Drug (IND) application. Also, *in vitro* studies, such as isolated hepatocytes or purified enzymes, might be used to assess the metabolic clearance of the lead candidate. When the drug enters the clinic, preclinical pharmacokinetics is then used to answer specific questions (e.g., does the compound show a drug interaction?). Hence, preclinical studies help in selecting the first dose in man, selection of the correct dosing regimen, and appropriate interpretation of toxicological studies. They also provide input in helping bridge historical toxicological data to new formulations of drug candidates. Preclinical MPK studies are GLP compliant.

The term *toxicokinetics* refers to the kinetics when compounds are administered to animal models at doses in the range of those used in toxicity studies, while *pharmacokinetics* refers to the kinetics of compounds given to humans or animal models at lower (i.e., pharmacological) doses (Fig. 6.20). Every compound that is identified as a potential lead candidate will undergo a battery of safety/toxicity screens prior to being considered as a NCE. Typical studies include genotoxicity, carcinogenicity, mutagenicity, ion channel safety (hERG potassium channel), reproductive toxicity, and target organ toxicity. The toxicokinetic support for these toxicology studies may help to determine the concentrations that cause toxicity, choose appropriate species for research questions, examine species variability, evaluate exposure-toxicity relationships, assess the safety margin, and define the therapeutic window.

Safety margin of a compound can be expressed as the ratio of drug exposure (C_{max} or AUC) at NOAEL doses in the most sensitive animal species to the corresponding parameter in human at a particular dose (Fig. 6.21). Allometric scaling can be applied to estimate human exposure if this is used for first-in-man dose selection. Modeling and simulation technologies can be used to generate the exposure if particular doses or dose regimens have not been tested in humans. Assumptions on PK linearity and others may be required. Because the safety margin is assessed across different species, the total drug concentrations should be converted to unbound fraction, if there is significant species dependency on protein binding. On the same plot, exposures that produce side effects either benign or serious, such as hERG interaction (toxicokinetic data), and exposures that produce the desired effects (pharmacological data) can also be presented. This kind of plot

Objectives

- Systemic exposure following single & multiple dosing (safety margin of parent & metabolites);
- Dose Proportionality
- Sex related differences in exposure

Types of toxicokinetic studies

- Single dose ranging toxicokinetic studies
- Four-week, 13-week, 26-week toxicokinetic studies in "two species"
- Two year carcinogenicity studies in rats & mice at 3 doses

Role of Toxicokinetics

- Selecting appropriate animal species for toxicity study
  - E.g., same metabolites as human & sufficiently high
- Assisting dose selection
  - E.g., dose non-proportional
- Comparing findings across species
  - E.g., species-specific toxicity
- Comparison of findings from toxicity studies employing different routes
- Relating exposure to toxicity findings
  - E.g., differentiate lack of toxicity from lack of exposure
- Assessing if plasma concentrations change over time course of dosing

FIG. 6.20. Toxicokinetics to Support Development

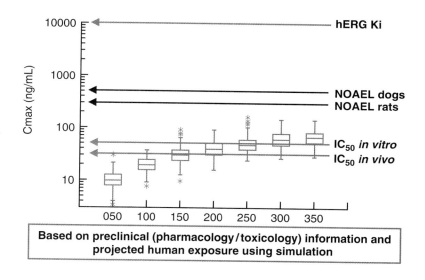

FIG. 6.21. Safety Margin

provides an integrated view on margin of safety and a means of dose finding based on animal pharmacology both *in vivo* and *in vitro*, toxicokinetic data, and human PK.

MPK plays a central role in discovery and preclinical screening phases to identify the ideal physicochemical (PC), bioavailability, biopharmaceutical, pharmacokinetic, and pharmacodynamic characteristics among the candidate compounds. Figures 6.22 and 6.23 present 12 representative possible problems with the *4 ADME areas*, including PC characteristics and their physiologic relevance to product selection in the product development process. Absorption issues revolve mostly around the compound's bioavailability, especially its variability. Distribution examines both protein binding in the blood and tissue effects, which can impact both efficacy and toxicity. Metabolism involves a compound's degradation or activation, including metabolites and their effect on drug interactions, efficacy, and toxicity. Elimination focuses on half-life and dosing impacts. By applying PC profiling, preclinical PK and metabolite screening, and safety evaluations early, it minimizes the probability of candidate failures in clinical development due to poor solubility and stability, lack of high permeability, absorption from the gastrointestinal tract, inadequate PK characteristics, short duration of action, metabolite(s), covalent binding, cofactor depletion, and so on.

A key question in the design of first-in-man studies is how to select an appropriate starting dose: too high a dose may lead to severe adverse events (AEs), and too low a dose may require many dose escalation steps before pharmacological evidence of activity is observed. Safety margin assessment

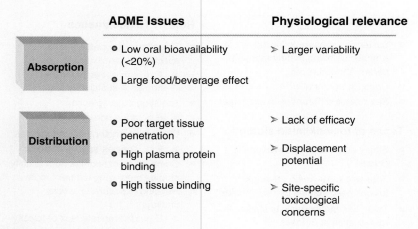

FIG. 6.22. ADME Factors as Development Issues -1

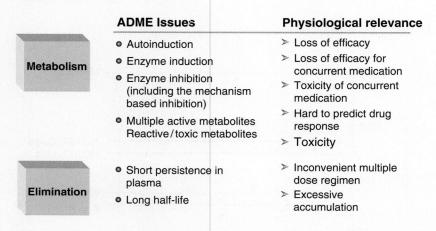

FIG. 6.23. ADME Factors as Development Issues -2

- Dose = CL* $C_{ss\ av}$*tau
- C_{ss} (steady state concentrations)
  - IC50 or EC50 or MIC..... = Cssav
  - EC_{50} in animals, when scaled across species, has shown remarkable correlation with negligible slope for either pharmacodynamics or toxicity, thus *in vitro* EC_{50} or values from animal studies can be used
- Tau can be based on several parameters but usually QD is dosing interval of Choice or 24 hours
- CL needs to be scaled across species. Common methods are:
  - Allometric scaling
  - Campbell method
  - *In vitro* metabolic prediction of CL

FIG. 6.24. First Dose in Man: Allometric Scaling -1

based on the ratio of exposure at NOAEL dose in animals and human exposure at a particular dose estimated according to allometric interspecies scaling may be a useful guide. Figures 6.24 and 6.25 enumerate key principles in determining the *first dose in man* [11, 12].

Interspecies scaling of PK data to predict human PK is based on similarities in physiology and anatomy among species. Allometric scaling can be conducted using the following relationship: $CL = Wt^b$, where the total clearance is scaled based on the body weights of various species. Similarly, volume of distribution can also be scaled, which is generally proportional to the body weight. Generally, the exponent, b, has a value of 0.75 for clearance and 1 for volume of distribution.

The second method that was used was the Campbell method where scaling method uses the body weight and the maximum life span. The projected dose can be calculated according to the equation Dose = $CL \cdot C_{ss} \cdot$ tau, where tau is the dosing interval (24 hours). Because the pharmacological effects have been shown to be similar across species, the *in vitro* IC_{50} can be used for the target concentration for the efficacious dose, while the concentration at the NOAEL (no adverse effect level) in toxicity studies can be used to predict the maximum tolerated dose in humans.

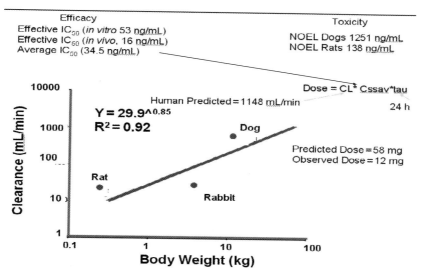

FIG. 6.25. First Dose in Man: Allometric Scaling -2

- **Single Dose Studies - Objectives:**
  - Safety and tolerability in healthy subjects
  - PK and its metabolites in plasma
  - Urinary excretion of the drug and its metabolites
  - PD markers of safety & efficacy
  - Initial insights in to putative human metabolites
  - Initial PKPD modeling
  - Can include effect of gender or food in these studies

- **Multiple Dose Study - Objectives**
  - Safety and tolerability in healthy subjects
  - Steady-state PK parameters such as fluctuation & accumulation ratio
  - Refinement of the PK, PD and PK-PD relationship

- **FIM studies: Oncology & HIV conducted in patient populations.**

FIG. 6.26. First in Human Study: Single/Multiple Doses

Usually, *single dose, first-in-man* (FIM) studies are designed as placebo-controlled, double-blind, randomized, parallel-group studies involving several groups of 8–12 healthy volunteers (males and/or females) that receive escalating doses (Fig. 6.26). Initial doses are based on allometric scaling with at least 1/10 to 1/20 of the NOAEL dose. Dose escalation is usually based on various methods including Fibonacci series or PK/PD driven, where the concentration of the next dose is predicted based on concentration of the prior dose and compared with a target for effective or safe concentration based on animal data. The studies evaluate safety, tolerability, pharmacokinetics, and pharmacodynamics in the first-in-man studies. The stop dose can be based on the maximum tolerated dose or the stop dose criteria based on the exploratory IND guidance by the U.S. FDA [13].

Multiple doses studies are of similar design, but their duration is usually based on the pharmacokinetics of the drug, so that steady state may be achieved on the anticipated duration of responses for the pharmacodynamic marker. These studies usually have 3–4 dose groups and the dose escalation and regimen based on single-dose study. Although these studies are usually conducted in healthy subjects and they can be extended to patient population. FIM studies for oncology and HIV should be conducted in patient populations because of the toxicity of the drugs and to accelerate development of the compound for the potentially life-extending drugs.

The objectives of mass balance studies include recovery of radioactivity in administered dose, excretion routes (urine vs. feces) of radioactivity in administered dose, and metabolite profile of excreta. Mass balance studies are usually single-center, open-label, single-dose studies after oral administration of the intended route in 6–8 healthy male volunteers [14, 15]. ^{14}C is the most common radiolabel used. The amount of radioactivity can not exceed 100 μCi and is based on the dosimetry calculation, taking into account the ^{14}C mass balance studies in two animal species and the animal quantitative whole-body autoradiography (QWBA) data.

In these studies, a series of samples of blood and excreta are collected to assess the distribution (in RBC and protein binding) and elimination of radioactivity after dose administration and to determine the PK. Plasma (blood), urine, and feces samples are collected for up to several days after dose administration, provided that discharge criteria have been met (i.e., all radioactivity is taken into account [>90%]). Metabolic profiling of plasma, feces, and urine is performed to determine the metabolic fate of the drug. The mean (±SD) ^{14}C radioactivity in plasma and blood over time and the mean (±SD) plasma concentration over time profiles of M100240 and MDL 100173 (active metabolite) following oral

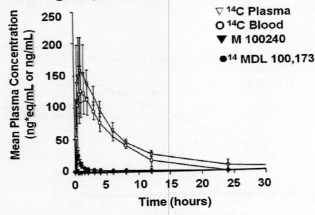

Shah et al. Am J Therap 2003;10:356-362.

FIG. 6.27. Mass Balance Study

- **Objectives**
  - For Innovator compounds BA/BE studies allow bridging the clinical data throughout drug development
  - For Generic drugs BA/BE studies are pivotal data for approval
- **General Features**
  - Two-way crossover design
  - "n" determined by reference treatment variability
  - Study population relatively homogeneous
  - PK sampling adequate to capture early and full exposure metrics
    - C_{max} & T_{max}: rate of absorption
    - AUC: extent of absorption
  - Analysis involves individual PK metric estimation
  - Bioequivalence criteria (Confidence Interval Methods)
    - mean ratio (T/R) and associated 90% CI of AUC & C_{max} are within 80% - 125%
    - Sustained release formulation should include C_{min}

FIG. 6.28. BA/BE Studies

administration of ^{14}C M100240 (25 mg/50 µCi) to 6 healthy male subjects for a *mass balance study* are presented in Figure 6.27.

Bioavailability and bioequivalence studies measure how much of the drug gets into the body and how fast is the absorption (Fig. 6.28). The pharmacokinetic parameter, area under the curve (AUC), explains the extent of absorption, and the PK parameter, C_{max}, explains the rate of absorption. Although T_{max} can also explain the rate of absorption, this parameter is not used for determining bioequivalanece. The role of BA/BE studies in product development differs for innovator versus generic drugs. That is, they are pivotal for approval for generic drugs, but they can serve as bridging studies for new formulations of innovator drugs.

The bioequivalence of the test formulation (test) versus reference formulation (ref) is assessed by examining the logarithmically transformed PK parameters (AUC and C_{max}) using an analysis of variance (ANOVA) model with subject as random effect and treatment regimen as factors. Point estimate and 90% confidence interval are calculated for the geometric mean ratio of the test to reference. If the 90% confidence interval for the geometric mean ratio falls within (0.8, 1.25), then the formulations are considered to be bioequivalent. For sustained-release formulations, the calculations will include the minimum concentrations as well. Other design issues include sample size (n), which is dependent on the treatment variability (more variability means more patients). Study populations are usually quite homogeneous to assist in reducing variability and permitting smaller sample sizes. PK sampling needs to cover early and later metrics for full exposure. Therapeutic equivalence is determined when, instead of pharmacokinetic parameters, clinical or safety end points are used in the calculations.

In the previous sections, most of the discussion of PK properties was limited to the behavior of the drug after oral administration with immediate release formulation, which is usually the most desired route of administration. However, as can be seen in the marketplace, there are various other routes of administration for a drug. Figures 6.29 and 6.30 review three alternative formulations, transdermal, extended release, and inhalation, including key features (advantages and design issues) and a few product examples. For these other formulations, the pharmacokinetics can be used to bridge the information of an existing formulation to develop a new formulation (e.g., extended-release formulation). The extended-release formulation may be desired for a drug with probably a short half-life requiring multiple administrations

Formulation	Key Features	Examples
Transdermal	- Type I – IV patches. - Evaluate release profile of drug from patch. - Conduct skin metabolism & penetration studies. - Compare BA/BE with respect to either IV or oral formulation. - Can bypass first pass effect thus altering metabolic profile.	Progestagel Estrogen Nicoderm Duragesic
Extended Release	- Eliminate multiple doses & increase compliance. - Site specific absorption & metabolism issues. - Drugs with short-half-life & poor absorption good candidates for extended release formulations. - Consistent & predictable performance without dose-dumping, in *vitro/in vivo* formulation correlation. - Effect of food can be substantial.	Albuterol (Volmax), Glipizide (Glucotrol), Oxybutynin (Ditropan)

FIG. 6.29. Alternative Formulation Evaluation -1

Formulation	Key Features	Examples
Inhalation	- Examine lung deposition, & gamma scintigraphy SPECT imaging, & charcoal block studies. - Correlate particle size of formulation with site of absorption. - Consider differences in DPI and MDI, solution and suspension MDI. - Account for extra variability. - Evaluate both oral and inhaled bioavailability; Ideal for drug to have low systemic exposure with low oral bioavailability, high protein binding to avoid systemic side effects. - Conduct lung metabolism studies - Long lung retention times & conduct binding studies in lung or evaluate lung residence time. - Inhaled drugs for systemic use should have low lung metabolism and no long-term toxicity in lungs.	Cortico-steroid Albuterol

FIG. 6.30. Alternative Formulation Evaluation -2

during the day. By improving the formulation, a more convenient and compliance-friendly once-daily or even less frequent formulation (e.g., bisphosphonates for osteoporosis) may be developed.

In order to avoid side effects of the drugs or inordinate first-pass metabolism, drugs may be administered to the target organ directly, such as in the case of inhaled drugs or ophthalmics. The use of inhaled corticosteroids in treatment and management of asthma have significantly reduced the systemic side effects, such as cortisol suppression observed after oral administration. Drugs with poor and variable oral bioavailability due to the first-pass effect can be challenging (e.g., selegiline). However the proposed use of selegiline in a transdermal patch can not only reduce the variability but also has led to the investigations for possible use of the drug in new indications such as Alzheimer disease [16].

Objectives of *dose proportionality* include dose-exposure relationship, changes in ADME in relation to dose, and accumulation of multiple doses. Dose proportionality of a dose-dependent PK metric implies that the surrogate measure divided by dose (e.g., dose-normalized AUC) is independent of dose (Fig. 6.31). An analysis of variance (ANOVA) can be performed on the log-transformed, dose-normalized surrogates (i.e., AUC and C_{max}), using dose as a fixed effect rather than a continuous variable. If a crossover design is applied, then a repeated measures ANOVA would be used. If the F-test for the treatment effect is not statistically significant, one

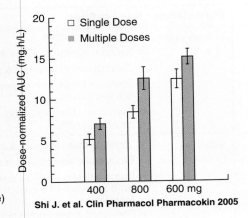

FIG. 6.31. Dose Proportionality Study

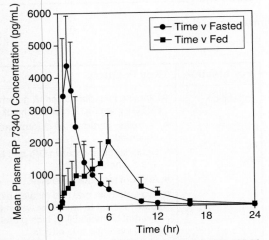

FIG. 6.32. Effect of Food
Reprinted with permission from Dustri-Verlag, Rockledge, FL.

would conclude dose proportionality. In the telithromycin example shown in the figure, the statistic test indicated that the AUC was deviated from dose proportionality at the doses tested, possibly due to a saturable first-pass metabolism of the drug. Another commonly used dose-proportionality assessment method is to use a power model, which assumes that log-transformed surrogate is linear to log-transformed dose: $Ln(S) = Ln(a) + b \cdot Ln(dose)$. The surrogate data after log transformation are fit to the power model using a mixed effect model. When b is tested not significantly different from 1, one could conclude dose proportionality. The graph shows a dose normalized AUC versus dose for both single and multiple doses for telithromycin (Fig. 6.31), which indicates some deviation from the dose proportionality [17].

Normally, early first-time-in-man studies offer a good opportunity to examine the widest dose range, while the targeted crossover study designs with carefully considered washout periods between doses are the best for evaluating within and between subject variability, thus, the most robust way to test dose proportionality. Dose proportionality is an important measure of the predictability of PK when dose adjustment is needed [18].

Administration of food may change the pharmacokinetics of a drug by possibly delaying gastric emptying, changing the gastric pH, increasing bile flow and splanchnic blood flow, and changing lumenal metabolism (Fig. 6.32). Food can also physically or chemically interact with a dosage form of a drug, and thus food can either increase or decrease the bioavailability or delay the absorption of the drug. The effects of food on BA depend on the physicochemical (solubility) and pharmacokinetic (site, rate, and extent of absorption, first-pass metabolism) properties of the drug and on the dissolution of the drug substance from the drug product. The information derived from the food interaction study can (a) optimize the formulations for early and mid-stage developmental compounds, (b) enable well-designed late-stage clinical trials, (c) provide prescribing options to physician for optimal patient compliance, and (d) avoid excess variability in drug responses or even toxicity. In the figure, the graph demonstrates exemplary impacts of food on drug absorption, with a C_{max} over twofold higher and a T_{max} of about 1 versus about 7 hours in the fasted versus fed states; [19].

Drugs such as ampicillin, aspirin, tetracyclines, and warfarin have reduced drug absorption, and drugs such as acetaminophen, diclofenac, digoxin, and valproate have delayed absorption. Drug absorption of diazepam, propranolol, griseofulvin, and carbamazepine are increased with administration of food, while absorption of oxazepam, tolbutamide, telithromycin, and propoxyphene remains unaltered. The bisphosphonates have a class label that requires patients take the drug first thing in the morning before any food and drinks. The drug in this class can form a complex by chelating calcium or other divalent minerals in the food or drinks, and the absorption of the drug is greatly dampened.

The most common effect of food study is a study evaluating the effect of a high-fat meal. This study is a crossover study where drug is administered to healthy male and female volunteers in fed or fasted state. Plasma samples are collected for 24 hours and the effect on the PK parameters, especially AUC and C_{max}, is evaluated using the bioequivalence criteria (however, a wider 90% CI, i.e., 70–143%, may be set for C_{max}). Other food-effect studies include time of administration of meal with respect to food study and special diet study such as for diabetics.

The *drug-drug interaction studies* are conducted to evaluate the PK and PD effects but primarily are designed to evaluate for metabolism-based PK drug-drug interaction, because metabolism changes by far are the most common of the mechanisms for DDI (Fig. 6.33). The following factors should be carefully evaluated when design a drug-drug interaction study: study population (healthy vs. patients), study design (crossover vs. parallel group; fixed sequence vs. randomized crossover), dose regimen (single or steady-state studies, dose and duration, timing of co-administration), mechanism of interaction (PK vs. PD; inhibition vs. induction), PK/PD characteristics (e.g., linear vs. nonlinear kinetics, presence of active metabolites, or delayed pharmacological response), wide or narrow therapeutic index, blinded or unblinded if a pharmacodynamic or safety outcome is to be assessed, and how the study will be interpreted in the product label. The bottom line is that the study should be able to maximize the probability for an interaction, yet still ensure the safety of the study subjects.

Several common *study designs* are used for DDI studies (Fig. 6.34). About 70% DDI studies used a one-way fixed-sequence crossover design by administering multiple doses of both victim drug and perpetrator drug to steady state. Such designs are best in mimicking the clinical therapy or clinical practice. The fixed sequence designs include the following, in which drug A is the theoretical victim drug and drug B is the perpetrator drug:

- Randomized crossover: Drug A (period 1) followed by drug A and drug B (period 2) or drug B (period 1) followed by drug A and drug B (period 2).
- One-sequence crossover: Drug A (period 1) always followed by drug A and drug B (period 2) or the reverse; this can be extended to 3 periods by giving drug B alone in period 2, and then drug A and drug B in period 3.
- Parallel design: Drug A in one group of subjects and drug A and drug B in another group of subjects (period 1).

Sometimes, both drug A and drug B can be victim drugs, or if the victim drug is not clear prior to the study, the design can be more complicated.

The bioequivalence criteria are used for AUC and C_{max} to show if there is a significant interaction. There is also a possibility of conducting a multiple probe or "cocktail" study that would evaluate the effect of the perpetrator drug on various CYP isozymes. Phase III population PK analysis is useful to confirm there is no large and unexpected DDIs. Drug interaction studies are important in drug development, especially

- **Objectives**
  - Effect of the NCE on the PK/PD of other drugs
  - Effect of the other drugs on the PK/PD of the NCE
- **General Features**
  - Healthy subjects, relative homogenous
  - Single (long $t_{1/2}$) or multiple dose (short $t_{1/2}$)
  - Dose to maximize probability for an interaction yet still ensure safety of subjects
  - Rationale for selecting interacting drugs:
    - Mechanistic understanding for potential interaction based on *in vitro*, preclinical and human MPK and safety data
    - Co-prescribe potential
  - Often conducted in parallel with Phase II or III trials
  - BE criteria for AUC & C_{max} to show if interaction exists

FIG. 6.33. Drug-Drug Interaction Study

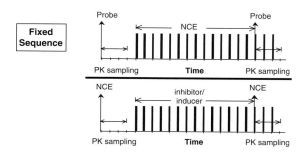

NCE – At standard or at highest relevant dose consistent with volunteer safety.
Probe CYP Substrate – Tolerated dose of probe that has adequate sensitivity for detecting activity of relevant CYP pathway.
CYP Inducer/Inhibitor – Standard dose used in similar studies in literature.

FIG. 6.34. Recommended Clinical Study Designs

	C_{max}	(90% CI)	AUC	(90% CI)
% Change in PK of telithromycin in presence of co-administered drug				
CYP3A4 inhibitors Ketoconazole	↑51	(↑12–105)	↑95	(↑50–152)
CYP3A4 inducers Rifampicin	↓79	(↓63–90)	↑86	(↑69–91)
% Change in PK of co-administered drug in presence of telithromycin				
CYP3A4 substrates Simvastatin	↑433	(↑134–654)	↑761	(↑661–874)
CYP2D6 substrates Metoprolol Midazolam	↑38 ↑162	(↑27–50) (↑53–302)	↑37 ↑511	(↑21–55) (↑278–1079)
CYP1A2 substrates Theophylline	↑17	(↑7–26)	↑17	(↑10–26)
CYP2C9 substrates R-warfarin S-warfarin	↑11 ↑12	(↑4–17) (↑5–9)	↑20 ↑5	(↑14–25) (↑1–9)
PGP substrate Digoxin	↑73	(↑59–89)	↑37	(↑32–42)

FIG. 6.35. DDI studies in telithtomycin program

where concurrent drug therapy is common in clinical practice, and they will improve product labeling and resultant use by clinicians to avoid problems.

A comprehensive *DDI program* was implemented in *telithromycin* clinical development. Telithromycin is a CYP3-PGP substrate, also a strong inhibitor to CYP3A, and a mild inhibitor to CYP2D6 (Fig. 6.35) [17]. The figure displays seven types of interactions with representative drug examples and includes the impact on C_{max} and AUC. For example, an interaction with midazolam on CYP3A results in a 162% increase in C_{max} and 511% increase in AUC of midazolam; another interaction with metoprolol on CYP2D6 enzyme results in a 38% change in C_{max} and a 37% change in AUC of metoprolol. Another interaction with digoxin on PGP efflux pump results in a 73% change in C_{max} and a 37% change in AUC of digoxin. Telithromycin was a significant advance in antimicrobial therapy with improved spectrum of activity and less drug resistance and hence marketed, even with many DDIs.

Based on the U.S. FDA guidance, a PK study in patients with impaired renal function is recommended when *renal impairment* is likely to significantly alter the PK of a drug and/or its active/toxic metabolites, and a dosage adjustment is likely to be necessary for safe and effective use in such patients (Fig. 6.36). Drug clearance during hemodialysis is another important renal MPK study. Additional rationale for renal workup is a drug with narrow therapeutic index or its metabolites. The graph in this figure shows the impact on plasma levels of sotolol with varying degrees of renal function [redrawn based on the data, Ref. 20].

Full design for a renal impairment study would include eight or more subjects in each of the categories of renal impairment, and no effect would be based on the bioequivalence criteria when compared with healthy control subjects in the study. A partial design is only conducting the study with any one of the categories and may also be considered, but it would impact the drug label with limited information for clinical drug use. Subjects are considered to have normal renal function if their creatinine clearance (CL_{Cr}) values are >80 mL/min. The subjects with CL_{Cr} >50 to 80 mL/min, >30 to 50 mL/min, and <30 mL/min are considered to have mild, moderate, and severe renal impairment, respectively. Subjects undergoing hemodialysis can also be included in the study as a separate group.

Based on the U.S. FDA guidance, a PK study in patients with *impaired hepatic function* is recommended if hepatic metabolism and/or excretion accounts for a substantial portion (>20% of the absorbed drug) of the elimination of a parent drug or active metabolite; if the drug and/or active metabolite is eliminated to a lesser extent (<20%) but its labeling or literature sources suggest that it is a narrow therapeutic range drug; if the metabolism of the drug is unknown and other information is lacking to suggest that hepatic elimination routes are minor (Fig. 6.37).

A full study design would be used to develop specific dosing recommendations across the entire spectrum of hepatic impairment, a study should be carried out in patients in the three Child-Pugh categories (mild, moderate, and severe), as well as healthy controls. For this study design to provide evaluable data, at least six subjects in each arm should be evaluated. A partial design may also be considered in some cases. This figure displays a concentration-time curve for ranolazine and the impact of liver disease on plasma levels at day three [21]. Moderate liver disease slows the clearance of

6. Metabolism and Pharmacokinetics

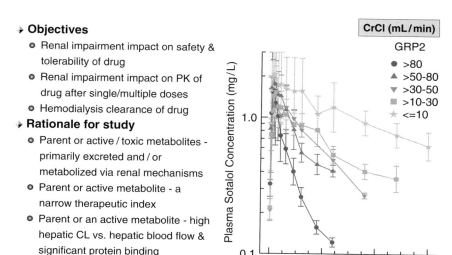

- **Objectives**
  - Renal impairment impact on safety & tolerability of drug
  - Renal impairment impact on PK of drug after single/multiple doses
  - Hemodialysis clearance of drug
- **Rationale for study**
  - Parent or active / toxic metabolites - primarily excreted and / or metabolized via renal mechanisms
  - Parent or active metabolite - a narrow therapeutic index
  - Parent or an active metabolite - high hepatic CL vs. hepatic blood flow & significant protein binding
  - Impact of dialysis is of interest

FIG. 6.36. Renal Impairment

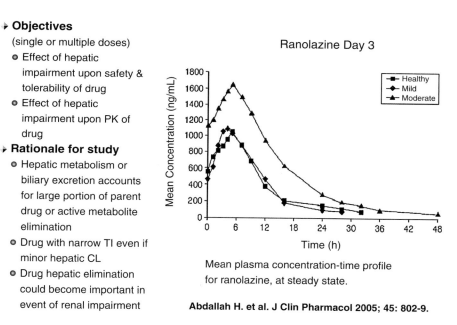

- **Objectives** (single or multiple doses)
  - Effect of hepatic impairment upon safety & tolerability of drug
  - Effect of hepatic impairment upon PK of drug
- **Rationale for study**
  - Hepatic metabolism or biliary excretion accounts for large portion of parent drug or active metabolite elimination
  - Drug with narrow TI even if minor hepatic CL
  - Drug hepatic elimination could become important in event of renal impairment

Mean plasma concentration-time profile for ranolazine, at steady state.

Abdallah H. et al. J Clin Pharmacol 2005; 45: 802-9.

FIG. 6.37. Hepatic Impairment (Reproduced with permission from Sage Publications, Thousand Oaks, CA.)

ranolazine and causes a higher C_{max} and AUC, but mild liver disease has no impact on hepatic clearance.

Age and gender are the most common sources of variability for the pharmacokinetics and pharmacodynamics of a drug and can be determined in a standalone study or via population PK/PD analysis (Fig. 6.38). A typical phase I study uses 3 separate groups of 12 or higher number of males and females (preferably of equal proportions): young (18–45 years), elderly (65–75 years), and elderly (>75 years), which is demonstrated in the graph [22]. Single or multiple doses are employed. Measurement of both PK and PD includes urine PK and creatinine clearance. Age and CL_{cr} can be correlated. ANOVA is used to determine the effect of age and gender on the PK and or PD parameters. Blood samples are usually collected for pharmacokinetic analysis up to 24 hours post oral administration. It is possible that age and gender may affect the PK and not the PD or vice versa.

Children cannot simply be regarded as "miniature" adults: they differ from adults and even from other children in regard to drug absorption, distribution, metabolism, and elimination. Furthermore, age-related differences in receptor binding characteristics are also evident, resulting in different drug

- **Common source of variability**
- **Study Design: Separate study**
  - 3 groups of 12 or more, males and females
    - Young: 18-45 years
    - Elderly (2): 65-75 & >75 years
  - Single or multiple doses
  - Measure both PK & PD including urine PK, CL_{Cr} and other covariates
  - ANOVA to determine effect of Age & gender on PK / PD parameters
- Can be also determined via population PK / PD analysis
- Age & gender may affect PK & not PD, or vice-versa

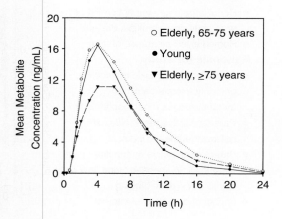

FIG. 6.38. Effect of Age and Gender (Reproduced with permission from AAPS Annual meeting 1998, Arlington, VA)

- Children may differ from adults
- PK/PD studies with relevant biomarkers central to bridge adult safety & efficacy data to pediatrics
- Appropriately applying a size adjustment approach critical in dose selection of trials & PK/PD modeling
- Limited sampling design coupled with mixed effects modeling (population approach) frequently used feature in pediatric trial & data analysis
- Necessity of dose adjustment for pediatric subpopulations should be judged based on whether or not PK difference will likely lead to clinically significant PD difference

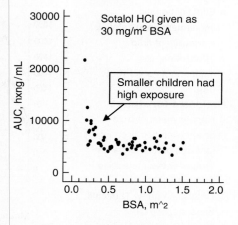

Shi J et al J Pharmacol Pharmacokin 2001

FIG. 6.39. Pediatric PK/PD Study

effects. *PK/PD pediatric studies* play a key role in clinical programs and are a central contributor to define pediatric dose adjustments specified in the product labeling. The graph (Fig. 6.39) in this figure shows substantial differences in AUC for different BSA (body surface areas in meters squared) for sotalol [23].

Populations in pediatric PK/PD studies frequently cover a much wider range in body size than similar studies in adults. Therefore, appropriately applying a size adjustment approach is critical in dose selection of the trials, optimal sampling design and PK/PD modeling for other covariates, and, ultimately, in dosage regimen recommendations. Limited sampling designs are a frequently used feature in population PK/PD analysis in pediatric populations. Sufficient methodology is now available to allow for the design of Dose-optimality or random sampling based schemes and validation of these schemes. Furthermore, reliable and unbiased results can be obtained using various Bayesian and nonlinear mixed effects modeling approaches, even though the data is sparse and unbalanced. Population PK/PD models, if used as bridging to recommend dose, should be carefully validated or evaluated.

The recent regulatory initiatives and policies have stimulated pediatric clinical studies resulting in improved understanding of the PK/PD of drugs prescribed in pediatrics.

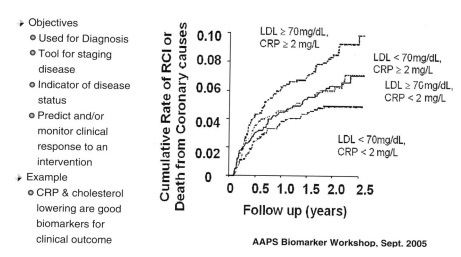

FIG. 6.40. Biomarker Development
Reprinted with permission from American Association of Pharmaceutical Scientists. Arlington, VA.

- Formalizes the Decision Making Process
  - Assumptions ⇒ Opinions ⇒ Decisions
  - Knowledge repository of disease, patient population, drug action, and trial designs
- Enhances Communication
  - (Clinical) Development Team
  - Regulatory Agencies
  - Senior Management

FIG. 6.41. Modeling and Simulation in Drug Development

The pursuit of relationships between systemic exposure and both response and toxicity specifically in pediatric populations represents the frontier in limited sampling design, population PK/PD modeling, and dose optimization. The integration of model-based techniques as a tool in these investigations is both rational and necessary.

Biomarker is a characteristic that is objectively measured and evaluated as an indicator of normal biologic processes, pathogenic processes, or pharmacologic responses to a therapeutic intervention. An example would be low-density cholesterol (LDL) and CRP (C-reactive protein) for heart disease, as represented in Fig. 6.40, which displays death rates over time with different levels of CRP and LDL [24]. Pharmacogenomics guidance further defines possible, probable, and known valid biomarker categories depending on available scientific information on the marker. Biomarkers are used to diagnose disease and predict response to therapies. Their development often is a stepwise progression throughout the research development process. At the preclinical stage, early methods are developed in animals, including correlation of animal response data with the marker, correlation of toxicity to drug exposure, and developing the bioassay parameters and variables for the biomarker. In phase II stage, the proof of concept should be accomplished, with data in healthy volunteers, clinical data in patients with the target disease (e.g., predictability of biomarker to disease outcomes, subpopulations of importance), and fine tuning of the test methodology and validation of the biomarker. In phase III, biomarker evolves to become a surrogate marker, necessitating full validation (phase III data for outcomes and biomarker correlation, full test criteria), full regulatory engagement for U.S. and worldwide approvals, and labeling considerations.

Recently, there is an increasing interest on implementation of *modeling and simulation* (M&S) to improve the efficiency of the efficacy and safety assessment of new chemical entities and to aid decision-making in preclinical and clinical development (Fig. 6.41). The contribution of pharmacokinetics/pharmacodynamics (PK/PD) in drug development was well established in the early 1990s. Since then, the exposure-response modeling and population approach has been expanded to clinical trial simulation. Companies now have large repositories of information for diseases, MPK data, PD data, trial designs, and patient populations, which can be used in the M&S. After 10+ years of practice of this technology, the value of M&S in expediting drug development and reducing the cost has been recognized. Acceptance by regulatory authorities for M&S is now available, as long as the negotiation for an NCE is done in advance and well accepted by the FDA in the development plan. Mathematical modeling, decision analysis, and simulation are powerful approaches that can be employed to enhance critical path drug development. The modeling and simulation need concurrence of the development team and senior management on its value and need to fit into the development plan and all the study results, followed by communication with and acceptance by regulatory authorities.

The focus of MPK package in a NDA is to address the FDA's approach of question-based review. The fundamental *regulatory expectations* are presented in Figs. 6.42 and 6.43. What is the dose versus systemic exposure relationship for drug and its metabolites? How are the responses (efficacy and/or

- What is the dose versus systemic exposure relationship for drug and its metabolites?
- How are the responses (efficacy and/or adverse effects) in relation to the dose and/or plasma drug concentration?
- How does exposure vary
  - In the presence and absence of intrinsic (age, gender, race, renal or liver function…)?
  - In presence and absence of extrinsic factors (food, concomitant drugs, smoking…)?

FIG. 6.42. Regulatory Expectations: Key PKPD Topics

Small Molecules:
- FDA Regulatory Guidances
  - FIM and Exposure Response
  - Population PK / PD and In vitro Metabolism
  - BA / BE and Pediatric
  - In Vitro In Vivo Correlation
- ICH Guidances
  - QTC

Biological Products:
- Manufacturing process is crucial.
- Combination of several conformations (e.g., isoforms) that cannot be distinguished by standard analytical methods.
- Metabolic breakdown primarily driven by breakdown of proteins.
- Production of antibodies, especially neutralizing antibodies
  - Can affect ADME, toxicology, and efficacy of compound
  - Show dramatic inter-species differences; thus, identifying animal models for toxicology and even pharmacology is difficult.

FIG. 6.43. Regulatory Expectations

adverse effects) in relation to the dose and/or plasma drug concentration? How does exposure vary in the presence and absence of intrinsic (age, gender, race, renal or liver function . . .) or in presence and absence of extrinsic factors (food, concomitant drugs, smoking . . .)?

For small molecules, FDA guidance primarily addresses the following six areas: FIM and exposure response, population PK/PD, *in vitro* metabolism, BA/BE, pediatric information, and *in vitro–in vivo* correlation. ICH guidances add QT_C. For biological products, manufacturing process is crucial. Combination of several related conformations (e.g., isoforms) cannot be distinguished by standard analytical methods. Metabolic breakdown is primarily driven by breakdown of proteins. Production of antibodies, especially neutralizing antibodies, can affect ADME, toxicology, and efficacy of compound and show dramatic inter-species differences; thus, identifying relevant animal models for toxicology and even pharmacology is difficult [25, 26].

Drug development is a sequential process—from discovery to preclinical through phase I to phase III and beyond including PK studies as well, as displayed in this *MPK summary* in Fig. 6.44. PK work involves *in vitro* work during discovery stage, followed by preclinical development with substantial animal research, next clinical phase I in healthy normal subjects, and then phases II–III in patients with the target disease. The role of metabolism and pharmacokinetics is to address the question of which compound should be selected for development among multiple candidates and how the compound should be dosed. The strategy of developmental value chain from early discovery to late-stage development is to develop and utilize new technologies *in vitro* or animal models that are predictive of human absorption, distribution, metabolism, and metabolism. The integration of pharmacokinetics/pharmacodynamics has recently become an important component of drug development programs to understand the drug action on the

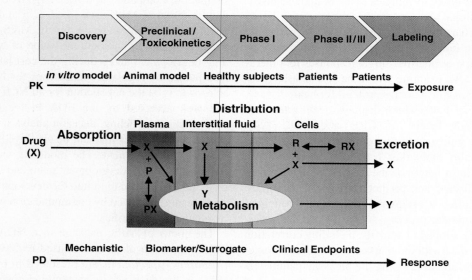

FIG. 6.44. Role of Metabolism & Pharmacokinetics

body. The ability to correlate drug exposure to effect and model it during the drug development value chain, from mechanistic studies in preclinical models, biomarkers in phase I, surrogate markers in phase II, and clinical end points in phase III, provides valuable insight into optimizing the next steps to derive maximum information from each study. PK/PD analysis and modeling and simulation is becoming increasingly important in drug labeling due to its potential for predicting drug behavior in populations that may be difficult to study in adequate numbers during drug development and due to its value in optimizing clinical trial designs.

References

1. Rohatagi S, Barrett J. Pharmacokinetic/pharmacodynamic modeling in drug Development. In: Applications of Pharmacokinetic Principles in Drug Development, Ed. Krishna R. Kluwer Academic/ Plenum Publishers 2003
2. Rowland M, Tozer TN. Clinical Pharmacokinetics, Concepts and Application, third ed. Williams and Wilkins, Baltimore. 1995
3. Li AP, Ed. Advances in Pharmacology: Drug-Drug Interactions: Scientific and Regulatory Perspectives. San Diego, CA: Academic Press. 1997, pp. 43, 189-203
4. Bjornsson TD, Callaghan JT, Einolf HJ, Fischer V, Gan L, Grimm S, Kao J, King SP, Miwa G, Ni L, Kumar G, McLeod J, Obach SR, Roberts S, Roe A, Shah A, Snikeris F, Sullivan JT, Tweedie D, Vega JM, Walsh J, Wrighton SA. The conduct of in vitro and in vivo drug-drug interaction studies: a PhRMA perspective. J Clin Pharmacol 2003;43(5):443-469
5. Natarajan C, Rohatagi S. Role of Pharmacokinetics in Preclinical Development. In: Industrial Pharmacokinetics, Eds. Bonate P, Howard D. AAPS Press 2004
6. Ho RH, Kim RB. Transporters and drug therapy: implications for drug disposition and Disease. Clin Pharmacol Ther 2005;78:260-277
7. Defining genetic influences on pharmacologic responses. Drug interactions table. Available at http://medicine.iupui.edu/flockhart/
8. Evans WE, Johnson JA. Pharmacogenomics: the inherited basis for inter individual differences in drug response. Ann Rev Genomics Human Genetics 2001;2:9-39
9. Kalow W. Pharmacogenetics, pharmacogenomics, and pharmacobiology. Clin Pharmacol Ther 2001;70(1):1-4
10. FDA Guidance. Bioanalytical method validation. U.S. FOA
11. Chaikin P, Rhodes GR, Rohatagi S, Natarajan C. Pharmacokinetics/pharmacodynamics in drug development: an industrial perspective. J Clin Pharmacol 2000;40:1428-1438
12. Guidance for Industry. Estimating the maximum safe starting dose in initial clinical trials for therapeutics in adult healthy volunteers. U.S. Department of Health and Human Services, Food and Drug Administration, Center for Drug Evaluation and Research, July 2005
13. U.S. Food and Drug Administration (CDER). Guidance for Industry, Investigators, and Reviewers: Exploratory IND Studies (*DRAFT GUIDANCE*). April 2005.
14. Shah B, Emmons G, Rohatagi S, Martin NE, Howell S, Jensen BK. Mass balance study of 14C-M100240, a dual ACE/NEP inhibitor in healthy male subjects. Am J Ther 2003;10:356-362
15. Rohatagi S, Wang Y, Argenti D. Mass balance studies. In: Industrial Pharmacokinetics, Eds Bonate P, Howard D. AAPS Press, 2004
16. Rohatagi S, Barrett JS, DeWitt KE, Morales, RJ. Integrated pharmacokinetic/metabolic modeling of selegiline after transdermal administration. Biopharm Drug Disp 1997:18(7): 567-584
17. Shi J, Montay G, Bhargava VO. Clinical pharmacokinetics of telithromycin, the first ketolide antibacterial. Clin Pharmacokinet 2005;44(9):915-934
18. Smith B. Assessment of dose proportionality. In: Industrial Pharmacokinetics, Eds. Bonate P, Howard D. AAPS Press. 2004.
19. Vaccaro SK, Argenti D, Shah BS, Gillen MS, Rohatagi S, Jensen BK. Effect of food and gender on the pharmacokinetics of RP 73401, a phosphodiesterase IV inhibitor. Int J Clin Pharmacol Ther 2000;38(12):588-594
20. Dumas M, d'Athis P, Besancenot JF, Chadoint-Noudeau V, Chalopin JM, Rifle G, Escousse A. Variations of sotalol kinetics in renal insufficiency. Int J Clin Pharmacol Ther Toxicol 1989;27(10):486-489
21. Abdallah H, Jerling M. Effect of hepatic impairment on the multiple-dose pharmacokinetics of ranolazine sustained-release tablets. J Clin Pharmacol 2005;45(7):802-809
22. Rohatagi S, Zannikos PN, DePhillips SL, Ferris O, Boutouyrie BX, Jensen BK. Effect of age on the pharmacokinetics (PK) and pharmacodynamics (PD) of a platelet GP IIB/IIIA antagonist, RPR 109891 (Klerval). 13th Annual Meeting of the American Association of Pharmaceutical Scientists, San Francisco. November 1998.
23. Shi J, Ludden TM, Melikian AP, Gastonguay MR, Hinderling PH. Population pharmacokinetics and pharmacodynamics of sotalol in pediatric patients with supraventricular or ventricular tachyarrhythmia. J Pharmacokinet Pharmacodyn 2001;28(6):555-575
24. AAPS Workshop. Integrated roadmap to biomarkers for drug development method validation and qualification. American Association of Pharmaceutical Scientists, September 8-9, 2005
25. Evens RP, Witcher M. Biotechnology: an introduction to recombinant DNA technology and product availability. Ther Drug Monitoring 1993;15(6):514-520
26. Carson KL. Flexibility—the guiding principle for antibody manufacturing. Nature Biotechnology 2005;23(9):1054-1058

7
Laws and Regulations: The Discipline of Regulatory Affairs

Elaine S. Waller and Nancy L. Kercher

Regulatory Authorities and the Laws	149
Protecting the Public Health	149
Advancing the Public Health	150
Helping the Public Get Accurate, Science-Based Information	150
Regulatory Development Strategies	157
Fast-Track Program	157
Priority Review	157
Rolling Submission	157
Accelerated Approval	158
Orphan Drugs	158
Rx to OTC Switch	159
Submissions to Regulatory Authorities	160
Product Review	165
Package Insert	167
Advertising	167
Postapproval Maintenance	168
Compliance/Quality Assurance	171
General References	177

The discipline of regulatory affairs integrates the scientific information obtained during the development of a drug product, the marketing opportunities and competitive landscape, protection of the intellectual property surrounding a drug product, and the compliance efforts of a pharmaceutical company. It requires knowledge and, most importantly, interpretation of the most current requirements of regulatory authorities around the world.

It is the responsibility of the regulatory affairs professional to represent their interpretation of the regulatory requirements to their colleagues in such a way that they constructively facilitate, not obstruct, the development of a drug product. The critical skills to being a successful regulatory affairs professional include ability to think strategically; being able to integrate the diverse aspects of drug development, whether they are related to chemistry, manufacturing, pharmacology, clinical, legal, or marketing. Additionally, it is necessary to balance patient care, science, and business needs of the product and company; educate internal staff, both scientists and marketers about the regulatory requirements and vagaries of product development and marketing; and effectively communicate and negotiate with regulatory authorities. For individuals who are more oriented to breadth of a broad subject (i.e., drug development and commercialization) versus the depth of an individual piece (i.e., toxicology) of that broad subject, regulatory affairs is an ideal discipline.

The chapter outline covers all the major aspects of regulatory affairs, with a focus on the environment in the United States and the Food and Drug Administration (FDA). Although the regulatory and intellectual property laws are different outside the United States, and some registration processes are different, fundamentally the discipline of regulatory affairs is practiced very similarly around the world. Having an understanding of the process and requirements in the United States provides an excellent model for the processes and requirements in the developed world. In response to the growth of global pharmaceutical and biotechnology companies, there have been efforts by major countries to harmonize regulatory requirements. Where appropriate, those harmonization efforts are discussed to provide a global perspective.

The chapter begins with the basis for the regulatory authority of the FDA and a brief orientation to the complex organization that constitutes the agency. It then follows a model chronology of the role of regulatory affairs during the drug development, approval, and postapproval process, starting with the development of a regulatory strategy designed to optimize the regulatory approval and marketability of the drug product. This is followed with a description of the process of submitting applications to the FDA for both the investigational phase and the approval phase of a product. The steps of the approval process and the interactions with the FDA during product approval are described. The regulatory obligations continue once a product is approved, as there continue to be regulatory requirements for as long as the product is made commercially available.

Underlying the entire process are requirements for compliance with FDA regulations in the way the pharmaceutical industry conducts its business, whether that be in the manufacture of a drug product, the way in which animal and human trials are conducted, or the commercial promotion of a product.

Finally, the chapter concludes with some of the enduring controversies in regulatory affairs. These controversies drive pharmaceutical companies and the FDA, whether independently or in cooperation, to try to address the issues that surface in these controversies in a balanced, risk-based approach.

Regulatory Authorities and the Laws

Regulatory affairs exists only because the federal government has elected to regulate the pharmaceutical and biotechnology industries. The role of the FDA and its regulatory authority is defined by the laws legislated by Congress. Therefore, in order to understand the basics of regulatory affairs, it is necessary to understand how the FDA is established under the law, how it interprets and applies the laws it is responsible for administering, and some of the key legislation that gives the FDA its authority. This section of the chapter will also provide an introduction to the efforts to establish global harmonization of some regulatory requirements, with the intent of reducing and/or eliminating some of the testing required to bring new products to market. FDA has been a leader in this global harmonization effort.

This introduction to the regulatory authorities and the laws includes a brief overview of the organizational structure of the FDA to better understand how it interacts with the pharmaceutical industry and clinical investigators. It also includes the major laws that grant, and so limit, the regulatory authority of the agency. It further explains how FDA interprets the laws legislated by Congress into federal regulations and guidance and the multiple means that the FDA uses to communicate to the regulated industry and the nation's public. Finally, the role of other federal agencies that interact with the pharmaceutical and biotechnology industries, whether in a regulatory capacity or not, is described.

The mission statement of the FDA is found in Fig. 7.1. This mission statement reflects the current mission of the agency, which has evolved since its inception. At the time the FDA was first established in 1906, the role of the agency was limited to the protection of the public health. As public demands and therefore legislative demands have increased on the FDA, their mission has expanded.

The FDA is responsible for protecting the public health by assuring the safety, efficacy, and security of human and veterinary drugs, biological products, medical devices, our nation's food supply, cosmetics, and products that emit radiation.

The FDA is also responsible for advancing the public health by helping to speed innovations that make medicines and foods more effective, safer, and more affordable;

and helping the public get the accurate, science-based information they need to use medicines and foods to improve their health.

FIG. 7.1. FDA's Mission Statement

Protecting the Public Health

The FDA is responsible for not only drugs and biologics, but their regulatory authority also extends to food, medical devices, veterinary drugs, cosmetics, nutrition products, and radiation devices. Thus, they are responsible for protecting the health of the nation's citizens across a broad spectrum, including such areas as labeling of the nutritional content on food, approving only drugs found to be safe and effective for human or veterinary use, the allergenic potential of cosmetics, and the certification of mammography equipment. FDA further interprets new laws from Congress and promulgates regulations and guidelines for the industry to follow.

From the perspective of the pharmaceutical industry, the agency is first and foremost involved in the evaluation of drugs and biologics, at both the investigational phase and later at the approval phase. No investigational drug or biologic may enter into a clinical trial without review by the FDA to ensure that adequate testing has occurred in animals that would predict a reasonable safety/efficacy profile worthy of exposing human subjects. Likewise, no drug or biologic may enter into commercialization without the review and approval of FDA.

It is the clear intent of FDA to approve only those products that have a risk/benefit ratio that is considered acceptable.

Advancing the Public Health

In more recent years, the FDA has become increasingly responsible for advancing the public health by working within legislated time-frames to speed the review of new drug products that have the potential to meet highly unmet medical needs. Additionally, the agency has streamlined some regulations with the intent of reducing the cost of drug development and hastening the introduction of generic products. In the 1980s and 1990s, the FDA came under significant criticism by both the public and the pharmaceutical industry for not acting quickly enough to make drugs for unmet medical needs available to patients who were in need of them. Comparisons were made to countries outside the United States, where often new and innovative drugs and biologics were commercially available years before they were available in the United States. During the same time period, the AIDS crisis in health care occurred with the urgent need for new products for this devastating disease. As a result of the public pressure, legislation was passed that made it necessary for FDA to work more quickly and interact with the industry more frequently, to facilitate the development, review, and approval of innovative drugs. As a result, U.S. citizens are now often among the first in the world to have access to drugs and biologics for unmet medical needs.

Helping the Public Get Accurate, Science-Based Information

The FDA is responsible for regulating the communications between the pharmaceutical, biotechnology, and medical device companies and the prescribing physicians, other providers, institutions, and patients who use their products. First and foremost, FDA fulfills this mission through approving the language in the package inserts of all drug and biological products, and regulating the educational and advertising materials for drug products. FDA communicates directly to the public through press releases and postings on its web site. These communications are usually directed at informing the public of the approval of a new product that serves an unmet medical need, or of a significant safety issue with a product on the market.

FDA also provides the executive and congressional branches of government with information about drugs on the market and in research, the processes and procedures about research, industry regulation, and communications to the public and health care community. The FDA works cooperatively with other federal agencies, (e.g., Office of the Surgeon General, Centers for Disease Control and Prevention, and the National Institutes of Health) to communicate directly with the nation's public on issues of public health that are aligned with drugs and biologics. For food safety issues, the FDA would be collaborating with the Agriculture Department and the Environmental Protection Agency for toxins introduced into foods.

On a limited basis, the FDA engages in scientific study with the goal of advancing the regulatory process. For example, they may study the limits of changes in drug formulations that have a real impact on the quality of a drug product. By knowing these limits, they have the data to define what changes a drug manufacturer may make in the manufacturing process of a drug product with, and without, prior agency approval.

The FDA resides in the executive branch of the federal government and is composed of certain offices, centers, and divisions as shown in Fig. 7.2. The FDA commissioner is selected by the president and is confirmed by the U.S. Senate. Thus, the agency's policies and initiatives, and even the occasional product approval or marketing withdrawal, are politically influenced. There are more than 9,000 people who work in the FDA organization, with oversight for items accounting for 25% of consumer spending. While the entire structure of FDA is quite complex, the following discussion highlights the primary organizational units that routinely interact with the pharmaceutical and biotechnology industries.

The Center for Drug Evaluation and Research (CDER) is responsible for the evaluation of all drugs under investigation; the evaluation and approval of new drugs; and the continual safety surveillance of drugs already on the market. The Center for Biologicals Evaluation and Research (CBER) has the same responsibilities for biological products. These two centers are being combined to gain efficiency and consistency in the review and approval of products, whether they are drugs or biologics, except for blood products and vaccines remaining within CBER.

Within CDER, there are several offices that have primary responsibility for the regulation of the industry. The Office of New Drugs is divided into smaller units, aligned into five offices covering 14 therapeutic areas (e.g., anti-infective and gastrointestinal), which have responsibility for investigational drugs and the evaluation and approval of new drugs.

- Office of the Commissioner
- Center for Drug Evaluation and Research (CDER):
  - Offices of New Drugs (5 Offices)
  - Office of Generic Drugs (OGD)
  - Office of Drug Safety (ODS)
  - Office of Biostatistics
  - Office of Pharmaceutical Sciences (Chemistry, Pharmacology, etc)
  - Division of Drug Marketing, Advertisement and Communications (DDMAC)
  - Division of Scientific Investigation (DSI)
- Center for Biologics Evaluation and Research (CBER)
- Center for Devices and Radiological Health (CDRH)
- Office of Regulatory Affairs (ORA)

FIG. 7.2. FDA Organization

Additional offices participate in the review and approval of new drugs. The Office of Biostatistics, the Office of Clinical Pharmacology and Biopharmaceutics, and the Office of New Drug Chemistry work with the five therapeutically aligned review offices providing expertise in their respective disciplines to evaluate drugs during both investigational and approval stages. There are separate offices for review and approval of generic drugs and for pediatric drug development.

Separate from the Office of New Drugs, the Division of Drug Marketing, Advertising and Communications (DDMAC) regulates the advertising and educational programs used by the pharmaceutical industry to promote their products to both prescribers and consumers of approved drugs, devices, and biologics. The Office of Drug Safety is responsible for adverse-event monitoring and safety assessment for marketed products and related interactions with the industry. Their review can lead to a product withdrawal after approval if new serious safety concerns arise during the general use of a product by the medical community.

The Division of Scientific Investigations has the responsibility for the inspection of clinical study sites, where trials on investigational drugs have been conducted. FDA inspectors from this division work with the Office of New Drugs to routinely inspect clinical study sites involved with pivotal trials for drugs being evaluated for approval. The Center for Devices and Radiological Health is responsible for the review and approval of medical devices (e.g., from surgical catheters to prosthetic hips). The center is also responsible for the regulation of radiation-emitting products (e.g., x-ray equipment). Finally, the Office of Regulatory Affairs (ORA) is the network of field FDA offices located across the country. The ORA has a wide range of responsibilities, including the inspection of pharmaceutical and biotechnology manufacturing facilities to the seizure of contaminated food.

The regulatory authority of the FDA has been shaped by many different laws since its origin 100 years ago. Figure 7.3 highlights those laws that are most significant in the establishment of the FDA's regulatory authority and have an important impact on the way FDA conducts its business today. Often, a health crisis in the United States creates a public outcry for legislation to change the regulatory authority of the FDA, to either protect the health of the public or make drugs more readily available.

The Federal Food and Drugs Act of 1906 for the first time added regulatory functions to the agency. Prior to this act, states exercised the principal control over foods and drugs. The Federal Food, Drug and Cosmetic Act of 1938 was a milestone in that it was the first law requiring that new drugs had to be shown to be safe before they could be marketed. This act remains one of the cornerstone pieces of legislation that the FDA still operates under today. The Public Health Service Act of 1944 established the regulatory authority of the FDA over biological products. The Kefauver–Harris Drug Amendments of 1962, named after the legislators who sponsored the legislation, required for the first time that drug manufacturers prove the effectiveness of their products in order for FDA to grant a drug product approval.

The Orphan Drug Act enables the FDA to promote research and marketing of drugs needed for treating rare diseases, defined as those having a prevalence of less than 200,000 in the U.S. population. This act provides financial incentives to the pharmaceutical and biotechnology industries to develop and commercialize drug and biologics products for patients whose diseases are not prevalent enough to be otherwise commercially attractive to the industry. The Drug Price Competition and Patent Term Restoration Act is oftentimes referred to as Waxman–Hatch legislation, named after the legislators who sponsored the legislation. This legislation expedites the commercial availability of generic products by permitting FDA to approve generic products without requiring repeat testing to demonstrate that the generic product is safe and effective.

The Prescription Drug User Fee Act of 1992 (PDUFA) requires pharmaceutical and biotechnology companies to pay fees for the FDA to review applications for new drugs and biologics and changes to approved drugs and biologics. The act also requires that FDA use these fees to pay for more reviewers to speed the review process. This act provides targeted periods of time for FDA to review an application and more accountability to meet these review deadlines. Today, the fee for a new drug or biologic product to be reviewed is more than $500,000, and typically the fee is raised each year. Another important provision of PUDFA in accelerating product approvals codifies some of the communications that occur between FDA and sponsors during the development of a drug to facilitate more rapid and efficient drug development and approval; for example, interim NDA deficiency letters from FDA during a review; a meeting upon the company request usually at key milestones such as end of phase 2, which will clarify development issues before the NDA either is filed or acted upon by FDA. This legislation has been so successful that both the government and the pharmaceutical industry have consistently supported its renewal.

The Food and Drug Administration Modernization Act of 1997 mandated the most sweeping reform of the FDA since

- Federal Food and Drugs Act –1906
- Federal Food, Drug and Cosmetic Act –1938
- Public Health Service Act –1944
- Kefauver-Harris Drug Amendments –1962
- Orphan Drug Act –1983
- Drug Price Competition & Patent Term Restoration Act –1984
- Prescription Drug User Fee Act –1992, 1997, 2002
- Food and Drug Administration Modernization Act –1997
- Pediatric Research Equity Act –2003

FIG. 7.3. Laws for FDA Regulation of Pharmaceutical & Biotechnology Industry the Food, Drug and Cosmetic Act of 1938. It provides many changes in the way FDA conducts its regulatory functions. Some highlights of those changes include the modernization of the regulation of biologics and streamlining the approval process for manufacturing changes of drug and biologics products. It also allows for drugs and biologics that can potentially treat serious and life-threatening diseases to be under a "fast-track" review process at the FDA. The Pediatric Research Equity Act of 2003 amended the 1938 Food, Drug and Cosmetic Act to authorize the FDA to require certain research on drugs used in pediatric patients be conducted by the pharmaceutical industry. Until this law, the pharmaceutical industry could elect whether or not it wanted to conduct trials with its products in pediatric populations. A benefit to the company for conducting pediatric product studies to their development plan and product labeling was extended patent exclusivity.

For legislative and therefore regulatory purposes, there are definitions of drugs, drug products, and biologics that form the basis of regulations (Fig. 7.4). These legislative definitions serve to clarify the authority of the FDA and what they are responsible for regulating.

The definition of a drug most commonly referred to is that of a substance "intended for use in the diagnosis, cure, mitigation, treatment or prevention of disease." This definition, along with the other definitions of a drug in Figure 7.4, serves to differentiate a drug substance from a drug product. The same drug substance can be found in many drug products (e.g., oral and parenteral drug products containing the same drug substance).

Biologic products are a subset of drug products and are distinguished by the biological manufacturing process. Drugs are sometimes differentiated by referring to them as "small molecules," and biologics are referred to as "large molecules".

The regulatory authority of the FDA starts with acts of Congress that grant them that authority and provide, at a high level, direction in how the agency is to exercise its authority. As mentioned in Fig. 7.5 two acts of Congress, the Federal Food, Drug and Cosmetic Act and the Public Health Service Act, are fundamental to the establishment of the FDA and its authority over drugs and biologics. Drug products are regulated and approved under the Federal Food Drug and Cosmetic Act, and biologic products are regulated and licensed under the Public Health Service Act. Over the years, several changes have occurred so that today, there are fewer regulatory differences between drugs and biologics.

Once an act is passed, it is the responsibility of the FDA to write regulations consistent with the legislative intent of the act, in order to enforce the statutes passed by Congress (Fig. 7.5). These regulations are legally binding, and it is necessary for the pharmaceutical and biotechnology industries to comply with these regulations or risk running afoul of the FDA from a compliance perspective. An example of a set of regulations is those requiring an Investigational New Drug Application to be submitted to the FDA before an investigational product can be administered to humans. These regulations describe at a high level the content needed in this application.

As a next step, the FDA writes guidance documents that serve to further explain how the industry can specifically do its work in order to be compliant with the regulations. These guidance documents are not legally binding, and the industry can elect to not comply with them, as long as they have good reason not to do so. The intent of the guidance documents is to provide the industry with as much of the FDA's thinking on a given topic as needed so the industry has an increased opportunity to satisfy the FDA's requirements the first time, rather than to guess what data the FDA will require. An example of a guidance document is one that describes in detail the necessary chemistry, manufacturing and controls data needed at the time of submission of an Investigational New Drug Application.

Any interested party or individual has an opportunity to participate in the writing of regulations by the FDA. FDA drafts regulations and has them published in the *Federal Register*, an official document of the federal government that is publicly

- **Drug:**
  - Recognized by official pharmacopoeia or formulary
  - Intended for use in the diagnosis, cure, mitigation, treatment, or prevention of disease
  - Substance other than food, intended to affect the structure or any function of the body
  - A substance used as a component of a medicine but not a device or a component, part or accessory of a device
- **Drug Product:**
  - Finished dosage form that contains the drug substance
- **Biologic Product:**
  - Any virus, serum, toxin, vaccine, blood, blood component or derivative, allergenic product, or analogous product applicable in prevention, treatment, or cure of disease or injuries. They are a subset of "drug products", distinguished by the biological manufacturing process.

FIG. 7.4. FDA Definitions of Drugs & Biologics

- Congress enacts Acts
  - Federal Food, Drug and Cosmetic Act
  - Public Health Service Act
- FDA interprets Acts
  - Regulations (legally binding)
  - Guidances (not legally binding)
- Industry/individual participation in rule making
  - Regulations proposed in Federal Register
  - Submission of petitions
- Code of Federal Regulations

FIG. 7.5. Congressional Acts – Regulations & Guidances

7. Laws and Regulations

available. At the time of publication, the FDA states that it will accept comments on the proposed regulations until a certain deadline. Anyone can respond, and the pharmaceutical and biotechnology industries typically do make comments. It is their opportunity to tell the FDA if their proposed regulations are going to have any unintended consequences on the industry and to seek clarifications to be written into the regulations. After this comment period, the FDA publishes final regulations in the Code of Federal Regulations, taking into account the comments received.

Any interested party or individual doesn't have to wait for FDA to publish regulations to make comments on how the agency does its work. At any time, they can submit a Citizen's Petition and ask that the FDA stop or start a particular practice. The pharmaceutical industry often uses this mechanism when they believe the FDA has overstepped its bounds or misinterpreted the law.

As for all federal government agencies, the regulations of the FDA reside in the Code of Federal Regulations (CFRs). The CFRs are divided into parts, and there are several parts that pertain to the pharmaceutical industry. The government makes the CFRs available in both bound paper volumes and online. A link to the parts of the CRFs that apply to the pharmaceutical and biotechnology industries is found on the FDA web site: www.fda.gov.

The FDA has a variety of tools with which it can communicate to the industries it regulates and to consumers. As mentioned in Fig. 7.6 one of the primary mechanisms by which the industry is informed of FDA's thinking is in the publication of *guidance documents*. These guidance documents are typically issued first in draft form and are open to public comment. Although guidance documents are not legally binding, they provide an excellent road map into the particulars of what FDA is most likely going to require of a sponsor. Guidances range from the acceptability of data from clinical studies conducted in foreign countries, to reporting of adverse drug events, to the process to be followed when requesting a formal meeting with the FDA. In many therapeutic areas, there are guidance documents for specific diseases and pharmacologic drug categories to guide sponsors in the clinical evaluations of their investigational products. More than 100 *guidances, clinical guidelines,* and *points to consider* have been promulgated by the FDA from 1977 to 2001 and are available from the FDA in pamphlets or online.

The FDA uses *advisory committee meetings* to provide a public forum for scientific and regulatory issues to be discussed. Advisory committees are composed of scientific experts from outside the FDA and are usually composed of academic and practicing physicians (majority of members), nurses and clinical pharmacists, and other technical experts such as statisticians, toxicologists, or epidemiologists. The core characteristics of members and the committees are recognized technical competence (primary criterion), personal integrity, commitment to public interest, objectivity, independence, and no conflict of interest. They meet on a periodic basis to discuss either a drug or biologic that is under review at the FDA (NDAs, ANDAs, prescription to nonprescription switch) or to discuss a scientific/regulatory issue that spans across a number of products. The advisory committee often will vote on whether or not a new product under review should be approved. This vote is advisory only and nonbinding on the FDA. These meetings are usually open to the public, and there is an opportunity for the public to comment on the issue being debated.

The FDA makes available, either on paper and/or at their web site, the *comments of FDA reviewers* on drugs and biologics that have been approved. These drug approval packages provide the basis of the judgment of the FDA scientific review team that a new drug or biologic has been found to be safe and effective. These drug approval packages help the industry to understand what development programs have been successful with other drugs or biologics that are similar to ones they are developing.

The compliance arm of FDA issues *warning letters* to the industry. These letters will be discussed further in the compliance section of this chapter and are recognized as important tools to communicate to the regulated industries on what they should *not* do. The industry reads and follows these warning letters to avoid their own compliance issues with the agency.

The FDA is also very visible at professional meetings, and the regulators speak frequently and formally from the podium on topics of recent interest: *podium policy*. Their presentations are oftentimes opportunities to understand how the FDA interprets certain processes, regulations, or guidances, and can provide opportunities to hear what the FDA is contemplating in upcoming requirements or regulatory controversies. Whereas podium policy is very informative, it is not legally binding.

Lastly, the agency maintains an excellent web site at www.fda.gov. This web site is very informative for both the

- Guidance documents:
  - Generated by CDER and CBER
  - Span scientific issues to administrative procedures
  - Guidelines, Point to consider
- Advisory committee meetings:
  - Public meetings sponsored by Divisions within CDER and CBER
- Drug approval packages:
  - FDA reviewer comments on data to support drug/biologic product approvals
- Warning letters:
  - Illustrate what is not acceptable
- Podium policy:
  - FDA speeches to outline very latest thinking
- www.fda.gov:
  - Information for both regulated industries and

FIG. 7.6. FDA Communication Tools

- **Health and Human Services:**
  - Food and Drug Administration (FDA)
  - National Institutes of Health (NIH)
  - Agency for Health Care Research and Quality (AHRQ)
  - Centers for Disease Control and Prevention (CDC)
  - Office of the Inspector General (OIG)
  - Centers for Medicare and Medicaid Services (CMS)
- **Other regulatory agencies:**
  - Consumer Products Safety Commission (CPSC)
  - Federal Trade Commission (FTC)

FIG. 7.7. Roles of Select Government Agencies

regulated industries and consumers and is a very convenient tool for the regulatory affairs professional to keep abreast of the latest at the FDA.

Administratively, the FDA resides in the Department of Health and Human Services (HHS) within the executive branch of the federal government. In addition to the FDA, there are a number of other agencies in HHS that may interact with the pharmaceutical and biotechnology industries (Fig. 7.7). The National Institutes of Health (NIH) is the steward of medical and behavioral research for the nation. Its mission is "science in pursuit of fundamental knowledge about the nature and behavior of living systems and the application of that knowledge to extend healthy life and reduce the burdens of illness and disability." The NIH works with both the FDA and industry to conduct basic and advanced research to determine underlying mechanisms of action of drugs and to advance the development of new clinical indications. The NIH can contact companies with research proposals, companies can contact NIH, or the FDA can request research to be undertaken. An example of NIH research is the study of important pediatric indications for approved products with no patent protection, when the sponsor does not want to develop a pediatric indication.

The Agency for Health Care Research and Quality's (AHRQ) mission is "to support research designed to improve the quality, safety, efficiency, and effectiveness of health care for all Americans." The research sponsored, conducted, and disseminated by the agency provides information that helps people make better decisions about health care. The AHRQ sponsored an assessment of the safety and efficacy of ephedra and ephedrine for weight loss and athletic performance enhancement. A systematic comprehensive literature review with meta-analysis was undertaken, as well as a review of FDA's extensive safety database. Conclusions were presented that led to further investigations and the eventual withdrawal from the market of products containing these ingredients.

The Centers for Disease Control and Prevention's (CDC) mission is "to promote health and quality of life by preventing and controlling disease, injury, and disability." CDC seeks to accomplish its mission by working with partners throughout the nation and world to monitor health, detect and investigate health problems, conduct research to enhance prevention, develop and advocate sound public health policies, implement prevention strategies, promote healthy behaviors, foster safe and healthful environments, and provide leadership and training. One very important task the CDC undertakes every year is identification of the likely leading strains of influenza so that vaccine manufacturers can produce a vaccine and have it available prior to the start of the flu season. In the case of a vaccine shortage, CDC attempts to coordinate with the manufacturers and distributors so that individuals at greatest risk are identified and receive the vaccine.

The mission of the Office of Inspector General (OIG) is "to protect the integrity of Department of Health and Human Services (HHS) programs, as well as the health and welfare of beneficiaries of those programs." The duties of the OIG are carried out through a nationwide network of audits, investigations, inspections, and other mission-related functions performed by OIG components for all agencies within HHS and also for government funded programs, such as the Veterans Administration's or Medicare's drug purchasing and related marketing practices. These investigations and inspections have uncovered pricing fraud, kickbacks and fraud with human growth hormones, drug diversion and substitution, and lack of protection of patient privacy during the conduct of certain clinical studies. Many of these cases resulted in very large financial settlements. Marketing practices for appropriateness of educational programming between the industry and health providers has become part of the purview of OIG because of its influence on product decisions. OIG has issued operating guidelines for the industry, which carry legal consequences if they are not followed.

The Centers for Medicare & Medicaid Services (CMS) "administers the Medicare program, and works in partnership with the States to administer Medicaid, the State Children's Health Insurance Program (SCHIP), and health insurance portability standards." Obtaining Medicare reimbursement, if possible, is an important hurdle after drug or biologic approval. Companies can meet with CMS prior to the design of the phase 3 trials to obtain early feedback on a design that would maximize the opportunity for reimbursement. In general, the study or studies must demonstrate that the product is safe and effective, and there is an improvement in net health outcomes, such as improvements in function, quality of life, morbidity or mortality. The product must be generalizable to the Medicare population and at least as good as if not better than similar products already covered under Medicare. Therefore, in the drug development plans for a product, in addition to proving safety and efficacy for regulatory approval, pharmacoeconomic and quality of life studies have become core studies in the plan for later marketing success.

Other important regulatory agencies in the health arena that are not directly under the HHS umbrella include the Consumer Product Safety Commission (CPSC) and the Federal Trade Commission (FTC). The CPSC is "charged with protecting the public from unreasonable risks of serious injury or death from consumer products under the agencies jurisdiction."

7. Laws and Regulations

Regulations applying to child-resistant packaging are governed by the CPSC. The Federal Trade Commission (FTC) "enforces consumer protection laws that prevent fraud, deception and unfair business practices." They are responsible for truth-in-advertising for over-the-counter drugs and monitoring health benefit claims for foods and dietary supplements. The FTC is also responsible for reviewing proposed company mergers to assure there is no possibility of unfair business practices or anticompetitive activities that could harm the consumer. As part of the merger process, the product portfolio of marketed products and products in the development pipeline are examined by the FTC to determine if there could be a monopoly in a therapeutic or pharmacologic category of products. After evaluating the potential acquisition of Immunex by Amgen, Immunex was required to divest Leukine® (sargramostim), which was purchased by Berlex®, prior to FTC approval. Amgen already had two related hematopoietic stimulating products, Neupogen® (filgrastim) and Neulasta® (pegfilgrastim).

Global harmonization of regulatory requirements was recognized as an urgent need in the 1980s, as the costs of drug development continued to escalate and the public began to demand that innovative products become approved and available in their country as soon as possible (Fig. 7.8). It was also at this time that the globalization of the pharmaceutical industry became more oriented to getting new drug products approved in as many countries around the world in as short a period of time as possible. The top 10 companies for product sales worldwide in 2004 were Pfizer, GlaxoSmithKline (British), Merck, Johnson & Johnson, Sanofi-Aventis (French), AstraZeneca (British/Swedish), Novartis (Swiss), Bristol-Myers-Squibb, Wyeth, and Eli Lilly.

As a result, parties from the regulatory authorities and the regulated industry around the world mounted an effort that came to be known as the International Conference on Harmonization of Technical Requirements for Registration of Pharmaceuticals for Human Use, or simply shortened to the International Conference on Harmonization, or ICH.

- International Conference on Harmonization of Technical Requirements for Registration of Pharmaceuticals for Human Use (ICH) initiated in 1990
- ...."brings together the regulatory authorities of Europe, Japan and the United States and experts from the pharmaceutical industry in the three regions to discuss scientific and technical aspects of product registration"
- "The purpose is to make recommendations on ways to achieve greater harmonization in the interpretation and application of technical guidelines and requirements for product registration in order to reduce or obviate the need to duplicate the testing carried out during the research and development of new medicines."

FIG. 7.8. Global Regulatory Harmonization

A process began to harmonize the data and process requirements for the successful development and approval of new drugs and biologics, such that a study conducted to meet the requirements of one country could be assured of also meeting the requirements of another country. This offered the opportunity to reduce the overall costs of drug development, as it became possible to reduce the amount of testing required.

The International Conference on Harmonization was officially initiated in 1990. The primary participants came from countries that represent the largest pharmaceutical and biotechnology markets in the world: Europe, Japan, and the United States. Organizations and smaller countries were invited as observers to the process. The parties involved in the ICH process have addressed such widely diverse topics as the testing of the stability of a drug product, the appropriate length of animal toxicology studies, and uniform definitions and requirements for reporting of serious adverse events that occur in clinical trials. Additionally, the parties addressed the format of documentation that can be submitted to regulatory authorities. Before the ICH process began, a pharmaceutical company had to conduct stability testing of drug products at slightly different temperatures and humidities, conduct toxicology studies for different lengths of time in different animal species, or conduct duplicate clinical studies in different countries due to concerns about ethnic differences. This duplicate testing was all to satisfy different requirements by different regulatory authorities. Although a large number of topics have already been addressed by ICH, the organization continues to address new topics and to reassess topics already addressed, all based on scientific evidence.

The regulatory authorities involved in the ICH process include the FDA and its counterparts in Europe (EMEA) and Japan (MHLW) (Fig. 7.9). The official observers of ICH include the World Health Organization, the European Free Trade Area (represented by Switzerland), and Canada. The major pharmaceutical trade organizations in the United States, Europe, and Japan coordinate the representation from the pharmaceutical and biotechnology industries, with the representatives being recognized experts in their areas in the industry. This ICH process has resulted in a greater appreciation and understanding of the challenges from both the perspective of

- Regulatory Authorities:
  - US - Food and Drug Administration (FDA)
  - Europe - European Medicines Agency (EMEA)
  - Japan - Ministry of Health, Labor and Welfare (MHLW)
- Industry representatives:
  - US - Pharmaceutical Research and Manufacturers of America (PhRMA)
  - Europe - European Federation of Pharmaceutical Industries and Associations (EFPIA)
  - Japan - Japan Pharmaceutical Manufacturers Association (JPMA)

FIG. 7.9. ICH Parties

- Expert Working Groups from the ICH parties developed guidelines addressing technical requirements for:
  - Quality
  - Safety
  - Efficacy
  - Common Technical Document
- Medical Dictionary for Regulatory Activities (MedDRA)
- FDA, EMEA, MHLW responsible for implementation of guidelines in their countries

FIG. 7.10. ICH Regulatory Harmonization Process

- Technical requirements for developing and registering new drug products containing new drug substances as they relate to:
  - Quality - those relating to chemical and pharmaceutical Quality Assurance
  - Safety - those relating to *in vitro* and *in vivo* pre-clinical studies
  - Efficacy - those relating to clinical studies in human subjects
- Common Technical Document (CTD) - organization of the common elements of a registration submission
- Medical Dictionary for Regulatory Activities (MedDRA):
  - international medical terminology for electronic transmission of adverse event reporting, both in the pre-and post-marketing areas, as well as the coding of clinical trial data
- www.ich.org

FIG. 7.11. ICH Topics

the regulators and the regulated industries. It has also resulted in greater informal collaboration among the regulators on product-specific issues.

Representatives from the regulatory authorities and the regulated industry have been formed into Expert Working Groups to address the technical requirements for the quality (manufacturing, product testing, and product formulation), safety, and efficacy of drug and biologic products (Fig. 7.10). Their work is based on available scientific and regulatory data that support appropriate guidelines for the development of drug and biologic products. For example, the Expert Working Group assigned to work on the acceptability of clinical trials in ethnically different populations evaluated available data on pharmacokinetic and pharmacodynamic differences, if any, of drugs in different populations to determine if ethnic differences were clinically significant.

In addition to addressing technical and scientific requirements, groups also addressed two administrative areas: the "Common Technical Document" (CTD) and a medical dictionary (medDRA), especially for adverse events, that could be used to report data to regulatory authorities. Once the technical requirements have been established and agreed upon by all the parties in the harmonization process, the FDA and its counterpart agencies in Europe and Japan are responsible for formally implementing the ICH guidelines in their countries. In the United States, the ICH guidelines are published as guidance documents, as referenced in Fig 7.10.

Topics for the ICH process were well developed in advance of the Expert Working Groups actually starting their work (Fig. 7.11). The technical requirements for developing and registering new drug products were grouped into three primary categories. The first of these categories, quality, is directed toward the development, manufacturing, and control of the actual drug product. Technical requirements in this area address such issues as stability testing, analytical validation, impurity profiles in both drug substances and drug products, quality of biotechnology products, and specifications for drug substances and drug products.

The safety category is directed toward preclinical testing, including both *in vitro* and *in vivo* animal testing of new drugs. Technical requirements addressed in this area include toxicity testing, carcinogenicity testing, reproductive toxicology, and toxicokinetics. The efficacy category is focused on the appropriate development of drugs during the clinical testing phase. Technical requirements in this category include the collection and reporting of clinical safety data, dose-response studies, ethnic considerations in conducting foreign trials, and studies in special populations such as geriatrics and pediatrics.

The ICH process has addressed one of the most frustrating registration issues for the pharmaceutical industry. Prior to ICH, each country had its own set of requirements for the formating and organization of documents to be submitted for review of a submission package by the regulatory authorities. Thus, sponsors found themselves reformating and reorganizing essentially the same data just to fit the requirements of different countries. With the development of the common technical document (CTD), pharmaceutical companies can now prepare one fundamental set of documents with one set of requirements for formating and organization. This set of documents can be submitted to all the regulatory authorities without making changes for each country. The CTD will be further discussed later in the chapter.

A Medical Dictionary for Regulatory Authorities (MedDRA) was also developed through the ICH process. This provides a single dictionary that is required for use for all adverse event reporting. Prior to the development of this international dictionary, pharmaceutical companies were coding the same adverse events to different medical terms, based on the medical dictionary required in each country. This created repeat, and often confusing work, for companies, and if anything, it served to obfuscate the interpretation of clinical safety data.

The ICH process has clearly advanced regulatory science around the world and has brought efficiencies to the development of drugs and the registration and reporting process. The ICH organization maintains an excellent and informative web site: www.ich.org.

Regulatory Development Strategies

The optimal regulatory drug development and registration strategy outlines rapid drug development, timely FDA review and approval, and expeditious market launch. The strategy should also take into consideration mechanisms for protection from competition whether through patent protection or market exclusivity provisions. Developing a regulatory strategy requires a multifaceted analysis that integrates all aspects into a development plan and timeline. With the product's clinical attributes as the basis for consideration, the analysis should address potentially viable indications, unmet medical needs, regulatory "opportunities," analyses of competitors' registration strategies (including approved products and those in development), medical practice guidelines for disease treatments of national societies, FDA advisory committee meetings and transcripts, and strategies for market protection through patents and market exclusivity. Competitive consideration dictates the labels of related products be carefully read, with the intent of developing a regulatory strategy that will lead to a competitive edge in the label. Key differentiating features could be the product's dosing regimen, administration, special populations, and monitoring parameters. The regulatory strategy for a first product in a pharmacologic class will typically differ from the regulatory strategy for those products that follow in the same class.

In general, the more life-threatening or severely debilitating the illness where no adequate therapy exists, the shorter the drug development timeline and the lower the hurdle will be for FDA approval of the product. The registration strategy for the development of a "me-too" product must take into account the expectations of FDA set by the approval packages of other products in the same drug class and the intensely competitive environment. Successful integration of these areas benefits the patients because they have access to innovative therapies as quickly as possible. The sponsor benefits because it can begin to receive a return on its investment as soon as possible. Regardless of the strategy, in order for it to be successful in the end, the data must demonstrate that the benefits of the drug product outweigh the risks in the target patient population.

The FDA has developed regulations and performance targets that accomplish a variety of goals including faster FDA review and approval timelines, incentives to innovators to study drugs in patients with rare diseases, mechanisms for taking appropriate prescription products to over-the-counter status, and pediatric indications for existing and new products (Fig. 7.12). This section specifically addresses some of these regulatory opportunities.

Fast-Track Program

The Food and Drug Administration Modernization Act of 1997 instructed FDA to specifically outline policies and procedures for "fast-track" drug products. The act states that "a drug designated as a fast track product is intended for treatment of a serious or life-threatening condition and demonstrates the potential to address an unmet medical need." There is FDA guidance outlining both when a disease is considered serious or life-threatening and when a drug can be considered to potentially address an unmet medical need. Unmet medical need is defined by FDA as well (e.g., drug effects serious outcome not seen with alternatives, improved effects for serious outcomes, benefits patients not tolerating alternatives, similar benefits of alternatives but avoiding serious side effects of alternatives). Sponsors may apply for fast-track designation at any time during drug development by presenting data that the new drug meets the criteria for treatment of both a serious and life-threatening condition and unmet medical need. Obtaining fast-track designation affords the sponsor an opportunity to access several programs to facilitate drug development and approval. These programs include early and frequent meetings with FDA, use of surrogate end points documented to be predictive of clinical benefit, the potential for priority review, "rolling" submissions for approval (discussed below), and/or accelerated approval. The FDA may require follow-up studies as part of the approval under fast-track status to confirm specific clinical issues of safety or efficacy.

- Fast track program:
  - To expedite development and review of drugs/biologics intended to treat serious or life-threatening conditions and with potential to address unmet medical needs; designated by FDA
- Priority review:
  - Faster FDA review times based on significant improvements compared to marketed products
- Rolling submissions:
  - Allows for incremental submission of reviewable modules of NDA/BLA to reduce overall FDA review time; designated by FDA

Fig. 7.12. Regulatory "Opportunities"

Priority Review

Priority review establishes a target of 6 months for FDA review of a drug product that "would be a significant improvement compared to marketed products in the treatment, diagnosis, or prevention of a disease." Designation of a priority review for a biologic product is stricter in that treatment must be for a "serious or life-threatening disease." A sponsor must request and justify a priority review at the time they make a submission for a drug or biologic approval. If priority review is not granted, the targeted FDA review time for a standard review is 10–12 months. Thus, a priority review provides an opportunity to have a product on the market 4–6 months earlier than a product with a standard review.

Rolling Submission

An opportunity exists to submit portions of the NDA submission for FDA review instead of waiting for the entire submission to be prepared. This approach must be requested by the sponsor and approved by FDA prior to submission. Rolling submissions are granted by the FDA only in situations where the public health can benefit from rapid product approval. In general, only complete portions of the chemistry and manufacturing section, the pharmacology and toxicology section, or the clinical section of the submission would be accepted in a "rolling" fashion. A rolling submission allows the FDA to review some portions of a submission while the sponsor is completing others, allowing for a faster overall FDA review time.

Accelerated Approval

Accelerated approval regulations apply to drugs developed with the potential to treat life-threatening and/or and severely debilitating diseases that provide meaningful benefit to patients over existing therapies (e.g., cancer) (Fig. 7.13). The FDA can approve the product based on adequate and well-controlled studies in which a surrogate for clinical benefit is the primary end point. The surrogate must be reasonably likely to predict a clinical benefit, such as survival or reversal of morbidity. For example, the FDA may allow the use of objective response rate as a surrogate for survival for a product used to treat solid tumors. If the surrogate relationship to the clinical benefit has not already been demonstrated in other studies, after approval, the sponsor may be required to conduct additional studies to demonstrate clinical benefit. If subsequent studies fail to show clinical benefit or the sponsor does not complete the studies in a timely fashion, FDA may withdraw approval of the product.

In the cancer example, the company must continue to perform the phase 3 type study with the definitive measure of efficacy, that is, to demonstrate improved survival. If the work is not done (or the follow-up study fails to document the full clinical benefit), the FDA can withdraw the product from the market more easily than usual. In 2004, Iressa®, gefitinab, was approved as a tyrosine kinase inhibitor of tumor cell growth for non-small cell lung carcinoma (NSCLC) based on the surrogate end point of significant reduction in tumor size and progression. At the end of 2004, one of three phase 3b/4 studies was reported to show no advantage over placebo for the definitive survival end point in a randomized double-blind parallel design. The company, AstraZeneca, informed the FDA within 48 hours of this result, sent out a dear doctor letter, disseminated results to the health care community, and stopped promotion of the drug. Withdrawal of Iressa® from the market awaited full analyses of all study results by the company and the FDA.

Orphan Drugs

In 1983, the Orphan Drug Act was promulgated, and the FDA Office of Orphan Drug Development was created and incentives were created for manufacturers to develop drugs and biological products (Fig. 7.13). There are estimated to be greater than 6,000 rare diseases and related conditions. A sponsor can request the FDA grant an orphan drug designation for a product to treat a rare disease or condition. The sponsor must demonstrate the disease is prevalent in less than 200,000 people in the United States, or if it is prevalent in more than 200,000 people, that upon commercialization, the sponsor would likely not be able to recover the cost of development. If orphan drug designation is granted, once the product is approved, the product will have 7 years of market exclusivity. The exclusivity does not allow a competitor to market the same drug for the same indication until the end of the 7-year exclusivity period. The Genzyme company particularly has used this regulatory approach in the rare, serious, and untreatable enzyme deficiency diseases; for example, Cerezyme® in Gaucher disease, Fabrazyme® in Fabry disease, and Aldurazyme® in Hurler syndrome. Serono used this process for Serostim®, growth hormone, for AIDS wasting syndrome. Amgen used this approach for Epogen® in anemia of renal disease because in the 1980s, the condition was documented to be within the 200,000 patient limit.

There is also an opportunity for a sponsor to receive tax credits on up to 50% of the dollars spent for certain clinical testing in the United States to develop the orphan drug. The FDA may offer financial grants to investigators or sponsors to defray some of the costs of developing an orphan drug. The orphan drug regulations are intended to provide incentives to pharmaceutical and biotechnology companies to develop drugs for treatment of diseases where there is little or no commercial incentive to do so. The product pipeline and commercialization of drugs for rare diseases has been affected significantly by the Orphan Drug Act. Prior to the act, there were 15 products approved for orphan disease. Today, there are more than 250 orphan drug products approved. Examples include Vidaza® (azacitidine) for the treatment of myelodysplastic syndrome and Clolar® (clofarabine) for the treatment of acute lymphoblastic leukemia. New molecular entities (NMEs) for orphan drugs also are approved faster by the FDA versus all other NMEs, about 10–12 months faster in 2000 to 2001.

- Accelerated approval:
  - For new drugs that provide meaningful therapeutic benefit over existing treatment for serious or life-threatening illnesses, using surrogate endpoint or restricted use provisions; designated by FDA
- Orphan drugs:
  - promoting the development of products that demonstrate promise for the diagnosis and/or treatment of rare diseases defined as prevalence in US <200,000; designated by FDA
- Rx-to-OTC switch:
  - OTC marketing of a product that was once available only by prescription; approved by FDA

FIG. 7.13. Regulatory "Opportunities"

Rx to OTC Switch

Dozens of products have been switched from prescription status to being available over-the-counter (OTC) without a prescription (Fig. 7.13). In order for a drug to be switched, the product must be used to treat or prevent symptoms in a disease where physician oversight is not necessary. The patient must also be able to self-diagnose the condition being treated. Sinus congestion, headache, pain, upset stomach, and itching are examples of symptoms a patient can recognize. Drugs that must be monitored carefully to assure efficacy or that have significant toxicities are not good candidates for OTC switch. A sponsor can petition FDA for an Rx to OTC switch or make a submission for OTC status. Petitions do not need to come from the sponsor but can be submitted by anyone. An insurance company petitioned FDA to switch Claritin® to over-the-counter status as a means to save money on reimbursement of prescription allergy medications. The family of ulcer medications (H-2 antagonists) has been approved for OTC use (e.g., Pepcid® and Zantac®) because of relative safety and the symptom changes will be discernable to patients, but the anticholesterol product Pravacol® was turned down in 2005 because of the lack of symptomology and the inability for patients to self-diagnose need and beneficial activity.

An optimal registration strategy is ideally developed by a multidisciplinary project team involving experts in basic research, pharmacology and toxicology, clinical development, statistics, marketing, manufacturing, regulatory affairs, and project management (Figs. 7.14 and 7.15). Typically, this project team is put in place when data are available from *in vitro* and animal studies suggesting the drug may be effective. If it is possible to develop multiple indications for the product, the indication that would lead to the most rapid approval and market launch is typically given the highest priority. The wording of the indication should be developed with input from the entire project team. The package insert from approved products with similar indications should be analyzed. A target package insert should be developed that compares desired claims/statements with those made in approved package inserts and products pending approval. This target package insert will serve as a guide to determine the types and design of studies for the preclinical, clinical, and chemistry and manufacturing development programs.

A review of competitors' drug development programs is extremely useful to determine which indication might be the lead indication for a drug candidate (Fig. 7.15). The FDA web site contains information regarding their review of competitors' applications, plus transcripts of advisory committees. The comments of FDA reviewers on a drug approval package become available after a drug product has been approved. Transcripts of FDA advisory committee hearings on competitors' products (if held) can usually be obtained.

It is also important to evaluate products currently in the pipeline that could be competitive with the one being developed. It is important that the evaluation of potential competitor products be evaluated for the science, the patent and market exclusivity situations. Preclinical and clinical data can be obtained at scientific conferences, as well as through internet/literature searches of periodicals, abstracts, and press releases. A wealth of information on a competitive product can be gleaned if that product is discussed at a public FDA advisory committee. Understanding why one product was approved, or another not approved, can provide valuable insights to aid in designing an optimal development and registration strategy.

The development plan must meet the regulatory requirements, which are legally binding. As an example, the Code of Federal Regulations designated 21 CFR 314 outlines application requirements for FDA approval to market a New Drug Application. The contents and format of the application, definitions of adequate and well-controlled studies, acceptance of foreign data and accelerated approval requirements are some of the many areas covered. Guidance documents, although not legally binding, provide far greater detail. As an example, the regulations are not specific as to the number of patient exposures required in an application for a product that is going to be chronically administered. There is a guidance document that outlines the number of patients overall that should be exposed to the drug, as well as the number of patients that needs to be followed for 6 months and for 1 year. Understanding this guidance will help ensure that the

- Define potential indications
- Develop targeted package insert
- Evaluate database and regulatory requirements of approved products:
  - FDA drug approval packages
  - FDA Advisory Committee meetings, transcripts, video tapes
  - Freedom of Information searches

FIG. 7.14. Develop Optimal Registration Strategy

- Collect regulatory information on products not yet approved:
  - FDA Advisory Committee meetings
  - Trade organizations, periodicals, abstracts, press releases
- Evaluate regulatory pathways with optimal critical path/market potential:
  - FDA regulations
  - FDA guidances
  - Podium policy
- Evaluate market exclusivity and patent extension opportunities
- Evaluate competitor patent status

FIG. 7.15. Develop Optimal Registration Strategy

development program is designed to study an appropriate number of patients.

Patent status of a new product is important for regulatory strategies along several lines of registration planning. The sponsor's new product patent will offer certain opportunities for exclusivity, but potential patent extensions in the future (e.g., new indications or formulations) need to be factored into the strategy as well. Also, the strategy needs to take into account the patent situation with the competitor's products, too.

In 1984, the Drug Price Competition and Patent Term Restoration Act was passed (Fig. 7.16). This act of Congress allowed generic companies to obtain approval of their drug without repeating all the testing required to demonstrate safety and efficacy of the brand-name product. The effect was to lower the hurdle for the development of generic drugs and to make them available as soon as the pertinent patents on the brand-name drug have expired. To maintain a balance between the commercial rewards for generic and innovative companies, under the act, the brand-name product is granted up to 5 additional years of patent protection. This additional patent protection is to compensate for the amount of patent life used up during animal and human testing, as well as FDA review.

Several opportunities for market exclusivity exist. A period of market exclusivity is one in which a competitor cannot market the same product, giving the innovator a market without direct competition. As previously discussed, an orphan drug is protected from direct competition for 7 years. If a new chemical entity is approved, the drug receives at least 5 years' exclusivity, even if its patent expires within this 5-year window, protecting it from generic competition. Three years' exclusivity is granted for new indications for an already approved drug product. Congress and FDA are keenly aware of the lack of approved indications for use of drugs in a pediatric population. As an incentive to generate data to support use in children, FDA grants 6 months of market exclusivity for new indications in children.

The Patent and Trademark Office of the federal government has responsibility for issuing patents. The FDA has only an administrative role to make information on patents and periods of market exclusivity readily available. Information regarding patent coverage, market exclusivity, and therapeutic equivalence for approved drugs and biologics appears in a booklet entitled "FDA Approved Drug Products and Therapeutic Equivalence." Because the publication has an orange cover, it has become commonly known in the industry as the Orange Book.

The FDA can grant a tentative approval to a generic version of a drug product that still has a remaining period of patent life or market exclusivity by the innovator. Generic companies are allowed to legally perform all the required development for a generic product approval, primarily bioequivalence and drug product formulation development, during the patent period of the innovator, resulting in accelerated market availability of generic products. A tentative approval is given when the FDA has completed its review and determined that all requirements for an approval have been met. A product with a tentative approval can be legally marketed as soon as the patent or period of exclusivity held by the innovator has expired.

Submissions to Regulatory Authorities

This section of the chapter is an introduction to the submissions to regulatory authorities that are required under the regulations. It covers both the timing of submissions in relation to the drug development cycle and the content of submissions. Although this section is focused on FDA requirements, there are similar submission requirements for other developed countries. Once the data and documents have been assembled for a submission to the FDA, they can be used for similar submissions in other countries. Investigational New Drug (IND) applications, New Drug Applications (NDA), Biologic License Applications (BLA), and Abbreviated New Drug Applications (ANDA) will be covered.

All sponsors of investigational drugs or biologics are required to complete the registration process before they can legally market their product (Fig. 7.17). If any trials are conducted in human subjects or patients in the United States, this requires that the sponsor submit an IND application to the FDA. Technically, the FDA does not approve an IND. If the sponsor has not heard from the FDA 30 days after submission of their IND, they may proceed with clinical trials.

- Patent term restoration:
    - Maximum five years of patent extension
- Exclusivity periods:
    - Seven years for orphan drugs
    - Five years for innovator products
    - Three years for selected changes in an approved drug product
    - Six months for pediatric studies
- FDA has administrative role in patent extensions and exclusivity:
    - FDA Approved Drug Products with Therapeutic Equivalence Evaluations (Orange Book)
    - Tentative approvals granted

FIG. 7.16. Patent and Exclusivity Interests

- Registration process
- Contents of Investigational New Drug (IND) application
- Applications for registration:
    - Contents of New Drug Application (NDA) and Biologics License Application (BLA):
        - Use of Common Technical Document
    - Contents of Abbreviated New Drug Application (ANDA)

FIG. 7.17. Submissions to Regulatory Authorities

Once a drug product has been developed and the sponsor believes it is ready for approval, they must prepare an application for its registration, which allows the sponsor to market the product. If the product is a new drug product, a NDA is submitted to the FDA for review and approval. If the product is a new biologic product, a BLA is submitted to the FDA for review and approval. The basic outline of the content of the NDA and the BLA are the same. Obviously, the content of the two submissions will differ based on one product being manufactured by a synthetic chemical process and another being manufactured by a biologic process. A supplemental NDA (sNDA) is used to change the labeling of a product, usually for new indications, formulations, dosing schema, and adverse experiences.

As discussed earlier in the chapter, under ICH, the format and outline of a CTD has been developed. As of 2005, sponsors can organize their submissions according to the older NDA/BLA outline. In the future, sponsors will be required to submit their registration packages according to the CTD format, and the submission must be made electronically. While legally the document is still considered a NDA or BLA, it is in the format of a CTD.

An ANDA is the appropriate registration package for a generic drug product. This package is a much smaller registration package than the NDA (hence the title "abbreviated"), because it is not necessary to repeat and report all the testing (e.g, toxicology studies, clinical trials) required for a new drug product. The focus is on pharmacokinetics studies ("bioequivalence"), manufacturing, and product quality. Currently, there is not a universal mechanism for the approval of a generic biologic product, primarily due to the complexity of the manufacturing process and the resultant impact on product performance in patients. However, the regulatory environment on that issue is changing, and it is believed that in time, there will be a legal mechanism for their approval.

Figure 7.18 illustrates the drug development process, with the interactions with the FDA added at the appropriate milestones during this process. Before an IND is submitted to the FDA, a pre-IND meeting with the appropriate review division of the FDA can be requested by the pharmaceutical company. A pre-IND meeting is not required but is highly recommended by FDA when a novel or innovative drug or biologic product is being developed. A pre-IND meeting, in which the development of the drug to date is discussed and the available data shared, provides a valuable exchange of information to assure both the FDA and the sponsor that the appropriate preclinical and chemistry work is completed prior to introducing the drug product into humans. This meeting can help avoid a "clinical hold" on the IND, a FDA regulatory action that will be discussed later in the chapter.

The sponsor is also encouraged, though not required, by FDA to meet sometime during the drug development process, typically in what's known as an "end-of-phase 2" meeting. It is at the end of phase 2 that a sponsor already has learned important elements regarding the performance of the investigational drug and is ready to launch into large, expensive, pivotal phase 3 trials, the results of which will be the basis for a drug or biologic product approval. Typically, the number of patients exposed to the investigational product will increase significantly in phase 3. For these reasons, it is in the best interest of the sponsor to gain feedback from the review division at the FDA on the remainder of the drug development program. This can help ensure that when the pivotal trials are completed, they will provide data considered scientifically necessary and complete for the drug's approval by FDA. During these meetings, the FDA is expected to comment on

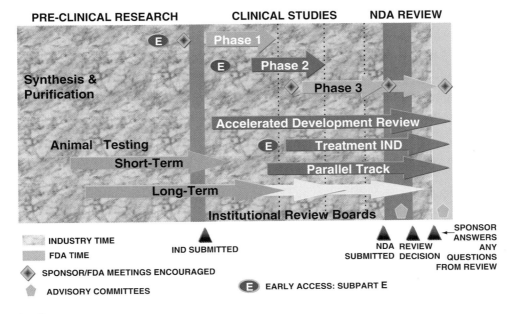

FIG. 7.18. Registration Process

the design of the proposed pivotal clinical studies, the selected doses and comparator drugs, and the adequacy of the preclinical and drug product formulation data needed to advance the drug into phase 3 of development. This is an appropriate forum for the sponsor and the FDA to respectfully disagree and debate scientific and/or regulatory issues.

Oftentimes, the next meeting with the sponsor and the FDA comes after the completion of the pivotal phase 3 trials and before the submission of the NDA. This is an opportunity for the sponsor to present, in general, the contents of the NDA. The main objective of the meeting is to familiarize the FDA with the anticipated data package and to discuss the format of the data so that it suits the FDA requirements for review. The FDA review division may have some special requests for data analyses and presentation that will help facilitate a rapid review.

As discussed earlier, after the application is submitted and during the FDA review process, a sponsor may be invited to present their data at a FDA Advisory Committee meeting. This is typically done when the FDA is creating new policy with the potential approval of a new drug or biologic or when the FDA believes the data warrant additional scrutiny. Advisory committees are covered more fully later in the chapter.

Although it is not a regulatory requirement that a sponsor to meet with the FDA during the development and review of a new drug or biologic product, it is highly recommended that the sponsor take advantage of every opportunity to discuss their data and plans with the FDA. This helps ensure that both parties are in agreement in principle on the plans going forward and helps to avoid unpleasant surprises during the process. And during the process, a sponsor would ignore the advice of the agency at their own peril. Most face-to-face meetings with the FDA include representatives from multiple scientific disciplines from both the agency and the sponsor. They are typically excellent opportunities to learn the multidisciplinary scientific and regulatory approach used by the FDA in their review of investigational drug products.

Though the above has described the typical face-to-face meetings with the FDA during drug development and approval, there are numerous interactions via phone, fax, and e-mail during the process. It is the responsibility of the regulatory affairs professional to be the single initial point of contact with the agency in order to manage the myriad interactions with the FDA, ensure that the sponsor is speaking with one consistent voice to the agency, and avoid pitfalls at a later time.

An IND application is required to be submitted to the FDA before a sponsor initiates any trial administering investigational drugs or biologics to humans in the United States (Fig. 7.19). Countries outside the United States have a process similar to the IND that is required. As stated in the IND regulations, the primary objectives in the FDA's review of an IND are to "assure the safety and rights of human subjects, and, in Phase 2 and 3, to help assure that the quality of the scientific evaluation of drugs is adequate to permit an evaluation of the drug's effectiveness and safety." If a sponsor company elects to conduct clinical trials with investigational drugs outside the United States, an IND is not required to be submitted to the FDA, though a sponsor may opt to submit an IND. Having an IND in place at the FDA gives the sponsor access to the advice and opinion of the agency and is very useful if the trials conducted outside the United States are intended to be used to support an application for a new drug or biologic in the United States.

- Purpose of IND
- Regulatory Requirements
- Contents of Investigational New Drug (IND) application:
  - Introductory Statement
  - General Investigational Plan
  - Investigator Brochure
  - Study protocol(s)
  - Investigator, facilities and Institutional Review Board data
  - Chemistry, manufacturing and control data
  - Pharmacology and Toxicology data
  - Previous Human Experience
- Types of INDs

FIG. 7.19. Investigational New Drug (IND) Application

The regulations written to regulate the IND process include a high-level description of the requirements for data in an IND. Most of these requirements are described in more detail in guidance documents issued by CDER and CBER within FDA and provide more instruction to sponsor companies on the requirements.

Once an IND is submitted to the FDA, the agency has 30 days to review the document. If the FDA reviewers are satisfied that the proposed clinical trial can proceed safely, the sponsor can assume that they can proceed once the 30 days have passed. If however, upon review, the FDA finds that critical data are missing, or that the clinical trial is not appropriate as proposed, the agency can place a "clinical hold" on an IND. The sponsor is not allowed to begin the proposed clinical trial until the concerns of the FDA have been satisfied and the agency has notified the sponsor that the clinical hold has been lifted. A clinical hold can take a few days to resolve (a relatively small change in the clinical protocol), or take months (an additional pre clinical study is required and the data submitted to the FDA before the clinical trial can begin). A pre-IND meeting, as described in the previous discussion, can help the sponsor avoid a clinical hold by thoroughly understanding the agency's requirements and incorporating these into the IND.

The contents of an IND application are usually straightforward and can be contained in a few hundred pages. Figure. 7.19 lists the main contents of an IND, as required by the regulations. The introductory statement and general investigational plan provide a general description of the drug or biologic product, and the general nature of at least the first clinical study proposed to be conducted, as well as the therapeutic indication of interest. The *investigator brochure* is a required document

7. Laws and Regulations

throughout the life of the investigational product, requiring periodic updating as additional data are obtained. It is basically the precursor to the *package insert* of the marketed product, informing all the clinical investigators of the available, pertinent information about the drug product's safety and efficacy. The protocol submitted is the first clinical protocol that will be followed in the clinic, assuming the IND is not placed on clinical hold. A brief description of the identity and credentials of the investigator(s) and institutional review board(s) responsible for the study are reported in the IND. The remainder of the document is the accumulated information on the chemistry, manufacturing and control data for the drug substance and drug product and the preclinical pharmacology, pharmacokinetic, and toxicology data.

Once an IND is submitted to the FDA, there is an ongoing obligation by the sponsor to maintain the IND. All new clinical protocols must be submitted to the IND before they can be initiated. Additionally, updated investigator brochures, information on additional clinical investigators, additional chemistry and manufacturing data, additional preclinical information, and any clinical data that have an impact on the evaluation of the safety of the drug or biologic product are required to be submitted in a timely fashion. If at any time the FDA deems it appropriate to stop the clinical investigation of the product, they may place an entire IND or a specific clinical trial on "clinical hold."

An IND must have a sponsor, who is legally responsible for the conduct of the investigations conducted under an IND. The most common sponsors are pharmaceutical and biotechnology companies and clinical investigators. An "investigator IND" is a mechanism for a clinical investigator(s) to conduct a clinical trial with an investigational drug product. Oftentimes, they are allowed to reference the pharmaceutical or biotechnology sponsor's IND for the same drug product, to provide the necessary chemistry, manufacturing and control data, and preclinical data to support the use of the product in humans.

IND sponsors can also file a "treatment IND" as a mechanism to provide new drugs not yet approved to patients with serious or life-threatening illnesses as well as illnesses where there is no alternative treatments. In order for FDA to allow enrollment under a treatment IND, some evidence of efficacy must have been demonstrated. Patients who are enrolled under a treatment IND are not eligible to participate in pivotal studies of the drug; however, safety and efficacy data are evaluated in the context of all clinical trials. Unlike a traditional IND, sponsors can require patients who are enrolled under a treatment IND to pay for the drug prior to commercial approval. Compassionate use, also called emergency use, does not require submission of an IND and is reserved for very rare life-threatening situations where there is not time to obtain IRB approval. In this situation, the sponsor and treating physician work closely with the FDA to exchange the necessary information so that the drug can be administered expeditiously.

- ▸ Purpose of applications for NDA or BLA
- ▸ Contents of New Drug Application / Biologic License Application:
  - Administrative information
  - Chemistry, manufacturing, and controls
  - Nonclinical pharmacology and toxicology
  - Human pharmacokinetics and bioavailability
  - Microbiology (for an anti-infective)
  - Clinical data
  - Statistics
  - Pediatric use
  - Samples and Labeling
  - Case report forms and tabulations
  - Patent information and certification
  - Financial certification or disclosure

FIG. 7.20. Applications – NDA/BLA

Once a sponsor has completed all the required phases of drug development, and in their opinion the cumulative data demonstrate that a drug or biologic is safe and effective, the next step is submission of a NDA for a drug product or a BLA for a biologic product (Fig. 7.20). Approval of either a NDA or BLA by the FDA is required before the product can be marketed and sold in the United States. Unlike an IND, which can be as small as a few hundred pages, a NDA or BLA is voluminous, in the tens of thousands of pages, and often surpassing 100,000 pages. Given the extent of the documentation required, it is no surprise that FDA will require in the near future that all NDA or BLA submissions be made electronically, which allows for easy and efficient navigation through a large submission and access to electronic data sets for analyses by FDA reviewers.

Once a sponsor submits a NDA or BLA, the FDA has 60 days to do a high-level review of the submission to determine if it has all the required elements and is organized appropriately to facilitate the review. If they find that the submission meets this threshold determination, the FDA officially files the submission and notifies the sponsor. In the event the FDA determines that the submission does not meet the threshold requirements, it will issue a "refusal to file" letter to the sponsor. The agency will not proceed with the review until the necessary changes to the submission are made by the sponsor.

The content of a NDA or BLA is established through the regulations. There are numerous guidance documents that provide the sponsor further instruction on the content, organization, and electronic formatting of a submission. Figure 7.20 provides a listing of the content categories of every new NDA or BLA. The submission is extensive in its requirements. It includes all the scientific data to demonstrate that the product manufactured is a quality product with adequate manufacturing controls; all the pertinent preclinical data to demonstrate the acceptable pharmacology, pharmacokinetics, and toxicology of the drug substance and product; reports of clinical pharmacokinetic trials; and reports from all clinical trials for the indication under review. Case report forms of patients who died or were withdrawn from the study are also included. All clinical safety data, regardless of the clinical study from which it was collected, must be submitted. The proposed package insert also is required in a NDA or BLA.

In addition to the scientific data, there are a number of requirements for administrative documents, including information about the patents that pertain to the product being reviewed, and financial disclosure of all clinical investigators that reveal any potential conflict of interest that might have influenced the outcome of clinical studies.

As mentioned earlier in the chapter, the International Conference on Harmonization (ICH) has developed a format for the organization of a NDA or BLA and their equivalent in countries outside the United States (Fig. 7.21). This format is the CTD. This format, now accepted by the United States, European, and Japanese regulatory authorities and those of other selected countries, greatly enhances the efficiency of the assembly of a registration package for submission in multiple countries. The content of the CTD has been outlined in modules, and they build upon each other.

Starting at the bottom of the pyramid, the CTD includes a module (no. 3) for quality (drug or biologic substance and drug or biologic product), a nonclinical module (no. 4) (animal pharmacology, toxicology, and pharmacokinetics), and a clinical module (no. 5) (human pharmacokinetics, clinical pharmacology, safety, and efficacy). These modules contain all the required data in its most granular form. These modules contain the data upon which all the data interpretations and conclusions regarding the safety and efficacy of the drug or biologic product are made. The data listings provided allow the FDA reviewer to conduct their own data review, for example, statistical analyses and pharmacokinetic modeling.

The next layer of the pyramid includes a distillation, summarization, and critical evaluation of each of the three modules in the bottom layer in the form of summaries or reviews. In this module (no. 2), the sponsor needs to really understand and fairly present their interpretation of the accumulated data. It is the sponsor's responsibility not only to highlight the positive aspects of the drug or biologic product but also to critically analyze any shortcomings of the product and what additional study should be done to better understand its appropriate use.

The top layer of the pyramid does not technically fall into the CTD, as there is not commonality in module 1 across all countries. The only commonality is that module 1 is reserved for the particular administrative aspects of a registration package required by each individual country. For example, in the U.S. NDA or BLA, module 1 is the appropriate section to include information about the U.S. patents that pertain to the drug or biologic.

Based on this CTD approach to the assembly of a submission, a sponsor can prepare modules 2–5 for submission to all regulatory authorities in the developed world. The only tailoring required for each country is module 1. The advantages of the CTD approach to the sponsor are obvious, and there are also advantages to the FDA and their counterparts. In the process of developing the CTD, the FDA was able to help design the structure of a submission that makes it very efficient for their review. By having all sponsors follow the exact same format for an electronic submission that FDA helped design, they are able to more easily navigate through a large number of submissions more quickly.

An ANDA is prepared when a sponsor is seeking registration of a generic drug product. A drug product is considered a generic if it is "identical in active ingredient(s), dosage form, strength, route of administration, and conditions of use"

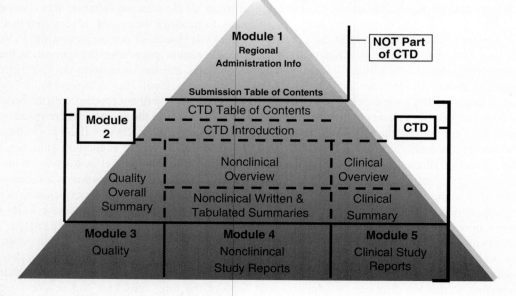

FIG. 7.21. Schematic of ICH Common Technical Document (CTD)

- Basis for ANDA
- Conditions of use
- Active ingredients
- Route of administration, dosage form, and strength
- Bioequivalence
- Labeling
- Chemistry, manufacturing, and controls
- Samples
- Patent certification
- Financial certification or disclosure

FIG. 7.22. Abbreviated New Drug Application (ANDA)

to an approved drug product. The sponsor of a generic drug product is not required to repeat the clinical studies necessary to determine that the drug is safe and effective.

The elements of an ANDA are listed in Fig. 7.22. The primary scientific elements of the ANDA include the chemistry, manufacturing and controls data that demonstrate that the sponsor has developed and can consistently manufacture a quality product. Data demonstrating the generic product is bioequivalent to the approved brand-name product also must be submitted unless the product is completely bioavailable, such as with an oral syrup. The sponsor must also submit the proposed labeling (package insert) of the generic product. This labeling must be fundamentally identical to the labeling for the approved drug product, including only those indications that are no longer covered by patents held by the sponsor of the brand-name product. The remainder of the ANDA is primarily administrative information.

A generic company may submit an ANDA well in advance of the expiration of all the relevant patents on the drug substance and drug product. In this case, under completion of their review, the FDA can give a "tentative approval." In the case of a tentative approval, the generic company has approval to market their generic product only after all the pertaining patents for the brand-name product have expired.

As mentioned earlier, at the time of this writing, there is not a universal regulatory mechanism for the review and approval of a generic biologic product. However, the FDA is carefully evaluating the requirements for demonstrating equivalence between two biologics products, and it is anticipated that there will be a regulatory pathway for the approval of generic biologic products (called "follow-on biologics") in the near future.

Product Review

The discussion of product review will focus on the FDA review and approval process of a NDA or BLA for a new drug or biologic product that has never before been marketed, as this is typically the most complex of FDA reviews. The FDA review process for a generic product is similar in principle but does not have the complexity of the review of a new drug or biologic product. The primary sponsor contact with FDA during product review is the regulatory affairs professional who has typically worked on the product during its development and is intimate with the contents of the submission. It is their responsibility to understand the overall review process and to effectively manage the communications between the FDA and the sponsor project team. It is also their responsibility to keep senior management of the company informed of the status of FDA's review. They are obligated to assure that information requested by FDA during the review is provided in a clear, complete, and timely fashion.

A submission of a NDA or BLA triggers a tremendous amount of work, both for the FDA and the sponsor (Fig. 7.23). The FDA has 60 days to determine if the application should be "filed" or if they will refuse to file the application. To determine if the submission should be officially filed, the FDA reviews the overall submission for all the required components, the sponsor's safety and efficacy claims in the draft package insert, the pivotal clinical studies to see if they generally support the claims, and a small number of clinical case report forms. If the FDA finds the submission lacking in required content or to be so poorly organized that it can't be reviewed, it will refuse to file the submission. At this time, the sponsor has no choice but to either address the issues or abandon the prospect of getting the product onto the commercial market in the United States.

It is typically during the review of a NDA or BLA that a sponsor and the FDA have the most frequent and intense communications that they will have during the development of the product. The number of critical activities to be accomplished within a compressed time frame require that both the agency and the sponsor have people dedicated to the review process. In order to facilitate the many interactions, both the FDA and the sponsor have a designated person who is primarily responsible for managing the liaison between the two organizations. Not only is this efficient, but it also ensures a smooth and orderly flow of information in both directions.

Once the application is accepted for filing, the various FDA review areas, including medical, statistical, pharmacology, biopharmaceutical, chemistry, and microbiology, conduct a detailed review of their sections. Typically, the FDA contacts the sponsor during the review to request additional information and/or clarification. The agency may meet with the sponsor toward the completion of their scientific reviews to discuss issues or discrepancies in interpretation of the data and advise the sponsor if an advisory committee is deemed necessary.

As the deadline for the completion of the review nears, the FDA reviewers meet to determine if the submission in its totality should be approved, including the findings from site inspections. If at this time they request the sponsor to submit a significant amount of new data, the approval of the product will be delayed. If there have been significant findings of noncompliance during any inspections, these issues need to be resolved prior to approval.

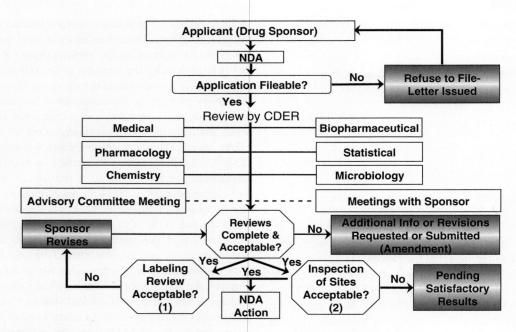

FIG. 7.23. Product Review with Regulatory Authorities

- Thoroughly understand NDA/BLA and literature data
- Interact frequently with reviewing Division
- Prepare for multiple audiences:
  - FDA, committee members, competition, & stock analysts
- Prepare
  - Committee briefing package
  - Primary and backup slides
- Study
  - Backgrounds of committee members
  - Past advisory panel hearings
  - Politics of the open, public forum
- Choose best presenters from company and investigators
- Practice, practice, practice

FIG. 7.24. Preparing for FDA Advisory Committee Hearings

At the same time the review team at FDA headquarters is reviewing the contents of the submission, FDA inspectors are working in the field. Inspectors will travel to selected clinical sites, manufacturing facilities, and perhaps animal laboratories to ensure that the studies submitted in the submission have been conducted according to FDA compliance requirements and that the manufacturing facility is capable of repeatedly producing a quality product.

Toward the end of the review process, FDA and the sponsor enter into negotiations on the final language of the package insert and the promotional materials that the sponsor wants to use to launch the product. As the package insert is the basis of what the sponsor can, and cannot, advertise and promote, these two activities are closely linked. Once the package insert has been finalized, barring any other issues, the application is ready to be approved by the FDA.

The agency may request that the data in a submission be presented to one of their advisory committees. This is usually done when the product represents an innovative class of drugs, or when the data present the agency with scientific issues that have significant regulatory implications. An advisory committee hearing adds significantly to the workload for both the agency and the sponsor during the product review process. However, it does provide an opportunity for the sponsor to become very familiar with the issues identified by the FDA, which they can hopefully help address.

The FDA has approximately 30 standing advisory committees aligned by product line (i.e., drug, food, biologic, or device) and therapeutic and pharmacologic categories (Fig. 7.24). Committees are composed of predominantly academic and clinical experts (physicians primarily plus nurses, pharmacists, and other technical experts), patient advocates, and industry experts. Credentials of committee members include recognized technical expertise; leaders in their field; independence from the company and product under review (or competitive companies); known reputation for integrity; and known for commitment to public interest. Although FDA may elect to seek guidance from a committee, the advice from the committee is not binding on the agency. Use of the committee process allows FDA to supplement its knowledge with expertise outside the agency and to provide a public forum as an educational tool for all sponsors involved in drug development.

When a sponsor receives even a hint that the data package may be presented to an advisory committee, they should begin preparing for a meeting. This includes preparing a briefing package for the committee and a presentation with primary and back-up slides. Key presenters may come from

7. Laws and Regulations

the sponsor or may be an external expert and should be chosen based on their thorough knowledge of the data in the application as well as the literature and their presentation skills. External medical consultants are often used to provide fresh insights on how the FDA and advisory committee may view the data. As the FDA review proceeds, the issues and questions that the FDA will raise to the committee will become clearer through frequent interactions with the FDA reviewers. The sponsor is well served by anticipating questions from both the FDA and the committee and addressing those in the sponsor presentation.

As mentioned previously, the advisory committee process is a public process. Attendees include FDA representatives, along with competitors, stock analysts, and the general public. The committee may be asked to vote on whether or not a product should be approved and to provide recommendations on additional studies that should be conducted. It is not uncommon to hear recommendations of advisory committees during the evening news or in the next day's paper.

In order to be fully prepared for an advisory committee meeting, it is incumbent on the sponsor to understand the backgrounds and areas of expertise or interest of the committee members. This will help identify in advance what issues they may have so those issues can be addressed in the briefing document and the presentation. Reviewing previous transcripts of the advisory committee hearings can also provide insights into potential issues. If there are political implications with the drug product under review (e.g., early drugs for the treatment of AIDS), it is necessary for the sponsor to understand the political environment that may be created at the meeting as advocates or detractors speak during the public forum section of the meeting. Finally, a well orchestrated advisory committee presentation requires many, many hours of practice by the sponsor to fully understand their data and to represent those data in a polished fashion.

Package Insert

The content and format of the package insert is outlined in FDA regulations (Fig. 7.25). The package insert must contain the following sections: description, clinical pharmacology, indications and usage, contraindications, warnings, precautions, adverse reactions, drug abuse and dependence, overdosage, dosage and administration and how supplied. Proposed labeling must be submitted at the time of the NDA/BLA/ANDA filing. The package insert must accurately reflect the data generated during drug development, and claims in the label must be supported by data. To better assist the FDA in substantiating the claims made in the package insert, an annotated version is provided in the submission. All information in the package insert must be cross-referenced to the appropriate section of the submission that supports the labeling statement. Depending on the type of product, a sponsor may elect to develop a patient package insert, which assists patients with product usage and also requires FDA review and approval.

As discussed previously, labeling negotiations between FDA and the sponsor are usually not initiated until the product review is substantially complete. It is typically the last action step on the critical path prior to FDA approval. The package insert forms the basis for what can, and very importantly, what cannot, be said in advertising and promotional material and used by the sales organization. Based on the importance of the wording in the package insert to the commercial success of the product, sponsors may often elect to continue to negotiate the labeling instead of opting for a more prompt product approval with less than desirable labeling. For example, a phrase like "arrests disease progression" versus "slows disease progression" will have significant impact on provider acceptance and the marketing strategy. The package insert creates the opportunities and limits for marketing a product.

Advertising

Consistent with the FDA's mission to protect the public's health, the agency has regulatory authority over the advertising and promotion of prescription drugs (Fig. 7.25). The intent is to ensure that promotions of drug and biologic products to prescribers and/or consumers are truthful, do not exaggerate the benefits, and fairly present the risks of the products. There are written regulations on what can and cannot be presented in advertising materials. It is a regulatory requirement that sponsors submit all their promotional materials to the FDA for review at the time of dissemination or publication.

Promotional material must not be false or misleading. The claims in advertising must fairly reflect the information in the package insert or additional scientifically defensible information. Examples of false and misleading claims are false statements regarding a competitor's product, a claim of unsurpassed safety not substantiated by data, or a claim of unsurpassed efficacy. The material also must contain "fair balance" of the benefits and the risks. The same scope, depth of detail, and prominence need to be presented for both in the form of words used and even type styles and sizes and colors. Because initial impressions are so critical to the lasting image of a product, it is incumbent on the sponsor to have the FDA pre-clear advertising materials used in the initial launch of a

- **Package insert (PI):**
  - Content and format established by regulations
  - Annotated product label is in registration submission
  - Negotiation of language is one of last tasks of submission review
  - Basis for language used in advertising copy
- **Advertising copy:**
  - Content and presentation established by regulations
  - Cannot be false or misleading
  - Must contain "fair balance" of benefits and risks
  - Advertising campaign submitted for FDA clearance

FIG. 7.25. Negotiating Product Labeling and Advertising Copy new product to ensure that the sponsor's advertising is compliant with the regulations. Negotiating this advertising copy with the agency can lead to intense interactions between the sponsor and the FDA. Educational materials used by the sponsor (e.g., publications, monographs, and symposia) must also comply with regulatory guidances, be within labeling claims, and are subject to FDA review.

Although there are countless exchanges between the sponsor and the FDA during the typical review process, there are certain critical milestone communications that transpire during the product review (Fig. 7.26). As mentioned earlier, once an application is submitted, the agency has 60 days to determine if it is substantially complete and be can officially filed. If it is complete, the sponsor will receive a letter stating the submission is filed and FDA will proceed with its complete review process. If a priority review has been requested by the sponsor, the review team at the agency will decide to grant the priority review or not and inform the sponsor accordingly. If a NDA/BLA application is not sufficiently complete, the applicant is sent a refuse-to-file (RTF) letter with the deficiencies outlined. Examples of deficiencies include an incomplete application form, inadequate English translations, inadequate organization of the NDA, or failure to submit sufficient information to evaluate safety and efficacy. Minor deficiencies that can be addressed and likely fixed during the review and interaction with the company will not lead to a RTF letter.

Abbreviated NDAs are handled slightly differently. If the ANDA for a generic product is complete, it is "received" and the applicant will be notified in writing. If it is not complete, it is not received and the applicant is usually notified by telephone.

When FDA finishes reviewing the application, they will issue either an approval letter or a complete response letter, the latter of which outlines the deficiencies in the submission. If the sponsor receives an approval letter, they are free to introduce the product into commerce in the United States, under the conditions outlined in the approval letter. If a complete response letter is sent to the sponsor, it will indicate whether there are major or minor deficiencies with the application. Minor deficiencies require a class I resubmission.

- Initial NDA, BLA, ANDA application:
  - Complete application
  - Refuse to File letter
- Action letters after application review:
  - Complete Response letter
  - Approval letter
  - Information Request letter
- Complete response to an action letter:
  - Class 1 resubmission – minor updates to the application i.e. draft labeling, stability data updates, minor reanalysis of data
  - Class 2 resubmission – major update, i.e. warrant presentation to an Advisory Committee

FIG. 7.26. FDA Communication During Product Review

Examples of a class 1 resubmission include certain safety updates, product stability updates, phase IV commitments and proposals, assay validation data, or minor reanalysis of data. These submissions are to be reviewed by the FDA within 2 months of their receipt. Major deficiencies require a class 2 resubmission. Examples of a class 2 resubmission include data from additional clinical trials or preclinical studies. These submissions are to be reviewed by the FDA within 6 months of their receipt.

Postapproval Maintenance

Once a NDA/BLA is approved and the product can be made commercially available, the obligations of the sponsor and the FDA continue as long as that product is on the market. If there are no significant safety issues that arise after the product is on the market, the maintenance of that product by both the sponsor and FDA is usually a routine process. Annual reports are provided by the company to the regulatory authority. Because a clinical program of an investigational product can never fully identify all the possible safety issues with a product, one of the most critical postapproval responsibilities of the sponsor and the FDA is the monitoring of the reported safety profile of the product when it is in general use.

The sponsor's obligations to support a product postapproval can be product-specific, in the form of postapproval commitments (studies), or general requirements that apply to all products (Fig 7.27). During the final phases of the approval process, the FDA may request a commitment from the sponsor to conduct certain studies after the product has been approved. Although the sponsor has some room to negotiate these postapproval commitments, with the approval of the product being held in the balance, most sponsors agree to conduct the studies proposed by the FDA. The specific commitments and time frames for completion are outlined in the approval letter. For every product that is approved, the sponsor is required to conduct safety surveillance, designed to capture and evaluate all reported adverse events, and submit these data to the agency. As the population exposed to the product is much greater once the product is available on the market and has broader demographics, concurrent disease, and concomitant medications, it is not unusual to see adverse events that were not observed in the clinical trials; or, the same adverse events may be observed but at a higher incidence or at a greater level of seriousness than observed during the clinical trials. Additionally, if the sponsor makes any major

- Post-approval commitments (studies):
  - Voluntary or required
- Post-approval safety surveillance requirements
- Changes to an approved application

FIG. 7.27. Post-approval NDA/BLA activities

changes to the product or to the package insert from what was initially submitted and approved, those changes need to be submitted and approved by the FDA.

As described earlier, during the final phases of the approval process, the FDA may require the sponsor to do additional work on the drug product after it has been approved. These postapproval commitments range from generation of additional safety and/or efficacy data to generation of additional data on the manufacture and control of the drug or biologic product (Fig. 7.28). During initial drug development, studies are conducted in a relatively small number of patients selected under strict inclusion and exclusion criteria with limited long-term safety data. Special patient populations, such as patients with concurrent diabetes or heart conditions, renal failure or hepatic failure, or pediatric and geriatric patients, may not have been specifically studied during the development program. Subpopulations within a general disease indication may need further exploration (e.g., at different disease stages or levels of severity). Due to the exclusivity provisions for studying products in a pediatric population, the sponsor may elect to do this work to obtain the 6-month market exclusivity that this work can afford. In addition, drug interaction information is likely limited at the time of product launch and may need to be further studied. Long-term safety studies may be required to determine whether there are unique safety issues associated with chronic exposure to the product. If the sponsor has agreed to postapproval commitments, then periodically they must advise FDA of the status of their work on these commitments. If the sponsor does not conduct the studies to satisfy these commitments in a timely fashion, the FDA may withdraw approval of the application.

Typically, the chemistry, manufacturing, and controls section of an approved NDA/BLA will require updating postapproval. For some products, real-time stability data must be generated to support extended expiration dating. The time the product is on the market is additional time the sponsor can study the stability of the product and determine its maximum shelf-life. Typically, the sponsor can extend expiration dating of the product based on additional real-time data that meets a FDA-approved stability protocol and change the expiration date on the product without obtaining FDA approval. It is required that those data be submitted to the FDA at the appropriate time.

For many products, FDA allows the sponsor to scale up the production of the product by 10-fold to accommodate the commercial demand, as long as the impact on the quality of the product is not different at this larger scale. The first lots made on a scale larger than what was approved in the NDA/BLA must be placed on a stability program and the appropriate validation data collected and available for FDA review.

Postapproval safety surveillance is critical for further defining and refining of the safety profile of a drug (Fig. 7.29). Data generated from clinical trials are from a small, tightly controlled subset of the overall patient population, and clinical trials are usually not large enough to detect rare adverse events. Vigilance on the part of the FDA, the sponsor, and the public must be maintained to ensure new and more serious adverse events are identified and promptly incorporated into the package insert so that prescribers and patients are aware of the risks associated with the product. Also, the most common reason for withdrawal of a product from the market by FDA action or voluntarily by a company is a serious adverse event situation often unanticipated based on the NDA safety file, especially of a cardiovascular or liver nature.

Sponsors are required to evaluate and report adverse events from all sources, including those reported to them in both the United States and foreign countries, the published literature, and postmarketing studies. If an adverse event occurs that is "serious and life threatening," it must be submitted to FDA within 15 calendar days of the receipt of the report by the sponsor. Once the drug is approved, the sponsor must submit all adverse experience reports, not just those that are serious or life threatening, and an evaluation of these reports, to the FDA every 3 months for 3 years. After 3 years, unless otherwise specified, reports must be submitted annually for as long as the product remains on the market.

FDA may determine that a drug can be approved but that the safety profile needs to be carefully monitored in patients

- Clinical:
  - Safety data in additional patients
  - Long-term safety data
  - Special patient populations (i.e. renal failure, diabetes)
  - Drug interactions studies
  - Pediatric trials and exclusivity
  - Geriatric studies
- Chemistry, Manufacturing, and Controls:
  - Long-term stability
  - Manufacturing data at scale
  - Validation data

FIG. 7.28. Post-approval Commitments

- Sponsor requirements
- Registry / risk management
- Med Watch:
  - Voluntary reports from health care professionals
  - Types adverse events to report:
    - Death
    - Life-threatening hazard
    - Hospitalization
    - Disability
    - Birth defects
- Actions based on new safety issues:
  - Medical Alerts
  - Revised Labeling
  - Boxed Warnings
  - Product Withdrawals

FIG. 7.29. Post-approval Safety Surveillance Requirements

who receive the drug. This can be accomplished through a registry, where each patient receiving the drug is "registered" in a database and followed periodically to collect additional safety information. A patient registry can also be used for a subset of patients (e.g., to monitor for birth defects in patients who take the product during pregnancy).

Another mechanism to uncover new adverse events is MedWatch. This is a voluntary reporting system established by the FDA whereby the general public can report serious adverse events, although most of the reports are received from health care providers such as doctors, nurses, pharmacists and dentists. MedWatch reports should only be submitted for serious adverse reactions. Serious adverse events are defined with the following criteria: 1, death (if an adverse event from the drug resulted in a patient's death); 2, life-threatening (if a patient was at risk of death at the time of the adverse event); 3, hospitalization (the adverse event requires a patient to be hospitalized or an existing hospitalization to be prolonged); 4, disability (if an adverse reaction results in a persistent or significant disability/incapacity); and 5, birth defects (a congenital anomaly/birth defect).

Depending on the severity of the adverse event profile, FDA, in conjunction with the sponsor, has several avenues available to alert health care providers, patients, and the general public of potentially serious adverse events. Medical alerts can be issued in the form of "Dear Doctor" letters. These letters are sent to physicians advising them of the new serious adverse event. The package insert can be revised to add the event or strengthen a warning regarding an existing event. If FDA feels an event needs to be prominently displayed, they can require that it be outlined in a black box in the package insert, a so-called "black box warning," or an adverse event could lead to a new "contraindication." Lastly, although rare, a drug can be withdrawn from the market either voluntarily by the sponsor, or as a mandate from the FDA, if the nature of the adverse event significantly changes the balance of risks and benefits to the patient.

After BLA/NDA approval, a plethora of activities can occur that requires submission of additional information to the FDA (Fig. 7.30). These changes can occur as a result of further experience in manufacturing and testing, new clinical information, the need for risk mitigation, or the desire to grow the market. The more significant the implications of the change, the greater likelihood FDA will need to review and approve it prior to implementation. Chemistry, manufacturing, and controls changes that occur after approval include such things as adding a new supplier of drug substance, adding a new finished product manufacturer and or packager, adding new vial/bottle sizes, changing the batch size, or changing the labeled storage conditions. Clinical changes could include new indications for the use of the product, new safety information, or new pharmacokinetic data in special populations. New information that impacts the product's package insert needs to be submitted, as well as advertising copy updated to reflect the new information.

- Types of changes:
  - Chemistry and manufacturing
  - Clinical
  - Labeling
  - Advertising copy
- Supplements to an approved application:
  - Prior approval supplements
  - Changes made at time supplement is filed
  - Annual Report changes

FIG. 7.30. Changes to an Approved Application

Major changes to an approved application require a supplement to the NDA (sNDA) or BLA describing the change and providing the necessary supporting documentation for FDA review and approval prior to implementation. Examples of these types of changes include broadening a drug substance or finished product specification, changing or adding a manufacturing facility where the facility is materially different from the approved facility, or any major change in labeling. These are called "Expedited Review Requested" supplements.

Clinical changes including new indications for the use of the product are major changes and require a supplemental NDA or BLA to be submitted. A new indication requires full phase 2 and 3 clinical studies for efficacy and safety. A new indication also may require preclinical work for safety and/or clinical pharmacokinetic studies. The registration strategy for Enbrel® (etanercept) included an original BLA submission for treatment of moderate to severe rheumatoid arthritis. After the initial BLA approval, supplements were submitted and approved for treatment of moderate to severe acute polyarticular-course juvenile rheumatoid arthritis, psoriatic arthritis, ankylosing arthritis, and chronic moderate to severe plaque psoriasis.

Less significant changes are submitted in a "Changes Being Effected" supplement. These changes can be implemented after the sponsor has submitted the supplement and before FDA approval. Examples of these types of changes include adding a specification that will provide added control to the manufacture of the product, changing or adding a new manufacturing facility that is not materially different from the approved facility, or changes in the package insert that strengthen instructions about dosing, precautions, warnings, or adverse reactions.

Minor changes can be reported in the annual report to the NDA or BLA. The sponsor is required to submit this report near the anniversary date of the product's approval. These changes include minor changes to the package insert such as grammatical changes, deletion of an ingredient that serves only to add color to the product, or an extension in expiration dating based on real-time data that conforms to an FDA-approved stability protocol.

Compliance/Quality Assurance

Compliance with the regulations, and the assurance that a company is in compliance, underlies all the regulatory aspects that have been discussed thus far in this chapter. The FDA requires that any drug or biologic product, whether at the investigational stage or being marketed, is manufactured, studied, and marketed in accordance with its regulations, and the FDA has substantial power to enforce their compliance to the regulations. Even the most innovative product for a highly unmet medical need cannot be expected to reach the market, or stay on the market, if there are serious compliance issues surrounding it.

Thus, a critical function of a regulatory affairs organization in any pharmaceutical or biotechnology company is quality assurance. It is the responsibility of quality assurance to both monitor a sponsor's adherence to the compliance requirements and to help the organization establish systems and processes that build compliance into the development (research), manufacture, and marketing of a drug or biologic product.

Compliance is generally divided into the six categories listed in Fig. 7.31. Like all FDA regulations, there is a great deal of interpretation of the regulations addressing these areas of compliance. FDA issues guidance documents to further explain to the regulated industry their views on compliance. Additionally, the FDA makes selected documents on their compliance findings of pharmaceutical and biotechnology companies available to the public. These documents are used by the industry as a judge of what is, and is not, acceptable to the FDA.

The compliance regulations apply to both a sponsor company and any service provider (e.g., contract research organization or contract manufacturer) that a sponsor company contracts with to conduct their work. The sponsor is ultimately responsible for ensuring that any service provider that they hire conducts their services consistent with the compliance regulations.

Good laboratory practices (GLPs) regulate the conduct of nonclinical laboratory studies that support or are intended to support NDAs or BLAs. The GLP regulations cover all aspects of a nonclinical trial that involves the use of animals and the laboratory practices associated to support animal studies. Good manufacturing practices (GMPs) regulate the manufacture, control, accountability, and documentation of drug or biologic substances and products manufactured for human use. These regulations ensure that products introduced into humans are of acceptable quality and that a product is appropriately labeled. The regulations also require adequate controls for the release and security of the products before they are allowed to go into the general marketplace. Good clinical practices (GCPs) address the conduct of clinical trials and the rights and safety of human subjects. These regulations cover the conduct of the sponsoring company, the clinical investigators, and the institutional review boards responsible for review and approval of the conduct of a protocol. The GCPs cover not only the technical aspects of a clinical trial but also the ethical considerations and informed consent process involved in inviting someone to participate in a clinical trial.

Certain electronic records and electronic signatures used to create and maintain records in support of GLPs, GMPs, and GCPs are subjected to regulation. The regulations that cover electronic records are in Part 11 of the Code of Federal Regulations and are commonly known simply as "Part 11" regulations. The intent of the Part 11 regulations is to ensure that electronic records of documents required under the other regulations are authentic and have the same basis of integrity as paper documents with handwritten signatures.

Lastly, the FDA regulates the advertising and promotion of marketed pharmaceuticals and monitors the industry to ensure they do not inappropriately promote products or indications that are not yet approved. FDA regulates all promotional aspects, from the direct-to-consumer ads seen on TV to what research publications are actively shared with prescribers.

The power of the FDA in the area of compliance is extensive. They have the right to inspect a sponsor company, a nonclinical or clinical investigative site, and a manufacturing site with or without notice. In the worst of offenses, they can bar a clinical investigator from conducting any future clinical trials, close down an institutional review board, seize manufactured product from being marketed, and deny or withdraw drug approvals. Therefore, the compliance regulations are to be taken seriously, as negative consequences can be significant.

Good laboratory practices are a set of regulations designed to establish standards for the conduct and reporting of nonclinical laboratory studies and to ensure the quality and integrity of nonclinical safety data submitted to the regulatory authorities (Fig. 7.32). It is not required that every nonclinical study (e.g., experimental pharmacology study) be conducted to the GLP standard. However, for studies that are considered pivotal to determining the safety of a drug in animals (e.g., carcinogenicity study), it is required that the study be conducted in compliance with the GLP regulations.

The GLP regulations have major categories, each of which describes the requirements that a sponsor must meet in that area. The first of these is the personnel who work in the laboratory facility and their responsibilities. Each study is to have a study director who is a scientist with the appropriate education, training, and experience. This person has overall responsibility for the technical conduct of the study, as well as for

- Good Laboratory Practices (GLP)
- Good Manufacturing Practices (GMP)
- Good Clinical Practices (GCP)
- Electronic Records (Part 11)
- Prescription drug advertising and promotion
- FDA enforcement activities

FIG. 7.31. Compliance

the interpretation, analysis, documentation, and reporting of results. Additionally, a quality assurance (QA) unit is required. It is the responsibility of the people in this unit to monitor each study to ensure that the facilities, equipment, personnel, methods, practices, records, and controls are in compliance. The QA unit must be independent of the personnel engaged in the conduct of the study, to ensure there is no conflict of interest in their reporting.

The regulations describe the required physical facilities needed to conduct a GLP trial. The regulations stipulate that appropriate animal care facilities, as well as laboratory space, be available to conduct the study. Equipment that is used in the collection or assessment of data collected in the study must be appropriately maintained and calibrated to ensure the equipment is generating accurate data. The laboratory must have standard operating procedures (SOPs) written, and in effect, that set forth the procedures to be done in order to ensure the quality and integrity of the collected data. In addition, it is necessary to document that laboratory personnel have been trained on these SOPs.

It is a tenet in regulatory compliance that "if it's not documented, it didn't happen." An inspector from the FDA can only inspect the documentation that supports the conduct of a study, as they are not present to observe the study as it is being conducted. Regardless of whether or not a procedure was done, if a procedure has not been documented, it is presumed that the procedure was not completed. Therefore, the documentation that allows an inspector to see that all regulations are being followed is absolutely critical.

The consequences of a pivotal nonclinical safety trial not being conducted in compliance with GLPs are substantial. In the worst cases, the FDA may disqualify a study if it was not conducted according to the compliance standards. In the case of a 2-year carcinogenicity study, a disqualification would be a serious blow to the ongoing development of a drug product, the timing of its submission for approval, and the loss of patent life.

The regulations for good manufacturing practices are intended to ensure that minimum manufacturing standards are established and adhered to, such that only quality products are produced and/or sold in the United States (Fig. 7.33). These regulations apply to all drug and biologic products intended for human consumption, without exception. The location of the manufacturing plant can be anywhere in the world. As long as the plant is producing product that will be sold in the United States, it is subjected to these regulations and to inspection by the FDA. These regulations apply to both the manufacture of the drug or biologic substance and the drug or biologic product.

The regulations set forth the minimum methods, facilities, and controls to be used for the manufacture, processing, packing, or holding of a drug product, to ensure that the drug has the identity and strength as labeled and meets the required quality and purity characteristics. The GMP regulations establish the minimum standards for the personnel manufacturing the product, along with the facilities and the equipment used to make the product. The regulations in these respects are extensive. Examples of some of the items described in these sections of the regulations include the protective gear worn by personnel during the manufacturing process, the air-handling systems in the manufacturing plant, and the cleaning of equipment that comes into contact with the drug product.

It is required that all components of the drug product, the containers into which it is packaged, and the labels used to identify the product are controlled at all times to avoid any contamination or use of the wrong component. Multiple layers of control are necessary during the actual manufacture of a product and once a finished product has been produced.

Once a product comes off the manufacturing line, it is the responsibility of a quality control unit to conduct the testing to determine that the finished product has the appropriate identity, strength, quality, and purity to meet predetermined specifications. A product is placed in quarantine until all the testing has been completed and the product is determined to pass all specifications. Once it has been determined that the product meets specifications, it is formally "released," which allows it to enter commercial distribution. If a product is produced that does not meet specifications, it must be quarantined indefinitely and an investigation conducted to determine the cause of the manufacturing failure.

- Intended to establish standards for conduct and reporting of nonclinical laboratory studies and to assure quality and integrity of safety data submitted to FDA
- Required for selected, but not all laboratory studies
- GLP elements:
  - Personnel, including Study Director and QA unit
  - Facilities for animal care, test article control, specimen control
  - Equipment maintenance and calibration
  - Appropriate Standard Operating Procedures (SOPs) for all aspects of nonclinical study conduct
  - Documentation of study conduct
- Study results can be disqualified if study not in compliance.

FIG. 7.32. Good Laboratory Practices (GLP)

- Intended to ensure that minimum manufacturing standards are adhered to, thereby helping to ensure the quality of drug products produced and/or sold in the US
- Required for all drug products intended for human consumption
- GMP elements:
  - Organization and personnel, building and facilities, equipment
  - Controls of components and drug product containers and closures, production and process controls, packaging and labeling controls
  - Holding and distribution, laboratory controls, records and reports, returned and salvaged product
- Misbranded and/or adulterated product

FIG. 7.33. Good Manufacturing Practices (GMP)

Documentation during all steps of the manufacturing process is required and critical. Documentation must be maintained on each manufacturing lot for the length of time that the product is presumed to be in potential human use. In the event of a problem identified after it has reached the commercial marketplace, these records will be carefully reviewed to determine if an error was made, and mistakenly overlooked, during manufacturing.

Unfortunately, even with all the many layers of control, manufacturing errors do happen on occasion. Oftentimes, a problem is reported to a sponsor, an investigation is conducted, and a determination is made that an error has resulted in a misbranded or adulterated product reaching the marketplace. Depending on the nature of the misbranding or adulteration, this can result in a drug recall to the wholesale level, pharmacy level, or, in the most serious case, to the patient level. In August 2004, the biotechnology company Chiron® identified bacterial contamination in some lots of their influenza vaccine and advised the FDA and the CDC of their ongoing investigations. Subsequently, the FDA inspected the manufacturing facility in the United Kingdom and announced that none of the 48 million doses of vaccine were safe for use because of significant deficiencies in quality control. Ongoing FDA oversight will be significant to monitor that the appropriate corrective actions have been taken to ensure that Chiron's manufacturing facility can produce safe and effective vaccines.

Good clinical practice (GCP) regulations are designed to both ensure the integrity of clinical data upon which product approvals are based and protect the rights, safety, and welfare of human subjects (Fig. 7.34). The FDA has written regulations for GCPs, and the International Conference on Harmonization has also written extensive guidance on GCPs. It is required that all clinical trials being conducted under a U.S. IND are conducted to GCP standards, regardless of the country in which the trial is being conducted. This includes clinical trials being conducted on either investigational or marketed products. There are no exceptions to this requirement.

A major component of GCPs includes the protection of the human volunteers who elect to participate in a clinical trial. The primary body responsible for protecting people who may potentially participate in the trial is the institutional review board (IRB). It is the responsibility of the IRB to evaluate the clinical protocol, available safety and efficacy on the finished product, and the informed consent. The purpose of this review is to ensure that the protocol is scientifically worthy of exposing volunteers to the products and procedures outlined, that the product is reasonably safe as used in accordance with the protocol, and that the patient is adequately informed of the risks they are assuming by participating. Most institutions (e.g., hospitals, academic centers) have an institutional review board that is responsible for the review of all clinical trials conducted in that institution. There are also commercial IRBs, who for a fee will conduct all the necessary work and assume all the responsibilities of IRB oversight. Commercial IRBs are typically used by clinical investigative sites that are not part of an institution. Regardless of the affiliation of the IRB, all are required to follow the appropriate GCP regulations and are subjected to FDA inspection.

The GCPs also require that clinical investigators are qualified by training and experience to do the work required of them in the conduct of a clinical protocol. The clinical investigator is responsible for seeing that the study is conducted according to the protocol, that the rights, safety, and welfare of the individual volunteer are maintained, and that the investigational drug product is controlled. The principal clinical investigator must sign a FDA Form 1572, in which they commit to conduct the trial according the applicable regulations and in which they assume responsibility for the conduct of subinvestigators and staff during the study.

The regulations further outline a host of responsibilities of the sponsor in the conduct of the trial. This starts first with the requirement to write scientifically sound clinical protocols. It would not be ethical to expose volunteers to the potential risks of the drug and procedures required in a clinical protocol if the protocol wasn't appropriately designed to allow valid scientific conclusions to be made. The regulations require that the sponsor monitors each clinical investigator and clinical site to ensure that the conduct of the protocol and the informed consent process is proceeding as planned and that the data being collected are accurate and have integrity. The sponsors are also responsible for keeping all investigators informed of any new pertinent safety data that might become available during the conduct of the trial.

As with all other compliance regulations, it is necessary that the IRB, the clinical investigator and staff, and the sponsor have adequate written documentation of all the procedures followed during the study so they can withstand FDA inspection. There are potential consequences for IRBs, clinical investigators, and sponsors if GCPs are not followed. The operations of an IRB can be closed down, a clinical investigator

- Intended to ensure integrity of clinical data on which product approvals are based and to help protect rights, safety, and welfare of human subjects
- Both FDA regulations and ICH guidelines govern conduct of trials involving human subjects
- Required for all clinical trials
- GCP elements:
  - Institutional Review Board review and approval
  - Informed consent
  - Qualified investigators
  - Adequate sponsor oversight
  - Scientifically sound clinical protocols
  - Documentation of study conduct
- Study results can be disqualified; investigators can be debarred

Fig. 7.34. Good Clinical Practices (GCP)

may no longer be able to conduct clinical trials, and/or the study results can be disqualified from use by a sponsor in support of a new drug approval.

Perhaps the most currently controversial set of compliance regulations is those regulating electronic records (commonly referred to as Part 11; Fig. 7.35). These regulations were written to ensure that electronic records and electronic signatures have the authenticity, integrity, and confidentiality of their equivalent paper records. The FDA wrote them in response to the reality that many records are kept electronically and those records need to be subjected to control so they cannot be altered, either intentionally or unintentionally. The scope of the regulations is limited to those records required under other regulations (e.g., GLP, GMP, and GCP records). They became controversial when their interpretation by both sponsors and FDA inspectors made them overly burdensome. The FDA is currently evaluating the regulations, with the intent of achieving the goal of integrity of electronic records without the unintended consequences wrought by the early interpretations.

The Part 11 regulations require that selected computer systems be validated. This validation procedure requires an analysis of risk and that an installation qualification and operation qualification be completed and documented. This validation is intended to demonstrate that a particular computerized system, be it hardware and/or software, functions in the way it is designed, preserves data integrity, and provides an audit trial of any changes to an electronic record. There must also be security, both physical and password protection, on systems to ensure that no unauthorized individual can make a change, either intentional or unintentional, to an electronic record.

For business purposes, a sponsor may want to have the capability to have electronic signatures of required documents. The regulations are written to ensure that an electronic signature is considered the legally binding equivalent of traditional handwritten signatures, which requires a sophisticated level of security on that electronic signature. And as true with all other compliance regulations, extensive documentation is required on not only the configuration of computerized systems but also risk analyses, validation, and security of these systems.

The regulation of prescription drug advertising and promotion by FDA is intended to achieve two main purposes: (1). ensure that the pharmaceutical industry does not promote unapproved drugs or unapproved indications of marketed products, and (2). ensure that prescribers and consumers are not misled through advertising that is false and misleading or lacks "fair balance" (Fig. 7.36). Every promotional advertisement or piece shown or given to prescribers (e.g., drug brochures or calendars) is submitted by the sponsor to the FDA for review. The FDA also has the authority to regulate all direct-to-consumer advertising for prescription drugs, including both written ads and those used in the broadcast media. Unlike other compliance regulations that are written to inform a sponsor on what they should do to be in compliance, many of the regulations for drug advertising are written to tell a sponsor what they can't do. It is not the intent of the FDA to stifle promotional creativity, which would result if they only dictated all the features of a drug advertisement.

Advertisements cannot be false or misleading with respect to side effects, contraindications, or effectiveness, nor can advertisements fail to present a fair balance between information relating to side effects and contraindications, and information relating to effectiveness. The regulations require a brief summary of the side effects and contraindications of the drug product to accompany each advertisement. For these reasons, drug ads in the written media are accompanied by a brief summary of the side effects and contraindications from the package insert, and direct-to-consumer ads have statements concerning the side effects and contraindications included in the voice-over of the advertisement. Under very limited circumstances, sponsors are allowed to advertise that a new product is coming to the market in the near future. For example, separate "coming soon" advertisements can state the name of the product or the disease to be treated, but no linkage of the two concepts can be made, and the two advertisements can never be used together.

The FDA recognizes that full exchange of scientific data on drug products and discussions of those data fulfill an important educational need. Educational activities that are deemed by the FDA to be independent from influence by the pharmaceutical company and nonpromotional in nature have not been treated as advertising or labeling and have not been subjected to the agency's regulatory scrutiny.

- Intended to ensure authenticity, integrity and confidentiality of electronic records
- Applies to records in electronic form that are used to fulfill FDA regulated records requirements
- Part 11 elements:
  - Validation of computerized systems
  - Security of electronic records, including audit trails
  - Electronic signatures
  - Computer systems documentation

FIG. 7.35. Electronic Records (Part 11)

- Intended to ensure that unapproved products are not promoted
- Intended to ensure a product is not misbranded through advertising that has false and misleading statements, and lacks fair balance
- Applies to all written and broadcast ads to both prescribers and consumers
- Applies to verbal exchanges between prescribers and pharmaceutical sales representatives
- Does not apply to non-promotional and independent scientific and educational exchanges

FIG. 7.36. Prescription Drug Advertising & Promotion

7. Laws and Regulations

- Types of GMP inspections:
  - Pre-approval inspection
  - Post-approval inspection
  - Surveillance good manufacturing practice inspection
- GCP and GLP Inspections:
  - Inspection of clinical sites
  - Inspection of animal testing facilities
- The inspection process:
  - May or may not be announced
  - Site receives "Notice of Inspection"
  - Inspector's observations –Form FDA 483
  - Establishment Inspection Report (EIR)

FIG. 7.37. FDA Inspections -1

- FDA inspection classification system:
  - No action indicated – NAI
  - Voluntary action indicated – VAI
  - Official action indicated – OAI
- Debarment list
- Disqualified/restricted/assurances list for clinical investigators

FIG. 7.38. FDA Inspections -2

In order to enforce the above-described FDA regulations on compliance, the agency is authorized to conduct inspections at any time to ensure that a sponsor, manufacturer, animal laboratory, or clinical investigator is following the regulations (Fig. 7.37). As FDA is responsible for ensuring that drugs and drug products are not adulterated or misbranded and are manufactured, packaged, tested, and distributed in accordance with good manufacturing practices, they conduct periodic inspections of facilities engaged in these activities. There are three types of GMP inspections: preapproval, postapproval, and surveillance good manufacturing practice inspections. A preapproval inspection is conducted when the NDA, BLA, or ANDA is under review at the agency. The purpose of the inspection is to confirm that the application accurately reflects the chemistry, manufacturing and control activities in the facility and that there are no significant GMP issues that need to be resolved prior to product commercialization. If there are significant issues, approval of the application will be delayed until corrective action is taken. After the drug product is approved, FDA will reinspect the facility to evaluate validation and testing data associated with manufacturing multiple large-scale commercial lots. They will also confirm that any changes to the process or test methods have been appropriately documented and, if required, submitted to FDA. Both of the above inspections are usually product specific. During a surveillance GMP inspection, a FDA inspector looks at the overall manufacturing facility, personnel, and operating systems and does not focus on one product. These inspections can be biannual or inspections "for cause." If the FDA receives product complaints such as product tampering or faulty packaging, the agency can send a team to do a "for cause" inspection to investigate the potential causes for the manufacturing problem.

The NDA/BLA preapproval process also includes clinical site inspections. Typically, several sites from the key pivotal trials are inspected to determine if the investigational product was appropriately dispensed, patients gave informed consent, study records were maintained appropriately, and there was appropriate clinical investigator oversight during the patient visits and evaluation of test results. The intent of the clinical site inspection is to ensure the studies were conducted in accordance with good clinical practices and that there were no fraudulent data reported. Case report forms can be checked against actual patient records to assess fraud. The FDA will also conduct an inspection of selected animal testing site(s) to ensure animal handling and testing was done in accordance with good laboratory practices.

Inspectors from the FDA can notify the company or site of their upcoming inspection or they can arrive unannounced. When an inspector(s) arrives, they provide the company or site with a "Notice of Inspection," which contains the purpose of the inspection. At the conclusion of the inspection, which can last a few hours to several weeks, specific and significant inspection observations are identified on Form FDA 483, with the most significant observations listed first. The inspectors will meet with key company individuals or the clinical investigator to review the results of the inspection, including the Form FDA 483 observations, prior to their final departure. The company or site is obligated to respond to the observations in Form FDA 483. An establishment inspection report (EIR) is issued, which provides a detailed summary of the overall inspection. Both the Form FDA 483 and the EIR are available to the general public, although certain proprietary information is redacted.

Once the EIR has been prepared, the inspection is given one of three classifications: no action indicated (NOI), voluntary action indicated (VAI), or official action indicated (OAI) (Fig. 7.38). A classification of NOI indicates there are no FDA findings that are objectionable and that there are no obligations by the company or site to institute any changes. The classification of VAI is assigned if inspection findings are not serious in nature, and particularly if the company or site has already voluntarily made corrections at the time of completion of the EIR. A company or investigator receives an OAI classification when the inspection findings are substantive enough to warrant regulatory or administrative action. Some types of regulatory actions include civil penalties, injunctions, product seizures, voluntary recall, or a warning letter.

Individuals who have committed criminal acts associated with the activities of drug development can be debarred. Types of debarment acts include falsifying data, lying to inspectors, enrolling nonexistent patients, and reporting false test results for manufacturing lots. All of these activities serve to seriously undermine the safety of the general public. Debarred individuals are forbidden from working for companies involved in drug development. The names of debarred individuals are public information. It is important that pharmaceutical companies be very familiar with this list so that they do not hire anyone debarred in any capacity in the company or as a consultant or investigator for the company. At the time of submission of the NDA/BLA/ANDA, the sponsor must certify that no debarred individuals have worked on the product's development, either within the company or as a consultant or investigator. If the company has used a debarred person, it can be fined up to $1 million and the debarred individual up to $250,000.

Clinical investigators who repeatedly or deliberately fail to comply with good clinical practices or submit false information during the conduct of clinical trials, if not debarred, may be disqualified from participating in future studies. Investigators who are "totally restricted" are not allowed to receive investigational drug, and investigators who are "restricted" can participate in the conduct of studies in accordance with the defined restrictions.

FDA has several formal enforcement avenues to deal with companies and individuals who are not complying with the appropriate regulations and statutes (Fig. 7.39). Warning letters, sent from the FDA to a sponsor or individual, outline compliance violations that need to be corrected immediately. For example, warning letters are typically issued when a company fails to adequately correct numerous manufacturing issues or is consistently using advertising and promotional materials that are misleading or lack fair balance, or when a clinical investigator is consistently not providing adequate oversight to patients in a clinical trial. If the infractions are related to good manufacturing practices, the FDA can use recalls and field corrections to remove potentially unsafe products from the market. A sponsor can initiate a voluntary recall of product, or the FDA can order a recall. The extent of a recall is determined by the potential seriousness of the health issue. Product may be recalled from the wholesaler and/or pharmacy and in the most extreme cases from the patient. Recall information is made public weekly in the FDA Enforcement Report. For products that pose a serious risk, a public warning will be issued via various media avenues such as radio and television news broadcasts, as was done in the case of Tylenol® tampering several years ago.

If the FDA cannot convince a company to voluntarily cooperate, they can petition a federal court for an injunction to legally mandate that a company cease making and distributing a product or products. An example of such a situation is products containing ephedra. After determining that the ingredient provided only short-term weight-loss benefits with potentially serious risks of heart ailments and stroke, FDA ordered ephedra manufacturers to cease distribution and have all products containing ephedra removed from the market permanently. Those companies that did not voluntarily comply were subjected to seizures and injunctions. By filing a formal complaint with a U.S. District Court, the FDA can also initiate enforcement activities in the form of a seizure of product. If granted by the court, a federal marshal can physically seize the product until the issue is resolved. In general, companies voluntarily comply with FDA directives, and seizures are not required.

A consent decree is yet another option the FDA has to enforce compliance. A consent decree is an agreement between the FDA and the sponsor submitted in writing to a court. Once approved by a judge, it becomes legally binding. Consent decrees usually occur when companies continually fail to correct deficiencies or meet correction deadlines despite multiple warning letters and FDA inspections. To date, it has been primarily used in cases of egregious manufacturing noncompliance. A consent decree is very serious. It binds the sponsor to precise ways of doing their work, with appropriate external and FDA oversight. It can take years of intense corrective work, independent outside audits, and FDA interactions to have the decree lifted.

Lastly, the FDA has the authority to suspend or revoke any product approval. The primary reason for a product suspension or revocation would be a serious safety issue with a product that a sponsor refused to voluntarily withdraw from the market.

The responsibilities of the FDA and sponsors are enormous. They both have a profound potential to impact the nation's public health, and they do so in a competitive marketplace, where public companies are held accountable to provide their shareholders with a reasonable return on their investment. Not only is the public health at issue but also the financial health of companies. Over the years, these tensions have given rise to controversies in regulatory affairs that are not easily resolvable (Fig. 7.40).

- Warning letters:
  - Letter to correct violations promptly or enforcement action could be taken
- Recalls and field corrections:
  - Action taken by firm to remove product from field or conduct a field correction
- Injunction:
  - Civil action to stop production or distribution of violative product
- Seizures:
  - FDA initiates seizure based on violation of the law. U.S. Marshal takes possession of goods.
- Consent Decree
- Product/license suspension or revocation

FIG. 7.39. FDA Enforcement Activities

- Balancing Priorities:
  - Speed of new product approvals versus sufficient safety assessment
  - Innovative pharmaceutical companies, versus generic companies, versus payers (health care)

FIG. 7.40. Controversies in Regulatory Science

One of the long-standing controversies is the balance between the speed with which a new product becomes available to patients who desperately need access to new therapies and the sponsor and FDA fully understanding its safety profile. The faster a drug is developed and approved, the less likely the safety profile is fully known. This puts the FDA in the position of judging whether it is better to approve a drug quickly or to wait until additional safety studies have been performed. This is also a dilemma for a sponsor. Whereas the sponsor would like to begin selling a product as soon as possible, they do not want to prematurely sell a product that may unpredictably cause harm.

Another long-standing controversy is that of balancing the interests of innovative pharmaceutical companies, generic companies, and the payors of health care. Innovative companies would like to have their investment in research and development be protected as long as possible, through patent protection and market exclusivity. Generic companies want to be able to compete with brand products as soon as possible. And the payors of health care, the largest being the federal government, would like the prices of pharmaceuticals to be as low as possible. The balance of these forces has led to intricate laws, which through FDA regulations try to effectively balance these two interests. As pressures continue to mount on the cost of health care, this issue will surely continue to be readdressed. The FDA is also making contributions to streamline the development of new products through reevaluation of the regulatory requirements. If the agency can lessen the regulatory burden on companies while not compromising on demands to demonstrate safety and efficacy, this should reduce the investment needed by sponsors to bring new therapies to the market.

The above examples are just a couple of many classic controversies that continue to shape the thinking and performance of both the FDA and sponsor companies. These interests are served through the enactment of laws, which lead to regulations, which lead to the discipline of regulatory affairs.

General References

1. Matthieu M (ed). New Drug Development: A Regulatory Overview, 6th Ed., Parexel International Corp., Waltham, MA, 2002
2. Matthieu M (ed). Biologics Development: A Regulatory Overview, 3rd Ed., Parexel International Corp., Waltham, MA, 2004
3. Available at www.fda.gov

8
Clinical Trial Operations

Carl L. Roland and Paul Litka

Introduction	178
Clinical Trial Conduct	181
Contract Research Organizations (CROs)	193
Selected Issues in Clinical Development	197
Summary	200
References	200

This chapter focuses on the activities and operations related to the conduct of clinical trials for new product development, applicable to both U.S. and non-U.S. trials as well. Although the focus of this chapter is on the development of new drugs (including biologics), many of the principles are also applicable to medical device development. Note, however, that while similarities in principle exist, device development is governed by separate U.S. federal regulations. Clinical trial operations are based on good clinical practices, best possible science, use of technology, best business practices, optimal management principles, people with expertise and interpersonal skills, the regulations and laws from governing agencies, and some good common sense. These eight areas, that is, skills and attributes, will be brought into the discussions in this chapter to best elaborate upon operations for clinical trials.

As described previously in Chapters 1–3, the development of a new product is complex, costly, and time-consuming. Therefore, effective planning is necessary. The proper conduct of the clinical trials required by the development plan is critical to the overall process of developing a new drug. This chapter will provide the reader with a broad understanding of what is required to conduct a clinical trial including the responsibilities of those working in the area of clinical operations, the major systems and processes, key outcomes, and relationships between the sponsoring company and the investigator(s), often at universities. Contract research organizations (CROs) can and do perform many of the roles and responsibilities of clinical research on behalf of a sponsoring company; they also are discussed. What and how the work is done in collaboration between a CRO and a manufacturer is addressed. Finally, a series of selected issues and controversies in clinical research in the industry are addressed.

Introduction

A clinical trial is a prospective study that compares the effect, safety, and value of an intervention (e.g., drug or device) against a control in humans. The historical basis for the clinical trial is often attributed to the study of scurvy reported by James Lind, M.D., in his "Treatise on the Scurvy" published in 1753 (Fig. 8.1). Dr. Lind reported on his experiment in 12 sailors with similar cases of scurvy aboard the ship *Salisbury*. Two sailors each were given a 6-day course of varying treatments, and Dr. Lind reported that the sailors who received two oranges and one lemon per day experienced the most sudden and visible good effects. In fact, he reported that one of them was fit for duty on the sixth day [1, 2].

Clinical trial operations are those activities related to the conduct of clinical trials in the product's development plan (Fig. 8.2). Clinical trials showing efficacy and safety are the central element of a marketing application (e.g., NDA in the United States or common technical document [CTD] in Europe). The conduct of a clinical trial is a complex integration of many activities requiring the coordination of a large number of individuals each with specific expertise. Conducting clinical trials can be further characterized as labor intensive, time consuming, and expensive; it is heavily

Historically, the first recorded clinical trial is that of James Lind, MD in his 1753 publication of "A Treatise on the Scurvy".

Clinical Trial is a prospective study that compares effect, safety, and value of an intervention (e.g., drug or device) against a control in humans.

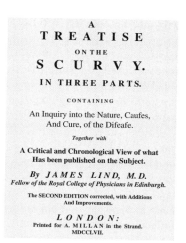

FIG. 8.1. Background – The Clinical Trial

- Clinical trial operations are activities, related to conduct of clinical trials in product's development plan that leads to a marketing application:
  - Involves numerous individuals, each with specific expertise
  - Labor intensive, time consuming, & expensive
  - Heavily regulated (FDA, HIPAA, OIG, NIH, etc)
  - Requires clinicians (investigators) & volunteers (patients)
- Drug development is not an exact science:
  - It is an effort to bring order & structure to process that is not necessarily orderly.
  - Demands use of SOPs & integrated management systems

FIG. 8.2. Background – Clinical Trial Activities

- End Goal must be Kept in Mind – NDA Completion & Approval:
  - Regulatory approval to market new drugs requires demonstration of safety & efficacy.
  - Clinical trials are critical to development process; hence, proper design and conduct of trials are critical.
  - Clinical trials drive development program.
- Health care systems today demand:
  - Evaluation of impact new drug has on cost of patient care
  - This necessitates that pharmacoeconomic studies are put into clinical development plans
- Each day it takes to bring product to market is revenue lost:
  - Therefore, planning is critical to entire drug development process, including clinical trial operations.

FIG. 8.3. Background – Goals

regulated by several government agencies in the United States and around the world (e.g., FDA, OIG, NIH, EMEA, MHLW); it requires experienced clinicians for investigators and willing volunteers for patients. Furthermore, the information collected will be intensively scrutinized and critiqued by the regulatory authorities in each area of the world and in countries. The marketing application must be approved by the Food and Drug Administration (FDA) or European Agency for Evaluation of Medicines (EMEA) or Ministry of Health, Labor and Welfare (MHLW) in Japan before the product may be made available to the public (see Chapter 7). Therefore, an understanding of the regulations and guidelines related to the conduct of a clinical trial is imperative. Drug development is not an exact science, and the regulations/guidelines do not address every issue that will occur. Developers must bring order and structure to a process that is not necessarily orderly. This order demands use of much structure and many controls in standardized forms, detailed standard operating principles (SOPs), guidelines, and integrated management systems.

As described previously, there are a number of different but interconnected disciplines involved during the development of a new product. Therefore, appropriate planning and effective communication within the project development team is important to ensure quality in an efficient manner. The clinical operations personnel are an important part of the development team, and the clinical trials are critical to the entire development process.

The overall goal in the development of a new product is to provide evidence that the new drug is effective and safe for public use (Fig. 8.3). The NDA is the marketing application that is required to seek authorization from the FDA to market a drug in the United States. The European equivalent of the NDA is a CTD, which is submitted to the EMEA. To complete an application and file it with the FDA or EMEA requires a number of clinical trials that provide the primary evidence for the safety and efficacy of the product (e.g., pivotal trials), as well as a program of supportive studies. Because the clinical trials are rate limiting to the completion of the NDA or CTD, they really drive the development plan. Clinical trials are becoming more complex in patient monitoring, larger in size, and broader in scope, plus they often include studies other than those of just safety and efficacy. Given the cost of health care and the expense of new products, the health care systems and payors of today demand more value from new treatments. Many development programs will include pharmacoeconomic studies to evaluate how the new intervention affects the overall cost of patient care. This need demanded by health care systems can add many more trials to the development plan with commensurate increases in costs and time. Often, many clinical trials are required to complete an NDA/CTD. Effective coordination may save years in filing an NDA/CTD. The time of development is important because each day a product is delayed from reaching the market is another day lost in the product's patent life—"time is money." Because of the complexity of a development program and the importance of time, planning is critical to the whole process including operational aspects of clinical plans and trials.

"Plan the work and work the plan." A plan is critical to ensuring efficiency in the complex research process (Fig. 8.4). Planning should be the first activity of the development and clinical team once the decision has been made to develop a new drug (i.e., file an IND or CTA). Thirty days after the IND has been filed with the FDA, the first clinical trial may begin if the FDA has not contacted the sponsor. The first clinical trial will need to be ready to begin on time and progress according to the planned schedule to ensure that all subsequent trials may begin according to schedule. Each trial at all phases (and its associated activities) fits into one plan. Therefore, the timing of the entire development plan begins with the first clinical study. The development team should spend an appropriate period of time planning the development strategy being sure to consider all aspects before the first study begins. As we will explain in more detail later, proper planning requires highly skilled resources, sufficient time, and proper follow-up to evaluate successes, delays, and failures so the plan may be modified as necessary. It is essential that members of the clinical team are part of this planning process.

The clinical trial is conducted according to the regulations and guidelines, as well as generally respected professional standards. Because the activities associated with the conduct of a clinical trial are regulated, they must be documented. An example of this documentation is the standard operating procedure (SOP), which describes in detail how each specific activity is to be completed (e.g., study monitoring). SOPs help assure quality is maintained throughout the conduct of the clinical trial and demonstrate that processes exist to ensure that the regulations are being followed and scientific integrity is maintained. A company may audit itself through its quality assurance department to ensure and demonstrate that the SOPs are being followed. A clinical study plan may contain 5–10 or more separate studies (phase 1, phase 2a and 2b, phase 3 and 3b, economic studies, phase 4). Each trial at all phases (and its associated activities) fits into one integrated sequenced plan. Follow-up on successes/delays/failures are key elements to planning and operations, and such corporate learning is a key success factor.

A wide variety of clinical activities should be ongoing while the clinical development plan is being drafted in order to have efficient start-up of the clinical studies. Some examples include review of studies completed to date (e.g., preclinical studies), completing the investigator brochure, identifying potential investigators and sites, determining timelines to complete trials, designing the trials, and writing the protocols including the statistical analysis section.

Each clinical trial is complex in study design and clinical operations. Trials require study designing, protocol writing, evaluation and selection of investigators and sites, and internal/external agreements. The design and study procedures may require special population group(s) and specialized equipment or techniques, respectively. The many specialized people will be addressed below, which are both internal and external resources.

Clinical trial operations will require a large number of people from different departments and from outside the company with specific expertise (Fig. 8.5). One study often will engage multiple sites and investigators. For example, a single clinical trial may include the following 20-plus different

- Planning - "Plan the work and work the plan":
  - A plan is critical to ensuring efficiency in complex process
  - Appropriate time is required to do it right!
  - Requires skilled resources
  - Each trial at all phases (and its associated activities) fits into one plan
  - Follow-up on successes/delays/failures are key elements to planning
- Need clear objectives for each trial. Goal is to demonstrate following:
  - Drug is safe and efficacious
  - Drug can be manufactured safely and reproducibly
  - Both research & marketing plans are fulfilled
- Clinical activities ongoing while development plan is being drafted:
  - Review of studies completed to date (e.g., pre-clinical studies)
  - Completing investigator brochure
  - Identifying potential investigational sites
  - Designing the trials
  - Determining timelines and costs to complete trials

FIG. 8.4. Background – Planning & Activities

Investigative Institutions:
- Patients
- Investigator Team
  - PI
  - Co-I
  - Coordinators
  - Investigational pharmacist
- Patient Care
  - Nurses
  - Laboratory & Radiology
- Support Operations
  - Research Office
  - Institutional Review Board
  - P&T Committee
  - Information Technology

The Company:
- Clinical Operations
  - CRAs – Monitoring
  - Clinical managers
  - Safety group
  - Biostatistics
  - Clinical Data Management
  - Metabolism / Pharmacokinetics
  - Clinical Research Organization
- Support Operations
  - Safety group
  - Quality assurance – auditing
  - Regulatory affairs
  - Medical writing
  - Information technology
  - Finance
  - Manufacturing – Clinical lots

FIG. 8.5. Background – Personnel in Clinical Trials

personnel at the institution and with the sponsoring company to perform all the functions dictated by patient care, the study protocol, company SOPs, and regulatory and legal requirements. All these research people need to be engaged but also directed and coordinated in the right ways at the right times.

In addition, some of the sponsor activities may be contracted out, which will present a number of further complexities. The aspect of clinical research organizations (CROs) will be discussed later in this chapter. The activities that each of these CRO personnel are responsible for must be coordinated and monitored by the sponsoring company to ensure that quality is maintained and timelines are met. Each clinical trial to be conducted will require good people, process management and leadership to ensure success.

For each clinical trial planned, it is critical that each study have clear objectives that meet the ultimate goal for the product, that is, its approval (Fig. 8.6). The goals of approval are to demonstrate that the drug is safe and efficacious, can be manufactured safely and reproducibly, and fulfills research, patient needs, and marketing plans. The goals as appropriate have a strong clinical focus. They require an understanding of the disease being treated, existing therapies (their benefits and limits), patient care and health care issues, and the product's potential benefit and value to the public 4–8 years before it is available. Therefore, it is essential that there is clinical representation, experienced in both research and patient care practice, on the project team who are involved in the development planning process. Ideally, a training and education session should be provided to clinical operations at the sponsoring company regarding this disease and therapies. Effective communication of information between all groups and within all groups of the project team(s) is critical to a successful project. Furthermore, the objective and study design need to be practical and match the needs of the development plan by answering key planning and operational questions. Five sample key questions should be asked during the planning process in assessing the fit of a study idea into the plan (Fig. 8.6).

Also, each clinical trial that is conducted should answer at least one critical question related to the product's characteristics as it relates to the development plan, for example, Avastin® for lung cancer: primary or secondary treatment, concurrent cancer chemotherapies, dose–dose frequency (daily–weekly), administration of dose (bolus/infusion), lung bleeding, tumor shrinkage versus survival, secondary cancers.

Clinical Trial Conduct

The product's characteristics are often referred to as the product's profile, which initially is the ideal list for a successful product. The profile will change over time as more information from research is uncovered, and the final profile for the NDA will ultimately be described in the product's package insert.

As described in Chapter 2, a product's profile is a realistic description of the new product's characteristics when it is ultimately marketed. Ideally, it should describe the key properties of the product such as efficacy, safety, and formulation information (Fig. 8.7). It may also contain information on the product relative to other similar products that will be based on active controlled trials to be done. Having an understanding of the product's ideal profile to meet patient care and the market needs is critical to the planning process and should be the first task of the project team. The profile is based on the disease, treatment needs, and any early product data, and it evolves over time as studies are done and information is available. Once the product profile is completed, the package insert may be drafted. The overall goal of drug development is to complete a package insert for the NDA (i.e., keep the end in mind). A package insert may be completed in conjunction with the product profile as it provides the framework for describing the product's characteristics.

The package insert (PI) provides the end users (physician, pharmacist, and nurse) of the product with information required to ensure the safety of each patient taking the product, along with patient selection (indication), proper dosing and administration (Fig. 8.8). It is required for all marketed

- Each clinical trial is complex:
  - Requirements: study design, protocol writing, selection of investigators & sites, and internal/external agreements
  - Other design and study procedures possible:
    - Special population group(s)
    - Specialized equipment or techniques
  - Number of specialized people involved
  - Internal & external resources needed
- Clear objectives & realistic design for each trial:
  - What is question to be answered?
  - Is question relevant to the development plan/product profile?
  - Is trial affordable?
  - Can we get it done, on time, on target?
  - Can the data be meaningfully analyzed and reported?
- Each clinical trial should answer at least one critical question:
  - Related to product's characteristics, as it relates to development plan.

FIG. 8.6. Background – Objectives & Designs

- The Product Profile, ideally:
  - Describes key properties of the product:
    - Efficacy, safety, formulation, & dosing & administration. May contain comparative information, e.g., product competitors.
  - Should be first task for a project team:
    - Overall goal of drug development is to develop a package insert for marketing the product.

FIG. 8.7. Product Profile

- The Package Insert (PI) - Goals:
  - Provide end user (MD, PharmD, RN) product information.
  - Is required by law for all marketed drugs.
  - Drives number and type of clinical trials required.
  - Drives indication, dosing, administration, & formulation.
- Information required in PI is described in CFR:
  - Usage Information:
    - Description
    - Clinical Pharmacology
    - Indications and Usage
    - Dosage and Administration
    - How Supplied
  - Safety Information:
    - Contraindications
    - Warnings and Precautions
    - Adverse Reactions
    - Drug Abuse and Dependence
    - Overdosage

FIG. 8.8. Package Insert

- Product profile & PI drive development plan.
- Full development plan:
  - Time & Events:
    - Clinical, manufacturing, pharmacology/toxicology, regulatory
    - Formulations, safety, QA-stability, package engineering
  - All studies for desired PI
  - Resource requirements
  - Budgets
  - Role of scientific advisory boards & focus groups
  - Publication plans

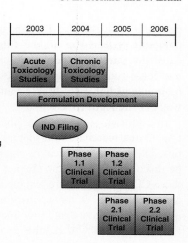

FIG. 8.9. Development Plan

drugs (i.e., required by law). The PI is a legal document affecting patient care. The planned PI drives indication, dosing, administration, and formulation to be used in clinical trials and also drives number and type of clinical trials required for the NDA/CTD in the development plan. The information required in the package insert is described in 21 CFR 201.56 and includes the 11 sections outlined on Fig. 8.8 [3]. The package insert may contain additional sections if appropriate and in compliance with the regulations. These sections may include animal pharmacology and/or animal toxicology, clinical studies (excepts of design and data), and references.

The clinical studies required to complete many of these sections are pivotal to the overall success of the development program. Once the proposed product profile and package insert are drafted and approved by management for a new drug, the clinical trials may begin to be planned and designed, along with additional nonclinical work as necessary. Therefore, the product profile and associated package insert will drive the development plan for the product and, ultimately, the clinical plan.

The development plan engages the whole company (all divisions, all staff involved at any stage in any way) and will contain all studies to be done (basic research and clinical), all resources, budgets, and potential publications (Fig. 8.9). Many representative departments in six divisions from across the example company are engaged and listed below. Any staff, budgets, and systems are included in the development plan. Input for the plan includes outside experts, who serve on company advisory boards and participate in focus groups where specific development ideas are discussed. Scientific and marketing information is shared. These boards and groups include experts with knowledge in the disease, drugs, health care environment, and payor situations. They provide invaluable reality checks for the company's clinicians, researchers, and marketers, as well as feedback on the products and plans of the company. The studies done by the company need to be published to communicate new findings to the medical community and public and receive peer review and acceptance based on good science. Especially with breakthrough products where existing knowledge is lacking, a full-blown education plan for the medical community is needed to establish the scientific basis for the new product and how it fits and exceeds existing therapy.

An example of a drug development plan is provided also on this Figure 8.9. A budget also would be associated with each activity. A basic development plan will contain the time, events, and responsible parties for the drug development activities for the following six company divisions (16 representative functional groups).

- Clinical trials (clinical operations, safety, medical affairs, pharmacokinetics (ADME), CDM, biostatistics, pharmacoeconomics)
- Manufacturing (formulations, package engineering, QC [stability])
- Research (pharmacology, toxicology)
- Regulatory and Legal
- Finance
- Marketing

Due to the complexity of the drug development plan overall and the conduct of each clinical trial, effective management (and leadership) of the project team is important to ensure that quality is maintained throughout and the timelines are met. This team includes representatives from all the departments that will be meeting team objectives and performing work in the plan. A typical project team may include individuals from the disciplines outlined on Fig. 8.10. The project team will work in their respective areas to complete the product profile, package insert, and the development plan. The team members should serve three roles: representing their departments on the team as communicators/liaison, performing the work with colleagues from their area, and decision makers within the team. Research groups, such as chemistry, pharmacology, and toxicology, are represented on the clinical project team to provide input from their research or follow-up on issues discovered

- Due to complexity of drug development & conduct of clinical trials, project management ensures that quality & timelines are met.
- Project team typically includes variety of functional groups with responsibilities for Development Plan and conduct of clinical trials:
  - Project management
  - Clinical research
  - Regulatory
  - Pharmacology & Toxicology
  - Chemistry, pharmaceutics and manufacturing
  - Safety group
  - Marketing
  - Biostatistics
  - Data management

FIG. 8.10. Project Management and Project Team

- Once development plan is created, clinical planning process may begin.
- This involves significant amount of management and planning in itself.
- Clinical plan is description of timelines associated with clinical trials and includes additional detail not provided in development plan.

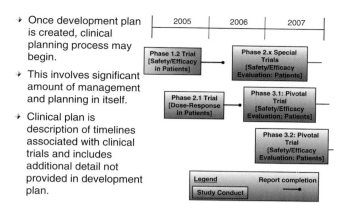

FIG. 8.11. Clinical Development Plans

during clinical studies. Manufacturing, stability, quality assurance, formulations, and package engineering are all groups that contribute to making sufficient and acceptable product available for clinical studies. Clinical operations will actually conduct the study with the outside investigators. Clinical-related groups such as biostatistics, data management, and medical writing support the study protocols, case report forms for study data, and report writing. Safety will analyze and report adverse events (AEs) about collected AE data. Regulatory oversees study operations and reports to guide protocols and studies with the regulations and NDA/CTD in mind, as well as liaison with the FDA as needed. Marketing provides research on marketplace data and helps create the best product profile and package insert to meet patient care needs and corporate marketing needs. They can serve as another perspective on provider and patient needs and disease opportunities compared to the competition.

The team will have a single leader, who may come from any department, calls and conducts the meetings, and fosters decisions. Team leaders are responsible to senior management and often are a separate management group within a corporate structure to foster planning and decision making independent from any one area. The project team will usually meet on a periodic basis (e.g., once a week) to track progress, discuss results, and modify the plan as necessary. Once the development plan is completed the clinical plan may be completed, and the clinical team may begin gearing up for the conduct of the clinical trials.

Within the drug development plan is the clinical plan. A clinical development plan is a description of the clinically related studies planned in order to assess the safety, efficacy, and value of the investigational product, leading up to an NDA/CTD. The usual studies in the clinical plan include efficacy and safety studies in phases 1 through 3, any special clinical trials, pharmacokinetic (ADME) trials, and pharmacoeconomic and quality-of-life studies. An example of a clinical development plan is provided in Fig. 8.11.

Note that the clinical development plan will include additional detail that is not provided in the overall development plan. In the examples provided above, the clinical development plan provides the timelines for the conduct of the trial and the time required to complete the report. It also provides a summary title describing each trial. As with the development plans, the clinical plan will also include budget information, but it is specific to each proposed study. It is broken down into its key components (e.g., per patient charges, overhead, lab charges). The above clinical development plan example also demonstrates the importance of planning to ensure efficiency. Looking at the phase 1.2 trial (Fig. 8.11), you will note that the conduct of the trial is completed, but the report is not when the phase 2.1 trial begins. The plan here may be that the project team has decided that sufficient pharmacokinetic information will be available before the report is completed to allow the initiation of the phase 2.1 trial.

To design and conduct a clinical trial, a thorough understanding of the disease area, the therapeutic area including competitive products, and the characteristics of the new drug known to date is essential (Fig. 8.12). In addition, a detailed understanding of applicable regulations, published guidelines, and generally accepted research practices (collectively referred to as good clinical practice; GCP) also is essential. Compliance with GCP ensures achieving three major goals of the clinical research: patient protection, acceptance of the study by the research community, and successful regulatory submissions around the world. Some key regulations and guidelines are provided in the Table 8.1; they have been promulgated by the FDA and found in the U. S. Code of Federal Regulations (CFR) and by the International Conference on Harmonization (ICH).

In addition, there is a large body of specific regulations, FDA guidances, "points to consider," and "guidelines" for clinical evaluations of many diseases (more than 100), as well as the full set of regulations available on the web at www.fda.gov Fig. 8.12 [4–10]. The particular value of these documents for

- Conducting clinical trial requires:
  - Thorough understanding of therapeutic area and characteristics of drug.
  - Also involves detailed understanding of applicable regulations, published guidelines, and Good Clinical Practices (GCP)
- Regulations and Guidelines to have as references include:
  - 21 CFR 50 Protection of Human Subjects
  - 21 CFR 56 Institutional Review Boards
  - 21 CFR 312 IND
  - 21 CFR 314 NDA
  - 21 CFR 201 Labeling
  - 21 CFR 202 Prescription Drug Advertising
  - ICH E2A-E Clinical Safety Data Management
  - ICH E3 Structure and Content of Clinical Study Reports
  - ICH E6 Good Clinical Practice: Consolidated Guideline
  - ICH E8 General Considerations for Clinical Trials
  - ICH E9 Statistical Principles for Clinical Trials

www.fda.gov for regulations and guidelines

FIG. 8.12. Clinical Trial Conduct

TABLE 8.1. Clinical Regulations and Guidelines.

Regulation or guideline	Title
21 CFR 50	Protection of Human Subjects
21 CFR 56	Institutional Review Boards
21 CFR 312	IND
21 CFR 314	NDA
21 CFR 201	Labeling
21 CFR 202	Prescription Drug Advertising
ICH E2A-E	Clinical Safety Data Management
ICH E3	Structure and Content of Clinical Study Reports
ICH E6	Good Clinical Practice: Consolidated Guideline
ICH E8	General Considerations for Clinical Trials
ICH E9	Statistical Principles for Clinical Trials

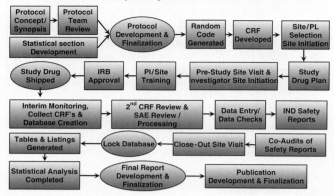

FIG. 8.13. Planning and Use of Process Maps

Principal and Co-Investigator	PI (principal investigator) is responsible for conduct of trial, including ensuring safety of participants (patients or volunteers). Co-investigator(s) assist principal investigator. Both principal and co-investigators listed on FDA Form 1572.
Monitoring Activites	Monitoring of investigational sites required under FDA/CFR regulations and performed by IND sponsor: Investigator following protocol, subjects being informed of study and providing informed consent, data collected accurately and recorded, drug being stored, prepared and accounted for, and study documents maintained.
Monitoring Site Visits	Pre-study site visit, Initiation site visit, Interim monitoring visit, and Close-out site visit.
AE's and SAE's	Any adverse event (AE) associated with use of drug in humans, whether or not considered drug related. Serious AE is any AE that results; death, immediately life threatening, persistent or significant disability/incapacitating, requires or prolongs inpatient hospitalization, or congenital anomaly or birth defect.

FIG. 8.14. Definitions/Terms -1

companies and investigators is in identifying the disease parameters, drug factors, and research practices that the FDA/EMEA/MHLW deem important in the NDA/CTD process. Companies are well advised to become familiar with and take this information into account when designing and conducting clinical trials.

Abiding by the regulations and following the guidelines require a number of processes to be adopted when conducting a clinical trial. Specific research activities must be done in an orderly and specific sequence of events. In order for the study to run smoothly, all of the many diverse participants need to understand the process, especially what the key events and timing are. The end point is a complete and successful trial, finishing in a final study report to put into the NDA. Therefore, the many activities required are often mapped (i.e., a process map) to aid in the management and conduct of this clinical study process. An example of 23 basic and standard activities, required to conduct a clinical trial, is illustrated in Fig. 8.13. Many of these processes will be explained further in this chapter. The process map provides a basic understanding of the activities required to conduct a clinical trial. The process map includes a number of terms and abbreviations that may be unfamiliar to individuals outside the industry.

Twelve standard terms that are important to understanding clinical trial conduct are included in Figs. 8.14, 8.15, and 8.16. First, the PI (principal investigator) is the person who is responsible for the conduct of a trial. Although they are most commonly a licensed board-certified physician, pharmacologists (Ph.D.), clinical pharmacists (Pharm.D.), and other qualified health professionals are permitted to be PIs along with a physician co-investigator. The PI major responsibilities are first ensuring safety of participants (patients or volunteers), reporting to the IRB, conducting the trial according to good clinical practices (GCPs) and the particulars of the protocol, and collaborating with the company's project team members. Co-investigator(s) is/are person(s) assisting the principal investigator. Both principal and co-investigators must be listed on FDA Form 1572.

Monitoring of an investigational site is required under the regulations and is performed by the clinical operations staff of the sponsor of the IND [7, 8]. Monitoring activities include ensuring the investigator is following the protocol (inclusion/exclusion criteria, patient assignment to study groups,

IND Safety Reports	Report that sponsor of IND must submit to FDA and all investigators, when any SAE occurs which is unexpected and "associated with use of the drug", within 7 days of any unexpected fatal or life-threatening experience and within 15 days for all other safety reports.
IRB/IEC	Institution based groups that serve important role in protection of rights and welfare of human research subjects, and have been formally designated to review and monitor research involving humans. IRB has authority to approve, require modifications, or disapprove research.
CRF	Paper or electronic form designed to capture data from a clinical trial.

FIG. 8.15. Definitions/Terms -2

assessments of efficacy and adverse event); subjects are informed of the study and are providing informed consent as required; the data collected is accurate and is being recorded as required; the drug is being stored, prepared, and accounted for as required; the site is notifying the IRB of the progress of the study as required; and study documents are being maintained appropriately. Monitoring site visits typically are done at several key milestones in the study timeline and consist of the following: prestudy site visit and initiation site visit ensure that the study staff and systems are prepared to follow the protocol and follow GCP; interim monitoring visits ensure all procedures and guidelines are being followed, answer questions from the site, and proactively audit for compliance; and closeout site visit is an audit-type visit to collect all data, inspect necessary records, and meet with investigators and staff.

Adverse events (or experiences) and serious adverse events are defined in the CFRs [9, 10]. An adverse drug experience is any adverse event associated at any dose in any treatment group with the use of a drug in humans, whether or not considered drug related. Blinded studies including placebo groups have these recorded. A serious adverse event (SAE) is any adverse experience occurring at any dose that results in any of the following five outcomes: death, immediately life threatening, a persistent or significant disability/incapacity, requires or prolongs inpatient hospitalization, and a congenital anomaly or birth defect.

As stated in 21 CFR 312.32, the *IND safety report* must be submitted to all the regulatory authorities by the sponsor of an IND, and the sponsor must notify all investigators when any SAE occurs that is unexpected (i.e., not listed in the investigator brochure) and "associated with the use of the drug," that is, "there is a reasonable possibility that the experience may have been caused by the drug" (Fig. 8.15) [9]. Of note, the sponsor must notify the FDA by telephone or facsimile within 7 calendar days of any unexpected fatal or life-threatening experience. All other IND safety reports must be submitted within 15 calendar days of becoming aware of the event.

Institutional review boards (IRB) and an *independent ethics committees* (IEC) are groups external to the sponsoring company that serve an important role in the protection of the rights and welfare of human research subjects. They have been formally designated to review and monitor biomedical research involving human subjects. Each university and health care institution involved in a study will have their own IRB, who each will approve a study. For multi-institution studies, especially with clinics and physician offices without IRBs, commercial IRBs are available to perform all the necessary functions. The regulations are the same no matter what IRB is used. Membership on IRBs is usually several physicians with appropriate experiences, a lay person, and an ethicist. In accordance with the regulations, an IRB has the authority to approve, require modifications, disapprove research, or stop an ongoing study. The PI and sponsor are responsible to report periodically to IRBs on progress in the study, and SAEs are reported as well [11, 12].

Case report form (CRF) is a paper or electronic form designed to capture all the data from a clinical trial. One CRF for a single patient can be 50 to 100 pages for one study. Please be reminded that the data is voluminous and includes, for example, a patient's demographic profile, checklists for inclusion criteria and exclusion criteria, drug administrations throughout the study, clinical assessments for efficacy at each time point (once or up to 12 months per protocol), clinical assessments for adverse event monitoring, AE report forms, and checklists for patient withdrawals from a study no matter what the reason. Technology is being used by some companies wherein data is remotely entered directly into the database via computers or personal data assistants (PDAs) by the investigator's staff. CDM group uses CRFs for data entry at the company.

Clinical data management (CDM) is the functional group within clinical operations that is responsible for the data derived from clinical trials, its storing, preparing, and entrance into databases (Fig. 8.16). Also, they usually create the forms (case report forms) wherein the data, for each patient throughout the study, will be recorded by the investigator and staff at the sites.

Electronic data capture (EDC) is used for collection and entry of patient information into case report forms by the staff at the investigators' sites. The staff uses varied computer/database systems, such as laptop computers or PDAs. The goal is improved efficiency of a very time consuming and labor intensive process, that is, data entry. Otherwise, staff enter data and information into paper CRFs, which in turn is entered into computer databases from the paper CRFs by the CDM staff of the company. The CDM group must still check the data entry for accuracy, consistency, and missing information. Issues limiting application of EDC are costs of implementation and comfort level and capabilities at the sites for this type of data processing by study staff [13, 14].

The *biostatistics* group will ultimately analyze these data. Biostatistics is the functional group that applies statistical methods to the data derived from a clinical trial. This group will help write protocols, particularly the statistical analysis section dictating how and what statistical methods will be used to analyze the data. Each type of data requires different types of mathematical tests. Also, their expertise is study design as well, and they will critique many other aspects to enhance quality of the data to be collected (e.g., appropriate comparison groups, frequency of testing, the amount of change in a disease parameter that is significant). They write a report summarizing all the study results, as well as present the results in appropriate tables and graphs. A database is secured (that is, *"locked"*) to prevent any further changes from being made. A lock will occur after procedures that are designed to verify the quality of the data have been carried out; for example, computerized edit checks, manual review of the data listings and tabular output, and audit of database output versus case report forms. When the database is locked, the formal statistical tests are ready to be performed, and final tables may be prepared for inclusion in the final report.

An *audit* is a formal examination and verification of the accounts and processes in the conduct of a clinical trial. Audits typically include examination of some number of CRFs with identification of any missing information and verification of data items against source documents such as medical records. Audits of CRFs are common in a monitoring visit to the investigator's site by CRAs. An audit of safety data is especially important to ensure its accuracy and to continually monitor safety of subjects participating in a trial. The group, quality assurance (QA), is a separate functional area in clinical operations responsible for conducting the formal audits after a study is completed by the sponsoring company. Regulations and company SOPs are used as references for the conduct of studies for the audits. Audits uncover protocol violations and study conduct issues before an NDA/CTD is filed. The regulatory authorities also routinely audit the trials (e.g., pivotal trials) at both the company and the investigative sites once an NDA is filed as part of their compliance function. A major audit problem in the conduct of a study can cause the FDA/EMEA to exclude the whole study from the NDA/CTD filing [15].

Due to the regulatory oversight and scientific nature of the clinical trial process, there are a number of required documents related to the conduct of a clinical trial. Four of the most important of these documents as they relate to the conduct of a clinical trial include the protocol, consent form, case report form (CRF), and final report. Many of the activities occurring within clinical operations are geared toward the completion of each of these documents.

Nine documents are presented in the next two figures (Figs. 8.17 and 8.18). A *protocol* is a detailed written description of the planned clinical trial. The protocol typically contains

CDM	Functional group within clinical operations that is responsible for storing, preparing, and processing data derived from the clinical trials.
Biostatistics	Functional group that applies statistical methods to data derived from clinical trials, creates data tables, and writes statistical report to be incorporated into study reports.
Locked Database	Database is secured ("locked") to prevent any further changes from being made. Analysis then can proceed.
Audits	Audit is a formal examination and verification of accounts and processes in conduct of clinical trials by sponsor and FDA. Audits of CRFs are common. Audit of safety data is very important to continually monitor safety of subjects in a trial.

Fig. 8.16. Definitions/Terms -3

Protocol and Amendments	Protocol is detailed written description of clinical trial, e.g., study objectives, patient selection, design, methods & procedures, and statistical methods that will be used to analyze data. Amendments are protocol changes & must be approved by IRB.
Consent Form	Consent form is provided to prospective study participants that describes benefits / risks, procedures, and study events subjects will experience if they agree to participate. Evidence of informed consent is provided by subject's signature on form.
Investigator Brochure (IB)	IB provides all available information on non-clinical (e.g., chemistry, formulation, manufacturing, pharmacology, toxicology) and clinical results to date. It is essentially package insert before drug is approved. IB must be provided to each investigator and IRB.
(DMP) Data Management / Statistical Plan	DMP is usually prepared by both data management & biostatistical staff before study starts and no later than unblinding of data. Plan describes how data will be managed (e.g., entered, transferred, cleaned) & statistical tests to be used in analysis of data.

Fig. 8.17. Documents -1

IRB Approval	Each protocol and consent form must be approved by an IRB. Documentation of IRB approval is required and must be filed.
Randomization Code	A code typically prepared by biostatistics that directs investigator as to which drug treatment each subject is to receive.
Statistical Report	Report of statistical methods used to generate and analyze data, plus all the results and analyses. It is integrated into final clinical study report.
Site Visit Report	Report of activities, results and accounts of each clinical monitoring visit (e.g., pre-study, initiation, interim, and close-out).
Final Report	Final Clinical Study Report describes study and all results from study and discussion and is submitted to the FDA. ICH E3 provides necessary guidance for completing this report.

FIG. 8.18. Documents -2

at least 10 commonly used sections, including a description of the disease and therapy background, rationale for this study, product description, study objectives, inclusion/exclusion criteria, dosing and product administration, methods and procedures, such as DBPCRPC study (double-blind, placebo-controlled, randomized or parallel, controlled study), patient assignment and randomization code process, and the statistical methods that will be used to analyze the data and report it in what tabular forms. The *consent form* is provided to the prospective study participant that describes a balanced reporting of the benefits/risks of this new treatment, alternative treatments, patient's rights to stop participating at any time, and the procedures and study events the subject will experience if they agree to participate. Evidence that the subject has provided informed consent is provided by the subject's signature on this form, as well as the signature of the individual administering the consent.

The *investigator's brochure* (IB) is the document that provides all available information on the nonclinical (e.g., chemistry, formulation, manufacturing, pharmacology, and toxicology) and any clinical results to date. It is essentially the package insert before the drug is approved. It changes over time as new information from studies is gleaned and added and must be updated on a regular basis. The IB must be provided to each investigator and IRB, including any changes over time.

The *data management plan* is usually prepared by both data management and biostatistical personnel before the study starts, most frequently and no later than the unblinding of the data. The plan describes how the data will be managed (e.g., entered, transferred, and cleaned) and the statistical tests to be used in the analysis of the data and its reporting.

Each protocol and consent form must be approved by an IRB (Fig. 8.18). Documentation of the *IRB approval* is required and must be filed with the investigator at each site and the company for each trial. The IRB file at the site must also contain the membership of the IRB. The *randomization code* is a coding process typically prepared by biostatistics that directs the investigator as to which drug treatment each subject is to receive. For multicenter studies, the randomization is centrally controlled with investigators calling the center for their patient's assignment to treatment groups. Often, this call center process for the code is automated as each investigator calls for each of their study patients. After a study is closed and the data is locked, a report of the statistical methods used to generate and analyze the data is integrated into a final clinical study report. The rationale for the selection of the particular tests to be used is provided. A description of the tables and graphs to be used is provided. This *statistical report* is integrated into the final clinical study report.

A *site visit report* provides the results and accounts of each clinical monitoring visit (e.g., prestudy, initiation, interim, and close-out). These visits and reports are conducted and written by clinical research staff from the company, also called "monitors or CRAS." The content of these reports includes, for example, the dates and times of visits, who was met with, what protocol violations were found, storage conditions of the product, what records and how many were reviewed, storage conditions of study records, and how were discrepancies found and rectified. The PI or at least the Co-I should be available for all site visits, especially because an assessment of the site and its staff is being made and also to address protocol questions.

The *final report* of a clinical study describes the protocol, especially design features, clinical conduct and results of the study, such as profiles of the patients (demographics and disease and treatment status), outcomes for the key end points, data analyses that were conducted for efficacy and adverse events, and conclusions drawn from the study. These reports must be submitted to the regulatory authorities in a timely manner. ICH E3 provides the necessary guidance for completing this report [22].

As important as the design of the clinical trial itself is the selection of the site(s) and investigator(s) in which the study will be conducted (Fig. 8.19). Some criteria that are important when selecting a study site and investigator include *qualifications and experience* of the investigators and staff (PI, Co-I, and especially study coordinators) and the *site's ability* to conduct the study in a clinically competent and GCP compliant fashion [16]. A prestudy site visit form will provide an outline for this evaluation. The *characteristics* of the personnel at the site that will be involved in the study include:

- Dedication: Will you be able to get in touch with the site when needed? Do the individuals enjoy their work and demonstrate a commitment to their work? Is quality exemplified in the work?
- Education: A curriculum vitae (CV) should be requested from each person who will be involved with the study. Do each of the individuals have a degree in their given profession? A physician (M.D. or D.O.) must be a co-investigator or principal investigator for medical ethics and legal reasons. Other researchers and clinicians in other disciplines can be a PI: Pharm.D., Ph.D., D.Nurs.
- Training: Are the medical staff board certified in their specialty? Has each person received specific training to perform their research job and has it been documented? Have they been trained in the regulations and guidelines related to conducting a study? Do they need training on the protocol? This is almost always necessary because of the novelty of the products and the nuances of the specific study that are important for trial conduct and appropriate data collection.
- Experience: How long has each person been doing their particular job and how long have they been involved with the conduct of clinical trials? What is their experience base with the target disease?

- Criteria important for selecting a study site and investigator:
  - Qualification and experience of investigators.
  - Site's ability to conduct study in clinically competent and GCP compliant fashion. Pre-study site visit will provide for this evaluation.
  - Characteristics of personnel at sites involved in study including:
    - Dedication
    - Education
    - Training
    - Experience
  - Availability of specific equipment that may be required by study
  - Geographic location (opinion-leader representation)
- Thought leaders or investigators:
  - Known to have committed significant time to studies of the disease at interest
  - Considered early on in development of drugs.

FIG. 8.19. Investigator/Site Selection & Training

The availability of *specific equipment* that may be required by the study is needed. Local laboratories that are up to national standards may be needed, but often samples are obtained, stored, batched, and sent to a central lab to ensure consistency and accuracy of testing, especially for any sophisticated tests, such as genetic analyses.

The *geographic location* of the site may provide some insight to the population that may be recruited at the site as well as a general idea of the cost. For example, sites located in more heavily populated areas (West and East Coasts) generally have a higher cost of living, and the cost of the study will typically be higher. How far is the site from the sponsor and the monitor? Is the site easy to access from the airport?

Do they have the patients in their practice or geographic area for the study, regarding not just numbers, which of course is very important, but also the type of patients in their diagnosis, severity, and availability for the study? Are there geographic, language, or other barriers to patient participation? Do the patients need assistance (financial, logistical) in participating in the study because of the extra time to perform all the tests?

Typically, *thought leaders* or experienced practitioners with many patients with the disease of interest are considered that are known to have committed significant time to the study of the disease at interest. A thought leader often will have a track record of studies done previously for drug development with other companies. A limit to too much thought-leader involvement as investigators is their usual location at universities, tertiary health care centers, that will have more sick and complicated patients that are not representative of typical patients and may not fit the study. Thought leaders are important to bring onto the project team early in the development of the product (e.g., before the IND is filed). A review of the literature and attendance at therapeutic-specific professional meetings will often dictate who the thought leaders are. These science-based individuals can assist in study design of existing protocols, suggest new studies that need to be done for the clinical plan, discuss competitive products, provide feedback on proposed product profile and package insert, provide access to other possible investigators, add credibility to your research program, and help understand any health care issues for this disease.

Training of investigators and staff is a common and important practice by clinical operations and medical affairs staff at manufacturers for multisite studies. Key training areas include the product, protocol, SOPs, data collection, the definitions and reporting for adverse events, and the investigators' responsibilities as stated in the regulations. Protocol compliance and data quality can be favorably affected by training investigators and their staff in these matters. Areas that can be problems to avoid at multiple sites, regarding excess variability, are the use of inclusion/exclusion criteria, monitoring (what, how, when), drug records and administration, and data collection with CRFs. Separate meetings for the education of the PIs, Co-Is, and study coordinators are a significant investment of time, money, and company staff that is well spent to result in a well conducted and completed study. The duration of such meetings is often 1–2 days at an off-site location to have an audience focused on the training.

Patient selection will be guided by the protocol and the availability of patients to the particular investigator (Fig. 8.20) [17]. It is common for many more patients to be screened and recruited but not selected to participate in the study. Many patients will not be selected due to the inclusion and exclusion criteria of the protocol, which are strict for pivotal trials and usually cannot be waived for individual patients. Some patients may decide not to participate due to the potential risks (e.g., adverse events, blood sampling, placebo groups) or requirements the study may pose on them. Long-term stays in care centers, high number of visits, long-duration studies, and extent of testing may lead to recruitment problems. Ultimately, a certain number of the patients recruited will volunteer to participate and will sign a consent form indicating that they have been informed of the study (including the benefits, risks, commitments). Once the study enrollment begins, it is critical to the success of the trial that the subjects be retained to complete the study. The investigator and the investigational site will influence the retention of study subjects. Trust in the investigator and staff and the facilities themselves will all influence subject retention in the study. The design of the trial will also influence this. Study designs that are demanding of the subject (e.g., require long inpatient stays, multiple visits, long studies, injectable products, frequency of product dosing, or too much invasive or time consuming testing) may lead to greater attrition during the study. Benefits that the study subjects may derive from the study may also influence the retention of subjects. If a subject does not feel that he or she is benefiting sufficiently from participation in the study, they may elect to withdraw. One benefit that a subject may gain is monetary, but it is critical that this benefit is not so excessive to be viewed as coercive. All of these factors must be considered when designing the trial and choosing the investigators and investigational sites. The added medical attention by clinicians and the free care given within the protocol are good incentives for participation and retention. The benefit to mankind in advancing care of this disease is a very good motivator for some patients to participate, but the investigator cannot oversell this situation as to be considered coercive. In the United States, the HIPPA privacy rules for personal information have complicated and added cost to patient recruitment, adding more time and process.

Independent review of clinical research by an *institutional review board* (IRB) or its equivalent is a universally acknowledged requirement for biomedical research to be conducted in human beings (Fig. 8.21). Documentation of the review and the IRB's approval is required before the clinical trial may begin. The IRB, usually referred to as an independent ethical committee (IEC) in Europe, is an independent body composed of medical, scientific, and lay community representatives that is responsible for protecting the welfare and rights of the research subject. The membership needs to involve experienced clinicians and seasoned clinical researchers with the time commitment to perform all their functions. IRBs confirm that the study is based on sound scientific principles and appropriate interpretation of data and that as designed it is capable of answering the scientific questions it poses with an

- Patient selection is guided by:
  - Protocol.
  - Availability of patients.
- Reasons patients may not participate in a study:
  - Do not meet inclusion/exclusion criteria of protocol
  - Patient elects not to participate due to:
    - **Risks and Excessive requirements**
- Subject enrollment begins:
  - When subjects sign consent form indicating that they have been informed of study (including risks) and agree to participate.
- Factors important to retention of study subjects (critical to success of trial):
  - Investigator and investigational site.
  - Study design. Studies requiring long interned stays or multiple visits may lead to greater attrition.
  - Subject benefits (e.g., additional medical attention or monetary).

FIG. 8.20. Patient Selection, Enrollment & Retention

- Independent body; medical, scientific, and lay community representatives
- Responsible for protecting welfare and rights of research subjects
- IRBs confirm:
  - Study is based on sound scientific principles & appropriate interpretation of data,
  - As designed study is capable of answering scientific questions it poses,
  - With acceptable level of risk to subjects
- Documents that IRB reviews and approves:
  - Protocol
  - Consent Form
  - Investigator Brochure
  - Advertisements
- IRB has ongoing safety oversight responsibility after study is approved:
  - IND safety reports
  - Periodic reports from investigator
- IRB be available to answer any questions a subject may have
- Types of IRBs:
  - Most IRBs are associated with hospitals or academic institutions
  - Commercial IRB or Central IRBs: duly constituted service business

FIG. 8.21. Institutional Review Boards

acceptable level of risk to subjects. The documents an IRB reviews to address these questions consist of:

- Protocol.
- Consent: The IRB verifies that the informed consent form (ICF) is
  - understandable to the subject population,
  - embodies the "Elements of Informed Consent" as stated in the U.S. Code of Federal Regulations [18] and the ICH E6 Good Clinical Practice: Consolidated Guideline [19], and
  - that an institutionally acceptable statement regarding the privacy of "protected health information" is incorporated (HIPAA compliance).
- Investigator brochure.
- Periodic safety report from the investigator and the company.
- Advertisements: The IRB also confirms that any compensation being offered to study participants is not coercive and that any advertising is not false or misleading.

In addition, the IRB has an ongoing safety oversight responsibility, both through review of adverse experiences through IND safety reports (which must be forwarded to the IRB by the principal investigator) and review of periodic reports by the principal investigator. The IRB also remains available to evaluate any complaints from subjects, which may arise in relation to their participation in the study.

Most IRBs are associated with hospitals or academic institutions. "Commercial" IRBs, also frequently referred to as "central" IRBs, also exist. These are duly constituted IRBs that operate as a service business offering review of clinical research. It is important to note that hospital, academic, and commercial IRBs are held to the same requirements under U.S. federal regulations and ICH GCP guidelines. In general, academic investigators are required to have their research approved by their institution's IRB. Some nonacademic hospitals may also require that research done by medical staff be approved by the hospital's IRB. In general, however, outpatient and multicenter studies conducted by nonacademic investigators may be reviewed and approved by a commercial IRB. Use of a single commercial IRB for review and approval of a multicenter study can yield substantial economies in effort, money, and time [11, 12].

In response to the need for ongoing independent monitoring of data from certain large and prolonged blinded studies of a variety of medical interventions in serious or life-threatening diseases, a role has developed for the *data monitoring committee* (DMC) also known as a data safety monitoring committee (DSMC) or data monitoring board (DMB) (Fig. 8.22) [20]. DMCs are discussed in a draft FDA Guidance (fda.gov/cder/guidance/index.htm). DMCs usually consist of individuals with the expertise in one or more of the disciplines of clinical medicine, medical specialties in question, preclinical science, clinical trials methodology and administration, biostatistics, and medical ethics. Sponsors may be represented but do not participate in meetings in which unblinded data are reviewed. DMCs periodically review data from a study, including

- Also known as a Data Monitoring Committee (DMC). These committees have developed in response to the need for ongoing independent monitoring of data from large and prolonged studies, typically in seriousor life-threatening diseases.
- Membership usually consists of experts in one or more disciplines (clinical medicine, pre-clinical science, clinical trial methodology, or medical ethics).
- Different and separate from an IRB.
- Examples include the safety concerns related to cardiovascular events for several NSAIDs.
- Studies can be stopped for unexpected adverse events or substantial efficacy during study conduct, usually at predetermined points.

FIG. 8.22. Data Safety Monitoring Committee

unblinded data as appropriate, from the perspective of whether safety issues, or clear failure of efficacy, or clear superiority of efficacy may warrant the premature termination of the study or modification of the conditions of the study. Of note, recent safety concerns relating to cardiovascular events for several NSAIDs were identified in large postapproval clinical trials by use of DMCs, as noted, for example, in Pfizer's statement on Celebrex [21].

It is important to recognize that while certain oversight activities of the IRB and DMC may appear to overlap, the role of the two entities is distinct. All human research requires IRB approval. The IRB discharges a broad set of responsibilities on behalf of the subject, only one of which is ongoing review of safety information, which is normally blinded. A DMC, on the other hand, is determined to be warranted for individual studies on a case by case basis and is chartered to periodically review data, usually unblinded, in order to provide the sponsor with recommendations regarding continuation with no changes, continuation with modifications, or premature termination of the clinical trial. If periodic checks of data are needed, for example for safety, such interim assessments need to be built into the statistical analysis of studies and formally described in methods, as they will impact the sample size calculations.

In the modern era of conducting clinical trials, computers have become an essential tool for completing the work. Word processors, database systems, statistical programs, scanning systems, and others are commonly used (Fig. 8.23). Therefore, it is essential that computer or *information technology* (IT) support is available throughout the process. The conduct of clinical trials will generate a large amount of data that is used to answer the questions for which the trials were designed. The data collected will come from a variety of sources and be of many diverse types including, for example, physical exams, laboratory analysis results from blood and urine samples, pathology reports, patient demographics, drug blood levels, drug records, patient or caregiver questionnaires,

- Computers have become essential tool in the modern era of conducting clinical trials:
  - Word processing
  - Databases
  - Statistical programs
  - Pharmacokinetic programs
  - Scanning systems
- IT support is essential throughout the clinical trial.
- Clinical trials generate large amount of data that come from many sources that must be collected in simple formats so coders can translate it into computer language.
- IT personnel collaborate closely with statisticians, clinical research, data management, and regulatory.

FIG. 8.23. Information Technology Support

- Responsible for ensuring quality of data generated from clinical trials.
- CDM works closely with IT, statisticians, clinical research, & regulatory.
- Responsibilities include:
  - Data entry
  - CRF development
  - Programming
  - Data cleanup
- Data entry is process wherein data is put electronically into database:
  - Entry from paper CRFs,
  - Transfer from one electronic source to another (e.g., scanned data or electronically captured data [EDC]).
- Programming and data cleanup:
  - Programming is required to build edit checks used to clean or scrub data. Data is checked for errors with aid of computer checks.
  - Data cleanup and corrections to data documented for clear trail if audited.

FIG. 8.24. Clinical Data Management

clinical tests (e.g., blood pressure), hospital/clinic charges, x-rays, electrocardiograms, and others. These all need to be collected in a simple format so coders can translate it into computer language. Case report form design needs input from IT groups. Some reports can be scanned directly into the database. Electronic data collection can be done directly at sites, but it requires proper training of site staff and is expensive for the equipment and setup. The data across all studies, which may cover 4–8 years and be thousands of records and CRFs, must be collected and entered into the database so it may all be pooled as necessary when the NDA is being prepared. One example of this is all of the safety data that will need to be pooled across all relevant studies for completing the NDA and package insert. With the technology expanding at a rapid rate as it has, making data consistent over a 4–8 year period of time can be a real challenge. IT personnel work very closely with the statisticians as would be expected. They also work with clinical research, data management, and regulatory within clinical operations but are involved in nonclinical areas. Some examples where IT support is needed include preclinical studies (e.g., toxicology and carcinogenicity) and computational chemistry (designing new drugs).

Clinical data management is responsible for ensuring the quality of the data obtained from the clinical trial (Fig. 8.24). As the name implies, this function is responsible for managing the data. This role will include formatting the data so it may be easily accessed when performing the appropriate statistical tests. Data management personnel will work closely with IT, statisticians, clinical, and regulatory during the clinical development program. Specific responsibilities within this group would include data entry, CRF development, programming, and data cleanup.

Data entry is the process by which the data is placed electronically into a system. This may include the entry of data from paper CRFs or the transfer of electronic data from scanned data or electronically captured data. Because the entry of data into one location from all clinical trials is essential to ensuring efficiency as the development program continues, it is important that clinical data management personnel are involved in the development of the CRF or the means used to collect the data (e.g., scanned CRFs or electronic data capture devices). For example, many clinical trials will require the collection of blood and urine for laboratory analyses. Many laboratories have the capability to report the results of the analyses in an electronic format. It is important to know that the sponsor of the IND has a system in place to accurately receive these data into its database. Furthermore, there will be a considerable amount of programming required to format the data received into a consistent and manageable format for further statistical analysis. Programming will be required to build a set of edit checks (described further below), which will be used to clean or scrub the data that has been collected from a clinical trial.

A large, time-consuming, and interactive responsibility of clinical data management is the process of data cleanup. Clinical trials will involve the collection of a large volume of information. For example, a single CRF (one patient) in a clinical trial may involve 20,000 fields of data. Therefore, a trial involving 100 patients will generate 2,000,000 pieces of data. It is highly likely that errors will have been made in the collection of this data even with the amount of on-site monitoring and training that has taken place. Many of the data fields collected can be checked for errors with the aid of computer checks (edit checks). An example is with blood pressure data. It is common to build programmed checks in which the entered data is checked against specific blood pressure criteria to identify possible errors that may exist. One type of edit check may read, "identify all systolic blood pressures (SBP) over 200 mmHg." This check would be run against all SBP data and the computer would identify all those patients and times that SBP was over 200. The entire collection of edit checks is run against the database several times until no further data is "kicked out" or identified. All data that is identified by an edit check is provided to the clinical team members responsible for the particular study sites so the data may be clarified and corrected as needed. Documentation of all corrections or explanation for no changes is required so that a clean trail is left should an audit be conducted in the future.

Once the clinical trial is completed, the results will need to be documented in an integrated *clinical study and statistical final report* (Fig. 8.25). These reports are the core documents for the NDA/BLA/CTD. The ICH E3 guideline provides the guidance for the format and contents of this report. In general, the format of the report will be as listed in Figure 8.25 [22].

An author of both the clinical and statistical reports will need to be identified early in the clinical trial. The lead author for the clinical report is the company's clinical leader for the product and project who often wrote the protocol as well, with significant collaboration with the lead biostatistician and input from clinical managers and principal investigators. The protocol and statistical report are the two source documents for the clinical report. It is not uncommon to turn over the responsibility for completing the report to a medical writer (often a separate function) who is responsible for managing the entire writing process. There is a considerable amount of material that must be incorporated into a final report, and much of it can be inserted from other sources. For example, many sections of the protocol can be pulled directly into the final report. As tables and figures are generated and the data is analyzed, new analyses and tables or figures may be necessary to more accurately demonstrate an important finding.

In our experience, it requires a minimum of 3 months and frequently more time after the end of a study to complete the integrated clinical and statistical final report. Much of this time is spent on entering the remaining data from the study, cleaning up the data, generating the final tables and figures, analyzing the results, authoring the report, and getting its review and approval. The report goes through a review process within the sponsoring company. The key questions are manifold: is the data reported clearly in tables and text?; is there data and information missing?; is the writing style scientifically sound and good English?; are the efficacy and safety summaries representative of the data?; are the design features adequately addressed to place qualifiers (limits) on the summaries?; does the data support the conclusions and recommendations? A key activity subsequent to finalization of the integrated report is publication in independent, refereed medical journals. Here the principal investigators lead the effort and work more proactively with the company's biostatisticians and clinical leaders to write the papers.

Clinical trials historically have been carried out outside of a sponsor's "home country" (i.e., "*global clinical research*") for a variety of reasons (Fig. 8.26). Many large pharmaceutical/biotech companies are now global operations in their basic research, research alliances, clinical research, and marketing of products. A measure of success (both research and marketing) for these companies now is the simultaneous approvals and launch of a product around the world. Cost-effective operations on a global scale include the following advantages: less demanding requirements for a Clinical Trial Application (the equivalent of an IND), less cost per patient, access to worldwide expert investigators, and access to more patients faster. In addition, "registration trials," which use local opinion leaders as investigators in multicenter trials, have frequently been used to facilitate drug approval in individual countries.

- Once clinical trial is completed, results must be reported in two parallel & linked final study reports (statistical and clinical reports).
- ICH E3 guideline provides guidance for format & contents for authors.
- In general, format contains following sections:
    - Introduction
    - Study objectives
    - Investigational plan
    - Study patients
    - Efficacy evaluation
    - Safety evaluation
    - Discussion and overall conclusions
    - Tables and figures
    - Reference list
    - Appendices
- Typically requires timeframe of minimum of 3 months to complete (often longer) after study has ended.

FIG. 8.25. Reports–Statistics and Clinical

- Clinical trials have been conducted outside the US for a variety of reasons including:
    - Less demanding requirements for a Clinical Trial Application (equivalent to an IND).
    - Less cost.
    - Access to expert investigators.
    - Access to patients.
    - To complete "registration trials" that utilize local opinion leaders to facilitate drug approval in individual countries.

- Drug development today is often carried out on a global scale:
    - High standards and GCP for grant of approval in multiple countries.
    - It is more feasible with ICH initiative.
    - It reduces costs associated with drug development.

FIG. 8.26. Non-U.S. Trials (Global Clinical Research)

Clinical product development is carried out on a global scale today. In past years, a separate development program might have been carried out in each of many different national environments. It is now generally accepted that a single development program conducted to a high and GCP compliant standard, with portions of it carried out in a number of different countries, will provide a basis for approval in multiple countries. The impetus for a single development program was initially economic, driven by the recognition that as clinical research became more complex, costly, and time consuming, carrying out a set of duplicative local development programs was unaffordable. As the International Conference on Harmonization initiative has progressed and as European drug regulation has become centralized, the single development program has become more feasible. The U.S. FDA embraces the ICH initiative and accepts international studies for U.S. approvals.

Sponsors conducting clinical research outside of their home country, at least in the industrialized world, can today reasonably expect that good clinical practice compliance is achievable and that, with appropriate study management, data from any study will be acceptable for multinational use. A practical consideration in international studies is the language challenges such that a consent form and case report form will need translation into the country's primary language; some English words or phrases may translate different meanings.

The overall *budget* for the clinical development of a product, including all studies and operations at the company and with outside resources (investigators and CROs), is usually about one-half of the total drug development budget (Fig. 8.27). The Tufts Center for the Study of Drug Development calculated this cost to be $450 million out of $802 million in 2001 [23]. An individual study may cost a company $250,000 to $5 million or much more depending on its size, duration, and complexity. Fig. 8.27 lists some of the internal and external budget considerations. Clinical operations consumes the largest portion of the overall budget, because of its labor intensive requirements, such as monitoring staff and visits, medical staff involvement and oversight, and protocol and report writing. CROs may consume a large portion of the budget, as companies may send out portions of work in the clinical area to absorb the major increases of work over specific time periods, especially to gear up for the large phase 3 trials, plus a CRO can perform patient assignment, statistical analyses, and more. As noted earlier, millions of data pieces are collected and need to be entered into databases and related work, which the CDM group performs. Drug has to be prepared in blinded fashion for the company and comparator groups. Safety data must be collected and analyzed on a continual basis keeping the medical safety group very busy. Basically, each department involved in any way in clinical research adds to the cost of clinical development.

The sites, where the studies are done, and their investigators have a budget for a list of work and services that they provide, for which examples are provided in this slide as well. The cost per patient and per site again depends on the size, duration, and complexity of the work for the site and may be influenced by other factors as well. An individual study budget can be very expensive. For example, as noted in the figure, a study involving 100 patients can cost $500,000.00 to $1.5 million to conduct. However some studies may be much larger. For example, some cardiovascular studies may involve 5,000 patients costing as much as $50 million. The institution or hospital may pass along patient care charges to the company, of course for any added work by the patient care staff for the study. Institutional overhead is demanded by institutions and is highly variable, often with universities demanding higher fees similar to government contracts, which have overhead as high as 50–75%.

Contract Research Organizations (CROs)

As the pharmaceutical industry has expanded and matured, industry managers have become increasingly sensitized to the need to operate the clinical operations area as a scientific business in relation to the maintenance of large in-house clinical development organizations (Fig. 8.28). In response to these concerns, contract research organizations (CROs) began to spring up during the 1970s and to play a progressively more prominent role in the industry over the subsequent three decades. CROs have offered industry clients the ability to outsource various activities as the need arises. The contract concept has been applied to manufacturing, development (preclinical and clinical), and business (e.g., sales, market research) functions in the pharmaceutical industry [24–26].

Many CRO companies now operate on a global scale, which fits well with the global operations of sponsoring companies. The research-related CRO size has grown to over $9 billion by 2003, out of a $40 billion overall budget; growth has been occurring at varying annual rates (from 3% to 23%).

- **Investigator Site Costs:**
  - PI / Co-I salaries
  - Coordinator salary
  - Lab fees
  - IRB fees
  - Drug control fees
  - Special procedure fees
  - Overhead (25%-65%)
- **Overall Cost Variables:**
  - Size of study (number of patients)
  - Duration of study
  - Complexity of design
  - Sophistication of design
- **Costs to Sites:**
  - Per Patient: $5,000 - $15,000
  - 1,000 patient trial: $10 MM

- **Company Costs:**
  - Investigator training & meetings
  - Patient enrollment
  - Site and CRF Monitoring
  - Auditing of sites
  - Drug Formulation, Manufacturing & Distribution
  - Data management (CRFs)
  - Report writing & Publications
  - Medical staff oversight
  - CRO expenses
  - Safety reporting
  - Regulatory affairs & FDA filings
  - Overhead
- **Overall Product Costs:**
  - Total: $900 MM – $1 B per drug
  - Clinical: 50% = $450 MM

FIG. 8.27. Budgeting

- Definitions
- Functions / Services
- Selection
- Management of CROs
- Characteristics

FIG. 8.28. Contract Research Organizations (CROs) Outline of Topics

- CRO Functions
- Unbundling
- Full service vs. Niche
- Academic CROs
- Domestic vs International
- SMOs
- CRO liaison/oversight

FIG. 8.29. CROs: Key Terms & Definitions

While certain of the principles in the selection and management of CROs are applicable to other functional areas, the discussion to follow is specifically directed toward CROs offering clinical development services. CROs are used for clinical trials I, II, and III for more than 60% of the projects in the industry. The top five multiservice larger corporate CROs in the United States with operations around the world include Quintiles, Covance, Parexel, PPD, and Kendle, who have collectively 50% of the CRO revenues and are growing [27].

Discovery research programs are, by their very nature, sporadic in productivity. As a consequence, compounds emerging from these programs enter and then progress through clinical development sporadically. Clinical development organizations may exist in a "boom or bust" environment of alternating under- and overstaffing relative to current clinical development needs. This problem can be compounded in the small or start-up company with a small or single product portfolio, where failure of a single project can render a previously overworked and understaffed clinical development organization idle overnight. Companies will work with both full-service CROs (about 80% of the time) and at the same time niche CROs with specific services (about 70%) to perform the various clinical development activities [27]. Five topics will be addressed in this chapter for CROs as listed in Fig. 8.28.

Before beginning a detailed discussion of CRO functions and of the selection and management of CROs, definition of several key terms and consideration of the concepts underlying them is useful: The following Fig. 8.29 summarize some of the key terms and concepts, as well as in the text below.

Unbundling is dividing up the clinical research work into manageable and discrete portions. Contracting out discrete tasks is carried out within a particular clinical development program to one or more CROs, frequently with other tasks being carried out by in-house staff. This is essentially a "mix and match" approach in which specific areas of deficiency within the sponsor's organization are complemented by the resources of the CRO. This is in contrast with the approach of contracting for completion of a complete development program. An example of the latter might be a contract with CRO X to conduct all aspects of a study from investigator identification through the writing of a clinical report. The former might be exemplified by a study in which five research business arrangement are made: (1) the sponsor identifies investigators and initiates and monitors some portion of the study sites, (2) CRO X initiates and monitors the remainder of the sites, (3) CRO Y is responsible for clinical data management, (4) an independent contract statistician carries out data analysis, and (5) a freelance medical writer writes the clinical report. Unbundling, by contracting out only where a truly complementary role for the CRO exists, can be the most cost-effective approach for the sponsor. However, this benefit may be offset by the complexities, cost, and risk of managing multiple vendors.

Certain CROs can justifiably claim to be *"full service,"* a popular marketing term meaning that they have at their disposal the resources to take on a development project in any therapeutic area at the preclinical phase and take it through to an NDA. In essence, the largest of these CROs resemble large pharmaceutical companies with all the same types of staff and systems, only without discovery research or a pharmaceutical sales force. The reader should note that working with a full-service CRO does not preclude "unbundling" as discussed above for a specific *(Niche)* function. Other CROs, so-called niche CROs, may limit the scope of their activities to a phase of development (e.g., phase 1), a therapeutic area (e.g., oncology, dermatology), or a specific domain of clinical development activity (e.g., study monitoring, data management, biostatistics or patient recruitment or assignment).

As of this writing, domestic CROs exist in virtually all national environments in which clinical research takes place. In addition, many CROs have multinational capabilities. In certain instances, they have the same scale as those of major multinational pharmaceutical companies, which can be particularly useful for a pivotal global clinical trial being done across several countries. The choice of a domestic or multinational CRO is driven mostly by the required geographic scope of a development program.

Historically, biomedical research has been heavily dependent on contract collaboration with academic investigators. These academic researchers often are the thought leaders for a disease or therapeutic area, lead the development treatment guidelines/standards, and are looked upon in the medical community for innovations. With the growth of the contract research industry, some academic medical centers have determined that, with the addition of significant organizational infrastructure, hiring of some research coordinators, and improving of the business operations of their research offices, existing talent (practitioners and investigators) could be organized into and marketed as a CRO offering services beyond the traditional sponsor–investigator collaboration. A new revenue source for the medical centers, that is, CRO type research contracts, has been another driver for these *academic CROs*. Potential attractions to clients might include improved access to academic sites, access to high-level investigators, and the external perception of academic excellence in their protocols, study operations, and study reports.

In the 1980s, the need arose to identify progressively increasing numbers of qualified clinical investigators as discovery research activity and thus the volume of clinical research studies increased. It was recognized that large numbers of individuals with excellent clinical medical backgrounds, but little or no familiarity with clinical drug development research, were practicing in the community. These independent practitioners have an advantage, that is, large number of "typical" patients to be enrolled in clinical studies. *Site management organizations* (SMOs) typically represent a collection of sites, may act as the intermediary in identifying qualified sites for a given study, act as a business representative for sites in contractual matters, educate the practitioners regarding research practices, and also provide to each site required guidance in setting up the infrastructure and systems needed to carry out GCP compliant clinical research. Some practice sites will need some added staff support and space for study work. The practitioners see benefits of SMO participation in bringing new and innovative treatments to their patients, advancing medical practice, and creating a new revenue source for the practice.

Some large pharmaceutical/biotech companies have a specific group dedicated to interacting with CROs *liaison/oversight*. Other companies may rely on different individuals, depending on the project, to interact with CROs. These groups or individuals are the primary liaison to the CROs in general, screen the potential CRO for the company, are familiar with who's who and what's what at the CROs, and lead the selection process and manage the overall relationship between the company and CROs. The primary liaison to the sponsoring company for a protocol or study conduct will usually be the clinical operations lead manager for a full-service CRO or the particular department lead for the product for a specific service, whether it's biostatistics, quality assurance, or regulatory affairs. A sponsoring company with a CRO department may benefit from a structure and process that provides for more fair, balanced, and knowledgable assessments of prospective CROs. The sponsoring company's CRO department will also be familiar with all the company's research programs and may try to create economies of scale by working with a particular CRO, if they can handle it, with several products.

As alluded to above, a sponsor may engage a CRO with appropriate resources to provide a narrowly defined or broad range of clinical development services (Fig. 8.31). These services may include creation of an IND, various regulatory affairs functions including initiating and maintaining IND-related contact with the reviewing division at FDA, performing any or all of the activities needed to conceptualize, execute, and report a clinical study or program of studies, and to create, file, and take an NDA through the approval process. A noninclusive list of 19 *potential services of CRO* is provided in Fig. 8.30.

The decision to engage a CRO can be driven by different considerations depending on the nature of the sponsor company. For fully constituted pharmaceutical companies, that is, companies with in-house staff representing the full set of skills required for clinical development, outsourcing has frequently been an *ad hoc* response to competing, unplanned for, resource demands at any stage in development. Examples might include the need to conduct an unplanned phase 1 mechanism of action study, receipt of an approvable letter from FDA conditioned on an additional unresourced phase 3 study, or the urgent need to do further source document monitoring against case report forms because of issues identified during NDA preparation. With increasing use of CROs by the industry, however, outsourcing by the larger companies has assumed a progressively more strategic role driven by long-term program and project planning and long-term staffing plans.

At the other end of the spectrum, the so-called virtual company, a business entity whose staffing may constitute a handful of managers who may have limited development experience, may require, at least for some stage in the company's development, outsourcing of the entire development process. For companies partially constituted by either chance or as the result of prospective planning, of course, task areas not represented in-house would be outsourced as a matter of practice.

Various business arrangements with a CRO may result in economies in both charges and administrative costs to the sponsor. Most common among these is a "preferred provider" scenario in which the sponsor places all work with CRO X for concessions on rates and does work under a single master services agreement rather than under multiple contracts.

There are numerous examples of CROs of all types doing outstanding projects of different types for industry sponsors.

- Investigators:
  - Recruitment
  - Training
- Advisory boards
- Patients:
  - Recruitment
  - Assignment
- CRF design
- IRBs
- Data entry
- Biostatistics
- Regulatory Services
- Monitoring:
  - Site visits
  - Telephone calls
- Lab services
- Electronic data capture
- Records retention
- AE reporting
- Drug Preparation, Dispensing, Records
- Audits @ site
- Protocol and Report writing

FIG. 8.30. CROs: List of Potential Services

Unfortunately, there are also examples of CRO–sponsor joint undertakings, which end in a less than satisfactory fashion for a variety of reasons (cost, quality, failure of execution, timeliness, etc.). In fairness, the successes and failures are both, in general, as much attributable to the management practices of the sponsor as to the qualities of the CRO.

The *selection of a CRO* or CROs is a critically important activity, which requires a systematic and reasoned approach (Fig. 8.31). Multiple CROs should be evaluated using a single set of prospectively defined evaluation criteria. The sponsor should begin with a top-level evaluation of each CRO in the following areas:

- organizational structure,
- overall quality and experience of management and staff,
- experience of CRO with the particular disease area,
- history of any regulatory actions including confirmation that no staff are debarred,
- financial stability,
- list of sponsor references,
- understanding of key elements of the contractual process,
- commitment by CRO of dedicated staff if possible,
- an understanding of the CRO's culture and values.

As an adjunct to more formal inquiry, the sponsor should work the network to see what is being said about the CROs under consideration (as with all rumor monitoring, this should be done analytically). Essential criteria in CRO selection has been captured by CenterWatch in 2000; strong reputation 41%, therapeutic expertise 59%, and ability to deliver patients 59% [29].

The above should result in the generation of a "short list" of CROs with the desired characteristics to support further evaluation. The importance of a culture and set of values consistent with the sponsor's cannot be overemphasized. Because the CRO will really be part of the sponsor company during execution of the contract, the ability of sponsor management and staff to create a shared culture within which to work is essential and unlikely to be successful if great dissimilarities exist at the outset. CROs on the short list should be then evaluated in detail by a sponsoring company assessment and visit team led by the CRO department with representation depending on the services desired. Are any involved in competing projects? Are there systems (SOPs, IT, a quality assurance function) in place that are adequate to support the project? Is training for assigned tasks performed and meticulously documented in training records? Is Staff turnover low? The CRO ideally should have had recent successful experience in the project therapeutic area or a related area. The sponsor must then very clearly define to CROs under consideration in a formal "request for proposal" (RFP) type document detailing what is needed of the CRO. The RFP needs to address this unambiguously at a truly "micro" level as the budget and timeline are directly determined from this. Responses to the RFP should be standardized in a sponsor-defined spreadsheet to facilitate comparison between CROs. The selected CRO should be subjected to an extensive QA audit by either the sponsor or a qualified independent entity contracted by the sponsor prior to contract execution.

CRO selection is a strategic activity. Bad choices made here will impact every aspect of a project and will jeopardize the ultimate success of the study and the NDA/CTD. CRO selection is a time consuming, often tedious process which produces large amounts of company data that must be rigorously analyzed (Fig. 8.32). Particularly in the case of the financial data, if you cannot really analyze it, don't be embarrassed to bring in someone from the business side who can. If the case of gross disparities in budgets from different CROs, step back and make sure that you have adequately communicated your requirements for the project and that all involved CROs have the same understanding of it, then ask them to re-bid. In making a selection remember that it is not only about price: "It is about value. value, which is a function of price × quality."

Once selected, CROs require intensive professional management by the sponsor from the top down to ensure successful business and research outcomes (Fig. 8.33). Sponsor *management of a CRO* is a demanding role requiring under-

- Good and Bad CRO Outcomes
- Evaluation Parameters (check lists):
  - People, experience base, & culture
  - Processes & systems
- Company:
  - Evaluation team
  - Visits to CRO
- Budgets:
  - RFPs
  - Standardized budgets
- QA Audits to Qualify

FIG. 8.31. CROs: Principles in Selection

- Strategic activity
- Time consuming
- Requires rigorous analysis
- CRO:
  - New business staff versus Coordinators & managers
- Disease area experience
- Reconciliation of disparities
- Price and Quality

FIG. 8.32. CRO Selection-Caveats

- Intensive
- Status reporting
- Communications and decisions:
  - Processes, systems, and people
- Responsibility
- Interfaces:
  - CRO and Company
  - PI and site
- Authority and chain of command
- Change authorization

FIG. 8.33. CROs: Principles in Management

- Culture
- Finances
- Conflicts of interest
- Processes and Systems
- Staff and Turnover
- Experience base
  (Disease areas, Study types, Years, & Type of companies)

FIG. 8.34. Characteristics of Desirable CROs

- Investigator Initiated INDs
- Patient Access
- Patient Expectations
- Publications
- Placebo Use
- Quality of Life Endpoints
- Compassionate Use
- Pharmacogenomics
- Outsourcing vs Internal Resourcing

FIG. 8.35. Selected Issues in clinical Development

standing of the development process, planning skills, and a well-developed set of human relations and organizational skills. A specific and designated liaison team at the company may be created for this oversight, communication, collaboration, and management role. It bears repetition that the CRO is *not* responsible for the study to the investigators, the institutions (sites), the IRBs, or the regulatory authorities; they are a vendor for services, and the sponsoring company is the responsible party. Companies with limited resources should be particularly careful to assess whether they have the ability to manage the CRO they engage.

The CRO must be assimilated into the sponsor company. CRO staff working in a company culture, which they are conversant and comfortable with, will work smarter and be more productive (Fig. 8.34). Assimilation requires time and attention from sponsor staff. Loci of responsibility and authority and chain of command must be absolutely clear. The frequency, format, and contacts for status reporting to the sponsor should be defined prospectively as part of the contract. Staff need to be dedicated on both sides over the whole study time frame, and kept the same, to enhance communication and coordination. Turnover upsets the relationships and must be minimized. Compatible systems must be available at the CRO; they must be worked out between the CRO and sponsor, must be integrated as much as possible, and must appear seamless to the investigative sites. Procedures for authorizing change from original contract work elements must be clear and must be followed without exception. The CRO to be successful and the company to be successful need CROs with sufficient years and types of experiences, regarding at least the regulations, therapeutic area, stage of research, research skills, and working clinicians focused multicenter studies.

Another important issue with CROs and a desirable characteristic is receptivity by the investigative sites. The CRO represents the company and gives them a good or bad name. The features that they find important are a collaborative team, supportive and responsive, professional monitors and staff and staff who are organized and prepared, according to a CenterWatch surveys of sites [30].

Selected Issues in Clinical Development

A collection of brief reviews of selected issues follows that can pose as controversies or extra challenges in clinical development (Fig. 8.35). Some topics, such as investigator-initiated INDs and compassionate use, have been perennially discussed within the pharmaceutical industry. Others, such as quality of life end points and pharmacogenomics, represent disciplines and concepts that are relatively new in their application to drug development. The purpose of this section is to introduce these issues and some of the key concerns surrounding them to the reader.

In an investigator-initiated IND, an investigator, typically an academician, who normally also serves as principal investigator on the studies under the IND/CTA, submits the IND under his or her own name (Fig. 8.36). Thus, the investigator is literally the sponsor of the IND. The role of a pharmaceutical company in this process generally may involve providing permission to reference existing drug master files and other regulatory submissions, including other INDs, and providing supplies of study drug. It is generally held that the advantages of allowing investigators to study new drugs under an investigator-initiated

IND consist of conservation of resources and a lower level of regulatory scrutiny. IND creation by a sponsor company is a significant undertaking in terms of time and expense due to the high standards that exist for the quantity and quality of data in industry-sponsored INDs. Investigator INDs typically are more rudimentary in nature and resources that are expended by the investigator rather than the company. Similarly, it is generally held that investigator-initiated INDs are subjected to a lesser degree of regulatory scrutiny and that potential issues such as clinical holds pending submission of additional data or changes in protocol design are less likely to occur.

On the negative side, studies carried out under the investigator-initiated IND are done with a total loss of sponsor control of key study elements, data analysis and interpretation, and handling of safety data. Whereas there are certainly instances where mutual good faith has allowed projects to proceed under these conditions, the risk of essentially giving a company asset to an investigator for some period of time is very real and must be clearly understood and balanced against the perceived advantages.

It is a generally held view among laymen that "patients with serious diseases are always looking for clinical trials to participate in" (Fig. 8.36). Seemingly reasonable corollaries to this would be that most of these patients do, in fact, participate in clinical trials and that, assuming a reasonably promising drug under study, recruitment of patients should not be difficult. The reality is quite different. For example, one study estimated that less than 2% of white adult cancer patients in the United States participated in a National Cancer Institute clinical trial in the common tumors from 2000 to 2002. Even lower participation was found among racial and ethnic minorities [31]. On a more parochial level, subject recruitment issues and challenges are a day-to-day reality for anyone involved in conducting clinical trials. Enrollment of patients into clinical trials of new drugs is affected by a variety of factors. Scientifically meaningful and medically safe evaluation of new drugs usually requires a highly defined study population, particularly in early patient trials. Thus, complex inclusion and exclusion criteria in study protocols markedly reduce the population available for enrollment.

Community physicians may be less eager to refer patients to clinical trials because either loss of patients or loss of control is a concern. Notably, in the one area of medicine, pediatric oncology, where participation in clinical trials is the rule rather than an exception, most clinical care is initiated and overseen through regional medical centers rather than by community-based physicians. Geographic access to centers is limited in many parts of the country. Study requirements (frequency of visits, testing, etc.) are frequently burdensome and, while appropriate in the context of the study, are far in excess of standard care. Concerns about random treatment assignment, especially to placebo, may exist.

Aside from traditional recruitment aids such as newspaper and radio advertising, other actions may aid recruitment. Working with sites within a health maintenance organization (HMO) or other organized network may maximize referrals. Making sure that there is awareness of a clinical development program by relevant patient advocacy groups and ensuring that studies are reflected in relevant databases, most notably the "Clinical Trials Data Bank," may also be helpful. Note that a sponsor of a clinical trial of a drug intended for treatment of a serious or life-threatening disease is required by federal regulations to submit information to this database, but that any sponsor may do so [32].

Patient expectations are heterogeneous and depend on a multiplicity of factors (Fig. 8.37). For healthy volunteer subjects ("professional subjects") participating in clinical pharmacology studies, the primary if not sole motivation is financial compensation, at times joined by some elements of altruism, curiosity, and desire to be with one's friends. On the other hand, for patients with life-threatening diseases, participation in a study of a new and promising drug may represent a last chance at receiving an effective medical therapy. Between these two extremes are an almost unlimited number of permutations of expected medical benefit, once again curiosity and altruism, economic benefit (free exams, free medication, stipends), and social benefits (meet people, get attention).

There are strongly held notions in academia, not without basis in fact, that suppressive *publication* practices are widespread in industry (Fig. 8.37). This has led to a requirement by several leading international medical journals that trials

- Investigator Initiated vs. Company INDs:
  - Controls in the design
  - Quality
  - Resources
  - Regulatory scrutiny

- Access to new drugs / Access to willing patients – Myth vs Reality:
  - Reasons
  - Outside help to find patients
  - Enrollment aids (advertisements, networks, databases, internet)

FIG. 8.36. Selected Issues & Controversies

- Patient expectations:
  - Heterogeneous
  - Spectrum from altruism to money

- Publications:
  - Industry vs. Academic perceptions
  - Some practical concerns
  - Publication policy for sponsors

FIG. 8.37. Selected Issues & Controversies

will need to be registered in a public clinical trial registry in order to be considered for publication [33]. There are equally strongly held notions in industry, one again with some merit, that less than rigorous publication practices (e.g., based on unreliable data or questionable analyses) are widespread in academia.

Attempts to publish based on incomplete or unreliable data (e.g., premature single-center publication from multicenter studies, publication prior to data lock, selective publication of data) or unsound analyses (post hoc statistics, invalid assumptions, etc.) clearly do not serve the interest of a sponsor or of any author. Efforts at preventing these occurrences are, if carefully presented, not likely to be construed as suppressive and more likely to be seen as facilitating timely dissemination of reliable data and analyses to the peer-review community. Review periods allowing for thorough consideration of intellectual property protection, if not unreasonable, are similarly accepted as standard practice and not ill perceived.

Sponsors should speak very clearly to their unequivocal support of the dissemination of scientific information, but build intellectual property protection (required review period by sponsor) and injunctions against inappropriate data use into study contract and protocol. The language should be carefully considered but once agreed should be adhered to without exception. In agreements with investigators especially at academic institutions, a company will require an opportunity to review and comment on a possible publication.

Randomized, double-blinded, *placebo-controlled* clinical trials, while generally regarded as the gold standard for scientific proof of the efficacy and safety of most new drugs, are limited in their application by ethical, scientific, and practical considerations (Fig. 8.38). In certain instances, while generally accepted effective standard treatment does exist, withholding it and using placebo may be acceptable, as in, for example, antihistamines in allergic rhinitis. In other clinical settings as, for example, in virtually all serious infections, the sequelae of withholding treatment would be medically unacceptable, thus mandating the use of a positive control of currently available approved therapy.

Use of an approved product as a positive control then poses the question of whether an approved treatment really works? This is less of a problem when it is possible to show that a new treatment is more effective than standard care. In that case, even if the assumption is that standard care is no better than placebo, demonstration of superiority for the test treatment is a compelling demonstration of efficacy. In the instance where equivalence to the standard treatment is shown, however, demonstration of efficacy is critically dependent on the robustness of the original demonstration of efficacy of the approved comparator. Note the concept of approval creep, i.e., showing equivalence to an approved product of dubious efficacy, when your product is approved, the next one down the line uses it as a comparator to gain approval and so on down the line [34].

Placebos will probably be scientifically appropriate and a regulatory necessity for a long time to come in many disease states. Uncritical condemnation of placebo-controlled studies as unethical leads one to the unpalatable alternatives of either approving drugs on the basis of inadequately controlled studies or not approving new drugs, neither of which serve the interests of either ethics or society.

It is to be hoped that new medicinal treatments will, by virtue of their being effective, have a measurable impact on health-related QOL (Fig. 8.38). Historically, the basis for new drug approval has accepted this hope as a reality by implicitly assuming that when biological evidence of efficacy exists, there will be health-related benefits to the patient. *QOL end points*, either as add-ons to a traditional efficacy study or as free-standing studies, have become increasingly important in supporting comparative product claims and drug pricing and reimbursement. QOL is not, however, necessarily different from efficacy. QOL may serve as basis for approval. The best examples of this are in the area of oncology. Traditional "biological" oncology end points, such as tumor response and time to event variables (especially survival), have historically served as bases for new drug approval. Although FDA Oncology Division guidances over the past 20 years have invoked QOL end points as an acceptable bases for approval, until relatively recently, sponsors have not sought approval based on these end points. Recent examples of approvals based on QOL end points include mitoxantrone in prostate cancer and porfimer sodium in esophageal and non-small cell lung cancer [35].

Compassionate use, also called treatment IND, is provision of an unapproved drug to a patient outside of a formal clinical trial (Fig. 8.39). In the usual course of events, the patient should be enrolled in a clinical trial if eligible and if an appropriate study exists. In general, the drug should have been shown to have provided direct benefit to the patient on an earlier study, or there should be extremely compelling medical reasons for believing that it will benefit the patient during compassionate use. The patient's disease should either be serious and progressive or highly symptomatic, and approved therapies should have been tried and found to be either ineffective or not well tolerated. Compassionate use requires notification to and approval by the applicable regulatory authority.

For the right drug and patient, compassionate use can meet an important humanitarian need. However, valid instances are

- Issues in use of placebo:
  - Nature of disease
  - Available treatment
  - Approval creep with active controls

- Quality of Life Endpoints:
  - QOL as basis for approval vs.
    QOL as "phamacoeconomic endpoint"

FIG. 8.38. Selected Issues & Controversies

- **Compassionate use:**
  - Definition
  - Uncommon
  - Ethical considerations
  - Practical impact
- **Pharmacogenomics:**
  - Metabolism
  - Safety
  - Efficacy

Fig. 8.39. Selected Issues & Controversies

uncommon. If many valid compassionate use situations arise, the sponsor might want to consider whether a drug might be suitable for accelerated approval. Poorly managed compassionate use programs can quickly get out of control, require inordinate amounts of time to administer, endanger patients, create legal issues, generate large amounts of data of uncertain reliability, and cast doubts on conclusions drawn from reliable data. Using promises of ongoing compassionate use to attract patients into short clinical trials is associated with all of the problems above, as well as with potential ethical issues.

Variation in response to drugs among individuals has long been recognized. One of the mechanisms for such variation is genomic variation in the population (Fig. 8.39) At the most readily observed level, this variation results in "responders" and "nonresponders" in clinical trials and clinical practice and also in the occurrence of drug toxicity in some treated individuals but not in others (called *pharmacogenomics*). More recently, individual variations in drug metabolism, particularly with respect to cytochrome P450 (CYP450) mediated liver metabolism, began to be systematically explored in an effort to understand the bases for individual differences in drug efficacy and toxicity. Numerous well-studied examples (e.g., dextromethorphan, tricyclic antidepressants, beta-blockers, narcotic analgesics) exist of interindividual variation in the expression of a CYP450 drug-metabolizing enzyme (phenotypic expression) resulting in varying systemic drug exposure and consequent clinical sequelae [36].

Advances in molecular biology, which had their beginnings in the early 1970s supported by powerful computer technologies, have made the identification of specific genetic sequences in an extracted sample of human DNA possible and allowed this technique to be carried out efficiently and economically on large numbers of samples. Thus, phenotypic variation may now be related to variations in genotype. Pharmacogenomics attempts to relate variation in individual drug response to genotypic variation. To date, much of the practical application of this discipline to clinical drug development has been in the area of drug metabolism. Genotyping may be used to efficiently identify and either include or exclude subjects with a given phenotype for a specific CYP450 enzyme. This approach may enhance safety or create a more homogeneous population for efficacy analysis.

Alternatively, DNA may simply be collected from any one of a number of tissue samples and stored for possible retrospective retrieval and genotyping should clinical findings from the study suggest that genetically determined variation in drug metabolism may have been operative. Storage of DNA samples for possible subsequent genotyping and linking to efficacy and safety end points, rather than purely metabolic end points, is an area of active investigation at this time. In concept, many facets of drug response could be genetically determined, although, in distinction to drug-metabolizing enzyme expression, these response phenotypes would be more likely to be based on multiple genes.

Summary

As discussed in this chapter, clinical trial operations encompass multiple medical and scientific disciplines. Successful clinical trial operations depend on extensive planning and appropriate application of diverse and extensive internal and external resources. Clinical trial operations are heavily regulated, and the conduct of a clinical trial is never "black and white." With any clinical trial, previously unrecognized issues and concerns may arise. Having an understanding of the regulations, pertinent guidances, disease pathogenesis and presentation, therapeutic area, clinical trial methodology, and the roles played by the various disciplines involved in clinical operations is essential to manage a clinical trial.

References

1. J Chronic Dis 1959;10:218-248
2. Bull History Med 1982;56:1-18
3. Code of Federal Regulations 21.201.56. Drugs. Labeling. General requirements on content and format of labeling for human prescription drugs
4. FDA Guidance documents for regulated industry. Available at www.fda.gov/opacom/morechoices/industry/guidedc.htm
5. ICH guidance documents. Available at www.ich.org/cache/compo/276-254-1.html
6. International Conference on Harmonization. Guidance on general considerations for clinical trials. Federal Register 62(242). Feb. 17, 1997
7. Code of Federal Regulations 21.312.53 IND. General responsibilities of sponsors. Selecting investigators and monitors.
8. Code of Federal Regulations 21.312.56. IND. General responsibilities of sponsors. Review of ongoing investigations.
9. Code of Federal Regulations 21.312.32 IND. Safety reports.
10. Code of Federal Regulations 21.310.305 Records and reports concerning adverse drug experiences on marketed prescription drugs for human use without approved new drug applications (CFR, Adverse Events)

11. Levine R. Getting past the IRB roadblock. Pharmaceutical Executive 2003;23(12):76-82
12. Anonymous. Working with institutional review boards and informed consent. DIA Forum 2002;(7):16-20
13. Successful clinical trials management. An imposing change (EDC). Pharmaceutical Executive Supplement, June 2004
14. King J. Electronic submissions. Is anybody ready? R&D Directions 2003;9(10):34-40
15. Anonymous. Introduction to audit programs. DIA Forum 2003;(1):24-27
16. Successful clinical trials management. Wanted clinical investigators. Pharmaceutical Executive Supplement, June 2004
17. King J. 10 ways to faster and easier patient recruitment. R&D Directions 2004;10(4):34-46
18. US Code of Federal Regulations CFR21 Part 50. Protection of human subjects
19. ICH E6 Good Clinical Practice: Consolidated Guideline (patient consent)
20. Schneider BA. Independent data and safety monitoring. DIA Today 2005;5(3):34-35
21. PR News Wire via Dow Jones, December 17, 2005
22. ICH E3 guideline. Structure and content of clinical study reports
23. Anonymous. Tufts Center for Study of Drug Development Outlook 2002 Boston. MA
24. Sentek K, Engel S. Special report on contract research organizations. The turnaround. R&D Directions 2002;8(8):44-94
25. Anonymous. Successful clinical trials management. Working with CROs. Pharmaceutical Executive Supplement, June 2004
26. King J, Jackson A. Adjustments along the way (CROs). R&D Directions 2003;9(8):38-55
27. Lamberti MJ (ed). An Industry in Evolution, 4th ed. Thomson CenterWatch, Boston, MA, 2004, p. 61
28. Lamberti MJ (ed). An Industry in Evolution, 4th ed. Thomson CenterWatch, Boston, MA, 2004, p. 72
29. Lamberti MJ (ed). An Industry in Evolution, 4th ed. Thomson CenterWatch, Boston, MA, 2004, p. 76
30. Lamberti MJ (ed). State of the Clinical Trials Industry. Thomson CenterWatch, Boston, MA, 2005, p. 177
31. (pg 79) (cancer audit of patients in trials)
32. FDA. Guidance for Industry: Information Program on Clinical Trials for Serious or Life-Threatening Diseases or Conditions. Available at http://fda.gov/cder/guidance/index.htm
33. Clinical Trial Registration: a statement from the International Committee of Medical Journal Editors. Lancet 2004;364(9438): xx-xx
34. positive controls and approval creep
35. J Clin Oncology 2003;21(7):1404-1411.
36. Bernard S. The 5 myths of pharmacogenomics. Pharmaceutical Executive 2003;23(10):70-78

9
Formulation and Manufacturing

Leo Pavliv and James F. Cahill

Dosage Form Decisions 202
Formulation Development 205
Early Manufacturing 214
Process Development 216
General References 221

The culmination of many years of research, development, animal testing, clinical studies, and mountains of paper or many gigabytes of memory eventually result in a patient taking a tablet, inhaling a powder, or taking some other formulation of a drug product to alleviate or cure a malady. This chapter focuses on the steps entailed to create the actual product, the formulation, which a patient takes for his or her malady, from initial concept through commercial sale.

There are many critical decisions to make in defining the desired product parameters, such as conducting the necessary studies to ensure the product is safe and effective, that it can be consistently manufactured at the desired scale, it is a cost-effective treatment, and the final form of the product is ready and acceptable for patient consumption. The choice of formulation will impact significantly all the other parameters, including efficacy, safety, manufacturing, and cost. Pharmaceutical scientists lead this effort at a company, working closely with the discovery scientists, clinical operations medical staff, manufacturing scientists and managers, marketing managers, and regulatory managers, among others, to devise the best possible formulation for patients and providers, at a reasonable cost for the company and the health care system.

This chapter addresses ten topics important in creating a product and its formulation as listed in above. The topic "dosage form decisions" reviews the possible product forms, product profile, route and dose, decision makers, changes during development, and post approval changes. "Formulation development" covers overall goals, active ingredients, preformulation issues, bioavailability and pharmacokinetics, and the seven product categories (oral, inhalation, injection, ophthalmic, topical, rectal, and vaginal). Six additional major issues in manufacturing and formulation of pharmaceutical products are discussed, including analytical development, process development, packaging, product stability, clinical supplies, scale-up issues, and commercialization requirements. The manufacturing and pharmaceutics groups at a company must consider all these issues for a successful IND/CTA and NDA/CTD, as well as acceptance by health care providers and patients and commercial success.

Dosage Form Decisions

One of the initial steps in the drug development process is to select a desirable product profile and create the dosage form that fits well the product and disease (Fig. 9.1). This profile is based on the physicochemical characteristics of the active ingredient, disease-related issues, how the product will be used by providers and patients, and marketplace issues (e.g., competition). Characteristics such as permeability, solubility, stability, safety, potency, half-life, and molecular size will strongly influence the product profile. If a product cannot be orally absorbed or is entirely metabolized by first-pass effect through the liver, then it would not be a good candidate for an oral dosage form. A low-potency product, requiring high doses to produce the desired effect, would not be amenable to injectable, topical, or inhalation delivery. Drugs that lack solubility in various solvents may not be good candidates for an oral liquid or a nebulized solution. A product with a very short half-life may require very frequent dosing, which may require a controlled release delivery.

- A desired product profile is planned early in development, depending on product characteristics and likely usage.
- What dosage form?
  - Oral - tablet, capsule, liquid, suspension:
    - Controlled, modified, or immediate release
  - Injectable - IV, SC, IM:
    - Prolonged or immediate release
    - Solution, lyophilized powder, emulsion, or liposome
  - Inhalation - nasal or lung:
    - DPI, MDI, nebulizer, spray (powder, solution, or suspension)
  - Topical - cream, gel, ointment, spray, foam, patch
  - Rectal/Vaginal - suppository, cream, foam, enema/douche

FIG. 9.1. Dosage Form Decisions – 1

The other key item in selecting the product profile is defining how the product will be used. A product intended for surgical anesthesia would be most amenable to an injectable dosage form that can be easily titrated to achieve the desired level of anesthesia. A product intended to treat heartburn would likely need to be self-administered as an oral dosage form. A product intended to be given predominantly to smaller children would be easier to administer as a solution or a suspension. Product use and the resulting product profile also may be dependent on the disease being treated. Diseases producing significant nausea such as cancer may need non-oral alternatives to ensure proper dosing. Asthma can be treated with inhalation products, depending on the mechanism of action, given the pulmonary site of disease and the accessibility of lung tissue to direct product administration. Diseases that harbor in specific tissues may benefit from special direct extravascular administration, such as intrathecal for some infections and cancers.

Oral dosage forms are the most popular products related to patient convenience and usually low cost of goods in manufacturing. They include tablets, capsules, various liquids, and suspensions. Oral products can often be developed as immediate release, modified release, or controlled release forms. Modified or delayed release products can better target an area of the GI tract that may be required for better stability, better absorption, a local action, or to reduce gastric irritation. Examples of delayed release products include Prevacid®, Ery-Tab®, and Cymbalta®. Extended release products such as Niaspan® and Zyban® can often overcome a drug with relatively short half-life and allow for a more convenient once or twice daily dosing schedule or in some cases can reduce side effects caused by high peak (C_{max}) blood levels that can be seen with immediate-release products. The patent life of a product can be extended substantially with a follow-on useful extended release formulation of an existing product. Procardia® was originally a very popular multidose per day oral product for hypertension (calcium channel blocker) that was going off patent. Procardia LA is an extended release product used once daily that offered more than another decade of patent protection, with major patient convenience and compliance benefits and protected sales for the company.

Injectable products are primarily designed for hospital or office-based administration by health care providers or when products are not amenable to other delivery modes. Most protein products, such as monoclonal antibodies, hormones, and enzymes, are unstable in the digestive tract due to degradation by the acid and proteolytic enzymes and either have no or very poor oral absorption. The great majority of these products must therefore be injected for an adequate therapeutic effect. Injectable products can often be developed as either immediate release (simple solutions) or as a prolonged release product. Examples of the latter include the technologies using suspensions, such as Depo-Provera® and Ultra-Lente insulin, and newer technologies using polymers, such as Lupron Depot® and Eligard®, liposomes, such as Ambisome®, Doxil®, and Depocyt®, and pegylated products, like Pegasys® and PEG-Intron®. These technologies, such as pegylation and liposomes, may also offer other product benefits such as reduced toxicity or increasing delivery to target tissues. Lyophilized products help to stabilize the active ingredient if it is not adequately stable in solution. The freeze-drying process affords reasonably good shelf-lives of products for commercial viability, but they will require reconstitution prior to administration and typically adds to the cost of the product. Emulsions, such as Diprivan®, may be required if the active ingredient lacks adequate aqueous solubility.

Products administered by inhalation typically are designed to deliver drugs systemically as in the case of products for anesthesia or for a more localized treatment such as a beta agonist or a steroid for asthma. The majority of these are in metered dose inhalers (MDIs), but nebulized solutions like TOBI® and Xopenox® and dry powder inhalers like Spiriva® and Serevent® continue to be developed to overcome some issues with MDIs. The inhalation route is being investigated as an alternative delivery to injections for insulin and other large molecules.

As with inhalation products, topical delivery is typically intended for a localized effect, although depending on the drug and formulation, systemic circulation may be achieved. Most creams, ointments, and gels deliver active ingredients to treat topical conditions such as dermatitis, psoriasis, or local bacterial, fungal, or even viral infections. Some exceptions include nitroglycerin ointment and testosterone gels. Transdermal products are engineered to deliver the active ingredient systemically, such as Duragesic® and Nico-Derm® over an extended period of time, ranging from 1 to 7 days.

The rectal and vaginal routes can be used for either local delivery to treat localized infections, hemorrhoids, constipation, fissures, or can also be used for systemic delivery in some cases. These routes can avoid first-pass effect and, although not typically popular in North America, can be of benefit in certain circumstances. In Europe, this route is more commonly accepted and employed.

Deciding on which dosage forms to develop should come out of the product profile decisions and is based primarily on four additional criteria: both provider and patient acceptances, physicochemical properties of the drug, and information from the absorption, distribution, metabolism, and excretion (ADME) studies seen in animals, which is dependent on the chemical composition of the drug itself (Fig. 9.2).

If the intended use and product characteristics do not match, the possible alternatives would be to modify the chemical composition to change the ADME conditions or to evaluate alternative uses of the product where there might be a better fit. Examples of chemically modifying an active ingredient include altering the salt form of the drug to modify solubility at desired pH conditions. More extensive modifications include creating a prodrug or in some cases an active metabolite to enhance absorption, improve activity, or minimize toxicity. Allegra®, fexofenadine, was the active metabolite of the drug terfenadine. Terfenadine was extensively metabolized by first-pass effect and when combined with certain other drugs that inhibited its metabolism caused potential cardiotoxicity. Xeloda® is a prodrug for 5-fluorouracil (5-FU). The prodrug allows for oral administration and is primarily converted to the active 5-FU within the tumor. Oral versus parenteral delivery not only improved patient compliance but also improved the activity of the drug.

The question of whether to change dosage forms during the development process is often debated. Typically, this should only be done if the benefits are significant and outweigh the potential delays and increased costs of doing so. The longer the decision is delayed, the greater impact it will have in the eventual product approval. Regulatory authorities almost always require that the final formulation to be marketed is the one that must be used in at least all the pivotal (phase 3) studies. Formulation changes at a minimum would require some pharmacokinetic studies, compared between the early and later dosage forms, but may require much more extensive studies, depending on their extent of change. It's possible that small changes in the formulation could result in different degradents or different levels of existing degradents that may need to be qualified in additional toxicology studies. Formulation changes could also increase absorption or result in extended profiles that may also entail additional toxicology studies or possibly additional clinical studies. The decision whether to change dosage forms should be made by all disciplines (of the company) that will be affected by the change including medical, development, regulatory, nonclinical, marketing, and finance. This group can assess and integrate the impacts that the change will have on extra study requirements, added product development work, manufacturing needs, patent issues, lost sales from any delays, improved patient benefits and marketability with an improved product, the approval time, and especially the costs associated with the change. They can then determine if those costs are warranted by the product improvements.

Making dosage form changes can be done postapproval and introduced as second-generation products, which is a major outcome of product life cycle management (Fig. 9.3). This will allow the initial product to be approved sooner allowing for a more informed decision making and thorough development effort for the improved product. Improving bioavailability or reducing variability can often be done with enhanced formulations. Saquinavir and cyclosporine are examples of initial formulations that exhibited poor bioavailability and were significantly enhanced in second generations, Fortovase® and Neoral®, respectively.

Decreasing adverse effects through reformulation can offer significant benefits. This can be accomplished in various ways, with a classic example of formulating amphotericin B liposomal products, Abelcet® and Ambisome®. Improving patient compliance through less frequent dosing can be accomplished by developing a controlled release product. Examples of this include Wellbutrin XL® and Concerta®.

Patent life can be extended in some cases. This was accomplished when nifedipine was formulated into the controlled release product, Procardia XL®. This change created a blockbuster with a number of years of exclusivity. Finding new indications for previously approved drug products may

- Match dosage form/route to what parameters?
  - Provider and especially patient acceptance
  - Physicochemical properties of active ingredient
  - Preclinical ADME data (bioavailability, t1/2, metabolism)
- To change or not to change?
  - As early in the development process as possible
  - Changing dosage form adds substantial time & cost to market:
    - Longer you wait the longer the delay and the greater the cost
  - Change done sequentially, in parallel, or post approval
  - Decision based on scientific data and cost vs benefit
- Who are Decision makers?
  - Input needed from medical, development, regulatory, nonclinical, marketing, and finance

FIG. 9.2. Dosage Form Decisions – 2

- What are Post approval changes (2nd generation)?
  - Improved bioavailability:
    - Saquinavir (Fortovase ®), cyclosporine (Neoral ®)
  - Decreased toxicity:
    - Amphotericin B liposomes (Abelcet ®, Ambisome ®)
  - Less frequent dosing:
    - Bupropion (Wellbutrin XL ®), Methylphenidate (Concerta ®)
  - New patent life:
    - Nifedipine (Procarida XL ®)
  - New indication:
    - Inhaled tobramycin (TOBI ®), topical tacrolimus (Protopic ®)

FIG. 9.3. Dosage Form Decisions – 3

entail a change in formulation or route of delivery. TOBI® is an inhalation solution to treat cystic fibrosis patients infected with *Brevundimonas aeruginosa*. Tobramycin was previously approved as an intravenous antibiotic. Protopic® is a topical cream developed to treat atopic dermatitis and was previously formulated as both an oral capsule and injection to prevent organ transplant rejection.

Formulation Development

Formulation development activities will be based on achieving the desired product profile and can be summarized by a series of goals (Fig. 9.4). The formulator will need to (1) create a dosage form that (2) delivers the active ingredient (3) to the intended site of action (4) in an amount required to achieve the desired effect with (5) minimal adverse effects (6) over the desired time course in (7) a consistent and reproducible manner. For example, if the product profile for a specific drug is an oral product taken by adults no more frequently than twice a day with blood levels to be between 50 and 400 µg/mL, the formulator will have a relatively clear path to develop prototype formulations. Based on the half-life of the drug, the formulator will be able to decide whether a controlled release or immediate release product will be required. Because adult patients will be self-administering the product, they can narrow the selection to a tablet or capsule. Initially, the target blood levels will be based on data from several different species of animals and later in development from human data. The formulator should have the commercial viability of the product in mind relatively early in the development of the product. This includes a product that can be easily and reproducibly manufactured with a reasonably low cost of goods. The product should be physically and chemically stable for a commercially viable period. This is typically between 2 and 4 years, preferably at room temperature, but other storage conditions may be required. It can take 6 months or longer from the time a product is manufactured until it reaches a pharmacy or hospital and this time needs to be taken into account.

The product should be as convenient to use as possible by the patient and/or the caregiver. A few examples of these include prefilled syringes, patches applied once a week instead of capsules taken two or three times daily, and antibiotics requiring a 3-day treatment period instead of a 10-day course. Differentiating the product from competitive products in some useful fashion should be considered throughout the development process. This is most important for products that offer a marginal therapeutic or adverse effect profile compared with other products in the same class. A premium price can be charged because of its good value to the health care system if this is accomplished.

Adequately characterizing the active ingredient is a critical and necessary step in developing a successful product (Fig. 9.5). Complete characterization is not technically required until the time of the marketing application, but a basic understanding is needed for the initial Investigational New Drug (IND) application. The more information that is available earlier in development can assist in avoiding later pitfalls. Basic characterization includes elucidation of the structure of the active ingredient. New chemical entities (NCEs) should be examined for stereochemistry, isomers, polymorphs, and very early in development salt selection. Structural modifications of the original drug and the impact on activity are needed to be assessed, especially for both potentially new improved products and patent protection against future competitors. Characterization of a biotechnology product can be more complicated and may include determining secondary and tertiary structures, degree of glycosylation, biological activity, isoform activities, impact of truncation of the molecule, amino acid sequence changes, pegylation, presence of neutralizing antibodies, and immunogenicity of the compound. These items can have a significant impact on both safety and efficacy of the compound and also possibly ease of manufacturing and patent protection.

From very early in development through commercialization, the impurity and degradation profiles of the active ingredient should be examined. Impurities and degradation products are supported by toxicology studies and later by human clinical data. Changes in these levels or differing profiles may require additional toxicology studies to ensure the safety of the product is not affected. Additional studies will add to cost and may

---

↳ **Goals:**
- Consistently deliver active ingredients to intended sites of action in amounts required over desired time course
- Make an easily manufactured, reproducible product that can be made at commercial scale & with low cost of goods
- Produce product that is chemically & physically stable for an acceptable shelf-life (2 to 4 years is common)
- Offer patients & providers convenient-to-use product
- Differentiate product from the competition
- Create product formulation with benefits allowing premium pricing for value provided

FIG. 9.4. Formulation Development

↳ Active Pharmaceutical Ingredient (API) characterization:
- A must for successful formulations
- A regulatory requirement:
  ➢ Basic understanding for IND and fully characterized for NDA
- Elucidation of structure (NCE and Biotech):
  ➢ Stereochemistry, isomers, polymorphs, salt selection
  ➢ Secondary and tertiary structure, glycosylation, biological activity, immunogenicity
- Impurities/degradents:
  ➢ Critical at all stages of development
  ➢ The purity used in toxicology studies supports clinical trials, which supports commercial use

FIG. 9.5. Formulation Development: Active Ingredient

- **Excipient compatibility:**
  - Identify potential interactions between API and excipients to decrease chances of future problems
  - Best to select GRAS components or excipients used in similar dosage forms
- **Solubility studies (pH and solvent):**
  - Helps determine potential suitable vehicles for many dosage forms
  - Select pH, buffer, and solvent for best stability of API
  - Balance between stability / solubility & route of delivery
  - Important whether in solution or solid dosage form
- **Forced degradation:**
  - Indicates how API will potentially degrade (hydrolysis, oxidation, light, etc)
  - Helps show analytical methods are stability indicating (critical at all stages of development)
  - Provides information for manufacturing process & packaging requirements

FIG. 9.6. Formulation Development: Preformulation

delay clinical development. Degradation products also may be an active compound and a source of a new follow-on second-generation product.

A very important part of the formulation development process is to conduct preformulation studies (Fig. 9.6). These studies are initiated prior to formulation activities, some of which need to be repeated during the development process when changes to the active ingredient manufacturing procedure occur. There are numerous excipients available to develop a product formulation, many providing differing benefits to the product. An *excipient compatibility* study should be conducted very early in the process to help narrow the selection of excipients to those where there is the least likelihood of a negative drug–excipient interaction. Excipients and other nonactive ingredients for tablets or capsules include, for example in Arava® tablets, magnesium stearate, talc, titanium oxide, polyethylene glycol, crospovidone, hydromellose, lactose monohydrate, povidone, starch, yellow ferric oxide, and colloidal silicon dioxide. Excipients provide a variety of benefits such as integrity of the tablet, ease in manufacturing, and stability of the active ingredient. Using excipients that are listed in pharmaceutical compendia, those listed as generally regarded as safe (GRAS) and in quantities used by other drug products will decrease concern that an untoward clinical effect will occur from the excipients. Use of novel excipients will likely entail safety testing on the excipient–drug combination and possibly on the excipients alone. Preservatives often are necessary to maintain the product's integrity and to provide antimicrobial activity, but they cannot interfere with product action or increase any toxicity.

A very basic but important preformulation study is to determine drug *solubility* and stability in different solvents and at different pHs (pH solubility and pH stability profiles). This applies to all dosage forms because an essential step in delivering an active ingredient is that it is in solution at some point in time prior to becoming available to deliver its therapeutic effect. For a solid oral dosage form or an injectable suspension it can occur after it is ingested or injected but it still must occur. Some products may have high solubility at a certain pH but poor stability at that pH. An intermediate pH may be necessary or stabilizing the product by other means may be required. Accelerated stability studies are done, often using higher temperature, early on to judge stability and related proper product storage conditions, as well as the potential shelf-life and reasonable time frame for use of a product (expiration dating).

One of the most essential early studies to conduct is a *forced degradation* study using elevated temperatures, oxygen (hydrogen peroxide), acid, and base. This is critical to ensure that the analytical method is stability indicating. It also helps determine by what mechanism the drug degrades and how sensitive it is to that condition. If the drug is rapidly oxidized, the formulator can look at adding certain excipients or modify manufacturing or packaging to minimize this degradation pathway.

The majority of patients prefer taking drug products orally over other potential routes (Fig. 9.7). To deliver an oral drug systemically, it must be bioavailable, which entails the drug going into solution, being absorbed by the GI tract, reasonably low liver metabolism of the active ingredient on the first liver pass, and then entering the systemic circulation. Acceptable *bioavailability* values will vary based on the drug, indication, required systemic levels for activity, and economics. For many drugs, a target bioavailability is at least 20%. However, in some cases, a bioavailability of less than 1% could be acceptable, as is the case with Fosamax® for osteoporosis. Formulation can significantly improve bioavailability in some cases, predominantly if the reason for the poor bioavailability is related to the drug's solubility.

- **Low bioavailability:**
  - No magic minimum values; Often driven by variability & economics
  - Fosamax ® deemed acceptable with <1%
  - Many drugs deemed not viable with <20% bioavailability
  - Formulation may be able to significantly improve bioavailability
- **Intersubject variability:**
  - Significant concern unless very large therapeutic window or dose titration is reasonable
  - Formulation often can minimize variability if absorption is factor
- **Intrasubject variability:**
  - Food effect, DDI, circadian effects
  - Formulation can minimize in some circumstances

FIG. 9.7. Formulation Development: Bioavailability

Other key bioavailability concerns include *inter- and intra subject variability*. Intersubject variability is a significant concern because prescribing and dosing information could be difficult if it varies too much from patient to patient. This concern may be reduced if the therapeutic window for the drug is wide. Also, wide patient–patient variability can be lessened by dose titration. Intrasubject variability is affected by a number of items, including food effect, drug–drug interaction, disease effect, and effects related to the individual's circadian rhythm. In either case, it's possible that formulation can improve bioavailability and minimize this variability.

A drug's pharmacokinetics is based on its chemical structure (Fig. 9.8). Drugs with a very short half-life and that require a sustained level for a therapeutic response pose a development concern. It becomes less practical for an individual to take a product more frequently than three or four times daily, and a target of once or twice daily dosing is becoming the norm. Two options exist to overcome a short half-life. The first option is to modify the chemical structure of the molecule either creating a prodrug, an active metabolite, or other chemical modifications, including adding polymers like polyethylene glycol to the structure. This option can be used for small molecules as well as proteins and peptides. The biggest drawback to this approach is that a new active ingredient is created, which requires going back to the beginning and conducting a full development program for the new entity.

In most cases, a more desirable option is to create a controlled release dosage form that releases a certain amount of the active ingredient over a prolonged time period. This approach is most amenable for oral products where different technology exists from osmotic systems to polymeric coatings. Glucotrol XL® is an example of the former and Avinza® Capsules are an example of the latter. Controlled release injectable formulations are less common but include suspensions, emulsions, implants, standard and polymeric liposomes, as well as drugs embedded in slowly dissolving polymers. Eligard® is an example of a controlled release subcutaneous injection delivery of leuprolide acetate with release from 1 to 6 months depending on desired profile. Lunelle™ is an example of a monthly injectable contraceptive suspension.

Drugs with very long half-lives are not typically a concern unless serious adverse events are associated with the drug or metabolites. There are no formulation strategies to reduce half-life, therefore leaving chemical modification as a viable alternative. This includes the possibility of developing an active metabolite if one exists.

Bioavailability and pharmacokinetic properties of the drug should be evaluated early in development, first in animal studies to determine initial formulations and then as part of the initial clinical trials to fine tune the desired parameters (Fig. 9.9). Formulation issues are identified once these properties are determined and compared with the desired product profile.

Bioavailability and pharmacokinetics are best assessed from complete ADME studies determining the drug's absorption, distribution, metabolism, and excretion. If a drug's *absorption* is thought to be inadequate, it is important to determine the cause as soon as possible. If it is related to solubility or rate at which it goes into solution, then that is something formulation may be able to overcome by several mechanisms. Determining the pH solubility profile for a drug and formulating it at an appropriate pH often can work with drugs that have an ionizable moiety. If aqueous solubility is still not adequate, another approach would be to use one of a number of pharmaceutically acceptable solvents and/or cosolvents, for example, sesame oil (Haldol®) and propylene glycol and alcohol (Nembutal®). Physical approaches such as particle size reduction and using an amorphous form of the substance may substantially increase solubility and/or the rate of solution. In some instances, selection of alternative salts can also enhance solubility. If none of the approaches work, then chemical modification may be required.

Poor cellular permeability is difficult to overcome without modifying the chemical structure. However, it's valuable to determine if permeability is active or passive and if passive

- Very short half life:
  - Formulation can often over come short half life
  - Can also structurally modify molecule:
    - Add/subtract chemical groups, prodrugs, active metabolites
    - This strategy creates a new active ingredient
- Very long half life:
  - Typically not as great a concern but can be if serious AE's
  - Formulation strategies not effective to shorten half life
  - Can structurally modify molecule
  - If long half life is desired can simulate with controlled release systems (e.g. weekly, monthly, or yearly dosing with special formulations of hormone therapy for prostate cancer)

FIG. 9.8. Formulation Development: Pharmacokinetics

- Establish bioavailability (BA) and pharmacokinetic (PK) properties and identify potential formulation issues:
  - ADME (absorption, distribution, metabolism, excretion)
  - Absorption:
    - Lack of solubility, permeability, or p-glycoprotein excretion out of cells
    - Solubility can be increased by adjustments of pH, excipients, solvents, amorphous solids, micronization, chemical modification, salt selection
    - Cellular excretion difficult to overcome but changing site of absorption, excipients, or "particle/droplet size" may be useful
  - Distribution:
    - Chemical modification (altering lipophilicity) typically required to alter distribution
    - Tissue targeting with certain formulations is feasible (e.g. liposomes) and is important for extravascular site of drug action or special tissues

FIG. 9.9. Formulation Development: BA & PK – 1

whether it is paracellular or transcellular. If cellular permeability is active, then it is possible to increase the amount absorbed by releasing the drug more slowly, not overwhelming the site of absorption. Formulation approaches are primarily limited to use of penetration enhancers, which the majority work by disrupting cellular membranes to differing extents. Using effervescence may enhance transcellular and paracellular pathways with a mechanism that has low toxicity. All agents intended to enhance penetration pose a number of potential clinical and regulatory concerns. By their nature, a penetration enhancer may not only enhance absorption of the drug but may also enhance absorption of accompanying unwanted agents. Agents that keep channels open for a longer period of time are theoretically more problematic.

Some drugs have adequate solubility, permeate through the cells, but are rapidly excreted out of the cell by P-glycoprotein. Chemical modification may overcome this cellular excretion although formulation approaches may also be useful. Certain excipients such as vitamin E-TPGS have been shown to decrease this excretion with some drugs. Another possible approach may be to alter the site of absorption to an area where cellular excretion is limited. In some cases, particle or droplet size may limit cellular excretion and may also modify the site of absorption. An alternative approach may be to deliver high doses of drug overcoming localized cellular excretion. This approach is likely more plausible for drugs where there is a reasonably large therapeutic window, the difference between effective and toxic doses.

Determining a drug's *distribution* is typically performed early in development. The studies initially are performed in animals to determine if there is a specific organ where the drug accumulates. Distribution also is determined for some products using radiolabeled drugs and assessing radioactivity in the organs. Distribution can be altered by chemically modifying the drug or in some cases through formulation techniques. Chemical modification can alter lipophilicity thereby altering distribution. Attaching ligands or antibodies to the molecule with cell specificity may result in a more precise approach, targeting a specific organ. Formulation can help target the drug to a desired site in some circumstances. For example, droplet or particle size of inhaled drugs will impact where in the lung the drug will deposit. Liposome size has been shown to impact which tissue a drug initially distributes into. Nanoparticles have been shown to more readily move into tumor cells than other sized particles thereby altering distribution. Paclitaxel is being attached to albumin as a nanopacticle, Abraxane®. Polyethylene glycol attached to a molecule (e.g., liposome for Doxil®) creates protection against immune reactions with less T-cell lymphocyte recognition and removal from the circulation.

Metabolism can have a significant impact on both therapeutic and toxic effects associated with the drug (Fig. 9.10). Metabolism can occur in many tissues, usually by degradative action of locally occurring enzymes, including the gut wall, skin, muscle, various organs, blood, and with most drugs

- Metabolism:
  - Significant source of concern due to drug-drug interactions (DDI), gut wall metabolism, extensive first pass effect, or toxic metabolites
  - Pharmacogenomic variations in population with drug impact on formulation decision or structure modification
- Excretion:
  - Product development strategies not typically used to alter excretion
  - If required, would likely have to modify chemical structure
  - Drug interaction or change in excretion in disease may impact decisions

FIG. 9.10. Formulation Development: BA & PK – 2

predominantly the liver. Drug–drug interactions often occur when drugs are metabolized by the same pathway or when one drug inhibits or enhances the pathway for other drugs. This could lead to either subtherapeutic or toxic doses of one or more of the drugs. Extensive first-pass effect in the liver may convert the active drug to inactive or possibly toxic metabolites. The predominant liver enzyme for metabolism is the family of cytochrome P-450 enzymes, and also the secondary level enzymes for glucuronidation or methylation. Extensive first-pass effect may be desired for a prodrug to be transformed into an active drug (e.g., codeine). Chemical modification and formulation strategies both can potentially reduce metabolism. Where a drug is extensively metabolized by first-pass effect, alternative routes of administration, such as sublingual, inhalation, rectal, or via injection, can overcome the initial degradation achieving therapeutic drug levels. For drugs that are metabolized by the gut, excipients that decrease enzyme activity or modify the local intestinal pH may have a beneficial effect.

The recent advances in pharmacogenomics have found metabolism to vary significantly in different patient populations. Genetic profiles allow for patient screening to see how people will metabolize certain drugs. These profiles can also assist in predicting if a certain drug will be effective. Herceptin® is targeted to treat breast cancer in patients who overexpress the HER2 Neuprotein. It has limited to no effect in breast cancer patients who do not overexpress this protein.

Although it is important to know which organs are responsible for excreting a drug, it is not common to try to alter the pathway. If the mode of *excretion* has to be altered due to organ status (i.e., kidney disease), then it would likely have to be done chemically by modifying the compound. A simpler and more common approach is to adjust dose to compensate for changes in excretion due to disease or other conditions. For example, Rapamune® dose is adjusted downward in patients with hepatic impairment.

Orally administered products, that is, tablets, capsules, powders, and liquids, are the most common dosage form (Fig. 9.11). Drugs administered orally are classified by two key parameters: solubility and permeability. A class 1 drug is both highly soluble and highly permeable. Drugs classified as

- Most common and desirable dosage form in the US
- Drugs categorized by solubility and permeability:
  - Class 1 (highly soluble and permeable):
    - Not likely to have bioavailability concerns due to formulation or manufacturing issues
  - Class 2 (low solubility and high permeability):
    - Formulation and manufacturing may affect bioavailability
  - Class 3 (high solubility and low permeability):
    - Formulation and manufacturing may effect bioavailability
  - Class 4 (poor solubility and permeability):
    - Can have significant bioavailability issues
    - Small changes can have a big impact

FIG. 9.11. Formulation Development: Oral Products – 1

class 1 are not very likely to have bioavailability concerns with relatively minor formulation or manufacturing changes. Also, when changes are made in most cases, dissolution studies are usually adequate to ensure the changes will not affect bioavailability. Class 2 drugs have low solubility and high permeability. In this case, both formulation and manufacturing changes may affect bioavailability. These types of changes may significantly improve or decrease solubility or rate of solubility, which can impact bioavailability. Class 3 drugs have high solubility but low permeability. It's possible that formulation and in some cases manufacturing changes may impact permeability and therefore bioavailability. Class 4 drugs have low solubility and permeability. Relatively small changes in manufacturing or formulation can impact bioavailability. These changes are most likely due to an effect on solubility, but it is possible that the change can also impact permeability.

When a product profile lists oral delivery as a key parameter, there are a number of factors that should be considered to better define which dosage forms should be investigated (Fig. 9.12). The first factor is to consider the patient, their age, and their ability to ingest the product. Tablets and capsules are standard adult oral dosage forms. If the product is intended for pediatric patients, a liquid or suspension is common and allows ease in swallowing and for easier dose titration based on patient weight. For example, the pain and fever treatments ibuprofen and acetaminophen offer different dosage forms for differing age groups. Both products have a concentrated form intended for infants where the product can be administered as drops. They also have a less concentrated form where a useful unit of measure is a teaspoonful or a measuring cup, allowing for relatively accurate dosing by the parent. For older children, they also have a chewable tablet and then move to a regular tablet or capsule as they reach adult dosing. Many other products, including over-the-counter and prescription products, have pediatric formulations consisting of liquids, chewable tablets, films, and other variations to ensure more accurate dosing and patient acceptance.

The geriatric population may also require specialized dosing. A growing number in this group have difficulty swallowing standard tablets and larger capsules. Options for this group include smaller sized capsules or tablets, liquids, as well as sublingual or fast-melt tablets that are easier to ingest. Elderly patients may also metabolize drugs differently, potentially requiring less frequent or lower doses. Considering this growing patient group may offer competitive advantages relative to other commercially available products.

After taking patient considerations into the equation, the physicochemical properties of the drug are a critical factor. In certain cases, these properties may override patient factors if technically they are not feasible. The preformulation work previously conducted guides the formulator in selection of compatible excipients. Final selection of excipients is then based on creating the desired dosage form dependent on the drug's properties. In most cases, the minimal amount of excipients should be used that allows the product to have the desired profile that is both easily and inexpensively manufactured. For example, if a chewable tablet was the preferred dosage form, mannitol would likely be one of the potential excipient choices, assuming the drug and mannitol were compatible. If the physicochemical properties of the drug allowed for a more efficient and less costly direct compression process, than a directly compressible form of mannitol should be selected. The formulator would then evaluate different amounts of mannitol and drug, selecting the least amount of mannitol that resulted in the desired taste, disintegration, size, shape, and ability of the tablet to be readily manufactured and stable.

Stability of the drug is also very important and can be improved by the formulation. If the drug is sensitive to moisture, a protective moisture barrier can be applied to a tablet. If the drug rapidly degrades in the acidic environment in the stomach or if it irritates the stomach, a pH-sensitive coating can be applied that dissolves at a more neutral pH seen in the small intestine.

The product profile should define the desired frequency of administration and release profile. Depending on the drug's observed pharmacokinetics and required dose, the formulator will determine if this profile is theoretically achievable. An immediate release drug is formulated without special coatings

- Factors to consider:
  - User:
    - Adult (tablet or capsule most common)
    - Pediatric (liquid or suspension common)
    - Geriatric (smaller capsules, liquids, sublingual, or fast melts)
  - Physicochemical properties:
    - Careful selection of minimally required excipients based on processing requirement needs for correct scale of equipment
    - Stability of API (can coat tablet to protect low pH in stomach)
  - Desired release profile:
    - Immediate release vs. delayed or extended release
  - Stage of development:
    - Often more efficient to start with simple formulation such as basic capsule & develop more complex delivery if required & warranted

FIG. 9.12. Formulation Development: Oral Products – 2

or excipients that would alter the drug's bioavailability. Typically, these products dissolve quickly and then are available to be absorbed depending on their permeability. Delayed release products are coated to prevent the tablet from disintegrating and drug from dissolving until a certain amount of time has lapsed or at a selected pH. The pH-sensitive coatings can target release in the proximal intestine where the pH is close 4 or 5 or closer to the colon where the pH is approximately 7. An extended release product releases drug over a desired time course. The time course is selected in part on the drug's pharmacokinetic properties, as well as the solubility of the drug and required dose. For example, if the drug's half-life is 6 hours and the dose is reasonably small to easily incorporate into a single tablet, then releasing adequate amounts of drug over a 12-hour period may provide for a desired once-daily dosing.

For cost- and time-efficient development, one should consider using a simple dosage form for the initial clinical trials and proof of concept study. A capsule with limited excipients can be quickly developed for less cost than a more complex coated tablet. Once the drug's pharmacokinetics is determined and the product profile is better understood, a parallel set of activities can occur with the eventual commercial parameters as the goal. One caveat is that the commercial formulation should be part of the clinical trials, either at a minimum in a separate pharmacokinetic study, preferably in the pivotal safety and efficacy studies.

Inhalation products are of growing use and interest not only for local delivery but also for systemic delivery of drugs as well (Fig. 9.13). Local delivery is intended for respiratory-related diseases such as asthma and cystic fibrosis and includes a variety of drug classes including steroids, beta agonists, anticholinergics, and antibiotics. One of the most significant advantages of inhalation products is being able to deliver a smaller effective dose compared with systemic delivery thereby minimizing side effects. Most drugs in these classes are potent and have significant adverse effects.

Inhalation products, including delivery to the lung, nasal and oral mucosa, bypasses first-pass liver metabolism, and therefore inhalation is a desired route for drugs intended for systemic delivery. Opioid agonists, 5HT agonists, vaccines, insulin, calcitonin, desmopressin, and other drugs that may face substantial issues with oral delivery are either available or may soon be available as inhalation products.

Inhalation products tend to be relatively complex because the formulation, drug delivery system/device, desired site of delivery, and physicochemical properties of the drug are all closely intertwined to develop an optimal system. For example, if the target is the deep lung, then the droplet size or drug powder size should be very small, approximately 2 to 8 μm in size, in order to reach the lower respiratory tree. Very small or very large particles may be swallowed, and medium-sized particles will go into the more proximal portions of the lung. Nebulizers and other devices impact the size, as does the size of the powder itself when delivered by dry powder inhalers. For nasal delivery or delivery to the oral mucosa for systemic absorption, a somewhat larger droplet size would be preferred to minimize the amount swallowed and lost to the GI tract. Finally, the patient or family member (e.g., mother for child) must be able to execute drug delivery fully and consistently, impacting the design of the inhalation system.

Most devices that deliver drug through a spray system deliver approximately 50 to 100 μL in a controlled pattern per actuation (Fig. 9.14). These systems can be activated by inhalation or by a propellant. Due to environmental concerns, most existing and all new devices using a propellant are fluorocarbon free. Formulations using these devices vary and can deliver powders, solutions, or suspensions. Many contain a preservative and can also contain an antioxidant, a buffer, and a propellant. Because these systems are complex, there are increased concerns regarding stability. As with all drug products, they must be physically and chemically stable, including any antimicrobial agent added as a preservative. In addition,

- Can be used for local or systemic delivery:
  - Local delivery includes steroids, beta agonists, anticholinergics
  - Can obtain undesired systemic levels in some cases
  - Desired systemic delivery for drugs like insulin, opioids, calcitonin, desmopressin, and others
- Lung, nasal and oral mucosal delivery avoids "first pass" effect
- Formulation, drug delivery system, & route of delivery based on:
  - Desired site of drug delivery and physical & chemical properties of drug substance
  - Optimum delivery to deep lung obtained with 2 to 8 μ sized particles
  - Local delivery with nasal and oral mucosa better using larger particles
- Particle size can be adjusted by controlling API size, droplet size, and/or delivery device

FIG. 9.13. Formulation Development: Inhaled Product – 1

- Spray devices designed for single or multiple dosages with controlled amount of drug (50 – 100 μl) sprayed in a controlled pattern.
- Spray can be activated manually or with a propellant.
- Typical formulations can include drug (powder, solution or suspension), anti-microbial preservative, antioxidant, buffer, and propellants.
- Long term stability concerns:
  - Chemical and physical stability of active and anti-microbial
  - Consistent dose throughout shelf-life
  - Extractables and leachables from device
  - Particle size must remain consistent:
    - Can be adjusted by selection of formula and device

FIG. 9.14. Formulation Development: Inhaled Product – 2

- **Formulation is based on several factors:**
  - Desired dose, route, and delivery procedure (i.e., bolus or infusion)
  - Physical and chemical characteristics of drug obtained from preformulation studies:
    - pH solubility/stability, compatibility with excipients and solvents, degradation route needed to define formula
  - Modified "ADME" parameters:
    - Pegylation & lipids / liposomes can alter distribution & clearance
- **Ingredients & parameters (pH, tonicity, buffer, preservatives) must be compatible with blood & tissue:**
  - Caution needed for drugs with limited solubility to ensure compatibility with blood and to avoid precipitation upon injection and change in pH or solvent concentration

FIG. 9.15. Formulation Development: Injectable Products – 1

the device must be shown to consistently deliver the desired dose through the product's expiration date. The device contains a number of parts necessary to deliver the drug, and potential extractables and leachables from the contact parts must be minimized and shown not to be a safety concern. Because particle size can change over time and a change may likely effect product safety and efficacy, this is one parameter that must be consistently monitored.

Injectable products offer several benefits over other routes (Fig. 9.15). They deliver the intended dose without first-pass metabolism, GI efflux, and other barriers faced by oral delivery. The formulations for these products are dependent on several factors. The dose, route (intravenous, subcutaneous, intradermal, or intramuscular), and delivery method (bolus or slower infusion) are essential elements. The feasibility of achieving the desired dose is dependent on product solubility for an intravenous delivery and on concentration and volume limitations for subcutaneous and intramuscular delivery.

As with other dosage forms, the drug's physicochemical characteristics will determine what types of formulations are feasible. The drug for an intravenous product must be solubilized in the formulation that could be in a standard aqueous-based vehicle, or in a miscible solvent, or in an emulsion, liposome, micelle, or using a cyclodextrin. The pH solubility information will help determine if an aqueous formulation is feasible. Solubility studies in solvents, such as alcohol, propylene glycol, or polyethylene glycol, that are acceptable for injectable products, may guide the formulator to a relatively simple nonaqueous or mixed solvent system. For example, Zemplar® injection to treat secondary hyperparathyroidism uses a propylene glycol and alcohol mixture to aid in solubility, creating an acceptable mixed solvent system. Solubility in surfactants, fatty acids, lipids, oils, and similar solvents may guide the formulator to look at emulsions, liposomes, micelles, or similar formulations. Even simple diluents can have significant impact on the delivery and stability of the product as well. Normal saline is acceptable for Leukine®, and dextrose 5% in water for Neupogen®, but not the converse.

Most injectable formulations do not alter the drug's own absorption, distribution, metabolism, or excretion parameters. Subcutaneous and intramuscular injectable formulations delivered as suspensions or in oils can sometimes be used to create a delayed absorption product (e.g., the once-monthly contraceptive Lunelle™ monthly contraceptive injection). The formulation of Injectable Risperdal® Consta™ employs microspheres to slow release of the product and achieve a prolonged action. Typical aqueous- or solvent-based intravenous formulations will not alter the ADME properties. However, emulsions or liposomes can effect absorption and to some extent distribution and metabolism and possibly excretion as well. Attaching polyethylene glycol to the drug, thereby creating a new drug substance, can affect each of the ADME parameters, especially distribution and clearance, extending half-lives significantly. This pegylation technique was successfully used in particular to develop biological products with less frequent dosing, Pegasys®, PEG-Intron®, Neulasta®, Macugen™, Somavert®, and Oncospar®. Each of these products has different ADME parameters than the non-pegylated version. Instead of thrice weekly interferon dosing in hepatitis C, weekly dosing is possible with Pegasys®. Neulasta® is given once after chemotherapy for neutrophil rescue versus daily dosing for 5–10 days of Neupogen®. Pegylation also was done for liposomal products to increase blood circulation times and to decrease the mononuclear phagocyte system (e.g., Doxil®).

It is important that any injectable formulation developed must be compatible with blood and tissue. These formulations should not lyse cells and should be minimally irritating upon injection. This is of specific concern where the drug is marginally soluble or very concentrated where either precipitation may occur or the formulation is hyperosmotic. The adjustment of the pH and the buffer system of the injectable solution can impact not only product stability but also the irritability of the injectable solution upon administration. A change from a citrate to a phosphate buffer for Epogen® reduced the local irritation upon subcutaneous administration.

An injectable product must be sterile, contain low levels of endotoxin (<5 endotoxin units/kg per hour in most cases), and contain low levels of particulates (Fig. 9.16). Each of these parameters is a very serious safety factor and must be stringently controlled. These three parameters are critical at all phases of development and must be controlled throughout the product shelf-life. Many recombinant proteins, monoclonal antibodies, peptides and other biotech drugs are not stable in solution and are heat labile. Almost all of these products cannot be heat sterilized and must therefore be sterile filtered and processed aseptically. In addition, to obtain an adequate shelf-life, a good number of these products must be lyophilized or must be stored refrigerated or even frozen.

Multi-use products require the addition of a preservative, unless the drug is self-preserving, to help suppress microbial growth with the multiple needle entries into the container system. Preservatives are often effective only within specific pH ranges and may become ineffective if they partition into

- Sterility, pyrogenicity, and particulates must be controlled, in addition to stability during shelf-life and administration.
- Most biotech products are peptides or proteins:
  - Limited stability in solution and cannot be sterilized by autoclaving.
  - Sterilization must be done by filtration (0.2 μ filters).
  - Stability can be enhanced by lyophilization.
- Impact of preservatives on product stability needs evaluation.
- Manufacturing controls are stringent at all phases of development:
  - To ensure sterile products with lowest feasible endotoxin levels.
- Need to conduct compatibility studies:
  - In diluents or reconstitution solutions.
  - To ensure product remains chemically & physically stable for during use.
- Need to determine stability and compatibility of final product with plastics, glass and rubber stoppers.

FIG. 9.16. Formulation Development: Injectable Products – 2

the lipid phase in emulsions, liposomes, or other biphasic formulations. Preservatives can also degrade over time and thereby become ineffective. The preservative effectiveness must be tested during formulation development and monitored to ensure the product remains acceptable for multiple uses. Preservatives include phenol, m-cresol, and benzyl alcohol. Selection of the preservative should include consideration of the intended indication and patient population. For example, benzyl alcohol is an effective and common preservative for injectable products. Benzyl alcohol also has a local anesthetic affect useful with subcutaneously administered drugs. However, benzyl alcohol has associated toxicities at higher levels and has been linked to a toxic syndrome in neonates. Product labeling should note that a product is contraindicated in patients with known hypersensitivity to benzyl alcohol. These compounds (e.g., benzyl alcohol) also can have an adverse impact on the stability of some active ingredients, especially proteins.

Many injectable products are diluted in common parenteral diluents, such as normal saline or 5% dextrose, prior to use and then infused over a specified time period. Compatibility studies must be conducted in different diluents to ensure the product remains chemically and physically stable for the period from preparation until it is used in the clinic. Stability of reconstituted lyophilized products should also be done to determine how long the product can be used after reconstitution. In a clinical setting, an intravenous product may be administered at the same time as other products, such that drug–drug compatibility in the intravenous (IV) solutions becomes an issue for testing during product development. For example, antibiotic and cardioactive products in the medical intensive care unit are common situations for combined IV administration.

Liquids, whether buffered aqueous solutions, solvents, or emulsions, are in contact with varied packaging components, including glass, rubber stoppers, and possibly certain plastics. Each of these components may leach undesirable substances into the product or the product may adhere into packaging materials and not be available. Red cell aplasia was associated with antibodies developed to Eprex®. Although not conclusive, these antibodies were believed to be formed by an excipient extracting out substances from the rubber plunger of the syringe and potentially acting as an adjuvant when the product was administered subcutaneously.

Ophthalmic formulations have many similarities to injectable products (Fig. 9.17). They must be sterile and compatible with the eye. However, because the ophthalmic products are not rapidly diluted and distributed as intravenous injectable products, they should be formulated at close to a physiologic pH and tonicity to not be overly irritating. In most cases, ophthalmic products are intended as multi-use products and therefore should be preserved. The same considerations regarding preservative effectiveness noted for an injectable product should be considered for ophthalmic products. However, there are several differences to consider for ophthalmic products.

Although most ophthalmic products are solutions or suspensions, ointments and gels may be more appropriate in certain circumstances. The residence time for most solutions is limited by the natural tearing process with the drug quickly washed away from the eye. Viscosity enhancers are often used as part of the formulation to increase residence time and therapeutic effect. The orifice of most ophthalmic containers is designed to deliver approximately 50 to 70 μL per drop. Only about half of that amount will be retained on the eye, the remainder is typically washed quickly away. Ophthalmic ointments and gel containers have a narrow orifice to allow a thin ribbon of product to be placed either directly onto the surface of the eye or into the lower conjunctival sac. The container system for solutions typically needs to allow compression by the patient for delivery of the drops.

- Similar formulation considerations as parenterals
- Required to assure compatibility with the eye
- Additional considerations include:
  - Containers designed to deliver 50 – 70 μl/drop
  - Only 20 – 40 μl of the drop will stay on the eye
  - Viscosity enhancers used to increase contact time of the drug to eye
  - Preservatives must be effective throughout shelf life
  - Formula compatible with pH and tonicity of the eye
  - Particulates and microorganisms must be controlled

FIG. 9.17. Formulation Development: Ophthalmics

Topical drug products are quite common and are used to treat a variety of local and, in some cases, systemic conditions (Fig. 9.18). One of the first factors to consider in developing a formulation is to determine where the desired site of action is intended, that is, topical, intradermal, or systemic. Anti-infectives, emollients, and sunscreens are examples of topical agents delivered locally to either protect or return the skin to its normal state. Many topical products are intended to deliver the drug through the stratum corneum and into the viable epidermal or dermal layers. A wide variety of drugs, including steroids, antivirals, antihistamines, counter irritants, and anesthetics are used to either treat skin diseases or relieve a variety of symptoms. Creams, gels, and ointments are different vehicles for topical drug delivery. Creams are oil in water or water in oil emulsions, while an ointment does not contain a water phase, and gels are typically aqueous based but may contain water-miscible solvents. The choice is based on a number of factors, including patient acceptance, the solubility of the drug in the vehicle, delivery of the active drug into the skin, stability, and other associated variables. Ointments also have an occlusive property that may aid in absorption. Absorption also can be enhanced for some topical drugs by using an occlusive dressing with the product.

Delivering drugs transdermally to obtain a desired systemic exposure is highly dependent on both the drug and formulation. Topical systemic drug delivery is subjected to metabolism within the skin but will bypass the more significant first-pass metabolism by the liver that can be seen with oral delivery. A variety of drugs have been delivered transdermally, including hormones, antihypertensives, opioids, incontinence medications, and nitroglycerin. Transdermal delivery can occur from creams or ointments but is more typical from patches. Patches are designed for a controlled delivery of drug from 24 hours to 7 days. Creams or ointments will only deliver drug systemically for a short time period, minutes to hours. Several drugs are now being delivered using iontopheresis to significantly increase penetration.

The drug's physicochemical properties will have a significant impact on delivery, whether it is intended for topical, intradermal, or transdermal delivery. Larger molecules typically have more difficulty penetrating the stratum corneum. Drugs larger than 1,000 MW tend to have poor penetration and those larger than 3,000 MW tend to have negligible penetration. Drugs that are more lipophilic tend to penetrate more readily into the lipid environment of the skin, and those that are more hydrophilic tend to poorly permeate through the stratum corneum. Penetration enhancers can be incorporated into the formulation to assist penetration through the stratum corneum and into the epidermal and dermal layers. Enhancers work by different mechanisms, some disrupting the stratum corneum, some through hydration, and others by penetrating into the hair follicle.

In formulating a topical product, interactions between the drug and vehicle, between the drug and skin, and between the vehicle and skin must be considered (Fig. 9.19). The first step to delivery is that the drug must diffuse out of vehicle; a partition coefficient exists for each drug in each vehicle that can suggest drug movement through the vehicle to the skin. The drug may preferentially bind to specific layers of the skin, which in some cases may be modified by the formulation. The remaining important interaction to consider is the vehicle and skin. Vehicles may hydrate the skin, which can increase permeability, or penetration enhancers can be incorporated into the vehicle, which may significantly improve absorption.

Several other critical factors must be considered in developing a topical product. The drug must be reasonably potent because only a limited amount of drug can be incorporated and applied onto the skin. Another important consideration is the state of the skin. Drug delivery can be significantly affected by abrasions or other wounds, psoriasis, sceleroderma, and other conditions that can increase or decrease permeability.

Rectal products can be used both for systemic or local treatment (Fig. 9.20). A relatively wide variety of drugs are delivered rectally, including steroids, antipruritics, pain medications, and anti-inflammatory agents. In much of the world, including North America, rectal drugs are not considered a primary method for systemic delivery, but it is quite accepted in certain European countries. The rectum contains veins entering both

- Five Factors to consider:
  - Desired site of action:
    - Topical – enhances barrier function of skin:
      - Anti-infectives, emollients, sunscreens
    - Intradermal – delivers drugs to viable epidermis or dermis:
      - Steroids, antivirals, antihistamines, anesthetics
    - Transdermal – delivers drugs systemically without first pass effect:
      - Hormones, antihypertensives, anti-anginals, opioids, incontinence treatments
  - Physicochemical characteristics of drug:
    - Size:
      - Penetration typically negligible when MW>3000
    - Lipophilicity:
      - Typically more lipophilic drugs will permeate into skin better
  - Can use penetration enhancers to increase penetration

FIG. 9.18. Formulation Development: Topical Products – 1

- Interactions:
  - Drug + vehicle:
    - Diffusion from the vehicle
  - Drug + skin:
    - Binding to specific layers
  - Vehicle + skin:
    - Hydration (increased permeability)
    - Penetration enhancement (alters skin structure)
- Other factors:
  - Required dose:
    - More potent compounds due to limitations of penetration
  - Skin disorders:
    - Wounds, psoriasis, scleroderma can significantly impact delivery

FIG. 9.19. Formulation Development: Topical Products – 2

- **Rectal Products:**
  - Can be used for local or systemic delivery:
    - Rectum has veins entering both hepatic and systemic circulation
    - Decrease first pass effect with delivery to lower rectum
  - Minimal fluid in rectum to dissolve drug, therefore dosage form is critical for more consistent delivery
  - Suppositories, creams, gels, and ointments:
    - Typical vehicle bases include fatty acids, water soluble glycols, polymers, and glycerinated gelatin:
      - selection based on drug solubility, stability, melting characteristics, and others
  - Enemas and foams:
    - Delivery of active can range from lower rectum to distal colon

FIG. 9.20. Formulation Development: Rectal Products

the systemic and hepatic circulation. This can make systemic delivery quite variable, especially drugs that are significantly metabolized in the liver. Dosage forms targeted to the lower rectum will have a significantly greater amount entering the systemic circulation because the veins entering the hepatic circulation are in the upper rectum. As with many other types of dosage forms, rectal delivery is quite dependent on the formulation. There is only a small amount of fluid in the rectum to dissolve the drug so dissolution of the drug is more dependent on the vehicle than for many other dosage forms.

Most of the same considerations described for topical products pertain to rectal and vaginal products as well. A key difference is that mucosal tissue is much more permeable than the skin. Rectal dosage forms, including suppositories, creams, ointments, and gels, often are used to solubilize the drug to enhance bioavailability. Where the drug is in suspension in the formulation, particle size may play an important role in efficacy. Common vehicles for rectal products include fatty acids, polymers, glycols, and glycerinated gelatin. The common component with these vehicles is they melt or dissolve relatively rapidly when inserted into the rectum. Selection of the vehicle is based on drug solubility, stability, and speed of dissolution/melting. Enemas and foams are also used for rectal delivery. These dosage forms vary considerably in being able to target the desired area, from lower rectum to distal colon.

Vaginal products are similar in composition to rectal dosage forms and include suppositories, creams, gels, ointments, foams, and douches (Fig. 9.21). Vaginal foams and douches are aqueous based and typically are formulated to have an acidic pH to maintain the natural microbial flora. If a formulation creates a hostile environment to the natural flora, the balance can be easily altered to allow growth of undesirable microbes. Certain drugs are formulated as vaginal tablets, which are very similar in both excipients and manufacturing techniques as standard oral tablets.

Most vaginal products are intended for local treatments, although systemic delivery is feasible. A major issue with systemic delivery is the potential for considerable variability in absorption due to changes in epithelial thickness due to age, menstrual cycle, and circulation to the tissue. Typical applications for vaginal products include contraception, lubrication, hygiene, induction of labor, and treatment of fungal infections.

Early Manufacturing

Analytical development begins at the initial stages of drug discovery because it is a necessity to be able to identify and quantitatively establish the compounds in question (Fig. 9.22). As the compound moves into animal testing and then into human clinical testing, the analytical methods are improved upon and must be able to ensure the purity, potency, and consistency of the product. The methods must be reproducible not only from time point to time point but also between individual analysts and on different pieces of analytical equipment, for example two different HPLCs. Both the FDA and International Conference on Harmonization provide guidance documents on general requirements for developing analytical

- **Vaginal Products:**
  - Can be used for local or systemic delivery:
    - Variable absorption due to changes in epithilium thickness and blood circulation with age and menstrual cycle make systemic delivery complex and uncommon
  - Typical applications include antifungals, contraception, lubrication, hygeine, and labor induction
  - Suppositories, creams, gels, and ointments:
    - Similar vehicle bases to rectal delivery
  - Foams, and douches:
    - Typically aqueous based, preferably with acidic pH to maintain natural microbial flora
  - Tablets:
    - Similar to oral formulations and manufacturing

FIG. 9.21. Formulation Development: Vaginal Products

- Critical component of drug development:
  - Ensures purity, potency, and consistency of the product
  - Must be reproducible
  - Process begins at onset of program and continues after marketing
  - Methods vary depending on the drug substance and drug product
  - International Conference on Harmonization (ICH) provides guidance on analytical method development
- Tests uniform to most products:
  - Assay (content, purity, impurities), appearance, and identity
- Additional typical product specific tests:
  - Oral products - dissolution, uniformity, hardness, friability, microbial limits, & moisture
  - Injectable products - sterility, endotoxin, & particulate matter
  - Topical products – preservative content

FIG. 9.22. Analytical Development – 1

methods from the clinical stages to the final validated methods to release the commercial product.

The methods themselves will vary widely depending on the type of product. Tests to identify the product, evaluate its appearance, and also to assay the product (measuring the content, purity, and impurities) are required for essentially all products. Specific tests for oral dosage forms usually include dissolution, ensuring the product adequately dissolves, content uniformity, establishing that each tablet or capsule contains the same amount of active ingredient, and others such as hardness, moisture, microbial limits, and friability. Injectable products also require the key safety tests ensuring the product is sterile, has a safe level of endotoxins, and has little or no particulate matter. These tests should be validated early in development because they are essential to patient safety. Topical products and other dosage forms that contain a preservative should have specific analytical methods to ensure the preservative content is at an effective level throughout the use period.

Physical analytical methods are relatively easy to develop and perform but nonetheless important to evaluate potential changes to the product (Fig. 9.23). Some of these methods include appearance, color, clarity, particulates, and weight or changes in weight over time. Chemical or biological methods may not be able to detect these types of changes that can impact the safety or efficacy of the product.

Chemical methods are usually the most sensitive to detect changes in the product and measure the content, purity, and impurities in the product, including potential changes over time. Various chromatographic methods, including the most common, high-pressure liquid chromatography (HPLC), are very sensitive, selective, and specific for the compounds in question. Titrimetric methods such as pH and spectrometry are quite common tools in drug development while acid–base, precipitation, redox, and complexation methods are somewhat less common and product specific. In some cases, complexation type methods are useful in determining levels of biological compounds using antibodies or similar technology.

The *United States Pharmacopeia* contains methods that the FDA typically accepts. Europe, Japan, and other countries have their own official methods, which should be used if developing products for those regions. Whenever possible, compendial methods should be used if they are appropriate for the product being developed. For example, when conducting sterility testing, the USP method should be used, if it is appropriate to ensure the product does meet the proposed requirements. The common thread is that the combination of all methods must assure the safety and efficacy of the product by ensuring the purity, potency, and consistency of the active ingredient and dosage form.

The *analytical tests for biological products* often are more numerous and complex than with traditional drugs (Fig. 9.24). Biologicals are complex molecules that are susceptible to many forms of degradation or alteration and often cannot be easily assessed for structure and purity. Changes can be chemical, for example, oxidation, reduction, methylation, demethylation, decarboxylation, proteolysis, hypo- or hyper glycosylation, loss of addition of disulfide brides, protein truncation. Changes also can be physical, for example, agglomeration, precipitation, or clumping. The result is a lengthy list of technically sophisticated analyses that must be done before any biological product is approved for marketing. Genes or plasmids (DNA), living cells, and media are the manufacturing components for recombinant DNA products. The integrity of genetic materials must be maintained throughout the manufacturing process, requiring specialized testing. Manufacturers may need to perform general chromosomal analysis (karyotyping), and also screen for genetic contaminants, such as mycoplasma, viruses, and cancer-causing genes (i.e., oncogenes). Throughout the manufacturing process, the stability of the inserted gene must be monitored.

Proteins are large and complex molecules whose structure can be analyzed with a variety of molecular probes. The primary structure can be determined by amino acid sequencing along with tests of protein stability (e.g., peptide mapping), but the more complex protein tertiary structures also require analysis. They include cross-linking, glycosylation, amino acid domains

- Physical methods:
  - Appearance, color, clarity, particulates, weight change, etc
  - Relatively simple but important
- Chemical methods:
  - Chromatography (high pressure, gas, size exclusion):
    - HPLC is one of most common stability indicating methods due to ability of method to be sensitive, selective, and specific
  - Titrimetric methods:
    - Spectrometric, pH, acid-base, precipitation, redox, complexation
- Compendial methods:
  - United States Pharmacopeia contains FDA accepted methods
  - Europe, Japan, and others have their own pharmacopeia

FIG. 9.23. Analytical Development – 2

- Complex Molecules:
  - Many degradation processes:
    - Chemical
    - Physical
    - Genetic
  - Tertiary structures:
    - Glycosylation, Cross-linking, Pegylation
    - Amino acid domains, 3D Folding
  - Bulk & Final Product, repeat testing for changes
- Final Product, Extra Tests:
  - Purity, Concentration, & Potency
  - Sterility
  - Stability
  - Heat & Freeze / thaw testing

FIG. 9.24. Analytical Development (Biologicals)

with specific actions, and 3D folding, requiring various specialized analytical tools, such as UV fluorescence, analytical ultracentrifugation, SDS-polyacrylamide gel electrophoresis (SDS-PAGE), and also HPLC.

Many protein analyses used to test bulk product are repeated at the final product stage to ensure that the purified recombinant protein in the vials has maintained its structure and activity, as it moved from purified bulk through formulation to finished product. Manufacturers also perform new tests on the final product for purity and sterility (e.g., bioburden, DNA, or endotoxin contamination) and for protein concentration and potency (e.g., bioassays, immunoassays). Stability testing of the final, labeled product in the vials is done. Long-term integrity and sterility are also monitored. Manufacturers test the final product under conditions of low temperatures, including freezing, and high temperature to evaluate possible denaturation under normal handling.

Process Development

A commercial manufacturing process should be designed to produce a pure, stable, safe product that is reproducible and cost effective (Fig. 9.25). This is significantly different than the early clinical scale where cost is much less important, as economies of scale are not possible and little effort has yet been applied to optimizing the process.

In process development, the equipment and processes must change in manufacturing as the requirements move from milligrams or grams to kilograms and metric tons, in order to meet the needs of moving from preclinical animal studies to human clinical trials and eventually to commercial requirements. Early small quantities of a typical small molecule are often made using standard chemical glassware up to a liter or so in size. At this stage, little effort is put into optimizing the process because the chemist's main responsibility is to produce a variety of different products as opposed to making a single optimized product. As the product moves into toxicology studies and initial clinical trials, the equipment would likely move to larger pilot-scale reactors at a 20- to 100-liter scale. At this stage, more thought is given to the process to begin improving yield but more so in being able to consistently reproduce the product and keeping impurities at acceptably low levels. Significant process development efforts must be underway during phase 2 clinical testing and often continue during the phase 3 clinical trials. The main focus at this stage is to develop an economical process while maintaining consistency preferably using available large-scale equipment. The type of equipment can be considerably different from that previously used and is often designed to be similar albeit at a smaller scale to the commercial scale equipment. By mid phase 2 or phase 3 clinical studies, the batch sizes can often be in the tens or hundreds of kilograms. Obviously, scale is highly dependent on the potency and frequency of dosing of the product, as well as the projected potential usage. For highly potent compounds, the commercial scale process may only be kilogram quantities while a compound requiring gram quantities for clinical efficacy may be at the 1,000-kg scale for later stage clinical studies. Biotech products undergo comparable process development activities. Scale will vary depending on potency and clinical and commercial requirements, but it often moves from 10- to 100-liter bioreactors at the early clinical stages to 10,000-liter or greater reactors at the commercial scale. In general, biotech products are much more prone to small process variations than small molecules, which can have a direct impact on purity, potency, and yield.

The purity profile of the active pharmaceutical ingredient (API) must be defined and used to establish the safety and efficacy of the drug at each phase of scale-up. Throughout the scale-up process, a link to the purity of the material used in animal and human safety and efficacy studies must be established. Changing processes or equipment can often result in qualitative and quantitative changes in impurities. Either substantially increased levels of impurities previously seen or the introduction of new impurities may require additional animal or human testing to qualify the safety and efficacy of the material.

Important or critical manufacturing parameters are established relatively early in the development process, such as processing temperatures, concentration of ingredients or raw materials, solvents, purification procedures designed to remove impurities, mixing procedures, and other physical and chemical items that could affect the purity and activity of the drug. These parameters can often be modified throughout the scale-up as equipment, manufacturing process, and control procedures, as well as assay methods, are modified or implemented to improve the efficiency of the process and increase the amount of material produced. One should always evaluate whether the improved process will have a negative impact on the potential impurity profile of the product and if so whether the benefit or the reduced cost will outweigh the additional cost and possibly time to implement the process improvement.

- Goal of manufacturing is to provide a pure, stable, safe product at high yield, reproducibly.
- Equipment and processing schemes will likely vary depending on batch size, based on market demands.
- Purity profile of API must be defined and used to establish safety and efficacy of API used in animal safety, human clinical trials, and commercial manufacturing:
  - Changing processes can often result in changes in impurities which may require additional animal safety testing to qualify new impurities.
- Critical manufacturing parameters:
  - E.g., temperature, raw material concentrations, solvents, removal of impurities, control of isomers, etc.
  - That affect the purity, efficacy and quality of API.
  - Must be established early and modified throughout development.

FIG. 9.25. Process Development: API – 1

9. Formulation and Manufacturing

The manufacturing process needed to purify most of the large and complex drugs produced through biotech procedures often requires special attention to impurities from host cells and adventitious agents from animal or human materials used in the process (Fig. 9.26). Possible contaminants such as host cell proteins, DNA, genetic mutations resulting in modified drug, viruses, and microorganisms must be considered as safety issues and controlled throughout the development phases. Processes are typically designed to use only raw materials and cell banks that have been screened for organisms or viruses that could infect the patient. In addition, the purification procedure must be designed with steps that have been demonstrated to remove or inactivate potential contaminants. These steps are especially sensitive to changes in scale-up and must be reevaluated with any change in the equipment, raw materials, or process. A seemingly small change in pH or buffer used during the purification process may allow contaminants previously removed to elute with the active ingredient and be present in the final product.

The purity requirements for the API can vary based on the route of delivery and subsequent "downstream" processing steps needed to produce the drug product. For example, the drug substance produced for products that are designed for parenteral use must be non-pyrogenic and have low microbial bioburden levels. During processing to obtain the final product, additional steps to sterilize the material are introduced to destroy or remove the organisms through filtration or autoclaving; however, bacterial endotoxins are not removed by these steps. The biological active ingredients produced through biotech processes generally must be sterile and are often produced so that the material can be used to produce the final drug product with no additional steps added to the process other than sterile filtration and filling. API for oral or topical delivery generally have a less stringent requirement for microbiological purity than parenterals in that endotoxin and sterility are not required and the potential for an immunological response to contaminating material is considerably less. The material must however be free of microorganisms that are objectionable and could cause infections in the patient, or that can grow in the product upon storage, or that can alter the stability, safety or efficacy of the drug product.

The container used for the API is selected to protect the material from conditions such as moisture, light, air, heat, and microorganisms that could affect the stability of the material prior to use in the preparation of the drug product.

The important steps in the manufacturing process that could affect the stability, purity, or efficacy of the product must be defined and controlled throughout the clinical development timeline and finalized prior to commercialization (Fig. 9.27). These items will vary depending on the characteristics of the drug substance, the formula, route of delivery, and the type of equipment needed. Typically, critical process steps include mixing or compounding times and temperatures; control of uniformity of dose to ensure that all of the product produced has a defined and consistent amount of drug; microbial and/or endotoxin control; and control of other environmental conditions such as air, light, or moisture. The preformulation, formulation development, and stability data will let the process development personnel know of potential concerns with the specific product. If the product is sensitive to oxidation, the removal of oxygen by nitrogen flushing and special packaging can be incorporated into the process.

Manufacturing of biotech products typically requires the use of live cells that need optimal conditions for cell viability and productivity. Often the manufacturing process is continuous with no clear distinction between what would be described as API or drug product other than one is in a bulk form and the other is in vials, ampoules, or prefilled syringes. The conditions to obtain acceptable levels of productivity and purity in a consistent manner throughout the process typically include media, buffers, pH, temperature, nutrients added to the growing culture, gas exchange, and mixing conditions. These items will vary for each cell line or culture.

It is important to establish a manufacturing process that can be scaled from the relatively small amount required for

---

- Special attention to API produced through biotech include:
  - control of impurities from host cells and adventitious agents from animal or human materials
  - Possible contaminants include host cell proteins and DNA, genetic mutations, viruses, and microorganisms
- Purity requirements for API can vary based on route of delivery and additional processing required to produce drug product:
  - API for parenterals must be nonpyrogenic
  - API produced through biotech processes often must be sterile and is often formula used for drug product
  - API for oral or topical delivery should be free of objectionable microorganisms
- Container for API is selected based on parameters that affect stability, such as moisture, light, air, and microorganisms

FIG. 9.26. Process Development API – 2

- Critical steps in manufacturing that affect stability, purity, or efficacy of product must be defined and controlled:
  - Mixing or compounding times and temperatures
  - Uniformity of dose
  - Microbial and / or endotoxin control
  - Oxygen, light, & moisture
- Biotech manufacturing typically uses live cells which require optimal conditions for cell viability and productivity:
  - Key items include media, buffers, pH, temperature, nutrients, gas exchange, mixing conditions, & time in cell culture
  - Conditions vary for each cell line
- Should develop a process that can be scaled from initial small numbers of clinical material to large scale commercial needs

FIG. 9.27. Process Development: Drug Product

- Packaging is an integral part of the drug product
- Can be associated with delivery of products:
  - Prefilled syringe, metered dose inhaler, single unit dosing
- Patient needs must be considered:
  - Multiple dose vs. unit dose packaging
  - Vials vs. prefilled syringes
- Outpatient products must be child resistant
- Should be tamper evident
- Several novel technologies being investigated to prevent counterfeiting (e.g., radio frequency ID)
- Ensure product is as stable as feasible:
  - Protect from contamination, moisture, and light if required
- Low cost of goods important:
  - For profitability and market competitiveness

FIG. 9.28. Packaging

preclinical and clinical use to large-scale commercial needs. Changes in formulation, dosage form (capsule vs. tablet), and manufacturing processes could affect the safety and efficacy of the product and will require additional testing or repeat of studies already conducted to demonstrate the comparability of the material.

Packaging is an integral part of the drug product (Fig. 9.28). For typical tablets or capsules, the packaging protects the tablets from light, humidity, and exposure to contamination in a cost-effective plastic bottle. In some cases, individual blister packaging may save the pharmacist time and offer a convenient reminder to the patient if they took their daily dose of medication. For a number of products, it not only protects the product, but packaging can also deliver the product, such as prefilled syringes and metered dose inhalers.

One of the main considerations in deciding on a packaging system is to consider the user, whether it is the patient, the physician, the nurse, the pharmacist, or other caregiver (family member). For example, an injectable product intended for emergency use in the hospital could be a standard vial, a kit containing a vial with diluent, or a prefilled syringe. Product cost will increase with more complex packaging, but a prefilled syringe may not only be more convenient and faster to administer in an emergency setting but may actually offer a cost savings when preparation time is considered. Some syringes are designed with auto-injectors or needleless systems to assist the patient in self-administration; however, these systems add to the cost of development and drug costs.

Another key consideration is using packaging that ensures the product has an acceptable shelf life. Not only must the packaging protect the product but also the packaging materials cannot leach unwanted substances into the product nor should the product bind to the packaging materials. Rubber stoppers in syringes or in vials can leach material into a drug product, and certain proteins may bind to plastics or to glass walls of a vial.

Packaging is important to not only help protect the medication from the elements, but also it should be designed to protect the package from tampering and should be child resistant to protect others from accidental exposure. All packaging intended to be given to patients in an outpatient setting must be child resistant, including in most cases investigational drugs during the clinical stages of development. There have been a number of instances of product counterfeiting, which endangers patient safety. New technologies are being implemented to allow for easier detection of counterfeit products including radio-frequency identification (RFID).

Labeling of information on the container system is yet another development challenge. Specific guidances for such labeling exist from regulatory authorities, including even type sizes, colors, and styles. Product names, usage, and storage conditions are commonly written on the bottles or boxes, along with NDC codes and bar codes. All labeling must be approved by regulatory authorities prior to use.

Packaging technology is quickly evolving with new polymers that are less prone to extractable and leachables while at the same time providing better protection to the drug. Even after approval, companies should continually look at ways of improving products, including novel packaging and delivery systems.

The *stability* of the active pharmaceutical ingredient and the drug product must be established to ensure the integrity of the material throughout the proposed shelf life (Fig. 9.29). This applies to materials used for preclinical safety studies, clinical studies, and for commercialization. The objective of stability studies is to determine the various conditions that cause degradation of the material and when possible provide packaging or adequate instructions for shipping, storage, and use to minimize or prevent degradation. An expiration date based on defined storage conditions and duration of storage must be established to assure the end user a product that is safe and effective. Data to establish the expiration date must be obtained with the actual product and final product packaging

- Must determine stability to ensure integrity of product during clinical studies and commercialization
- Stability testing must be done with identical packaging to clinical and/or commercial packaging
- FDA has guidance document for drug substance and drug product stability requirements
- Desirable to have at least a two year expiration date for commercial products
- Stability and compatibility in different diluents (e.g., normal saline, D5W) is required
- Examine temperature extremes (e.g., accelerated conditions, freeze-thaw, humidity)

FIG. 9.29. Product Stability

that will be used either in the clinic or with the commercial material. Variations in the dose, product purity, formulation or composition, and size of the final container can affect the stability of the product. For example, a product that is sensitive to moisture, light, or oxygen could degrade if the container was modified slightly and the seal protecting the product were to become less effective.

The regulatory agencies have provided a number of guidance documents for stability requirements. Exposure to extremes in temperature, light, humidity, and time must be evaluated to determine their effect on the product during shipping or storage. The documentation collected with several batches of material must be included in marketing approval applications. Stability and compatibility in different diluents (saline, D5W, and others used with the delivery of the drug) and containers is required. In addition, compatibility with other drugs that may be used in combination with the material should be evaluated and described in the package insert. Most products should have at least a 2-year expiration date for commercial products.

The purpose of drug products produced for use in the various phases of clinical evaluation can vary, known as *clinical supplies* (Fig. 9.30). For example, phase 1 clinical trials are designed to demonstrate the safety and pharmacokinetics of the material to help guide future studies in determining potential effective doses for subsequent studies. Phase 2 clinical studies use the data from the phase 1 studies to select the dose range and frequency of dosing, looking for a proof of concept of efficacy. The phase 3 clinical studies confirm efficacy and safety on a larger scale. The formulation and clinical trial material will likely vary for these different phases, keeping it simple up to a "proof of concept" study, and then becoming more complex as may be required for the later phases.

The primary concerns in evaluation of material used in early phase clinical studies from a chemistry and manufacturing perspective is that of safety, establishing a purity profile, and demonstrating that the product contains the required amount of drug. Although the material must be stable throughout its use in the study, long-term stability is not needed at this stage of development. Because of the limited number of subjects and duration of the initial studies, simple dosage forms can be used in order to speed the entry into the clinic. For example, a powder in a bottle that can be dissolved immediately prior to use or a simple capsule formula can be used for oral delivery of drugs that will ultimately become tablets or more complex controlled release products. Many biotech products that may be unstable in the liquid form can often be used as frozen liquids instead of the more complex and costly lyophilized products.

Clinical trials often have specific requirements that are not needed for the commercial product. Placebos or comparator products are often needed for use in blinded studies. These materials can require similar development efforts in formulation development and stability evaluation as the drug. They are often made look-alike by over-encapsulation with excipients, placement of a tablet into a capsule, or compounding or coating the material as a tablet with colored coatings. If novel or unproven excipients or additives are included in a formulation, vehicle controls that do not contain the drug are often needed to distinguish the safety characteristics of the drug from the excipient. Special packaging such as blister packs and use of containers with devices to monitor use can often improve compliance and assist in drug accountability to ensure proper use.

The final formulation that will ultimately become the commercial product generally must be used in the phase 3 or pivotal clinical studies. Any change to the product may require additional studies to reestablish the safety and efficacy of modified product.

As the *scale* of the manufacturing operation increases in size to accommodate the amount of drug required for large-scale clinical studies and for commercial production, larger capacity or high-speed processing equipment is typically required (Fig. 9.31). Small changes in excipient amounts or

---

- Early clinical trials are often done with simple dosage forms (powder in bottle, simple capsule, frozen liquid) to speed entry into clinical research
- Clinical trials often have specific requirements not needed for commercial product:
  - Blinded studies may require tablets placed in to capsules or tablets with colored coatings
  - Comparator products may need to be made or over-encapsulated
  - Placebo products that look like the active and/or comparator need to be developed
  - Special packaging to better ensure patient compliance may be required (e.g., blisters, bottles with use indicators)
- Final formulation usually required for phase 3 trials

FIG. 9.30. Clinical Supplies

- Changes in batch size of API or Drug Product often require significant processing and formulation changes due to:
  - Required higher speeds or larger processing equipment and tanks can alter critical parameters, such as heat distribution, mixing characteristics, and time that affect purity of material
  - Excipients and formulation could be modified to improve processing characteristics, such as flow of powder for compressing tablets, coating, or drying
  - Chemical reactions can be altered because larger volumes of liquids often take longer to obtain processing temperatures
  - Longer processing times can cause degradation of unstable materials, and can increase potential for bacterial growth in liquid materials
  - Aseptic processing, autoclaving, and lyophilization affected by batch size and duration of filling operation

FIG. 9.31. Scale-up Issues – 1

the addition of certain excipients could be modified to improve the processing characteristics required for the manufacturing equipment. Properties such as increased flow of the powder used for compression, changes in drying conditions, or type of blender to ensure uniformity are typical items to consider in the scale-up of tablet or capsule formulations. Modifications in the equipment, manufacturing process, or formula can alter critical manufacturing parameters such as heat distribution, mixing characteristics, and drying time. Sometimes changes do not affect the purity of a drug but may alter its crystal structure or other properties that can impact the safety and efficacy of the product. Chemical reactions can be altered because the increased volumes or mass of material often takes longer to obtain required processing temperatures and also longer times to cool the product. The increase to larger batch sizes of liquid parenteral products that are terminally sterilized by autoclaving can be exposed to significantly more heat simply due to the increase in time required to reach sterilizing temperatures, and then to cool down. The additional heat can affect the stability and purity profile of the material unless properly controlled. Products that are sensitive to heat and must be aseptically filled and lyophilized are especially prone to scale-up issues due to increased processing times prior to lyophilization and the need to modify the lyophilization cycle to establish acceptable and consistent levels of moisture in the finished product. The additional exposure to heat during processing and filling, the exposure to potential inadvertent microbial contamination during filling, and the potential for vial to vial variation in drug and moisture levels must be thoroughly evaluated and controlled during development and scale-up.

The effect of any processing change on the purity profile and potency of the product must be reestablished by demonstrating equivalence of material produced with the modified process to that of the original material (Fig. 9.32). This typically entails comparison of physical, chemical, and biological attributes of the drug as well as the stability characteristics of the API and drug product. In some cases where the changes in processing are extensive and specifications may be altered, additional preclinical or clinical comparability studies may be required to adequately demonstrate the equivalence of the modified to the original material and/or to establish the impact the changes have on the safety and efficacy of the drug.

Processing changes can also have an impact on the productivity or yield of the drug or biological material. This is especially the case with biotech products where the active ingredient is expressed through growth of microorganisms or cell cultures with subsequent purification from complex mixtures of media and host cell components through chromatographic procedures. A balance between the cost and purity of the drug must be established, with the caveat that the efficacy and safety cannot be compromised.

Even after the drug substance and drug products have been developed and demonstrated to be safe, effective, and stable, there are several significant issues that must be completed before *commercialization* of the product can occur (Fig. 9.33). Manufacturing, quality control, packaging, and distribution facilities must be built or established through expansion of company facilities and/or agreements with contractors well before the product is ready to launch. The decision to scale up for commercial product use must occur years before a product is approved, in order to put into place construction, people, and processes (with testing and validation). The cost is substantial, in the tens to hundreds of millions of dollars, and has a considerable amount of risk given it takes about 2–5 years to build out a manufacturing capability, which often is done prior to knowing how large of a market the product will take in or even if it will be approved at all. Agreements with contractors are designed not only to address the cost and production related issues but also to clearly define the responsibilities of the various parties to provide the quality assurance and control aspects of the manufacturing and distribution of the product. Several batches of material must be produced at the selected sites using the procedures and packaging components previously established and defined in the documents provided in the marketing applications to the various regulatory agencies. These

- Effect of the change on the purity profile of the product must be re-established:
  - If a process change occurs, new material must be compared to original material using physical, chemical, and possibly biological assays
  - Additional preclinical or clinical comparability studies may be required depending on the degree of change
  - Changes may not effect purity but may effect crystal structure or other parameters that can impact safety and efficacy of the product
- Effect can change stability of API and drug product, which must be re-established
- Purity and productivity (yield) of biotech products can often be impacted by changes in scale

FIG. 9.32. Scale-up Issues – 2

- Additional steps needed at end of development process to get products on to market include:
  - Set up commercial manufacturing and quality agreements with contract manufacturers
  - Manufacture validation and launch batches:
    - Validation batches demonstrate the manufacturing process is under control and produces quality material reproducibly
  - Pre-approval Inspections (PAI) of manufacturing and testing facilities by Regulatory Agencies to:
    - Verify documentation in marketing application
    - Assure material is produced in compliance with Good Manufacturing Practices (GMPs)
    - Ensure validation of equipment, manufacturing and control processes, and test procedures

FIG. 9.33. Commercialization Requirements – 1

qualification or validation runs are designed to demonstrate that the product can be produced on a regular basis in a defined manner to meet the established specifications designed to demonstrate its purity, potency, and safety.

Regulatory agencies will conduct preapproval inspections (PAI) of the manufacturing and test facilities to evaluate compliance with good manufacturing practices as well as to verify the accuracy and adequacy of the documentation in the marketing application. During the inspection, a detailed examination of the facilities, flow of materials through the facility, equipment, operating procedures, batch records, test procedures, validation and monitoring procedures used to ensure process control, qualification of personnel, and documentation submitted in the application will occur. Documentation and the rationale used to validate various manufacturing processes such as washing, sterilization, and depyrogenation of equipment and containers; mixing and compounding; aseptic filling and lyophilization will be evaluated. The PAI is often scheduled at a time that will allow observation of the manufacturing and control operations.

Distribution facilities along with procedures for shipping and returning or recalling material must be established and implemented (Fig. 9.34). Meetings with wholesalers of products who will store and distribute products are needed well before approval, including major end-users who serve as their own wholesaler (e.g., Walgreens pharmacy chain, or Premier hospital system), so that they will maintain the integrity of the product through its distribution channels. This situation is especially needed for unique formulations and packaging. Specialty distributors may be needed for product administered in specialized settings (e.g., Epogen® in dialysis centers). Sometimes at approval and launch of a product, the manufacturing is not fully geared up to meet patient demand. Then, special distribution systems are needed to ensure that the patients most in need of the product will receive it first. This problem occurred with the biological product Enbrel® for rheumatoid arthritis. Final labeling, including package inserts, advertising materials and all packaging materials, must be approved by the regulatory agency prior to distribution of the product. Often, unique packaging presentations such as those used for patient samples, additional dosages, or products designed with logos printed or imprinted on the drug are needed. Appropriate manufacturing and control documentation as well as stability data similar to that used to support the marketing application are required.

General References

1. Wells JI (ed). Pharmaceutical Preformulation, 2nd ed. 1999. Taylor and Francis Ltd, London, England.
2. Gibson GS (ed). Pharmaceutical Preformulation and Formulation. A Practical Guide from Candidate Drug Selection to Commercial Dosage Form. 2002. CRC Press, Boca Raton, FL.
3. Ranade VV & Hollinger MA. Drug Delivery Systems, 2nd ed. 2004. CRC Press, Boca Raton, FL.
4. Felton L. Pharmaceutical Manufacturing. In Gennaro A (ed). Remington The Science and Practice of Pharmacy, 21st ed. 2005. Lippincott Williams & Wilkins, Philadelphia, 691-1092.
5. O'Connor RE, Schwarts JB, and Felton LA. Powders. In Gennaro A (ed). Remington The Science and Practice of Pharmacy, 21st ed. 2005. Lippincott Williams & Wilkins, Philadelphia, 702-719.
6. Ando HY, Radebaugh GW. Property based frug design and preformulation. In Gennaro A (ed). Remington The Science and Practice of Pharmacy, 21st ed. 2005. Lippincott Williams & Wilkins, Philadelphia, 720-744.
7. Crowley MM. Solutions, emulsions, suspensions, and extracts. In Gennaro A (ed). Remington The Science and Practice of Pharmacy, 21st ed. 2005. Lippincott Williams & Wilkins, Philadelphia, 745-775.
8. Akers MJ. Parenteral preparations. In Gennaro A (ed). Remington The Science and Practice of Pharmacy, 21st ed. 2005. Lippincott Williams & Wilkins, Philadelphia, 802-825.
9. Lang JC, Roehrs RE, Jani R. Opthalmologic preparations. In Gennaro A (ed). Remington The Science and Practice of Pharmacy, 21st ed. 2005. Lippincott Williams & Wilkins, Philadelphia, 850-870.
10. Block Lh. Medicated topicals. In Gennaro A (ed). Remington The Science and Practice of Pharmacy, 21st ed. 2005. Lippincott Williams & Wilkins, Philadelphia, 871-888.
11. Rudnic EM, Schwartz JB. Oral solid dose forms. In Gennaro A (ed). Remington The Science and Practice of Pharmacy, 21st ed. 2005. Lippincott Williams & Wilkins, Philadelphia, 889-938.
12. Ding X, Allani AWG, Robinson JR. Extended release and targeted drug delivery systems. In Gennaro A (ed). Remington The Science and Practice of Pharmacy, 21st ed. 2005. Lippincott Williams & Wilkins, Philadelphia, 939-977.
13. Sciarra JJ, Sciara CJ. Areosols. In Gennaro A (ed). Remington The Science and Practice of Pharmacy, 21st ed. 2005. Lippincott Williams & Wilkins, Philadelphia, 1000-1034.

- Additional steps needed at end of development process to get products on to market further include:
  - Establish distribution sites and shipping/packaging procedures
  - Manufacture drug and print final labeling/inserts to be available for initial distribution (FDA must approve all labeling)
  - Obtain approval from FDA for brand name
- Different dosages or packaging presentations such as patient samples may be needed:
  - Stability of product in all packaging is required prior to distribution

FIG. 9.34. Commercialization Requirements – 2

10
Commercial Division

Thomas Lytle

Challenges	222
Framework	224
Commercialization Process	228
Practical Approaches	235
Commercial Responsibilities	237
References	239

Successful products require substantial, consistent investment and a multifaceted approach to development and commercialization. Multiple strategies, real product-specific advantages that differentiate a product from key competitors, and the marketing muscle necessary to achieve share of voice are key to long-term success. Effective prelaunch planning and execution is required to shape and prepare the market and to ensure rapid adoption at launch. Companies are constantly faced with a series of choices. How those choices are made will ultimately determine the future success of the brand. There is substantial cost to developing and marketing potential blockbusters, and early commercial analysis and input is critical to the process. Prelaunch investments can be significant. Through a partnership involving research, clinical development, and commercial, we can hope to optimize the decision-making process, reduce the risk of investment, effectively shape products and their markets for launch success, and focus on areas of opportunity.

The concept of "backwards planning" is an important overall best practice for leading pharmaceutical companies and is a core principal in all that is done in commercialization. It means that you start with a clear objective, focused plans, and a vision of what you think the product can be. You may not always be able to achieve what is in your vision, but starting there, and then planning for what needs to be done to get there, can be an effective way of the development of the best possible product profile and then the best possible launch, ensuring future success.

Launching products is a journey. This chapter focuses on the steps toward successful commercialization. It discusses what "best-practice" companies have done and outlines some of the key steps in the product development, commercialization, and launch process. In any organization, someone needs to be accountable and responsible for results. The commercial team plays a key role in ensuring that the product profile meets the needs of the market, collectively, the disease, patients, providers, gatekeepers, payors, health care system, and manufacturers, and also that the necessary plans and programs are in place for a successful launch and commercialization.

The chapter addresses six areas that comprise the commercial division's involvement in developing new products and their launch. *Challenges* are discussed for developing and commercializing new products and also the global environment, since worldwide operations are the necessary *modus operandi* in business including health care and the pharmaceutical industry. *Frameworks* are suggested for assessing potential opportunities for diseases and products and then building successful brands, which are products associated with a particular company within a market for a disease. The *commercialization discussion* offers lessons learned, along with key elements that comprise a potentially successful launch. The *launch* of a product, that is, the marketing and sales of the product by a company following product approval by a regulatory authority, involve both the prelaunch time activities for about 2 years before approval and the postlaunch time activities for about 2 years after approval. A myriad of options exist in the research and marketing for a new product. Therefore, *practical approaches* are important in commercialization for setting priorities, making choices, and leveraging insights and inputs. Finally, the *responsibilities* of the commercial division are covered particularly for product development and launch roles.

Challenges

Best-practice companies consider two distinct sets of activities in developing a product for commercialization: What has to be done to shape the product (Fig. 10.1) and what needs to be done to shape the market (Fig. 10.2). Most of this work with *product shaping* is done in the prelaunch period before

Focus Activities on Successfully Shaping the Product

- Market research – what to do and when
- Selecting a trade name
- Pricing and the value proposition
- Commercial input to label design
- Competitive analysis
- Manufacturing
- Distribution
- Dosage forms
- Clinical research program
- Sustainable competitive advantage

FIG. 10.1. Best Practice Companies – The Product

Focus Activities on Successfully Shaping Intended Markets

- Pre-launch activities
- Identifying unmet medical need
- Competitive weaknesses
- Publications
- Education (Disease & Product)
- Though leader support
- Creating awareness, interest, trial, and demand
- Laying the foundation & creating the need

FIG. 10.2. Best Practice Companies – The Market

product approval. The collaboration with R&D for product profiles and study ideas can occur even during the research phase more than 5 years ahead of approval time to assist basic scientists in their animal studies and early formulation work. A range of activities are key to driving each area, and both phases of the launch process are important to a product's success. Successful products don't just happen, rather they are the result of thoughtful planning, customer insight, and identifying a need that can be satisfied by the proposed product. Market research particularly assists R&D in more fully understanding the disease from the eyes of the providers and outside research community, regarding unmet needs for a disease and therapies and then the extent of the opportunities. You can identify subsets of patients without appropriate or sufficient care. Number of addressable patients can be identified for the new product based on its profile. Special care circumstances can be elucidated, as far as sites of care, chronicity, and concurrent treatments. Cost of care can be documented. In competitive analysis, the current products may have problems with too low efficacy, unacceptable side effects, too much drug interactions, unreasonable dosing or administration, or poor formulations. Study ideas can be generated for the clinical research from focus groups. This learned knowledge from market research helps the marketing group at a company support other key development issues; acceptable formulation alternatives for the pharmaceutics group, cost of care and pricing of a product, manufacturing and distribution in amount and timing, package engineering for product labels, and regulatory group for labeling needs and opportunities. Product name needs to be selected that fits the disease and market and that is acceptable to the health care community. A memorable name assists the company in sales and even product acceptance. A bottom-line issue for shaping the product is to create a competitive advantage that is sustainable over time and in the face of competition.

Shaping the market occurs throughout the R&D period before a product's NDA/CTD is even submitted (Fig. 10.2). For some novel products with whole new mechanisms of action for a disease and a first-in-class product, the prelaunch work in shaping the market can start while phase 1–2 studies are being done 5 years or more prior to marketing. Best-practice companies focus activities on not just understanding the disease and market but also on how the market is changed or can be changed by a product. Many examples exist over the past several decades up to the present: H-2 antagonists for ulcer disease instead of antacids and surgery, thrombolytic biologic enzymes in heart attacks, antimicrobial drug cocktails for HIV infections, and cancer therapy cures, an unheard of opportunity a decade ago. Yes, it is the novelty of the product, but successful uptake (rate) and improved patient care also is based on a host of roles that a company supports. Education of the medical community about the product and its disease care opportunity is needed. Payment policies are put into place for the new product. Thought leaders for a product are given the advance knowledge and experience through research, beyond the few involved in phase 2–3 studies, which can be just 20–100 centers out of all the universities and medical centers around the world.

An analogy of a good gardener and nursery fits the pharmaceutical commercialization situation. They choose and cultivate the land to be planted, discusses the opportunities that exist with the customers for novel plants, advise on and select the appropriate plants for the locale and that meet customers' needs and preferences, sell the plants and support materials (soil, fertilizer, weed kill, and more), and plant and care for the flowers and shrubs. A company helps create the awareness, interest, and then demand for a product for patients and the health care community through their research, publications, education, and advertising. If a product already exists, a company will identify limits of these products, which are weaknesses to be exploited for the new product in advancing the care of patients. They lay the foundation and even create a need often related to the unmet medical need, the transformation of care with a novel product, the impact on the health care system (vis-à-vis, pharmacoeconomics and quality of life), and the benefits and risks of the new product. Perhaps, one outstanding example is the severity and impact of anemia on dialysis and cancer patients. The nephrologists and oncologists, respectively, were treating serious and fatal diseases, with anemia being considered a relatively minor issue. Amgen and Johnson & Johnson for dialysis and cancer, respectively, did the research to demonstrate the severely

- What should be global?
- What should be regional?
- When should we start?
- When should we invest?
- Is it ever too early?
- How do we integrate other functions?
- Who takes the lead?
- Who should do what?
- Who's accountable?
- What's our positioning?
- How do we select a name?
- How should we determine pricing?
- How will we get reimbursement?
- What's the sales forecast?
- Etc……

FIG. 10.3. Familiar Questions

- You can never start too early
- You can never invest enough
- Think globally and recognize regional differences
- Work in teams…get everyone involved
- Insights create value
- Positioning is everything
- Get close to your customers
- Think value and not just price
- Work with payers early on with reimbursement plan
- Have a plan (global and regional)
- The fundamentals still matter
- Execution, execution, execution

FIG. 10.4. Answers to Questions . . .

debilitating impacts of anemia on quality of life and the benefits of the novel product, epoietin alfa, well beyond the initial expectations of providers. New research methods and outcomes had to be created. Much education of the health care community, providers, payors, and government was needed. New payment policies needed to be created for a whole type of therapy. Now, epoitin is the standard of care, and patients' lives are dramatically improved through the collaboration of health care providers and the companies.

Anyone who has been assigned the task of bringing a product to market, organizing a launch team, or launching a new indication will have more *questions* than answers at the beginning of the process (Fig. 10.3). This situation is good. Knowing what to ask, who to ask, and what to do with the information is key to a successful process. This chart raises several of these questions (14) and draws attention to some of the more critical decisions and choices that organizations must make as they prepare for launch. People questions are particularly important for optimal functioning. Role clarification for departments, cross-functional teams and processes, and accountability all contribute to the process of product development and launch success. Questions for scope of plans and work equally need attention for global versus regional research and marketing, and single versus multiple indications. Timing issues have a huge impact on an organization's spending levels and timing, when to start and when to invest. Marketplace questions are manifold; name selection, pricing, reimbursement, and forecasting.

Despite the many tough questions, there are some possible *answers* based on experience and what the best practice companies have learned (Fig. 10.4). You can never start too early and probably can never invest enough. Both global and regional approaches are combined. Global opportunities being addressed are the norm through expansion or partners. Regional differences in culture, medical practice, and opportunities must be identified. Strive to work in teams to get the job done with defined responsibilities and decision making. Talk with your customers and listen to what they have to say. The insights you draw from their comments can help you differentiate your product and deliver product benefits and minimize product risks that are meaningful to them. The health care community wants value for their investment in new products and technology and is willing to pay a premium price if appropriate for the situation. Don't overlook the fundamentals in planning and operations for people and programs and for the science and marketing. Remember that plans are great (global and regional are needed), but it's the execution that really matters and makes the difference.

Framework

To be successful, organizations need to be aligned around a clear set of *goals and objectives* (Fig. 10.5). Most importantly, remember why you are in business: first and foremost to serve and help patients, and second to provide better solutions to your customers. "Customer" has become an expansive term in the pharmaceutical and health care world, for example, beyond the focus of providers and payors to include patients, health professionals (nurses and pharmacists), gatekeepers (managed care organizations, prescription benefits managers, group purchase organizations), societies, scientists, distributors, press, investors, government, lobbyists, and the public. Manufacturers are science-based for-profit businesses with stockholders, such that you have revenue targets, profit targets (bottom line contributions after research investments and operating expenses), necessitating world-class commercialization and needing a stream of successful products—blockbusters. The industry is an ethical and heavily regulated industry, such that you must follow the laws, regulations, and guidelines of many groups representing many constituencies

- Grow our business
- Increase revenues
- Improve the contribution to the bottom line
- Develop new products
- Have world class commercialization
- Comply with rules and regulations
- Launch a stream of "blockbusters"
- Know & support the customer (providers & payers)

FIG. 10.5. The objectives should be clear . . .

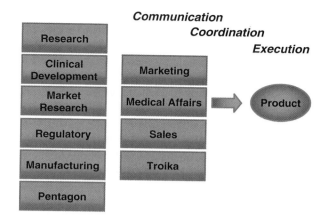

FIG. 10.6. Creating the next blockbuster is a complex process

that impact product development and usage (acronym soup: FDA, OIG, IRS, MMS, HHS, PhRMA, AMA, ASHP, IRBs, NIH, P&T Cs, MCOs, PBMs, GPOs).

Creating the next blockbuster is a complex process (Fig. 10.6). There are many ways to map the process. This chart reflects the various parts of the organization with important roles in the commercialization process. There are clearly many more and you should look carefully at the company's organization to see who does what, who is accountable, and where things get done. As the product moves through the development process, it is ultimately marketing, sales, and medical affairs who have the responsibility of creating the brand, creating its image, and executing the plan. This *troika* is responsible for commercialization, although distinct in their roles and organization, and needs to coordinate and communicate closely with each other and then execute. The rest of the organization (*the pentagon*) listed on this slide play principal roles in the commercialization plan, and they must also communicate, coordinate, and execute, especially in concert with the troika, for successful commercialization.

The pharmaceutical and biotechnology industries are unique, regarding the science/business/health care environment affecting access to, use of, and payment for products, and launching a product is not easy. The *hurdles* are high and have been increasing over the past 20 years (Fig. 10.7). Customers want to know if a product works and how it works; whole new disease targets and mechanisms of action are being developed in a world of burgeoning and novel science, especially biotechnology. Just because it works does not mean that it can be approved; regulatory hurdles have been rising with more studies, larger studies, more sophisticated testing, and more regulations, such as risk management. If it is approved, that does not mean that anyone needs or wants it; educational demands by providers and patients, phase 4 studies, and more publications are now paramount in product acceptance. If the medical community needs it, that does not mean that it will be reimbursed; value must be established further through added pharmacoeconomic studies, and gatekeepers need to be convinced. If it can be reimbursed, that does not mean that the company has the will to win and the resources to make it happen; the cost of research and launch for a new product has grown to more than $1 billion, leading to major industry consolidation and many partnerships. Despite all of these growing hurdles, good products succeed, and the industry does a lot of good for patients and their quality of life.

The pharmaceutical industry is about developing better products, improving the ways we treat and manage disease– *better outcomes* (Fig. 10.8). The environment is becoming increasingly challenging through increased pressures to control costs, increased regulation, and competition from a variety of sources. Achieving our objectives will require continuous improvement in many areas; product areas (product, their value, access), marketing (message and positioning, value proposition, customer insights), customers (relationships, communications, information, value proposition), operations (global coordination, collaboration and teamwork, delivery, communications), and government (compliance, communications, value proposition).

In some industries, the traditional *"4 Ps"* still dominate the *basics of marketing* (Fig. 10.9). Get the right product in the right place, at the right price, with the right level of promotion, and products will sell. For us, however, there are at least 6 additional "Ps"; somebody has moved our cheese, that is, the pharmaceutical world has been and is changing in complexity and demands. We need to have the right message for providers and payors, which may be different for each group but must be consistent and integrated. Policies at the national and local level impact our ability to access customers, and

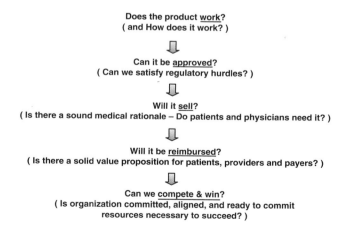

FIG. 10.7. Developing and successfully launching products – hurdles

- Products
- Value proposition
- Message and positioning
- Communications
- Delivery
- Access
- Relationships
- Customer Insights
- Information
- Global Coordination
- Compliance
- Collaboration and Teamwork

FIG. 10.8. The future will require better outcomes...

Somebody Moved Our Cheese

The 4 "P's"		At least 6 more "P's"
→ Product		→ Providers
→ Price	+	→ Payers
→ Place		→ Policy
→ Promotion		→ People
		→ Processes
		→ Performance

FIG. 10.9. Basics of Marketing

these policies are constantly changing. People still matter as do the relationships that are built with all the varied customers: patients, providers, payors, policymakers, principal investigators, and professionals at the company. Strong processes help us get things done in an effective and efficient manner, and we are expected to perform, financially and as individuals, meeting the expectations of management, the medical and health care community, the public, regulators, and even investors.

The *challenge* of bringing new products from an R&D perspective is increasing in complexity (Fig. 10.10). The environment of the 1970s and 1980s was very different than the environment of today. Today, a new set of pressures have increased the complexity, the challenge of new technology, need for speed to market, the size of the opportunity related to developing and commercializing new products, and the aspiration of the employees and companies for blockbusters and medical breakthroughs. This slide lists 14 of these pressures.

This *next* chart reflects the annual *investment in research and development* that has been made since 1993 by the government in NIH funding and by the industry segments of biotechnology and drugs (PhRMA) (Fig. 10.11). The industry has invested about two-thirds of the research dollars annually over this decade. The investment has continued to rise, more than 100% in less than a decade to more than $66 million in 2001. The cumulative annual growth rate (CAGR) is 11% and 13% in the 4 years 1998–2001 [1].

But fewer products (*new molecular entities;* NMEs) have been approved, which means that sponsors must make the most of each approved product and must be extremely diligent and thoughtful in the choices and decisions they make during the development process (Fig. 10.12). NMEs are the novel products, advances in therapy, which have gone down dramatically. Even excluding the excellent year of 1996 with 53 products, the rate for NME approvals has been about halved, while research spending by the industry has increased from $18.4 million to $46 million (150%) [2].

Organizations have different *strategies* to the science and business of pharmaceutical development and commercialization (Fig. 10.13). The approaches can all work, but they have substantial impact on the internal company operations of research and marketing activities and especially the type of products and the advances in health care to be achieved. If you liken the process to a baseball game, some companies rely on a series of base hits, doubles, and triples with the occasional home run. Others go only for the home runs; they may let quite a few pitches (products or opportunities) go by, but when they see the one they want, they go for the fences. Most companies, however, go for the hits, but are ready for the occasional home run. When they see the opportunity, they are ready and they know what to do with it to win the game. What approach does the organization take that you are working with

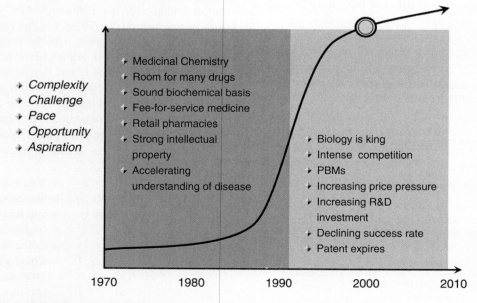

FIG. 10.10. Challenge Profile of Therapeutics Industry

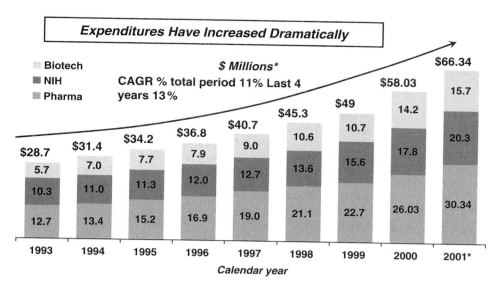

FIG. 10.11. Total Industry R&D Spend

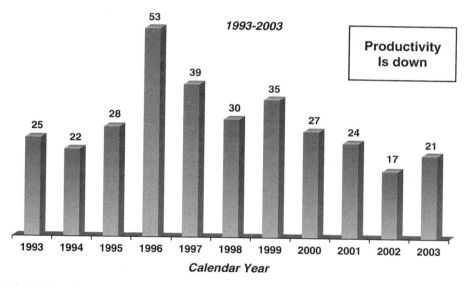

FIG. 10.12. New Molecular Entities Approved
Source: U.S. Food and Drug Administration 2002.

"Winning the Game"

- A series of singles, doubles and triples – hope for the occasional home run

- Wait for a great pitch - go for the home run

- Rely on a series of solid base hits - but plan to hit a few home runs

FIG. 10.13. Organizations have Different Strategies

or for? This is a question worth debating, because the answer and choice is very important to portfolio management and a broad range of other strategic and operational decisions.

The *traditional approach* for pharmaceutical product development and marketing in the 1970s through the early 1990s is outlined in Fig. 10.14 with seven operating principles usually employed. So if the world has changed and a new approach is required, what is a good way to approach it? *"Backwards planning,"* or starting where you want to finish represents an innovative approach. If you know where you want to end up (the finish), then start there and work backwards to outline what you need to do to get there. An example of this practice is to write out the ideal product profile with competitive advantages,

Environment Changes - Marketing Approaches Need to Change

Traditional Approach: ➡ **Backwards planning:**

- Driven largely by R&D & Regulatory
- Little commercial input-often late stage
- Approval - the focal point
- Value in modest product improvement
- Fewer hurdles
- More promotional latitude
- Marketing muscle to drive share

- Start where you want to finish
- Assess market needs & opportunities
- Recognize realities of new environment
- Continuous commercial input is essential
- Develop the "ideal" label early
- Develop product and design trials to help get you there
- Build competitive advantage

FIG. 10.14. New Environment Requires New Approach

and then the package insert, for a particular target disease with a product of interest that has been discovered or licensed in by the company, and perhaps only animal studies show some promising results. This ideal product profile will drive the studies needed to create the data for this insert (the end product), as well as market research and marketing planning for a future launch. The seven new practices for the backwards planning approach are outlined on this slide.

Commercialization Process

The chart in Fig. 10.15 highlights the *best practices for* phases of commercialization and overlays them with the phases of development and approval. Premarket activities (shape the market) can begin in late phase 2 and continue well after launch. Activities to shape the product (that is, the studies and early market research) start very early (phase 2 and before) and continue throughout the life cycle of the product. When a new indication or new opportunity is identified, the cycle virtually begins all over again. Planning for the future, looking for life cycle extension opportunities, and focusing on shaping the product and the market maximize the value of the brand to the organization.

Successful organizations work effectively in teams and leverage the skills, knowledge, and talents of each individual and group. But successful organizations also know that someone needs to be accountable, and someone needs to be in the lead. In best-practice organizations, leadership transitions over time based on the stage of development or commercialization of the product. In the early phases, the bulk of the knowledge rests with the scientific and clinical development organization, so it makes sense for them to be accountable and to take the lead role in the process with strong commercial input at key decision points. As we learn more about the brand (the product and market), and it gets closer to approval and launch, the emphasis shifts to commercialization. The commercial team takes the lead with strong R&D input. Strong collaboration and teamwork are the keys to successful product and market evolution and effective transitioning.

Brutal honesty and dealing with reality are keys to the development and commercialization process (Fig. 10.16). Understand four parameters: what you really have and what it can do ("the product" and "the outcomes"), what you can really say ("the label", that is, the package insert), and what the market really wants and needs ("the opportunity"). Collectively, these four parameters comprise your product "brand." This objective assessment drives further product development, forecasts, and market planning. The research and development managers and the sales and marketing managers together must be openly and completely self-critical with each other about their own product, the supportive studies, and the marketplace information. Many companies will bring in outside consultants at key development steps who are experts in the therapeutic and disease field to critique and advise. Only then can the best decisions and plans be made for successful development, approval, and commercialization.

Is there a difference between pharmaceuticals and *biotechnology*? The basics are the same, as represented by

Understand interface between R&D & S&M for commercialization

- Commercial planning starts early
- Need to shape products & markets
- Leadership changes over time

| PreClinical | Phase 1/2 | Phase 3/3B | Approval & Launch | Phase 4 |

Shape the Product (R&D Leads)

Shape the Market (Marketing Leads)

Commercialization

FIG. 10.15. Best Practice Companies – R&D with S&M

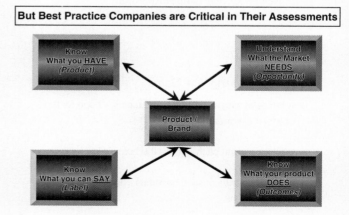

FIG. 10.16. The Brand-4 Parameters

10. Commercial Division

Same/Similar:
- Planning
- Strategy
- Positioning
- Processes
- Marketing
- Execution
- Market research
- Required skills
- Fundamentals
- Discipline

Different:
- Products
- Mode/mechanism of action
- Manufacturing
- Channel
- Reimbursement
- Market data
- Value proposition
- Image
- Promise

FIG. 10.17. Biotechnology... What's Different?

the 10 activities and behaviors listed in Fig. 10.17, but there continues to be some key differences between the two categories. These differences are narrowing as traditional pharma companies venture into biotech and as biotechs get bigger and look more like big pharmas. The titles aren't important, but the differences can be. The products differ in how they work and how they are administered. The mechanism of action can be unique and targeted to more specific sites or diseases. Manufacturing is a substantially more complex and costly challenge and should not be taken for granted; Is there capacity? can you scale up? and is the necessary experience available? Channels may be different if products require refrigeration or any special handling or are used in specialized health care settings. Reimbursement may take on different characteristics and challenges depending on the price point of the product and how and who administers it. There may not be any market data as traditional audits may not provide the data you need for your specific market. The value proposition with patients, providers, and payors is changed for biotech products because of their novelty in mechanisms, properties, and often advance in medicine. The image of the biotech product remains very positive, and the promise that these products will be able to do things not previously possible continues into the 2000s.

What separates the great companies from those who try, but just never seem to make it? Recent articles have identified *benchmarks* that are evident in the companies who are the best at developing medical advances, building businesses, and building brands. Three key operating principles are listed here in Fig. 10.18, dealing with scientific excellence, skillful commercialization, and operational excellence, especially alignment and teamwork. The three principles lead to a visionary type of outcome, changing the game to advance patient care, and establishing market leadership. Of greatest interest is the option that they don't look at the work as it is, but look at it as it could be, and they change the rules to give their products the best chance of success [3].

These same leading *benchmark* companies also do certain things better than other companies or their competitors (Fig. 10.19). There is a sense of urgency, a will to win, they

Three Key Operating Principles (Best Practices):

- Medical and scientific excellence and creativity to establish the rationale for the product
- Highly skilled commercial personnel to prepare market for new brand, accelerate market adoption, & maximize share of voice at launch
- Seamless organizational alignment and cross-functional teamwork to minimize time at launch and maximize medical and commercial impact

Change the rules to give the new brand the best chances for commercial success – control the game and capture the market

FIG. 10.18. Benchmarks for Blockbuster Launches
Source: Pharmaceutical Executive, Blauvelt B, Feb 2003

What the successful do BEST → Shifting the Life Cycle

What the successful do BEST	Shifting the Life Cycle
Sense of urgency	Assemble a dream team
Outstanding PR	Write marketing messages now
First & flexible	Find influence points
R&D productivity	Pre-empt the competition
Branding (products & markets)	Anticipate regulatory risks
Communication & Teamwork	Plan patient strategy
Business intelligence	Develop formulation strategies
Functional expertise	Differentiate portfolio products
Regulatory strategies	Establish critical alliances
PreLaunch investment	Support messaging with data & science
Supply chain management	
Global infrastructure	

FIG. 10.19. Benchmarks for Blockbusters
Source: Blauvelt B., Pharma Exec. Feb 2003; Daly M. Lolassa; Pharm Exec Supp, March 2004

are first, and they are flexible. You can read the remainder of the list of these 22 best practices identified in this publication in *Pharmaceutical Executive*. But these are companies that take time to understand the key elements of what it takes, and they develop the competencies and organizational capabilities to be the best [3, 4]. Another article highlighted these activities (10), as essential to shifting the life cycle of a product so that you can start earlier, sell more, and, more importantly, sell longer. They all point to careful planning, using the best science, understanding the health care marketplace, taking a long-term view, and executing on the fundamentals of what the best practice companies do.

Commercial organizations can make the greatest contribution when they know what they want, what the customer wants, and what the key ingredients for the success of the brand need to be. A core philosophy for the relationship and

Start with a Core Philosophy

- Customers are KEY
- Insights create value
- Positioning is everything
- Develop & implement strong processes, methods, & discipline
- Early and continuous commercial input is key to successful product development
- Remind developers that study designs (phase 2 and 3) create framework and limits of the package insert
- Effective use of information creates competitive advantage
- Think globally

Fig. 10.20. Commercial Input Into Product Development

interactions between the *commercial division and the R&D division* is represented on Fig. 10.20. Communicating these suggested eight principles to their partners in R&D can enhance the success of the overall process at an earlier stage, development of the product, and then later, commercialization. Some principles focus on collaboration or communication, impact planning, engage investigators–providers–payors (the customer), and ultimately what marketing can say or do. A critically important principle for R&D is that the study designs for phases 2 and 3 and their results are the information used by the company and the regulatory authorities in creating the labeling, the package insert, comprise the data for the major publications, and are the limits of any later promotions.

What about the *team*—who should be on it, and who needs to be represented (Fig. 10.21). The answer will vary with the project, but be sure to create a core team that represents the critical functions and stakeholders in the organization. The core team that meets regularly is suggested to include 8–10 groups; research and clinical research, marketing (commercial leads) and market research, medical affairs (phase 4 trials and pharmacoeconomics), regulatory affairs, global representation from the key functional areas (not necessarily one person), and project management to help pull it all together. Additionally, an extended team exists with groups brought in at key time periods to provide information, perform development or launch roles at the best times, and help make some key decisions, such as manufacturing, finance, and legal (patents). Don't forget about your education and advertising agency, because you need to get the creative process started early to create a brand voice, identify it, and give it personality. The communication and education occurs often years before approval of a product, especially for novel products in previously poorly treated diseases wherein the new product is not known and yet will advance medical practice (e.g., anemia in cancer and epoietin alfa, HC Co-A enzyme inhibitors and cholesterol).

Planning is paramount, involving the right key players is absolute, and the best operating principles and structures are critical, but when "the rubber hits the road," the work needs to get done on target, on time, within budget, vis-à-vis, *"flawless execution"* (Fig. 10.22). The people in development and marketing who sit on the teams and their operational partners in their departments have to deliver to achieve any level of success. Flawless execution of the plans and meeting objectives

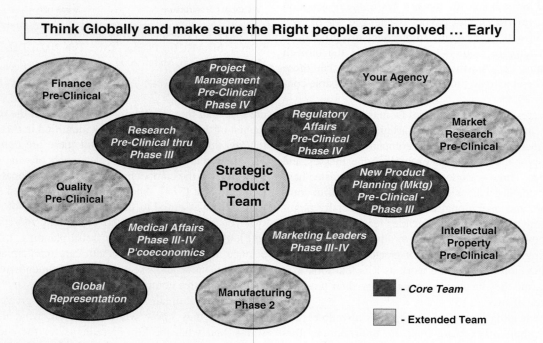

Fig. 10.21. Key Product Team Members

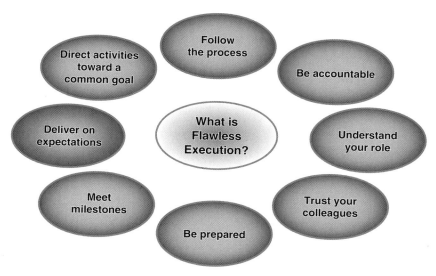

Fig. 10.22. Execute Flawlessly

is represented by eight behaviors on this slide. The hallmarks of flawless execution are preparation, common goals, understanding roles and milestones, trust, following processes delivery, and accountability.

The *label* defines what you have and what you can say in the future (Fig. 10.23). The label is primarily the package insert. The document is based on the science, summarizes the data supporting the product, assists in product usage by providers and patients, provides all necessary precautions and warnings, and is a business document, that is, a marketing tool that directs what can be said and done. In addition, please note that the competition will read your label in great detail and plan their strategies based on the weaknesses in the label. Start with the label you want, and backwards plan to do the work to get there ("backwards planning"). The product team, especially clinical research, medical affairs, and marketing, need to work closely with regulatory to guide them in the negotiations with the regulatory authorities on the content and wording of the label. In hypertension, a study may demonstrate that blood pressure is "lowered significantly" or "substantially" or "greater than the comparator product" or "returns to normal"; all the phrases are related but they carry much different impact. This figure lists ten considerations in designing the label. Besides the features, outcomes, and proof (data) as expected, the label needs to address benefits; differentiation from competitors on clinical and possibly pharmacoeconomic terms. The customer needs, that is, providers, patients, and payors, are considered in labeling phraseology. Of course the studies, their design, and data will dictate how and what can be stated.

Methodical *market research* is an important element of the prelaunch, product development, and postlaunch processes (Fig. 10.24). Through good research, you can gather the insights you need to shape the product and markets and meet the stated or latent unmet medical needs. This chart may help in understanding just some of the questions you will want to answer through market research and advisory boards. Start early and

Start Where You Want to Finish

- Product features
- Product benefits
- Clinical trial design
- Customer/market needs and wants
- Sources of differentiation
- Platform for promotion
- Outcomes
- Economic differentiation
- Sustainable advantage
- Proof

Fig. 10.23. Developing the Ideal label…

Customer Insights Can Create Value

- Identify the "customer (s)"
- Unmet needs – new opportunities
- Understand the buying process
- Recognize barriers to adoption
- Decision processes and influencers
- Know thought leaders (research and practice)
- How to ensure access
- Drivers of value
- Patient needs and behaviors
- Communication and promotional challenges
- Competitive perceptions and positioning

Fig. 10.24. Market Research

Insights Create Value

Phases in the Process:	Analyzing Results:
• Explore products & attitudes	• Perceptual mapping
• Product positioning	• Gap analyses
• Concept development & message testing	• Trade off analyses
• Creative executions	• Market segmentation
• Ad testing	• Market simulation modeling
• Analysis	• Measures and metrics
• Adjustments	

FIG. 10.25. Market Research Plans ...
Source: Adapted from: Fox S. Market Research 101, Mar 2001

invest in market research to understand your customers and markets. The market in health care is a complex environment as you observe in this chart, involving medical needs, access, buying process, the thought leaders, decision process—who makes it and who influences it, barriers to adoption, patient behavior, competition, and what drives value perception. Market research seeks out, analyzes, and reports the information needed to gain these insights.

There are many sources and references to help you develop the right *market research plan* for a product. This chart in Fig. 10.25 draws on a published article to outline some of the traditional phases of the market research process and what you should expect to accomplish at each step along the way (methods of analysis) [5]. We will elaborate on a few of these phases and analyses. Attitudes of the providers, decision makers, influencers, and payors should not be assumed, and market research will provide collective qualitative and some quantitative data that can significantly impact how a product can be studied by the company and then how it can fit into a market. Attitudes can involve, for example, current or future product use, medical needs, product characteristics, competitive products, barriers to care or product use, reimbursement, and value. Product positioning needs to be tested with customers and thought leaders, that is, based on the product's likely characteristics, the current or possible future market, and competition; marketing creates a framework around a product that presents it best case for its use. Message testing involves the marketing team and ad agency creating key phrases or sentences ("messages") about the new product for awareness or prescribing impact that need to be evaluated for their reception and impact. Market segmentation as the name implies is done by market research to understand where and how a product likely will be used; for example, hospitals versus outpatient clinics and offices, or the various classes of products that are used for a disease, or payors—fee-for-service versus managed care or private insurance versus government versus out-of-pocket (patient). Trade-off analysis designates potential benefits of a product, possible formulations, possible cost structures, and examines combinations and how one combination is traded off in the minds of the customer versus another. You may find a primary driver for product choice through this exercise. Market research will use representative groups of customers and thought leaders in focus groups and/or surveys and/or existing databases to help conduct the research.

That same article discussed some of the *tools of the trade* in developing and executing a market research plan (Fig. 10.26) [5]. This is an area where you should seek good advice to make sure the marketing types are asking the right questions, talking to the right audiences, and doing the right type of research for your product or situation. Three sets of goals and tools for market research are suggested as shown in this table for development, prelaunch period, and launch time periods.

Development	Pre-Launch	Launch
• Preliminary product concept testing	• Promotional development	• Attitude, trial and usage tracking
• Forecasting research	• Preliminary message testing	• Formulary adoption
• Competitive landscape	• Consumer reactions	• Program acceptance tracking
• Market needs	• Pricing studies	• Patient acceptance & usage
• Patient needs	• Reimbursement assessments	• Response to promotions
• General exploratory research	• Product concept refinement	• Refine promotional campaign
• Economic considerations	• Attitude and usage baseline studies	• Competitive response
• Dosage/delivery preferences		

FIG. 10.26. Market Research -Tools of the Trade
Source: Adapted from: Fox S. Market Research 101, May 2001

- Forecasting
- Simulations
- Competitive assessments
- "War-Gaming"
- Leverage market research
- Continuous refinement
- Life cycle planning
- Reality

Fig. 10.27. Model the Product and Market

Allow me to point out a couple of key issues and tools at each of these three stages. Product concept testing helps R&D understand what is desired by the medical community, how their product fits, and what needs to be done in research. Dosage and delivery preferences need to be obtained to guide research (pharmaceutics group) in formulation development; obviously, this market research needs to be done very early, even at preclinical stage. In prelaunch, pricing studies are needed for that decision; it involves much product data and market information to reach the best price for the value.

Reimbursement needs to be understood and ideally worked out before product approval. Promotional ideas need broad input from the development team, regulatory and clinical, and testing with customers at least for medical appropriateness and marketing impact. At launch and thereafter, market research especially will track a variety of key parameters, such as attitudes, formulary adoption, program acceptance, provider and patient acceptance, and impact of programs.

With your market research in hand, you can begin to *model* your market and how your product might fit (Fig. 10.27). The goal here is to develop the "best scenarios" based on what you know and to understand the ranges (higher and lower) around what you believe to be the best case. War gaming can be a good way to test your assumptions and to see what the "other guy" can do and might do to slow uptake of your product or to position you in a way that is advantageous to them. A marketing and product team must be sure to inject a real dose of reality at every step and ask the tough questions. Critical self-analysis is indispensable; the other companies certainly will defend their product by finding and emphasizing any weaknesses of the product, the label, the publications, and the promotions. Again, a few of the processes identified on this chart will be elucidated. Forecasting is research into foretelling how much and how a product will be used, which appears to be a crystal ball exercise. However, patterns of existing product use, other product launches, good thought-leader and practitioner input, and marketing experience will collectively create often a pretty reasonable picture of future usage. A major principle in marketing is to use the market research data continuously and refine what is being said (messages) and being done (education and promotion).

Companies, providers, patients, payors, and gatekeepers all talk about *outcomes*, but what outcomes are meaningful to the markets and customers (Fig. 10.28)? The answer may be that it depends. Consider the traditional measures of improved outcomes for a disease or therapeutic category, but also look for areas that can differentiate your product and create a more meaningful outcome than the competition, or existing therapy. The outcome needs ideally to be important, relevant, and differentiating for its acceptance by the medical community and for marketing advantage. Consider also what it will take to prove the outcome you want to claim, which will guide the R&D organization in the research plans: efficacy (expand beyond type and extent to disease progression, length of stay, functionality, or quality of life), safety (avoiding or reducing a side effect), convenience and compliance (dosing frequency, formulation, packaging like needle-less syringes), avoidance of other treatments (either drugs, tests, or surgeries), and economic (competitive price for value, overall cost of care).

When everything else is done, you now need to set a *price* for your product (Fig. 10.29). It's an important decision, with many things to consider. The customer must be understood in the purchase decision, and each one has a different set of criteria, patient versus provider versus gatekeeper versus payor. If there is competition in your segment, will you be the same

Important, Relevant, and Differentiating

- Clinical
- Safety
- Economic
- Length of stay
- Compliance
- Quality of life
- Disease progression
- Avoid or eliminate other products/procedures
- Functionality
- Convenience for provider or patient

Fig. 10.28. Outcomes to be Sought

- One of most important decisions you will make
- Start your research early
- Are you setting a price or establishing value
- Thorough competitive analysis
- Understand all aspects of reimbursement
- Can you create value through:
  - Better outcomes and / or
  - Better formulations
- What data do you have?, and what will you need?
- Consider international component
- Contracting and distribution policies
- Think both long and short term
- Understand impact on your customers

Fig. 10.29. The Pricing Decision is Critical

price, higher, or lower? Think both short-term and long-term at the beginning, because choices early will obviate some choices later. What about discounts? Discounts are part of all product relationships with customers, such as patient coupons, wholesaler stocking discount, and GPO or MCO discounts for group purchases. If your product is first in class, how do you establish a value for what you are offering? What will payors think, and which payors? How does the reimbursement work for care of and drugs for the target disease? International pricing also must be done for the multinational companies, which is the norm now. Even though the health systems vary across the world, we are a world without borders in health care and the knowledge of providers, payors, and the public. A company must do its homework, consider the options, do the models, and make the decisions. As stated earlier, market research plays an important role in the modeling, assessment of competition, and market (customer and health system) tolerance. They will identify what we know and what data we need, and then go get it [6].

What's in a name? Everything! Names carry power beyond the words for all types of products and businesses; think what Maytag™ or Tylenol® or Nike™ or Herceptin® means to you about the product or company beyond the base use and benefit of the product. Use a scientific process in *name selection*; again, market research is important (Fig. 10.30). Do your homework, do it early, and select a name that will mean the most to your target audiences and will form the basis of creating a distinct identity for the brand. This chart offers several important considerations in name selection, such as avoiding confusion, considering international meanings, performing a trademarks search for conflicting names, and use professionals who are experts with this work (research, name impact), all of which comprise best practices in choosing a product name.

Developing and commercializing a product is a journey that begins very early in discovery. What processes do you

- Apply the "science" of naming your product
- Leverage good research and let the "professional" naming and branding companies help you
- Apply best practices
- Be certain to check for conflicting trademarks, look-a-likes, sound-a-likes, and similar names
- Avoid names that could be confusing, or difficult to say and write
- Consider international impact
- Have a back-up as your 1st choice might not be accepted
- Understand the naming and name approval process
- Consider establishing your name very early

FIG. 10.30. Selecting a Name … Creating a Brand

have in place to ensure a collaborative effort that leverages the knowledge and experience of the key stakeholders? What we have learned, however, is that even in the very early stages, thoughtful *commercial input* can play an important role in helping organizations make the right decision about which compounds to advance, what formulations to develop, and where to focus development efforts. This diagram in Fig. 10.31 suggests some of the opportunities (six examples) when and for what the commercial team can assist R&D in developing the best possible molecules in the best way for an optimal product outcome and approval.

Best-practice companies have a clear process, and the commercial team submits a formal assessment and point of view at each phase of the decision-making process. Five such *decisions* in product development are presented in this simplified chart in Fig. 10.32, including the time frame (when), the decision to be made (what), and the deliverable (marketing's specific inputs). The commercial team–development team relationship is an iterative process requiring frequent periodic communications

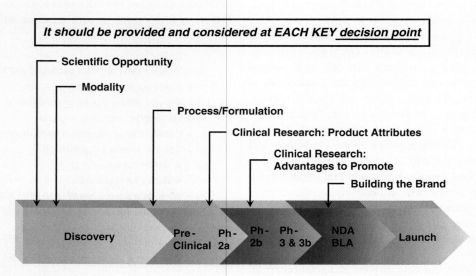

FIG. 10.31. Commercial Input in Development

WHAT →	WHEN →	DELIVERABLE
Initial Opportunity Assessment	Target Identification	Preliminary Statement of Commercial Interest
Initial Promise Assessment	Pre-clinical Evaluation	Statement of Commercial Interest
Secondary Promise Assessment	Phase 2	Formal Marketing Needs Report
Proof Assessment	Early Phase 3	Updated Needs Report
Conviction Assessment	End of Phase 3	Preliminary Commercial Plans

FIG. 10.32. Commercial Input to Product Development Decisions

and proactive collaboration. As products move along in the process, more data is available, and more is known about both the product (study results) and the market. What were estimates early in the process become commitments later in the process. Inputs are transparent and open for debate and discussion within the development/commercialization team.

Life cycle management (LCM) starts with the discovery of a new molecule in research and continues until the product is no longer used or sold (Fig. 10.33). The core idea of LCM is to maximize the value of the product to the medical community and the company over its full life span through proactive cross-functional planning and execution at each stage of evolution. Hence, the product team that led to launch functions onward for years to generate all the follow-on new developments for the product. This chart reviews seven operating principles for LCM for our consideration. LCM is one best practice of successful companies and engages the whole company as you can observe in the chart (e.g., law department for patents, pharmaceutics for new formulations, clinical research for the new indications and publications, manufacturing and process engineering for improved and less costly production, medical affairs for phase 4 research and a continuing stream of publications).

Think Globally & Take a Long Term View

- Aggressive development of new claims structure, indications, data / publications, formulations, delivery
- Look for fast follow-on opportunities
- Plan for providing new news to the field regularly
- Protect intellectual property aggressively and over time
- Plan out two-four years to anticipate and proactively address marketplace dynamics
- World class phase 4 execution capability
- Look for second generation molecules early with new features

FIG. 10.33. Life-Cycle Management . . .

Building Blocks in Commercialization Process

Structure (Global Marketing)	◆ Efficient and consistent commercial global input ◆ Alignment of R&D with S&M ◆ Clear roles & responsibilities, yet aligned ◆ Partnership of R&D & S&M throughout LCM

Governance (Review Boards)	◆ Clear requirements for decision making
	  • Durable, transparent, & timely decisions
	  • Who, What, How, and When defined
	◆ Better and faster cross-market and cross-function decision making

Processes (Development & Commercialization Process)	◆ Deliverables and accountabilities defined ◆ What is required (timely and complete data) ◆ Documented, Planning, and Communication ◆ Resource allocation follows decisions

FIG. 10.34. Leverage What Has Been Learned

Practical Approaches

If you consider the literature and experiences of many in what makes the difference in effective processes, here are some *lessons learned* (Fig. 10.34). The themes are fairly clear; clear roles, defined responsibilities and accountabilities, partnerships across the organization, communication, global views, defined decision process, durable decisions, and resources to support he decisions.

Some companies have also developed structures and processes to ensure commercial readiness and success. Groups like global marketing can become a focal point for cross-market planning and representing the commercial voice. Review boards can become focal points for decision making and resource allocation, and clearly defined processes can help ensure that each function knows its role and what the organization expects.

Below, seven practices are spelled out, as well as summarized in this chart, for lessons learned that are building blocks for successful development and commercialization.

1. A consistent commercial and scientific partnership must exist throughout the product life cycle.
2. The commercial organization must communicate product and market needs with a consolidated global view.
3. Clear roles and accountabilities must be established with well-defined deliverables for seamless workflows and handoffs.
4. Decisions must be durable, transparent, and timely.
5. Decision makers and decision points must be clearly defined and adhered to.
6. Resources must follow decisions with trade-offs made across the organization.
7. R&D, clinical development, and commercial must be fully aligned at key decision points.

There are many ways to assess what it will take to assess an opportunity and ensure that everything possible is done to ensure success. *OPPC* is one such approach (opportunity,

Opportunity	**Recognize market opportunity:** Competitive weakness or unmet medical need (stated or latent)
Promise	**Define the product promise:** Unique selling proposition, benefit, or improvement
Proof	**Provide evidence:** Comparative data, thought leader support, or indications
Conviction	**Commit to success:** Investment, urgency, expertise, or follow through pre and post launch

FIG. 10.35. OPPC – Use a common language

promise, proof, and conviction), displayed in Fig. 10.35. The goal here is to ensure that the organization has a way to communicate with a commonly understood language and to discuss their opportunities. The definitions presented and questions asked in this process are the basics in the process, as shown in this chart. These four questions will be elaborated further below.

Assessing the *opportunity* helps us understand where the need may be in patients and the health care system, the competitive situation, and what the size of the market could be. Consider all 10 factors regarding unmet medical need and unsatisfied market, outlined in this chart in Fig. 10.36.

The *promise* addresses the role the product will play in the market, the patient benefits, and answers the question of why this product versus something else (Fig. 10.37). What is really new and significant? A compelling rationale that can be communicated clearly and succinctly goes a long way toward positioning the product and laying the promotional foundation. Market insights help examine the possible fit of the product for the target disease ("the opportunity space"). The type and extent of the promise also can help decisions to take the time to create an optimal or ideal profile (e.g., with two,

---

How attractive is disease and/or marketplace for a new product?

- Unmet medical need:
  - Lack of effective treatments
  - Current therapies woefully inadequate
  - Large undiagnosed or under-treated population
- Unsatisfied market:
  - Competitive weakness:
    - Efficacy & Safety
    - Tolerability
    - Convenience
    - Value
  - Impact on outcomes unclear

FIG. 10.36. The Opportunity

Why Should I Use This Product vs. What I Use Today? What's New?

- Succinct statement of compelling reason to use product
- Important patient benefit
- Leverages key market insights:
  - Creatively examine possibilities of product and assess new opportunity space for product
- Consider need for optimal profile versus speed

FIG. 10.37. The Promise

three, or four indications), or accelerate the speed to market with a safe and effective, but more focused profile.

Today more than ever, *proof* is required to support the claims you make around the product (Fig. 10.38). The proof of course is the studies and the data from them. Get as much as you can into the label from the research. Look for areas of differentiation from existing products and treatments. The stronger the sources and the data, the better the situation is. Strength of the data and sources includes the best study designs for the disease and product, the optimal number and size of studies, the best investigators and influential institutions, and solid publications in the best journals. Engage your opinion leaders in this process early; their belief in the uniqueness and value of the product will influence usage and uptake. They are sought after for their therapeutic opinions by practitioners and other university experts.

Does the organization have the will to win? If everything has been done, there is an opportunity, you have a compelling promise, and the proof is there, it then takes *conviction* to make it happen, described as follows (Fig. 10.39). Be prepared to spend early (12–24 months prior to launch) and continuously throughout the commercialization process. The investment is not only in the marketing (education and promotion around the product) but also the clinical research for new indications, expanded research opportunities with the indication (phase IV), and pharmacoeconomics (outcomes). Ensure a sense of urgency for this new product that will increase motivation, engagement, and follow-through. Look for and create alignment across the organization for effective

---

Is the Promise Supported by Data?

- Package insert claims
- Comparative studies proving differentiation on efficacy, and/or safety
- Publications in referred journals
- Outcomes studies
- Credible value proposition
- Key opinion leaders' belief in uniqueness
- Weigh need for strong proof versus speed

FIG. 10.38. The Proof

10. Commercial Division

How Competitive & Aligned is Investment in Product? Is There a Proper Sense of Urgency?

- Focus on key drivers of business
- Serious investment 12-24 months prior to launch & long-term
- Dedicated public relations support
- Dominant or at least competitive share of voice:
  - Launch incentives
  - 200 day field sales execution plan
  - Heavily target early for rapid adoption
  - Expand treatment to broader audience of physicians / patients
  - Target managed care for rapid adoption
  - Key opinion leader driven adoption process
- Consider the role of the patient
- Ongoing investment in clinical research & added indications / outcomes

FIG. 10.39. The Conviction

operations and launch. Focus on the key business drivers for this disease and market. This chart lists six specific marketing activities and two others at launch that comprise the aggressive plans and action items associated with successful launches.

Commercial Responsibilities

Launches in the past were characterized by heavy upfront investment and as rapid uptake as possible (Fig. 10.40). Historically, how a product performed in the first few months postlaunch was a predictor of how that product would perform over its life. The diagram in Fig. 10.40 presents sales data in market share or sales dollars over time for a variety of products [7]. This diagram (left frame) documents and summarizes 70 actual product launches in the 1980s and 1909s (with six representative ones displayed) and supports this historical launch theory. The trajectory of sales at launch follows the four launch parameters presented earlier; opportunity, promise, proof, and conviction (spend). That may still be true in some categories, but the new reimbursement world and other environmental factors may radically change the product adoption process. Forecasts should reflect how a product will be adopted and should reflect how barriers to adoption will be addressed and when. Newer product launch models need to be considered, and the diagram (right frame) demonstrates the results of three very different product launches for Lipitor®, Viagra®, and Celebrex®, which were considered major health care innovations and represent more recent huge blockbuster products.

The times are changing, so a company must be sure to understand how quickly the market will and can adopt your product

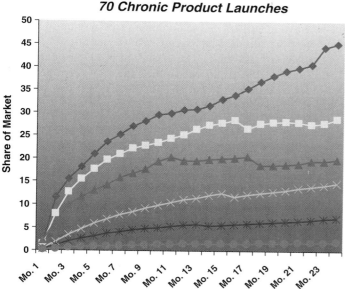
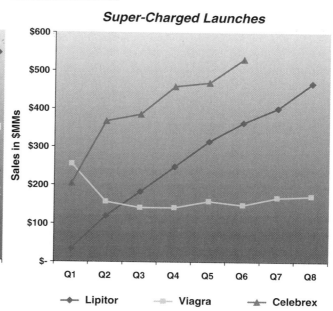

- Launch curves never cross
- Fate of launch determined in first few months
- Trajectory at launch comes from opportunity, promise, proof, & pre-launch / launch spend

- Aggressive pre-launch activities created pent-up demand

FIG. 10.40. Old Paradigm

Uptake may be slower than you think
- Reimbursement requirements
- Slower adoption
- New competition
- Impact of guidelines and formularies
- New "hurdles"
- Manufacturing and ability to supply
- Thought leader support
- CME requirements

Key Point: Understand what must happen to ensure adoption & how quickly it might occur…reflect actual uptake in your launch forecast

FIG. 10.41. New Paradigm

Applying the Traditional 4 Ps in Launch

Product:		Place:	
Label	Positioning	Channel	Distribution
Packaging	Advantages	Patient Flow	Office model
Promise	Unmet need	Formularies	Support
Outcomes	Manufacturing	Access	Market research

Price:		Promotion:	
Value proposition		Pre-launch	Publications
Reimbursement		PR	DTP/DTC
Discounts	Competition	Field force	Medical Affairs
Rebates	Economics	Proof	Medical Education
		Training	Targeting

FIG. 10.42. Launch Products & Build Brand Success

(a new paradigm) (Fig. 10.41). With the reimbursement environment, it may take longer than you think. Know what those steps are, what the barriers to adoption are, what you can do prelaunch, and what can only be done postapproval. Do the forecasts and set expectations accordingly. Educational practices are changing in what can or cannot be done with whom, how, and when, which alters thought-leader engagement and continuing medical education (CME), and possibly again altering product uptake. Formularies and guidelines have more impact than ever before, and they could slow product usage until these gatekeepers and thought leaders change them to incorporate the new product; a potentially time-consuming step requiring new approaches. Make sure manufacturing is geared up to meet the demands at the beginning and how they might change. Marketing research and forecasting must work closely with manufacturing. Manufacturing may need to start gearing up 5 years ahead of a launch or change because of infrastructure and cost requirements, staffing, and scale-up technical issues. Enbrel® from Immunex could not meet the medical need and product demand for a couple of years because of manufacturing scale-up issues. Lunesta® was not available at all for months after its regulatory approval. *Key Point:* Understand what must happen to ensure adoption and how quickly it might occur, and reflect actual uptake in your launch forecast.

A company should not be bashful about using checklists. The process is complex, and the company should not only worry about what they have done, but also what they may have overlooked or forgotten to do. Develop checklists for the specific product and its situation; then use them. This approach may be helpful, and it is centered around the 4 "Ps" of marketing, product, place, price, and promotion, as listed in this chart in Fig. 10.42 along with more than 30 issues and action items. The whole company focus is on *launching new products and successful brands*, and it engages R&D, L&R, and F&M with S&M; that is, research and development, legal and regulatory, finance and manufacturing, sales and marketing, respectively.

This list in the chart in Fig. 10.43 is a good bit more exhaustive (about 40 activities and action items), and buckets many of the same items around the three phases of the launch process. A comprehensive list for each product for launch should be done as represented here that works for the disease, product, and company. Regarding the time period for all these activities, the launch of a product is from 1 to 2 years before approval to at least 1 year postapproval.

Some summary thoughts are offered for the road to commercialization and the development process particularly for the *role of the commercial team* at a company (Fig. 10.44). There have been many lessons learned about what it takes to create blockbusters, or to maximize the value of a molecule in the marketplace. Some common principles apply. It's the role of the commercial team to take the lead, build a collaborative environment and process, and to be accountable for the results. The commercialization team, as a reminder, truly needs to consist of not just marketing but also research, development, manufacturing, medical affairs, regulatory, and market research as the core with many others contributing substantially at different times. Eight best practices are presented here in this chart for this team. The practices should (will) lead to products for unmet patient needs, of value to the health care system, that are paid for, and help companies be successful (yes for profit) business enterprises in the business of advancement of health and well-being of the public.

In summary, *start where you want to finish*, and then plan backwards to make it happen. Understand what "finished" will look like, and know what has to be done to get there. This chart in Fig. 10.45 summarizes 11 keys to success in product development and commercialization; key words include team, listen, strive, label, proof, value, commercial input, creative, and brutal honesty. The destination is well worth the journey.

Pre-Launch	Launch	Post-Launch
• Shape the product	• Launch "Event"	• Customers
• Shape the market	• Training	• Tracking
• Reimbursement	• Field readiness	• Feedback
• Value proposition	• Introductory ads	• Adjustment
• Publications	• Plan roll-out	• Competitive response
• KOLs	• Key customers	• Medical meetings
• Field plans	• Speakers	• Phase IV trials
• Label	• Distribution	• KOLs
• Phase IIIb & IV trials	• Contracts	• Marketing programs
• Positioning	• Phase IV trials	• Implementation
• Global coordination	• KOLs	• Adjustments
• Launch planning	• DTC/DTP	• Metrics
• Pricing		
• Coming soon		
• Sales training		

FIG. 10.43. The 3 Phases of Launch – Best Practices

Both the New Product Development and Commercialization Process

- Leadership
- Accountability
- Represent the voice of the customer and market
- Objective forecasting
- Winning product profile
- Help set expectations
- Support product development team
- Be accountable

FIG. 10.44. The Role of the Commercial Team

Backwards Planning - Some Keys to Success

- The commercial team plays a key leadership role
- Start early
- Work as a team – collaborate
- Listen to the market and customer
- Strive to create multiple advantages
- The label is key
- Develop strong proof with publications
- Credible value proposition
- Continuous commercial input into decision making
- Early creative input
- Brutal honesty in assessments and decision making

FIG. 10.45. Summary: Start Where you Want to Finish

References

1. PhRMA Member Companies' R&D Expenditures for 2001 is estimated
2. U.S. Food and Drug Administration 2002. IDEC's Zeralin (Feb.), Gilead's Hepsera (Sep.), Seron's Rebif (Mar.), Amgen's Neulasta™ (Jan.), Pfizer's Vfend (May), Novartis' Zelnorm (July), Astrazerer's Faslodex (April) plus 4 others
3. Blauvelt B. Benchmarks for blockbuster launches. Pharmaceutical Executive, Vol. 23, pg. 53, Feb 2003
4. Daly, M. Lolassa; Pharmaceutical Executive Supplement, Vol. 24 March 2004
5. Fox, S. Market research 101. Pharmaceutical Executive. Vol. 21, May 2001
6. Gregson N, Sparrowhawk K, Muaskopf J, Paul J. Pricing medicines: Theory and practice, challenges and opportunities. Nature Drug Discovery 2005;4(2):121-130
7. Anonymous. Product Launches Report. Personal communication. 2001

11
Medical Affairs and Professional Services

Ronald P. Evens

Introduction 240
Departmental Issues 241
Medical Information Services 245
Medical Science Liaisons 252
Medical Communications 256
Outcome and Pharmacoeconomics Research 265
Late-Phase Clinical Trials 267
References 272

Introduction

Novel drug, device, and biologic pharmaceutical products are being developed across a host of human diseases to extend current medical treatment and improve patient care. The pipeline of drug development requires years of tireless scientific effort, health care provider relationships and support, patient dedication, public awareness, and business acumen to move a molecular entity from initial discovery, through the layers of clinical trials, and if successful, through negotiated regulatory approval by the U.S. Food and Drug Administration (FDA).

The medical affairs unit, also named professional services at some companies, is a substantial effort by brand-name companies to provide services to and support for the providers, patients, and institutions that use their products, helping to ensure appropriate and safe use. Novel products also have substantial regulatory requirements to be fulfilled, as well as an ethical need for health care support of providers and health care systems for the safe and appropriate use of the new product. The medical affairs unit is a major component of companies to meet these customer and regulatory demands.

About 3.5 billion prescriptions were written in 2004. By the year 2010, increasing "baby boomer" demand and rising drug development costs are expected to increase drug costs from 10% to greater than 14% of total national health care expenditures following hospitalization and physician services in costs [1]. In 2002, pharmaceutical companies spent greater than $30 billion and shouldered the majority proportion of drug development costs in the United States [2, 3]. In medical affairs area, the phase IV spend was $1.6 billion in 2002 and has been growing substantially (37% from 2000 to 2002).

Under federal law and Food and Drug Administration guidance, pharmaceutical companies shepherd promising therapeutic candidates through preapproval hurdles and vigorously develop postapproval life cycles to maximize marketability before loss of patent protection. A tremendous amount of research is done and published over 10 years for various indications, safety, efficacy, formulations, pharmacokinetics, manufacturing, and more. Generic intrusion into the market at the time of patent expiration can significantly depreciate the profitability of a brand manufactured drug product. Generic drug industries typically do not perform all the research and reporting requirements for an IND/CTA or NDA/BLA/CTD and do not possess novel drug development technologies, personnel infrastructure, or extensive educational costs, as compared with fully integrated pharmaceutical companies. As an example, the first bottle of generic Captopril® (from Bristol-Myers Squibb) retailed at approximately $800 USD/bottle. Six months later with several generic brands advancing into the marketplace, the price of the generic fell to less than $13.80 USD/bottle, and consequently brand drug purchasing was reduced [4]. Within narrowing profit margins, generic pharmaceutical industries tend to minimize other health care value added services and do not invest budget into pharmaceuticals of tomorrow [5–7]. Medical affairs groups usually are skeleton operations at these generic companies in contrast with the medical affairs units at major pharmaceutical houses, which is the focus for this chapter.

Chapter 11 offers an overview of a medical and professional affairs department in supporting pharmaceutical product development. As with any large corporate entity, there are numerable variations of organizational structure and specific job responsibilities, so this chapter will focus on the basic requirements found most commonly in the industry. Medical and professional affairs (medical affairs) is responsible for accurate and appropriate communication of drug- and disease-specific information to internal and external customers, conduct of late-phase, postmarketing clinical trials and pharmacoeconomic trials, collection and reporting of product adverse events, implementation of educational programs, support of product launch, scientific congress support and peer-reviewed publications. The staff is composed of health care professionals, practitioners, and researchers, predominantly in the medicine and pharmacy fields with varied advanced educational degrees (Pharm.D., M.D., Ph.D., and M.S.N.) and support staff.

Medical information is the role of providing information to mostly health professionals and patients about the company's products upon request. Medical field professionals (liaisons) are health care professionals (Pharm.D., M.D., Ph.D.) employed by a company and geographically based across the country focusing on medical education and clinical research. Marketing communications is a group that supports marketing groups with education program development and key opinion leader development. Outcome and pharmacoeconomics research involves practice-based and cost-based research that explores the full value of a product's impact on the full health care system beyond safety and efficacy of the product. Late-phase clinical trials are a major research activity in the life cycle planning for a product, continuing the clinical research into the periapproval time period and thereafter, addressing many product usage questions with an ever expanding number of investigators.

Departmental Issues

Many specialized departments and procedures are involved in the drug development process throughout a product's life cycle; the medical affairs department focuses on late-stage issues working across the organization to fulfill its mission, including collaboration with research & development, legal and regulatory affairs, manufacturing, project management, sales and marketing, finance, and human resources, *a FIFCO, fully integrated pharmaceutical company* (Fig. 11.1). Research & development (R&D) covers from discovery to market place; research and development of traditional small molecules and some biologicals is a process that can average approximately 10 to 15 years. Research and development is responsible for interfacing with preclinical research and performing first-in-man studies to elaborate phase III human clinical trials of safety and efficacy. Medical affairs (MA) is

FIG. 11.1. FIPCO & Medical Affairs Interactions

provided data and studies by R&D that are used in patient, provider, and payor communications. MA uncovers new key usage questions that may require follow-up research by R&D involving dosing, stability, or formulation issues. Pre- and post-product approval laws and pharmaceutical guidances are monitored by the legal and regulatory (L&R) department. This department is the main interface with FDA regulatory agencies and helps set regulatory strategies, submissions, negotiates product insert labeling, postapproval maintenance including additional clinical trials and safety surveillance, reviews postmarket advertising and promotions, and oversees general manufacturing procedures. Medical affairs works closely with L&R to ensure compliance of educational programs and services to providers and that health care entities follow ethical and the various legal guidelines, as well as fulfilling some postmarketing requirements for clinical trials adverse reaction monitoring.

The global organization (EU, USA, and Asia) involves health care approval agencies, patient populations, pharmaceutical marketplaces, and manufacturing technology that can vary significantly across the globe. International presence of a pharmaceutical company often requires personnel familiar with local registration and requirements of pre- and post-drug approval. Often, multicenter clinical trial designs require large patient pools and must reach internationally or to special populations (e.g., HIV, genetics) for statistically significant end points. Medical affairs' roles and services need coordination across the globe for consistency with providers and regulatory agencies, plus avoiding duplication of effort (e.g., letter responses to common medical questions). *FDA good manufacturing practices* ground pharmaceutical development and commercial product manufacturing [8]. Consistency in batch manufacturing, packaging, labeling and storage, working out product complaints, and availability of clinical trial supply are all important contributions of the manufacturing department in their interface with medical affairs. The department of sales and marketing supports the drug development process through analysis of the potential patient marketplace, setting strategies for drug launch and life cycle management, and working closely to maximize ethical use of pharmaceutical products. Such roles also support medical affairs in their work with customers using the new products. Finance group supports medical affairs similar to the rest of the organization with budgeting and expense analysis and reporting. The human resources group supports medical affairs in recruiting, hiring, and training of staff.

The department of medical and professional affairs is a service-oriented cog in the rotating wheels of drug development. Medical professionals are typically viewed as "medical or product content experts" and are involved across the organization from medical support in early drug discovery and licensing acquisition to postapproval surveillance.

Across the industry, whether small molecule, biologic, or device manufacturer, the medical and professional affairs department's responsibilities start before product approval in preparation of the product launch regarding phase 4 trials, working with clinical research and other research groups on monographs and publication needs, creating medical letters and inquiries, collecting adverse events, and developing educational programming, all of which continues throughout the years of marketing the product (Fig. 11.2).

The medical affairs department is the primary clinical interface of the company with the external world and provides most of its work and outcomes to customers of the company's products, that is, providers, patients, payors, and institutions, as well as collaborating with and supporting the marketing and sales division. These sets of dual roles also are a dilemma from a legal and regulatory perspective and all the associated operating guidelines for the industry, regarding the need to be independent, fair balanced, medically focused, unbiased, and separate from the commercial interests. Furthermore, upon request by providers, the company through medical affairs can provide both "on-label" and "off-label" information.

Besides all the L&R issues, credibility of the clinical staff and their services (e.g., medical letters and educational programs) with the health care community necessitates the creation of *firewalls* between medical affairs and the commercial operations to ensure the separation and independence (Fig. 11.3). Staffing, hiring, and training of the medical and pharmaceutical personnel is done separately with human resources from the commercial division. Although the whole company is a for-profit commercial entity, the performance goals and related compensation should not be tied to product sales, using instead for example numbers of trials initiated, number of inquiries, satisfaction of customers with the services, and benefit of educational programs on learning and appropriate product use. Physical location in a distinct and separate area is a useful goal for independent operations. Organizational structure is important with reporting lines separate from sales or marketing, reporting into R&D, or L&R, or directly to the company president. Operating guidelines should be spelled out and demonstrate the independence, fair balance, and lack of bias. For example, for letter writing, no marketing review is entertained, and positive and negative data are included. Clinical trials should have specific and distinct review and approval processes, along with a goal to publish the work, pro or con. Educational programs have a host of industry, regulatory, and professional guidelines to follow, which will be discussed later and should be spelled out in operating guidelines. A compliance officer for educational

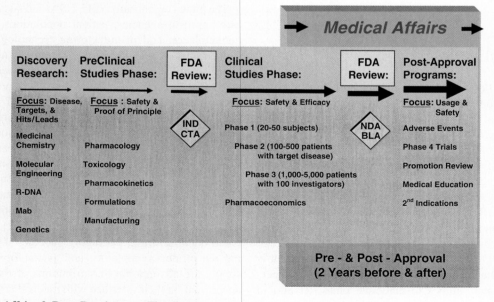

FIG. 11.2. Medical Affairs & Drug Development Timeline

- Staffing, hiring, and training
- Goals & Compensation
- Location
- Organizational structure
- Operating guidelines & SOPs:
  - Letters
  - Educational programs
  - Clinical trials
- Budget
- Support of legal & regulatory

Focus: Separation, independence, & patient care yet integration, within a medical business operation

FIG. 11.3. Medical Affairs: Organization & Firewalls

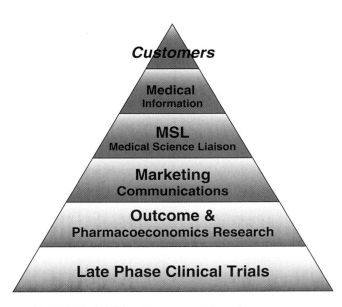

FIG. 11.4. Medical Affairs: Departmental Functions

and customer support programs should exist. The budgeting process always will depend on the profitability of the company regarding the availability of funds, but budgets for medical affairs must be separate from marketing and sales. Finally, the legal and regulatory groups need to work collaboratively with medical affairs in the execution of the MA programs, as well as providing oversight. Medical affairs and the company will be perceived, and be in reality, a medical consultant to providers regarding the company's products and have the credibility and capability to perform this role for the medical community with such firewalls [5].

The department of medical and professional affairs is a *service-oriented function* supplying product information and diseases state literature to support a wide variety of internal and external customer needs in a pharmaceutical company, represented in Fig. 11.4 by the five groups who all serve the various customers of the company. Medical professionals typically staff the department and are viewed as "medical or product content experts" and are involved across the organization. This section will overview the customer, who the customers are, how MA relates to product launch and thereafter, and the interface of MA with external customers and internal customers, especially both R&D and S&M. "Customer" is a business term that is appropriate even for this health care context, as the products of a company are used by individuals, and as such they are customers of a company. They will need assistance in the proper use of the products by, and the best information from, the health professionals who are the content experts with the product.

Medical and professional affairs is responsible for interacting with a variety of *external customers* regarding product and disease state information (Fig. 11.5). Drug support is typically based on information contained within the FDA-approved product package insert (PI), the voluminous data held by the company related to an NDA/BLA/CTD and later studies, and the published scientific literature. Communication between the pharmaceutical industry and external customers is essential and may include an information exchange with purchasers (patients, providers, managed care organizations, pharmacy benefit managers, CMS); patient care guideline and policy developers (P&T hospital formulary members, compendia audiences, medical and professional organizations); public relations (licensing and acquisitions, public investors) and drug regulatory bodies (FDA, EMEA, Health Canada). The following is a list of customers and examples of information that might be requested from a medical affairs professional.

- Patients and family members—Disease state information and patient package insert regarding product(s); provide information on samples and indigent program and special product access;
- Providers inclusive of health care professionals (physicians, pharmacists, nurses, dieticians, and social workers)—"On-" and "Off-label" use of product; and triage access to clinical trials;

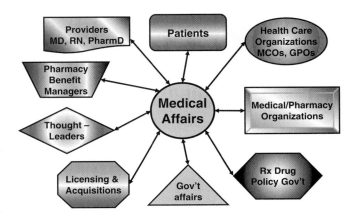

FIG. 11.5. Customers (External)

- Major medical associations—National, regional, and local medical societies may request drug information for basic provider education, background understanding for authorship of disease state treatment guidelines, and educational programming;
- Health care organizations—Safety and efficacy product information; outcome and pharmacoeconomic data; access to indigent care programs; research funding;
- Pharmacy benefit managers—Safety and efficacy product information; outcome and pharmacoeconomic data; packing and product storage information; access to indigent care programs;
- Regulatory authorities (FDA)—Postmarketing safety and surveillance reports; audits done by these agencies for these spontaneous reports; medical education and medical letters roles; product information for agencies related to policy or reimbursement;

External communication is typically delivered by two types of pharmaceutical professionals that have similar yet distinct roles: internal medical affairs professionals (headquarter-based or contract service organizations affiliated with the company) and field-based medical liaison professionals [9, 10].

Medical and professional affairs is responsible for interacting with a variety of *internal customers* regarding disease state and product information (Fig. 11.6). Drug support can be based on product-specific information, disease states, therapeutic categories, health care systems (organization and operations), phase 4 trials, competitor publication activity, or literature reviews to assist evaluations of future therapeutic interests. Communication within a pharmaceutical company is essential and may include an information exchange with field personnel (sales representatives, national account managers, and medical field professionals), product support departments (safety and quality assurance departments), and public liaising departments (investor and public relations, government affairs, legal and regulatory, licensing and acquisition). The following is a list of customers and examples of information that might be requested from a medical affairs professional [10, 11]:

- Sales representatives and national account managers—Disease state and "on-label" product information;
- Regional medical liaisons—Disease state and "on-" and off-label" product information; product data and studies held by the company, for example, stability, early phase 2 data on new indications, drug interactions, pharmacogenomics; competitor literature; access to phase 4 trials;
- Government affairs—Drug information for disease state and patient care guideline development;
- Development managers and project planning—Customer inquiry reports; customer contacts for research ideas and availability, phase 4 study needs;
- Quality assurance—Product or packaging inconsistency reports (complaints); stability issues;
- Legal and regulatory—Consolidation of scientific literature to comment or defend guidance to industry proposals;
- Investor and public relations—Disease state or product labeling information; competitor marketplaces;
- Safety department—Scientific literature reporting of adverse reactions; triage of patients and providers with adverse events;
- Licensing and acquisition—Disease state and product competitors on- and off-label use; competitor marketplaces.

The *launch* of a new product by a company is a seminal event in its history, "*the sine qua non*" of success in the industry, especially for advancing patient care with a novel product and for a product with blockbuster potential. The medical affairs department plays key roles in the product launch (Fig. 11.7). For true innovations, a pent-up demand exists from the medical community and public of course for the product but also significantly for information about the product and its impact on the target disease, as well as its safety. The sales and marketing division leads the effort at a company but the whole

Fig. 11.6. Customers (Internal)

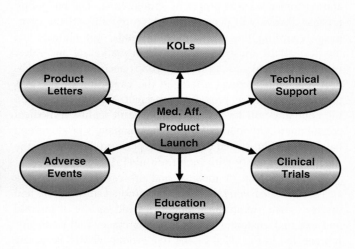

Fig. 11.7. Product Launch: R&D ◇ MA ◇ S&M Interfaces

organization is engaged. Besides the publications and package insert, which would be widely and publicly available, R&D holds the huge volume of clinical data filed with the FDA and often prior to publication, especially for safety, efficacy, dosing, stability, and formulation, including subset analyses. Medical affairs is the company interface to the medical community and others, especially patients, providers, and payors for this information, and must prepare well in advance of a launch and anticipate customer needs for medical and product questions with answers ready to go in letters. Adverse event monitoring and reporting will be needed. Key opinion leaders (KOLs) will need to be reached out to with educational materials; please be reminded that an NDA may engage 20–100 medical centers, but many experts in the field of study for the new product will not have had the opportunity to research the product, yet their patients will expect them to have knowledge about it. In the life cycle of a product, continued clinical research within the current indication is a key way to reach out to KOLs and needs to be ready prior to launch. Marketing will need technical support with the company publication and medical literature in crafting both the educational and promotional materials. Basically, medical affairs usually will need about two years before a product is approved to prepare their staff and all the services and materials for launch.

Medical Information Services

Medical professionals typically staff the department and are viewed as "medical or product content experts" and are involved across the organization. These individuals receive regular scientific and literature training around peer-reviewed literature, journal publications, therapeutic patient care guidelines, abstract and poster publications from major scientific congresses, customer interaction techniques, and FDA requirements for postmarketing safety and surveillance. Customer interactions are governed by two predominant regulatory policies that are strictly enforced: (1) Food and Drug Administration Modernization Act (FDAMA "Safe Harbor") for responding to "off-label" questions [12]; and (2) Health Insurance Portability and Accountability Act (HIPPA) to protect patient confidentiality [13, 14]. Medical and professional affairs through their medical information group serves a critical role as an acceptable group at a company to provide disease state and product information, whether "on-label" or "off-label," however only upon an unsolicited customer request. The scope of services from a medical information unit includes medical and product inquiries, pharmacy and therapeutics committee support, adverse event reporting, support of field sales staff, educational seminars, and medication use evaluations. Pharmaceutical medical professionals typically staff a "customer contact center" for which the telephone and e-mail product queries can be answered [11].

The pharmaceutical industry has evolved into one of the most highly regulated industries by government and health care and professional entities in the world. The U.S. Food and Drug Administration (FDA), European Agency for Evaluation of Medicinal Products (EMEA), and Japan's Ministry of Health, Labor and Welfare (MHLW) approve pharmaceutical drugs based on the clinically proven parameters of safety and efficacy. Prescribing information is required for approved pharmaceutical products, and the specific language of approval is contained in a product package insert (PI). Pharmaceutical industry supported activities including sales and marketing efforts are considered promotional in content and are rigorously restricted to language contained within this "on-label" PI information. External customers including patients, providers, hospital systems, managed care organizations, group purchasing organizations, and medical societies are becoming increasing more sophisticated and demand the latest clinical information, often "off-label." The demand for medical information is amplified by the structure and delivery of medical care in the United States and around the world, as well as the public demand through ready access to modern technologies, such as the Internet and mass media. In order to meet customer demands, the pharmaceutical industry and government regulatory bodies are reengineering to better accommodate specific "off-label" information requests beyond the package insert [11].

Overall, the most significant function of the *medical information* department is to support postmarketing information needs of health care providers and patients and represent the product package Insert (Fig. 11.8). Pharmaceutical medical professionals typically staff a "customer contact center" for which telephone and e-mail product queries can be answered. As the primary "customer voice," pharmaceutical management will often call upon the medical information team to better understand customer reception, needs, and inquiries about the company's products and services. In times of vacant sales territories, these internal medical professionals may also be available to help answer clinical questions from field customers. A typical company will answer about 130 inquiries per day in a typical 8-hour day, which is about

- Customer contact center
- "Voice of Consumers"
- Inquiries (1* telephone, e-mails, fax, & written):
  - (1* providers, 2* patients, & 3* many others)
- "Standard Letter" generation
- Represent product Package Insert (PI):
  - FDA approved product PI
  - Patient package insert (PPI)
- Pipeline product information
- FDAMA "Safe Harbor"

21 CFR 99 - Dissemination of Information on Unapproved/New Uses for Marketed Drugs, Biologics, and Devices

FIG. 11.8. Medical Information Services

32,500 per year, with a staff of 19, according to a survey of eight companies published in 2000 [15]. Telephone was the primary means of receipt (65%) and 24% electronically; responses were telephone in 48% and written in 50%. The amount of time on the telephone was stated to be almost 8 minutes per call in one survey [16]. Automation is usually used by companies to receive and distribute the call to an appropriate individual in 60% of companies surveyed. A description of an automated medical information system for the DuPont Company was published back in 2000 [17].

"Standard Letter" Generation

To more efficiently and consistently respond to commonly asked customer questions, medical information may develop a frequently asked question (FAQs) document used for internal training and also formatted into standard letter responses that can be mailed to customers upon request. Commonly asked customer questions are often anticipated and developed into standard responses and drafted in advance of a product launch, along with support from the product package insert, published scientific literature, and occasionally clinical or registration data on file at the pharmaceutical company. Standard letters typically incorporate relevant reprints of support clinical literature and serve as a main written form of customer communication. Because of potential wide distribution, standard letters often require approval by several departments including corporate legal and regulatory affairs and R&D representatives. Standard letters individualized to the inquiry being dealt with are commonly sent to the caller/inquirer, about 73% of the time in one survey [16]. Reprints are provided about half of the time. The inquirer was satisfied with the written response 70% of the time, in comparison with 63% satisfaction with the verbal response.

Product Package Insert

FDA-approved product labeling originates from years of rigorous safety and efficacy clinical trials. Labeling must include the established name, proprietary name (if any), adequate directions for use, and adequate warnings. The agency considers the approved product labeling, sometimes called the full prescribing information, to be adequate directions for use and adequate warning. This mound of evidence is compiled into a product package insert document and is highly negotiated word-by-word between the regulatory authorities around the world and pharmaceutical company's legal and regulatory department [18, 19].

Initial product labeling becomes available at the time of drug product entry into the marketplace with regulatory guidance in the United States under CFR Title 21 Sections 10, 100, and 200. Language in the product package insert outlines validated product claims and quarantines pharmaceutical industry market and sales promotion to within these boundaries. "On-label" is a frequently used industry term referring to information contained with an approved product package insert. "Off-label" information is slang terminology that refers to nonapproved uses of an approved and marketed pharmaceutical product for which scientific data is published or available at the company.

After initial approval, life cycle management and subsequent clinical trial development is implemented to maintain clinical marketplace preference before patent expiration. Labeling changes may be sought to strengthen information and product package insert language around new indications for a known molecular entity, expanded patient populations, new dose or drug forms, or advanced quality of life or long-term safety and/or morbidity and mortality data. The regulations in the United States [18, 19] lists examples of types of labeling materials under CFR: "Brochures, booklets, mailing pieces, detailing pieces, file cards, bulletins, calendars, price lists, catalogs, house organs, letters, motion picture films, film strips, lantern slides, sound recordings, exhibits, literature, and reprints and similar pieces of printed, audio or visual matter descriptive of a drug and references published (for example, the Physician's Desk Reference) for use by medical practitioners, pharmacists, or nurses, containing drug information supplied by the manufacturer, packer, or distributor of the drug and which are disseminated by or on behalf of its manufacturer, packer, or distributor are hereby determined to be labeling as defined in the FD&C Act."

Patient Package Insert

Pharmaceutical manufacturers may also gain approval for distribution of a patient package insert [20–22]. This version of the product package insert is developed for a patient consumer in language appropriate for a non–health care practitioner. The patient package insert is considered an extension of the product labeling and may be distributed to patients when the drug is dispensed from a pharmacy. Key information about the drug is described in common vocabulary and includes drug benefits, risks, how to recognize risks, and dosage and administration. Patient package inserts are required for certain drugs like oral contraceptives [23], estrogens [24], progestational drug products [25], otherwise it is considered a voluntary tool.

In addition, the Drug Information Branch within CDER produces Consumer Drug Information Sheets containing general information about newly approved prescription drugs based on the package insert. Unlike the more complex wording of a package insert, the CDER information sheets are provided for patient reading and understanding. Contained within the FDA web site is also access to the complete drug package insert. Only information about drugs approved since January 1998 appears on this page. More information on a specific drug can be found at the FDA Web site [26].

Under DDMAC guidance, nonapproved, off-label scientific information must be contained away from and separate to marketing and promotional activities [27]. Access to drug and disease state information and large publication databases is key to successful, timely, and accurate search and retrieval of scientific information.

Pipeline Product Information

Pipeline products will be undergoing their research, preclinical, PK/PD, and clinical trials 5 or more years prior to approval, and the medical community and public desire access to information about these investigational agents, all of which is considered off-label. Regulation of this information is intended to limit what can be done with investigational agents, especially given that the efficacy and safety of the products has not been established, and of course has not been approved by regulatory authorities. MA needs to work with R&D and management to respond to such inquiries with medically and ethically appropriate responses [27].

FDAMA "Safe Harbor"

Under the Food and Drug Administration Modernization Act (FDAMA) of 1997, a pharmaceutical medical information professional can respond to a question on the nonapproved use of an approved and marketed drug if specifically requested by an external customer [12, 26, 27]. The Food and Drug Administration Modernization Act, enacted November 21, 1997, amended the Federal Food, Drug, and Cosmetic Act relating to the regulation of food, drugs, devices, and biological products. The FDA Modernization Act (FDAMA) addressed several areas of improved efficiency regarding drug approval, including fast-track drug approval, consumer-accessible databank of clinical trials for serious and life-threatening diseases, dissemination of pharmacoeconomic information to formulary and pharmacy benefit managers, monitoring of postmarketing study obligations, and dissemination of information on off-label uses of drugs [28].

Under the "safe harbor" of FDAMA, only peer-reviewed publications and textbooks are approved materials for distribution under this act and should be accompanied by a bibliography of other publications on the new use and a current approved product package insert. If a book including some off-label information is proactively disseminated, it must be identified as "off-label" use if other similar products are being used and if there are any financial arrangements between the pharmaceutical company and journal authors.

Pharmaceutical organizations can utilize "safe harbor" dissemination of information provided a manufacturer commits to performing clinical studies on the new use with the intent of filing for FDA approval of the use and the following variety of conditions are met including:

- concerns a drug or device that has been approved, licensed, or cleared for marketing by FDA;
- is in the form of an unabridged reprint or copy of a peer-reviewed scientific or medical journal article, or an unabridged reference publication, about a clinical investigation that is considered scientifically sound by qualified experts;
- does not pose a significant risk to the public health;
- is not false or misleading;
- is not derived without permission from clinical research conducted by another manufacturer;
- includes certain disclosures (e.g., that the new use has not been approved by FDA), the official labeling, and a bibliography of other articles relating to the new use.

In 1998, the Washington Legal Foundation challenged the authority of FDAMA in the court system as restrictive and unconstitutional, limiting free speech. FDA has since clarified FDAMA in the *Federal Register* as a "safe harbor" for the dissemination of off-label uses; however, considerable interpretation can be made and most of the conservative pharmaceutical industry tends to cautiously weigh the risk versus benefit of proactively applying FDAMA to sales and promotional material.

Medical Inquiries (Off-label) to Sales Representatives

Drug products or indications that are not yet approved for marketing cannot be promoted by sales representatives in any way, as they are not yet approved to be safe and effective. Actually, they can say and do nothing proactively in regard to an investigational product, or for an approved product for an unapproved indication. If they receive a question that is off-label, that is, an unapproved use, they must not answer the question even if published literature is available. They can refer the provider to the company's medical affairs or professional services department to answer the question. Companies can answer off-label questions and even send letters and literature, as long as the provider asks the question, and the appropriate home office professional only is engaged in the interaction. Sales representatives are also contacted to be a source of research grants and drug product for drug studies for on-label and off-label ideas, but for the latter they can only refer the provider to the appropriate staff at the home office for any further discussion. Finally, sales representatives are used by companies to identify potential investigators for studies, based on their knowledge of the researchers and providers in their region, but again, they cannot make a contact with the potential investigator to discuss any issues related to off-label, unapproved issues.

Besides anchoring external customer inquires for drug information, the medical information postmarketing team also supports the following *other responsibilities*, many of which directly relate to launch of new products as well as continuing support over the life of a product (Fig. 11.9). The roles

- Responsibilities:
  - Scientific publication tracking
  - Scientific committee member
  - Scientific meeting support
  - Disease state management
  - Research questions to be addressed
  - Patient registries
  - Web-site (products and questions)
- Triage:
  - AEs ⇨ Safety; Complaints ⇨ QA; Patients/Trials ⇨ Development
  - Reporters ⇨ PR; Investors IR; Lawyers ⇨ Legal; Reimbursement ⇨ Specialists; Competitive intelligence ⇨ Marketing

FIG. 11.9. Medical Information: Other Responsibilities

described below for the medical information group in some companies are performed by the medical communications group in medical affairs.

Scientific Publication Tracking

Across a pharmaceutical company, and especially in a global marketplace, a significant amount of organization is called upon to manage the content and timing of peer-reviewed and medical congress publications [29]. Typically, a publication planning committee includes members across medical information, clinical development, pharmacoeconomics, biostatisticians, medical writing, and sales and marketing. Publications are generally adapted to include key scientific and marketing messages consistent with data available and are tailored to educate a target audience on a drug product and/or disease state message.

As an example, the data sets would be managed to publish:

- Randomized, controlled, peer-reviewed scientific clinical trials
- Literature manuscript review or meta-analysis
- MUE or DUE evaluations
- Pharmacoeconomics
- Medical congress abstract or poster
- Case reports
- Letters to the editor

Formulary Material Development

An important step in drug commercialization, especially in the hospital setting, is to gain hospital formulary approval for inpatient and/or outpatient use. To better assist drug information hospital pharmacists and members of pharmacy and therapeutic committees (P&T) with product information, medical information often develops a "product information" packet. A "formulary kit" typically contains in written or electronic format several sections including product package inserts, product information sheets, large-print and annotated prescribing information, compendia monographs, quick reference dosing cards, supporting clinical literature reprints, frequently asked questions, and business reply cards for customers to request mailing of any "off-label" or additional during information.

Scientific Meeting Support

Medical congresses present regular key educational and marketing opportunities that allow face-to-face interactions with potential customers or patients. Medical societies routinely hold large symposium meetings that allow promotional and academic exchange of information with customer providers, medical trainees, benchtop and clinical trialists, policy and government representatives, vendors and contract research organizations, investors and news reporters, and other related and supportive business enterprises. To fully support customers, a pharmaceutical company may staff promotional and/or scientific booths with sales representatives, account managers, government representatives, and pre- and clinical development and medical information specialists. If medical society regulations allow, disease state or drug product speakers and symposiums may be developed and supported during medical congresses. Investigators or research groups may also be invited to present key scientific or pharmacoeconomic data during "poster sessions." Acceptance of an abbreviated research synopsis, commonly called an "abstract," can lead to an opportunity to present further key research findings. Poster sessions allow an investigator an opportunity to develop an 4' × 6' poster of his or her research and discuss or defend results in a peer-to-peer environment. Often, and especially in accelerated clinical research environments such as oncology or HIV, abstract and poster data can serve as a significant preview to early trends in clinical trial data and are therefore inherently available to the public domain prior to final manuscript generation.

A few large scientific medical congresses held in the United States include the five listed below, wherein more than 10,000 physician and other health care providers participate.

- American Heart Association (AHA)
- American Society of Nephrology (ASN)
- American Society of Health System Pharmacists (ASHP)
- American Society of Clinical Oncology (ASCO)
- Heart Failure Society of America (HFSA)

Disease State Management

The pharmaceutical industry is beginning to take an active role in total disease state management to lower the overall burden of health care costs. Beyond standard physician detailing and in answer to customer requests for more in-depth information, provider and patient education is taking a more active role as a communication tool. Programs focus on conditions such as asthma, diabetes, depression, and heart disease and include clinical practice guidelines, provider and patient intervention techniques, monitoring of patient laboratory results, and treatment outcomes. Educational seminars often offer value-added services typically at no additional cost to health care organizations or consumers. Patient education can be provided in the form of non-branded brochures, newsletters, consultation with health care professionals (HCPs), or non-branded web sites. Such programs require substantial advance planning and financial investment from pharmaceutical companies and often target eventual cost savings and lessened use of health care resources. More and more disease states are recognized as interconnected syndromes with relationships and ties to other medical conditions. For example, the close relationship of heart disease and chronic kidney disease is commonly referred to in advisory circles and published literature as the "cardio-renal syndrome." This is perhaps one example of the human body's adaptive goal for survival that causes physiologic attempts at "mass-balancing" diseased systems. Pharmaceutical companies are beginning to focus on medical providers and reimbursement partnerships to better understand preemptive identification, treatment of existing disease, and realignment of normal body systems while targeting lower total health care expenditures [30–32].

Pfizer is one of the leading contributors to customer disease state management partnerships. *Pharmacy Today* reported an innovative arrangement with the Florida Medicaid system, in which Pfizer promised to achieve $33 million in cost reductions over 2 years in return for inclusion of all of its products on a new restrictive formulary in the sunshine state [33]. If the deal works, it could be a model for other Medicaid systems, third party payers, and pharmaceutical companies to follow in addressing the challenge of escalating prescription drug prices. The Pfizer arrangement uses many of the principles of pharmaceutical care, repackaged under the banner of 'disease management,' to achieve cost reductions through decreased emergency department visits and hospitalizations, all while keeping Medicare beneficiaries healthy by using cutting-edge pharmacotherapy.

According to a July 9 front-page story in the *Wall Street Journal*, Pfizer plans to achieve the cost reductions through disease management, focusing primarily on 12,000 patients who are high utilizers and have chronic diseases such as diabetes, asthma, or heart disease[34]. Some 60 nurse case managers will use computer software designed for chronically ill Medicaid patients to encourage patients 'to take their medicines dutifully, follow diet and exercise regimens, and have regular checkups,' the *Journal* reported. Florida's Medicaid budget for the fiscal year that began July 1, 2001 is $9.7 billion, and Governor Jeb Bush and state legislators were looking for reductions of $650 million during that year's legislative session. As Pfizer began negotiating a deal that would exempt it from a restrictive Medicaid formulary, the united industry front that had previously defeated such proposals crumbled and the proposal became law in late May. Pharmaceutical manufacturers have two choices under the statute: either compete for places on the formulary by making cost concessions (in the form of rebates) to the state or offer cost-saving services through arrangements that guarantee fixed amounts to the state. Thus far, only Pfizer has chosen the latter option.

Patient Assistance Programs

An important industry-wide practice is the provision of pharmaceuticals free of charge to patients who are medically indigent and cannot afford their medications [35]. Millions of dollars worth of medications are provided annually. The medical information area often coordinates this process, wherein patients need to provide their personal financial data to establish eligibility. Usually, income and expense statements are provided to be reviewed and validated with a minimum income requirement below the poverty level.

Research Questions to be Addressed

As the "voice of the customer," medical information is the primary touch point for customer inquires and can offer insight into needs for future clinical trials. Frequently asked questions from customer groups raise questions on how to fill voids of postapproval drug information. This type of customer inquiry and frequency can serve as a valuable tool in launching new clinical trials.

Special Patient Registries

For many pharmaceutical products, an opportunity to gather preapproval safety and efficacy information in pediatric and pregnant patients may be limited. Typically, the risk of conducting investigator drug trials outweighs the benefits of advancing pharmaceutical science, or these special patient populations are not initially considered as primary beneficiaries of an approved drug therapy. Therefore, as available, the medical information (MI) department will reactively collect customer case reports of drug exposure in these populations. Information collected in these types of "registries", including drug exposure, fetal outcome, and adverse events, may allow better clinical trial design and extrapolation of safety and efficacy recommendations [36,37].

Communication Triage

Questions come into a company from many sources and need information of many types, separate from product and medical questions from the medical community and patients. MI serves a triage role for the following types of questions and the referral department or group is noted as well.

- Adverse event reports for completion of postmarketing safety and surveillance reporting to FDA; Safety group;
- Ordering of a product from a valid purchasing agent; Customer services;
- Media inquiries of any type; Public relations;
- Investment or stock inquires; Investor relations;

- Legal and licensing inquires; Law department;
- Complaints about a product regarding the packaging, the vials or bottles, functionality of a unit, or stability; Quality assurance group in manufacturing;
- Reimbursement questions including special drug access to indigent patient programs or billing code questions; Third-party reimbursement specialists;
- Reports on competitive intelligence on marketing activities of other products, clinical trials, or reimbursement programs being offered by competing products or marketplace influences; Marketing group.

After approval of a product, prescriptions written by providers, and patient use, some patients will experience adverse events that may be expected, according to product labeling, or unexpected in type, frequency, or severity. The patient, family member, or provider (physician, pharmacist, or nurse) will report these findings to a company, which is obligated to receive, investigate, and report on these *adverse events* (Fig. 11.10). Three codes in the *Federal Register* are applicable for adverse reaction reporting to guide a company in compliance for marketed drugs in unapproved areas [39], for drugs postmarketing [40], and for biological postmarketing [41].

During the preapproval drug development process, hundreds to a few thousands of patients are exposed to a new molecular entity under the rigors of clinical trials environments (patient selection, specified doses, close monitoring, and tight controls). As an approved drug enters the public domain, postapproval side-effect monitoring and adverse event reporting become extremely important and are required under law. Commercialization increases drug exposures dramatically and makes product available to an "all-comer" patient population outside the control of well-designed clinical trials. In these highly variable and extensive usage situations, adverse events will occur that could not have been anticipated. The FDA uses postapproval surveillance to assist updates to drug labeling, reduce medication errors, develop new methods to provide patients with adequate medication information, and, on rare occasions, to reevaluate the approval of a marketing decision. See the May 1999 FDA report, "Managing the Risks from Medical Product Use: Creating a Risk Management Framework," for a report of current and recommended premarketing/postmarketing risk assessment processes and postmarketing surveillance programs [42].

To maintain patient privacy during the safety reporting process, the FDA requests a unique code number identifier be assigned to each patient case report, preferably not more than eight characters in length. The applicant should include the name of the reporter from whom the information was received. Names of patients, health care professionals, hospitals, and geographical identifiers in adverse drug experience reports are not releasable to the public under FDA's public information regulations. Records are maintained for a period of 10 years on all adverse drug experiences known to the pharmaceutical applicant, including raw data and any correspondence relating to adverse drug experiences.

Industry Requirements for Postmarket Safety Surveillance

Postmarketing adverse event surveillance information should be obtained or otherwise received by the pharmaceutical company from any source, foreign or domestic, including information derived from commercial marketing experience, postmarketing clinical investigations, postmarketing epidemiological/surveillance studies, reports in the scientific literature, and unpublished scientific papers [43–45]. All reported adverse event terms are coded using a standardized international terminology, MedDRA (the Medical Dictionary for Regulatory Activities).

- **Adverse drug experience.** Any adverse event associated with the use of a drug in humans, whether or not considered drug-related occurring from drug product in professional practice; overdose whether accidental or intentional; abuse; withdrawal; and any failure of expected pharmacological action.
- **Serious adverse drug experience.** Any adverse drug experience occurring at any dose that results in any of the following outcomes: death, a life-threatening adverse drug experience, inpatient hospitalization or prolongation of existing hospitalization, a persistent or significant disability/incapacity, or a congenital anomaly/birth defect. Important medical events that may not result in death, be life-threatening, or require hospitalization may be considered a serious adverse drug experience when, based on appropriate medical judgment, they may jeopardize the patient or subject and may require medical or surgical intervention to prevent one of the outcomes listed in this definition.
- **Unexpected.** As used in this definition, refers to an adverse drug experience that has not been previously observed (i.e., included in the labeling) rather than from the perspective of such experience not being anticipated from the pharmacological properties of the pharmaceutical product.

- Post-Marketing Safety & Adverse Event Surveillance:
  - Requirements of industry:
    - 15 Day "Alert" and follow-up
    - Periodic quarterly
    - Post Marketing studies exempt
  - "Medwatch" FDA form 3500
  - HIPPA:
    - Access to medical records. Notice of privacy practices
    - Limits on use of personal medicalin formation
    - Prohibition on marketing. Stronger state laws
    - Confidential communications
    - Complaints
  - Safety department interface for MI
  - Risk management

FIG. 11.10. Medical Information & Safety (Adverse Events)

There are several types of adverse event reports required by the FDA depending upon the severity and source of the information:

- **Postmarketing 15-day "alert reports."** The applicant shall report each adverse drug experience that is both serious and unexpected, whether foreign or domestic, as soon as possible but in no case later than 15 calendar days of initial receipt of the information by the applicant. The applicant shall promptly investigate all adverse drug experiences that are the subject of these postmarketing 15-day alert reports and shall submit follow-up reports within 15 calendar days of receipt of new information or as requested by FDA. If additional information is not obtainable, records should be maintained of the unsuccessful steps taken to seek additional information. Postmarketing 15-day alert reports and follow-ups to them shall be submitted under separate cover.
- **Quarterly periodic adverse drug experience reports.** The applicant shall report each adverse drug experience not reported under "Alert Reports" at quarterly intervals, for 3 years from the date of approval of the application, and then at annual intervals. A narrative summary and analysis of the information in the report and an analysis of the 15-day alert reports submitted during the reporting interval; a mandatory FDA Form 3500 for each adverse drug experience not previously reported; and a history of actions taken since the last report because of adverse drug experiences (for example, labeling changes or studies initiated).

Information is collected in the Adverse Event Reporting System (AERS), a computerized information database designed to support the FDA's postmarketing safety surveillance program for all approved drug and therapeutic biologic products. Reports in AERS are evaluated by a multidisciplinary staff of safety evaluators, epidemiologists and other scientists in the Center for Drug Evaluation and Research's (CDER) Office of Drug Safety to detect safety signals and to monitor drug safety [43]. As a result, the FDA may take regulatory actions to improve product safety and protect the public health, such as updating a product's labeling information, sending out a "Dear Health Care Professional" letter, or reevaluating an approval decision.

Postmarketing study adverse events do not apply under this unless the applicant concludes that there is a reasonable possibility that the drug caused the adverse experience. Adverse events that occur during clinical studies are to be reported to FDA as specified in the investigational new drug/biologic regulations. For applicable regulations and industry guidance on mandatory reporting for drug/biologic manufacturers, distributors, and packers, go to the FDA web site for MedWatch [44].

MedWatch

MedWatch is a voluntarily reporting process for serious adverse events, product problems, or medication errors suspected in association with an FDA-regulated drug, biologic, device, or dietary supplement [44, 45]. The FDA Form 3500 should be used by health care professionals and consumers for voluntary reporting of adverse events noted spontaneously in the course of clinical care, not events that occur during IND clinical trials or other clinical studies. Voluntary forms may be completed via online, telephone (800-FDA-1088), fax (800-FDA-0178), or U.S. post.

A serious adverse event is any undesirable experience associated with the use of a medical product in a patient and involving one of the following:

- **Death**: Report if the patient's death is suspected as being a direct outcome of the adverse event
- **Life-threatening**: Report if the patient was at substantial risk of dying at the time of the adverse event or it is suspected that the use or continued use of the product would result in the patient's death. Examples: Pacemaker failure; gastrointestinal hemorrhage; bone marrow suppression; infusion pump failure that permits uncontrolled free flow resulting in excessive drug dosing.
- **Hospitalization (initial or prolonged)**: Report if admission to the hospital or prolongation of a hospital stay results because of the adverse event. Examples: Anaphylaxis; pseudomembranous colitis; or bleeding causing or prolonging hospitalization.
- **Disability**: Report if the adverse event resulted in a significant, persistent, or permanent change, impairment, damage or disruption in the patient's body function/structure, physical activities, or quality of life. Examples: Cerebrovascular accident due to drug-induced hypercoagulability; peripheral neuropathy.
- **Congenital anomaly**: Report if there are suspicions that exposure to a medical product prior to conception or during pregnancy resulted in an adverse outcome in the child. Examples: Vaginal cancer in female offspring from diethylstilbestrol during pregnancy; malformation in the offspring caused by thalidomide.
- **Requires intervention to prevent permanent impairment or damage**: Report if you suspect that the use of a medical product may result in a condition that required medical or surgical intervention to preclude permanent impairment or damage to a patient. Examples: Acetaminophen overdose–induced hepatotoxicity requiring treatment with acetylcysteine to prevent permanent damage; burns from radiation equipment requiring drug therapy; breakage of a screw requiring replacement of hardware to prevent malfunction of a fractured long bone.

Product problems should also be reported to the FDA when there is a concern about the quality, authenticity, performance, or safety of any medication or device that may have occurred during manufacturing, shipping, or storage. Issues may include suspect counterfeit product; product contamination; defective components; poor packaging or product mix-up; questionable stability; device malfunctions; and labeling concerns.

Medication error reports include marketed human drugs (including prescription drugs, generic drugs, and over-the-counter drugs) and non-vaccine biological products and devices. The National Coordinating Council for Medication Error Reporting and Prevention defines a medication error as "any preventable event that may cause or lead to inappropriate medication use or patient harm while the medication is in the control of the health care professional, patient, or consumer. Such events may be related to professional practice, health care products, procedures, and systems, including prescribing; order communication; product labeling, packaging, and nomenclature; compounding; dispensing; distribution; administration; education; monitoring; and use." Examples may include miscommunication of drug orders, which can involve poor handwriting, confusion between drugs with similar names, misuse of zeroes and decimal points, confusion of metric and other dosing units, and inappropriate abbreviations; or incomplete patient information (not knowing about patients' allergies, other medicines they are taking, previous diagnoses, and lab results).

In 1992, the FDA began monitoring medication error reports that are forwarded to FDA from the United States Pharmacopeia (USP) and the Institute for Safe Medication Practices (ISMP). The agency also reviews MedWatch reports for possible medication errors. Currently, medication errors are reported to the FDA as manufacturer reports (adverse events resulting in serious injury and for which a medication error may be a component), direct contact reports (MedWatch), or reports from USP or ISMP.

See regularly updated information at the FDA web site for following reports. Safety Alerts for Drugs, Biologics, Devices, and Dietary Supplements; Safety-Related Drug Labeling Changes; FDA Safety-Related Information Recalls; FDA Enforcement Report, Medication Errors, Drug Shortages, Biologic Product Safety Information and Recalls & Withdrawals; Dietary Supplements Warning and Safety Information; Medical Devices: Safety Alerts; Public Health Advisories, and Notices; and FDA Patient Safety News [44].

The MedWatch to Manufacturer Program (MMP) is designed to expedite the transmission of serious voluntary adverse event reports from FDA to licensed drug and biologic manufacturers participating in the program.

HIPPA

The Health Insurance Portability and Accountability Act of 1996 (HIPPA) was put into public law in 1996 to protect the privacy of patient information [13]. On April 14, 2003, the Department of Health and Human Services (HHS) set new HIPPA standards to provide patients with access to self medical records and more control over how personal health information is used and disclosed.[1] The new privacy regulations ensure minimum federal standards for safeguarding the privacy of individually identifiable health information and sets a national floor of privacy protections for patients by limiting the ways that health plans, pharmacies, hospitals, and other covered entities use patients' personal medical information.

Key provisions of these new standards include:

- Access to medical records
- Notice of privacy practices
- Limits on use of personal medical information
- Prohibition on marketing
- Confidential communications
- Complaints—complaints can be made directly to the provider or health plan or to the Health and Human Services Office for Civil Rights (OCR; http://www.hhs.gov/ocr/hipaa) or by calling (866) 627-7748.

This HIPPA law carries heavy criminal and civil penalties. The Office of Civil Rights may impose monetary penalties up to $100 per violation, up to $25,000 per year, for each requirement or prohibition violated, and criminal penalties can range up to $50,000 and 1 year in prison for certain offenses; up to $100,000 and up to 5 years in prison if the offenses are committed under "false pretenses"; and up to $250,000 and up to 10 years in prison if the offenses are committed with the intent to sell, transfer, or use protected health information for commercial advantage, personal gain, or malicious harm.

HIPPA brings special challenges and additional cost to pharmaceutical drug safety surveillance and research. HIPPA protects patients from inappropriate use and disclosure of patient information. In many cases unique patient identifiers and third-party research firewalls are required to utilize large health information databases, such as hospital discharge and medical records, epidemiological databases of disease registries, and government reimbursement compilations, which often include other vital and health statistics. For information on how the privacy rule may affect specific research areas, see the companion pieces to this booklet: Health Services Research and the HIPAA Privacy Rule; Repositories, Databases, and the HIPAA Privacy Rule; Clinical Research and the HIPAA Privacy Rule; Institutional Review Boards and the HIPAA Privacy Rule; and Privacy Boards and the HIPAA Privacy Rule.

Medical Science Liaisons

Medical field professionals (also commonly known as *medical science liaisons*) are scientists and clinicians with a company and who are geographically located across the United States or global marketplace (Fig.11.11). They are responsible for key opinion leader development. Depending upon department alignment with a pharmaceutical organization, liaisons support most aspects of medical affairs postmarketing activities including medical information, speaker development, medical

---

http://privacyruleandresearch.Nih.gov/Qc_03.asp.accessed12/12/2004.

- Medical health care professionals geographically based near key customer accounts
- Corporately aligned based on therapeutic area, product, or account type
- Focused on Key Opinion Leaders (KOL):
  - Local clinical grants
  - Access to company and investigator initiated clinical studies
  - Investigator meetings
  - Scientific literature and education
  - Speaker updates and training

FIG. 11.11. Medical Science Liaisons Functions

congress support, publication planning, hospital formulary approval, basic product and disease state education, and technical customer presentations. The variety of job functions also may include clinical study site identification and patient recruitment and retention support, investigator recruitment, liaison for phase 2 through 4 studies, and advocacy development of regional thought leaders and health care influencers. As an extension of the corporate office, medical field professionals are subjected to conduct of business under regulatory guidelines such as FDAMA and HIPPA [46–49].

Medical field professionals departments are typically structured in one of two ways that allow very different product support:

1. An extension of sales and marketing departments, in which the job functions strictly support on-label disease and product information only; or,
2. An extension of research & development with marginal association with sales and marketing departments. This latter official reporting structure allows fair, balanced, off-label response under FDAMA "Safe Harbor" and is commonly involved in earlier phase product development, selection of clinical investigators, and study site evaluations.

Medical field professionals are geographically based near key customer influencers and academic accounts across the United States or global marketplace. They are typically aligned with corporately defined therapeutic areas, product and disease state, or specialized to account type (payor, academic, federal, pharmacoeconomics). They are assigned to specific individuals to work with or specific institutions.

To better support local, regional, and nationally recognized key opinion leaders, liaisons leverage and match professional relationships with pharmaceutical programs to maximize mutually interesting opportunities in education and research. In a recent *DIA* publication, Schneider described that liaisons answer technical questions, establish professional relationships, develop programs to increase awareness of current and future use, and foster new R&D opportunities [49].

There are a variety of tools available that liaisons leverage to interest a key influencer in learning and working with a company on a disease state, product, or patient population related directly or indirectly to a pharmaceutical product.

- Local clinical grants: As funding grows tight in academic settings, local clinical grants are generally welcomed to help support departmental grand round speakers and visiting professorships. Legal and regulatory review is often involved to be sure local grants are appropriate and do not risk being construed as a method of coercion or inappropriate influence to buy or support pharmaceutical drug products.
- Access to company and investigator initiated clinical studies: Research remains the cornerstone to clinical trial data generation and establishes best medical practice. Key influencers may be interested in joining a company-initiated multicenter trial or submitting a research hypothesis for funding or drug supply.
- Support investigator meetings: Investigator meetings are typically well attended by primary investigators, study coordinators, and supportive clinical trial personnel. Key influencers may be engaged to provide a disease state or clinical trial–related presentation. Often liaisons attend these venues to help welcome key opinion leaders and support staff, offer clarification to aspects of the clinical trial, or gather interest in investigator-initiated subset analysis or follow-up studies.
- Scientific literature and education: Updated published information and educational materials are provided for the caregivers or patients in a practice or academic site.
- Speaker updates and training: Updates to clinical trial or disease state information are often shared with speaker bureau members in thought-leader meetings by pharmaceutical liaisons.

A *key opinion leader (KOL)* can be defined as an influencer of current medical practice and capable of inspiring peers to improve patient outcomes through research, scientific literature, clinical or experiential information (Fig. 11.12). These KOLs are the most important interface for the MSLs in the

---

KOL: Influencer on current medical practice and capable of inspiring peers to improve patient outcomes through research, scientific literature, clinical, or experiential information

- Current
- Communicator:
  - Publications
  - Teaching
- Researcher
- Clinical Experience:
  - Disease state expert
  - Amgen & Competitors
  - "Hands-On" use
- Location:
  - Universities & Health centers
- Fellow & early physician development
- Influencers:
  - Peers
  - Guidelines & Policy
  - Formulary
  - Medical associations
  - Academic affiliation
  - Large community or national medical directors

FIG. 11.12. MSLs & Key Opinion Leader (KOL)

medical community. Typically, key opinion leaders generally serve as a role model to other health care practitioner peers and spearhead advancements in patient care and publication, generate research, or influence policy or reimbursement committees. KOLs can be readily recognized at all health care levels and may include patients, physicians, pharmacists, medical trainees, nurse practitioners, dietitians, social workers, laboratory technicians, societal influencers and committee heads, policy and guideline committee members, formulary and P&T participants, and are usually located at large community or academic settings.

KOLs are good communicators and influence others through publications or teaching. Generally, KOLs can offer reflection on the status of health care as a disease state expert, and in most cases, he or she can offer experiential testimonies from clinical trial or hands-on practice care of company and/or competitor products. They often are important advisors to both the R&D and S&M groups regarding medical/pharmaceutical practice and research.

Influence mapping and identification of key opinion leaders is established during phases 1 and 3 of the drug development process (Fig. 11.13). In concert with phase 1, it is paramount to identify national and regional influencers that can help guide registrational trial program development, conduct solid and timely clinical research, generate leading disease state publications, help draft treatment guidelines, and begin cultivating potential speakers to help leverage clinical trial product use immediately post drug approval. KOLs identified at this early stage of product development may be involved with one or more of the following activities with a company:

- Registration study site identification
- Phase 2 trial participation (proof of principle)
- Phase 3 trial participation (registrational trials)
- Global publication planning
- Advisory boards for R&D (target populations, study designs)
- Advisory boards for S&M (market research and product profile development)

During the later stages of drug development and prior to product launch, KOL mapping is again revisited to help identify more regional and local early adopters of similar drug therapy. Postapproval phase 3B/4 studies often require large numbers of patients, study investigators, and clinical trial sites. Leaders from early clinical trials are typically identified to help conduct later phase investigator meetings, serve as a sounding board for marketing messages and medical education initiatives, and advise on commercialization tactics and payor strategies.

KOLs identified at this stage may be involved with one or more of the following activities:

- Phase 3B and 4 studies
- Primary investigators and clinical trial sites
- Investigator meetings
- Medical education strategy/tactics
- Payor strategy/tactics
- Guideline development
- Phase II, exploratory studies

This is a *strategy flow diagram* in Fig. 11.14 that offers an example of medical science liaison *integration* into disease management and product support. In this example, high-level science interactions are planned after a national physician speaker training meeting. Liaisons help further develop targeted speaker platform and group moderator skills, revisiting important clinical literature and product registration studies, and provide ongoing literature updates on relevant topics.

In tandem, liaisons continue to identify principal investigators for future company-sponsored research and bring forth investigator-initiated clinical trials in support of brand product or marketing messages. Often, motivated key opinion leaders wish to publish self-experience or create publication overviews of key leanings from involvement in clinical trials and extrapolate on how a compound may fit into current medical care. If needed, liaisons help leverage statistical or medical writing support to assist these types of publication efforts.

Figure 11.15 displays required *professional degrees* and related *titles* of MSLs in a six representative pharmaceutical companies. Advanced professional health care academic degrees generally are required at the doctoral level for two needs; credibility with the KOLs and also for ideal technical preparation of MSLs in level of knowledge, research skills, and practice experience. The level of the position is at least manager up to director level. Such levels recognize their potential contribution to the company, enhance their credibility with the KOLs, and permit recruiting at a sufficient salary level to obtain appropriately talented individuals. The university setting in pharmacy and medicine is from where the MSL will be recruited.

Just as departmental reporting structures differ, the measurement of MSL *professional influence* (activities and effectiveness) varies across the pharmaceutical industry (Fig. 11.16). If liaison teams are structured through sales and marketing or research and development divisions, activities cannot be tied to promotional activities or product sales figures. The rationale is appropriate independence from marketing for regulatory reasons, that is, to perform off-label work and sponsor

Work Stream: Operating Principles and Guidance
Timing of influence mapping and the frequency?

Lead Optim. → Pre-clinical → Phase 1 → Phase 2 → Phase 3 → Filing → Launch → Post-launch

- Registrational study site ID
- Global publication planning
- Advisory boards
- Guideline development

- Phase IIIB and IV studies
- Identify PIs & sites
- Investigator meetings
- Medical education (Strategy/Tactics)
- Payers (Strategy/Tactics)

FIG. 11.13. Global KOL Strategy & Influence Mapping

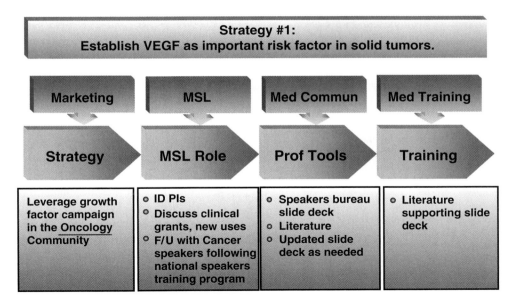

FIG. 11.14. MSLs & Strategy Integration – Flow Diagram

Company	Degree	Acronym	Title
Abbott	PharmD, RN	MSL	Medical Science Liaison
Amgen	PharmD, MD, PhD	RML	Regional Medical Liaison
AstraZeneca	PharmD, MS	MIS	Medical Information Scientist
	PharmD, PhD, MD, MS	MDS	Market Development Scientist
	PharmD, PhD, MD, MS	MML	Medical Marketing Leader
Bayer	PharmD, MD	CRA	Clinical Research Associate
	BS	SAL	Scientific Affairs Liaison
Bristol-Myers Squibb	PharmD, PhD, MD	OMLM	Oncology Medical Liaison Manager
	PharmD, PhD, MD	MSM	Medical Science Manager
Eli Lilly	PharmD, MD, RN, MS	RMS	Regional Medical Scientist
		MSLM	Medical Science Liaison Manager

FIG. 11.15. MSLs Commercial & Scientific Functionality

acceditable educational programs, and for professional reasons, that is, to demonstrate to KOLs that their work is research and educational, fostering access and collaboration. This slide represents both qualitative and qualitative benchmarking used to measure field activities.

The Pharmaceutical Research and Manufacturers of America (PhRMA) represents companies and chief executive officers of the leading U.S. research-based pharmaceutical and biotechnology companies devoted to inventing medicines for longer, healthier, and more productive human lives. In 2002, under pressure from policymakers and the public over the cost of drugs, *PhRMA* instituted a new *code* that defines how drug industry representatives should interact with physicians and other health care professionals (Fig. 11.17) [50]. By taking the lead in addressing this issue, PhRMA identified ways that individual pharmaceutical companies can conduct their marketing practices and avoid concerns about inappropriate influence on the prescribing practices of physicians. The code covers business practices, for example related to clinical research, consulting agreements, business entertainment, and continuing medical education (CME) activities. The code has been voluntary accepted and widely distributed within the pharmaceutical industry, the physician community, and major public media. PhRMA has made a significant effort

- **Qualitative or Subjective Professional Influence:**
  - Clinical trial support
  - Increasing KOL advocacy ratings
  - Quality of sales force support
  - Hcp feedback award exposure to medial education
  - KOL Publication planing
- **Quantitative Assessment:**
  - Investigator initiated trials, protocols submitted; publications
  - Investigators identified & signed up for trials
  - Number of sites reviewed for clinical trials
  - Frequency and reach of educational programs
  - Number of speakers developed; Number of speaker programs
  - Formulary approvals
  - Number face-to-face KOL meetings
  - Increase in prescriptions post educational programs
  - Product knowledge tests
  - KOL customer surveys

FIG. 11.16. MSLs Professional Influence

- Covers business practices ranging from consulting agreements to entertainment to CME activities:
  - Interactions should ensure that health care professionals have latest, most accurate information available regarding prescription medicines.
  - Items given to health care professionals must not be of substantial value ($100 or less).
  - Gift certificates redeemable for medical books on patient care could be provided if they are not of substantial value. "Open ended" gift certificates are not allowable under PhRMA guidelines.
  - Marketing of products should ensure that products are used correctly for maximum patient benefit.
  - Adherence to PhRMA code is voluntary and enforcement measures are not included; however highest leaders of pharmaceutical company management are committed to the change.

FIG. 11.17. CME and PhRMA Code of Practice

to ensure there is a broad awareness and understanding of the principles.

The Office of Inspector General (OIG) of HHS promulgated its version of industry regulations that covers product purchases by government agencies and related transactions and relationships, "Compliance Program Guidance for Pharmaceutical Manufacturers," in April 2003 [1, 51, 52]. Fraud and abuse are the foci of legal action by the OIG. These latter activities, that can be related to product purchases, can have a major impact on clinical research and education of health professionals. Government agencies include Medicare and Medicaid programs and also the Veterans Administration and the military establishments. Basically, the OIG looks for undue influence on health care professionals (HCPs), particularly physicians and pharmacists and their administrative entities, who are involved in product decisions through any mechanism. The OIG makes recommendations to manufacturers about their relationships with HCPs and institutions and that the "PhRMA Code of Interactions with Healthcare Professionals" should be used and followed. It is not a "safe harbor" in the IOG legal language but will reduce risks of any problems for fraud and abuse. The OIG is aware that this guidance can have a chilling impact on good working relationships between manufacturers and HCPs regarding education and research; good faith efforts to comply and use of the PhRMA code will substantially reduce risks.

Arrangements between manufacturers or their vendor companies and HCPs include education, research, or clinical practice activities. Heightened scrutiny and criminal or civil action will occur if the following questions are answered with a significant suspicion of undue influence. Does the remuneration or inducement to a HCP influence clinical decision-making or the integrity of the formulary process? Can the arrangement with a company possibly create over utilization or increase inappropriate utilization? Can the arrangement raise concerns of patient safety or quality of care? Are the educational grants or research grants too closely linked to product marketing? Does the service provided by a company have a substantial independent value to the buyer (hidden discount)? What influence does the physician have on the business of the manufacturer? Is remuneration tied to volume or value of business, e.g., prescription activity? Is the payment (education grants, consulting fees, advisory fees, or research grants) within fair market value? Is there a written agreement (desirable) for the services provided by the HCP to the company, identifying need, nature of services, and fair market value? Are any gifts including entertainment, recreation, travel, meals, and payments of less than $100.00 in value and not tied to product decision-making and related activities? The manufacturers and the health care entities, and HCPs as well, must police themselves and incorporate appropriate operating procedures, education of staff, compliance programs, and investigations.

Medical Communications

As with traditional drugs, postmarketing requirements for biologicals include adverse event reporting, research to document commercial usage patterns, and clinical trials to evaluate extended patient populations. To better understand formulation or product safety, phase 4 trials often extend for years after a drug or biological product is formally FDA approved. Above and beyond the intrinsic value of a product alone, manufacturers and third-party vendors develop continuing medical education (CME) and education-oriented advertising and promotions that can serve as a "value add" to customers and help cement a health care partnership in any particular pharmaceutical marketplace. Promotional programs and materials such as monographs, books, slides, reprints, patient education pamphlets, and grants for in-services

11. Medical Affairs and Professional Services

- Publication planning of R&D trials
- "Compendia" support; R&D trials and data
- Sales & Marketing support:
  - Thought-leader programs:
    - Speaker training
    - Advisory boards
    - Investigator meetings
  - Scientific meeting support; R&D data identified & interpreted, especially competition
  - Scientific & sales tool development – technical review
  - Field sale straining – scientific information

Fig. 11.18. Medical Communications: Responsibilities

or demonstration projects are considered "on-label" activities and often require CBER screening and review.

The medical communication groups *responsibilities* support drug commercialization through the development and conduct of education symposiums; as medical advisor to speaker training, advisory boards, investigator meetings, and pharmaceutical compendia generation; publication planning; scientific meeting support and competitive intelligence gathering; field tool development and training resource (Fig. 11.18) [53–56].

The marketing team needs a medical and pharmaceutical resource to support their efforts, which can be the R&D organization to obtain company data and especially medical affairs through the communications group to help use and interpret data. Publication in peer-reviewed journals is an integral part of biomedical research. Frequently, publication planning meetings join together members from research & development, medical affairs (medical communications group), sales and marketing, outcomes research, and medical field professional liaisons. Advancement of clinical practice and maintaining timely publication on rationale for new product or disease state treatment in the best possible journal are the goals of any successful publication planning team. Clinical advances, safety issues, pharmacoeconomic outcomes, disease management algorithms, or simply updates to clinical trial data can foster physician discussions on a drug product or clinical marketing message and perhaps develop interests in clinical research. Publication timing is critical, especially in parallel with initial drug launch activities, such that sufficient information is available at launch to both create interest in the medical community, guide product use, and provide the sales organization with publications for dissemination to providers and payors. The publication plan needs to take into consideration the authors and their associated institutions, the journals (prestige, specialty vs. general medicine or pharmacy), the number and type of publications, original research and review articles, appropriate coverage of the new product's attributes and advances in patient care, pharmacoeconomic issues, on-label and off-label information,

potential impact on P&T committee decisions and treatment guidelines, regulatory issues, and timing of the publications (a steady stream) [11, 29, 57].

As "content and information experts," medical communication (MC) professionals are asked to travel in support of product and disease state marketing campaigns at major medical meetings and scientific congresses. Under DDMAC guidance, nonapproved, off-label scientific information can be provided upon customer request and must be separate from sales and promotional activities. Medical congresses often serve as excellent venues to unveil major clinical trial results and collect competitive information on drug products supporting similar disease states. Speaker meetings, scientific advisory boards, and investigator meetings are excellent venues to share and gather opinions on clinical trial results. As "content experts," medical communications can offer unique interpretation of company-sponsored publications and trial results.

To further assist dissemination of important clinical trial results, medical communications develops and trains medical field professionals on tools and scientific slide presentations and also advises on production of sales materials. For marketing assistance, medical communications often is responsible for collection and interpretation of competitive clinical data at medical congresses introduced at poster sessions, vendor booth activities, and evening symposiums. The marketing group is supported by the technical input from available medical information into both the marketing plans and the promotional and educational materials. Also, MC reviews and approves marketing materials to ensure accuracy of information, proper abstracting in context, and proper medical context for the marketing messages and the materials.

Compendia approval and access to monograph information are important information issues for hospital formulary consideration. Companies need to work on getting their product information into five key *compendia* listed in Fig. 11.19 that are used by health systems, in addition to the *Physician's Desk Reference*, which contains a collection of package inserts. Compendia information often extends drug information beyond approved package insert labeling to published nonapproved clinical trial literature and case reports. Medical communications shepherds the compendial applications for the company, collects clinical trial information to support monograph development, and assists education of compendia personnel should additional information be needed by them, when contact is permitted. Often these

- **Orange Book**
- **United States Pharmacopoeia (USP)**
- **USP Drug Information**
- **Martindale – The Extra Pharmacopoeia**
- **ASHP – Drug Information**

Fig. 11.19. Medical Communications (MC): Compendia

compendial organizations do not permit live contact with the company, except for the application and submission of the data. Independence is the goal of the compendial organization adding credibility for the monographs they generate for government agencies and other payors and institutions. Practitioners and payors need to be aware that a major time lag occurs between the date of approval and appearance of a product in the compendia, as much as 1 year or more, which can adversely impact patient care because compendial inclusion may be a criterion for reimbursement for product use with payors.

"The Orange Book"

The hard-copy Orange Book edition is published annually usually by March. It is a listing of drug products (original innovators and generic) with related information that have been approved for safety and efficacy through December 31st of the previous year. After the annual publication, there is about a 3-week lag after the month of approval for the monthly Cumulative Supplement publication to be published. The supplement contains the current month and previous months' cumulative changes since the annual edition was published. The Electronic Orange Book Query enables searching of the approved drug list by active ingredient, proprietary name, applicant holder, or applicant number. The data is updated with the publication of the annual edition or cumulative supplements. The current edition of the Orange Book and its monthly cumulative supplements are available from the U.S. Government Printing Office: Superintendent of Documents, Government Printing Office, P.O. Box 371954, Pittsburgh, PA 15250-7954. The telephone number to charge your subscription is 202-512-1800 or toll free 866-512-1800. The cost is $110.00 annually.

United States Pharmacopeia Drug Information®

USP DI® Drug Reference Guides are compiled and reviewed by USP advisory panels comprising more than 800 volunteer health care experts. Impartial, consensus-based, and reliable, USP-DI drug information is subjected to external/public review by hundreds of individuals from medical, pharmacy, nursing, and dental schools; state and national medical and pharmacy organizations; consumer groups; government agencies; and manufacturers. Features include U.S. federal government–recognized resource for drug utilization review (DUR), unlabeled use reimbursement, and patient education; current, accurate, in-depth data thoroughly reviewed, evaluated, and updated by clinical experts; a variety of electronic licensing opportunities enabling easy integration into a variety of applications; data relied upon and respected among international health care organizations and educators. The Centers for Medicare and Medicaid Services in the U.S. government use the USP-DI as a primary source for information, including off-label indications, and they will pay for such usage if the data appears in this compendium.

United States Pharmacopeia

United States Pharmacopeia–National Formulary (USP-NF) offers official national standards for drug substances and dosage forms (USP) and pharmaceutical ingredients (NF). Monographs provide information on molecular and structural formulas, molecular weight, chemical name, Chemical Abstracts Service (CAS) Registry Number, description, packaging and storage, identification tests, and chemical and physical properties. USP provides the latest FDA-enforceable standards of identity, strength, quality, and purity for prescription and nonprescription drug ingredients and dosage forms, dietary supplements, medical devices, and other health care products. It includes tests, analytical procedures, and acceptance criteria. The main edition of USP-NF is published every November and becomes official January 1 of the next year.

Martindale: The Extra Pharmacopoeia

A comprehensive British publication that provides well-referenced information on pharmacology, toxicology and therapy of foreign and U.S. drugs (prescription, nonprescription, obsolete and investigational drugs, and natural substances). It includes nonapproved indications and chemical and physical data.

AHFS–Drug Information Essentials

The American Society of Health-System Pharmacists publishes their drug reference book, *American Hospital Formulary Service Drug Information*, on a regular periodic basis in order to publish up-to-date information. A rigorous evidence-based independent editorial process is used by ASHP staff; the medical pharmaceutical literature is used. On-label and off-label information is provided in efficacy, dosing, and side-effect areas. A series of product monographs is created with the following standard sections; classification and brands, uses, dosage and administration, cautions, interactions, pharmacokinetics, stability, actions and spectrum, advice to patients, and preparations.

Nearly half of physician continuing medical education (CME) is funded by the pharmaceutical industry at a price tag of nearly $1.1 billion in 1999 according to the Accreditation Council for Continuing Medical Education. For new products, CME is an indispensable component of the launch process to educate the medical community, and much work is done in the periapproval time frame (years −2 to +1 of launch). Because of all the new laws and regulations circa 2000 and PhRMA guidelines, the medical communications group takes the lead in developing and/or coordinating the medical education programs, of course consistent in general with the marketing plans. CME is a key component of product development; the clinical trial information is worth very little until a company can maximally reach the thought leaders and potential prescribers. From the industry, two important sources of information on therapeutic products for health care professionals are (1) activities (programs and materials) performed by, or on behalf of, the companies that market the products; and (2) activities, supported by companies, that are otherwise independent from the promotional influence of the supporting company. Specific criteria exist to assess the *appropriateness* of an *educational event* (Fig. 11.20), and reflect societal, professional, government, and industry guidelines.

The number of educational events is a bit staggering. In 2003, PhRMA companies sponsored 454,038 educational events, costing $2.5 billion, including varied locales, for example, dinner meetings (no.1 event), physicians' offices, hotels, hospital, research facility, and conventions [58]. More than $100 million was spent in 2003 on education by seven companies (Pfizer, GlaxoSmithKline, Merck, Lilly, AstraZeneca, Novartis, and Johnson & Johnson). The differentiation from independent educational events and more promotional yet still educational events are listed in this figure. Physician participation has been estimated to be four per month for promotional educational meetings. About 50–60% of physicians attend consultant meetings, and about 25% attend dinner meetings. Prescribing behavior was changed or the relevance of the meeting was rated high by physicians for 45% of participants in consultant meetings and 30% of dinner meetings in a 2003 survey [59].

Although both independent and promotional educational events provide valuable and sometimes vital information to health care professionals, the programs and materials performed and disseminated by companies are subjected to the labeling and advertising provisions of the Federal Food, Drug, and Cosmetic Act, whereas the truly independent and nonpromotional industry-supported activities have not been subjected to FDA regulation. CMS observes that CME can include certain types of information that cannot be promoted by drug sales representatives, such as off-label uses of marketed drugs and efficacy information on investigational drugs. Pharmaceutical companies have been very strong supporters of effective strategies to educate physicians. A wide variety of CME approaches are available, such as advertising in professional journals, meeting with individual physicians, and sponsorship of organized CME.

The FDA recognizes the important role accrediting organizations can play in ensuring that industry-sponsored educational activities are independent and nonpromotional. The agency also recognizes the importance of avoiding undue government interference in postgraduate and continuing education for health care professionals, as the agency seeks to ensure that company promotional activities meet applicable legal requirements. Thus, the agency will continue to work with major accrediting organizations to monitor company-supported educational activities conducted by their accredited providers.

Written agreement between the CME provider and the supporting company is encouraged and should reflect responsibilities for designing and conducting the activity, and that the activity will be educational, nonpromotional, and free from commercial bias. Although not required, a written agreement, coupled with the factors described in Figure 11.20, can provide valuable evidence as to whether an activity is independent and nonpromotional [12, 50–53].

Advisory board meetings are commonly held with KOLs for the company to obtain expert feedback about research plans, products, and marketing strategies, plans, or materials. Key questions of high importance can be addressed affecting future research opportunity or marketing success: Is the research or marketing medically appropriate? What is current state of therapy or research for a particular disease? Is the study design appropriate for the question being posed? Is the

---

Criteria for Independence and Educational and Non-promotional Value

1. Control of Content and Selection of Presenters
2. Disclosures: company funding; relationship of provider, presenter, or moderator & supporting company; discussions around unapproved uses of products
3. Focus of Program: intent of company and provider - independent and non-promotional, focus on educational content, & free of commercial influence/bias
4. Relationship Between Provider and Supporting Company
5. Provider's Involvement in Sales
6. Provider's Compliance to Standards of independence, balance, objectivity, or scientific rigor
7. Multiple Presentations of same material
8. Audience Selection: intended to reflect marketing goals (e.g., reward high prescribers) vs meaningful medical discussion
9. Opportunities for Discussion
10. Dissemination of Information beyond program
11. Ancillary Promotional Activities: sales presentations or promotional exhibits
12. Complaints regarding attempts by supporting company to influence content

FIG. 11.20. MC: CME Assessment Issues

program targeted at the right patients or providers? What information is missing? As stated previously, the PhRMA code [50], OIG guidance [51], and the personal services and management contracts safe harbor in the Code of Federal Regulations [60] all must be followed for the legal, ethical, and professional benefits of the health care provider serving as a consultant and expert and the company. A map for operational and legal success in using KOLs as advisors has been proposed for advisory boards: (1) establish business purpose (why, who, what, where, and how), (2) develop the content (expectations, information, experts, feedback), (3) submit program for approval (legal, medical, and regulatory at the company), (4) establish feedback mechanisms (expectations, guides, surveys, forms), (5) follow-up (corporate and advisors), (6) outcomes (modifications of research, messages, strategies, and tactics; base of advisors, investigators, faculty). The key focus is the feedback being solicited, tabulated, and actually used by the company [61]. Participants need to avoid any conflicts of interest between their pharmaceutical company consultation and their work in their practice in patient care, their research, their health care company relationships, and any societal or professional endeavor (e.g., guideline work, committees in societies, committees at their health care institution) [62].

A key role for medical affairs department, usually in the medical communications group, is preparing the information dossier for a new product to obtain formulary approval for its use at an institution or health system, such as managed care organizations, prescription benefit managers, hospitals/medical centers, or health insurance companies. In 2000, with updates in 2002 and 2005, the *Academy of Managed Care Pharmacy (AMCP)* has created a very specific submission dossier in format and content, as well as process, to assist health organizations and manufacturers in the decision making for drug coverage and for adding new products to institutional or system formularies of approved drugs for use (Fig. 11.21) [63]. The goals are ensuring more efficient process, identifying and developing necessary expertise in health systems, focusing on value of a product, improving economic models, and developing trust between industry and pharmacy department managers.

The information requested by the institution is exceptionally complete as represented on Fig. 11.21, including off-label information. Although AMCP expects the dossier to be done and ready at launch, the institution must make an "unsolicited" request to a manufacturer for this formulary dossier in order to obtain the off-label information. The off-label information for an institution or health system helps minimize surprises in product usage beyond the approved labeling, which is a common occurrence in clinical medicine. Information on the competitive product situation is also requested. The figure lists the three major sections of a dossier, that is, the detailed product description, the complete clinical and outcomes studies and data section, and the economic modeling section. Spreadsheets are a key component of the AMCP format for the dossier, wherein the clinical studies for efficacy and safety are summarized in a table and a separate one for the outcome studies. An overall summary focuses on overall value and cost to patients and the health care systems. Also, confidentiality is a major concern for a company for off-label data and studies, which may compromise a future regulatory submission or help a competitor already on the market by being aware well in advance of data on a new product still under study. The details of the submission are well reviewed in a June 2005 supplement to the *Journal of Managed Care Pharmacy* [63].

Another major element of the AMCP formulary dossier is the added focus on outcomes and value, requesting very specific types of data, study types, and economic models. The good news for a company is they know what to expect for questions in these areas in advance of approval and what has to be generated. The studies are in more realistic settings of product use wherein normal phase 3 trials usually are not performed because of all the scientific and regulatory requirements of the pivotal trials. The bad news for a company is that much of this work is asked to be done before approval, adding cost and possibly more time to the drug development process for expensive studies in various settings.

Clinical development for product approvals creates clinical information based on the company-sponsored studies, which is used by the marketing division to promote the product, especially the claims of efficacy (indications), and the safety profile. The FDA has an oversight role for the appropriateness of this information used in advertising, to ensure public safety and avoid misleading claims or information being disseminated. Many companies will have the advertising materials preapproved before use, but it is not required to be preapproved. The role of medical communications is to work with the marketing team as technical experts (an advisory role) in content of promotional materials and technical reviewer (an oversight, quality-control role). However, marketing teams can be creative on how they use the information with their advertising agencies, and sometimes the material may contain either exaggeration or inadequate information in their

- **Product Section:**
  - Product description (detailed overview, full labeling, on- & off-label indications)
  - Place of product in therapy
  - Pharmacogenomics
  - Comparative product profiles
- **Clinical Information Section**
  - Safety & efficacy trials
  - Off-label uses
  - Effectiveness and other data
  - Outcomes studies
  - Evidence table spread sheets (2) for both efficacy and outcomes studies
- **Modeling Report:**
  - Health consequences & costs combined
  - Cost-effectiveness analysis
  - Perspective
  - Time horizon
  - Discounting
- **Product Value & Overall Cost Section**
- **References & Appendicies**
- **Formulary Process:**
  - Requested by Institution
  - AMCP format required
  - Available at launch
  - Company contact information
  - Confidentiality

FIG. 11.21. MC: Formulary Submission (AMCP)

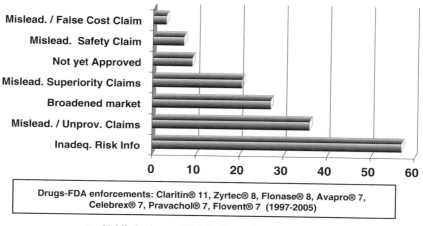

Fig. 11.22. MC: Product Promotion & FDA Enforcement
Source: USA Today May 31, 2005, 4A

interpretation of the data. This newspaper report in 2005 on *drug promotions and FDA enforcement* highlights some of the problems that the FDA had investigated and then sent out regulatory letters to the respective company to correct the false, misleading, or inadequate promotional information that had been used and needed to be corrected or stopped (Fig. 11.22) [64]. In 2000 to 2004, 220 FDA letters were sent to drug companies that cited them for violations of the marketing rules for their promotional materials. The most commonly offending products also are listed in this slide; asthma drugs were four of the seven offending drugs. Inadequate safety information (57%) was the top single problem. Efficacy related claims were more common when you note the variety of ways claims were exaggerated; 36% unsubstantiated or misleading claims, 27% too broad of a market, 23% unproven or misleading claims of superiority, and 9% marketing for unapproved use. The FDA has increased the penalties over the past few years, requiring major corrective action more frequently that can include corrective advertisements or "Dear Doctor" letters to all potential prescribers. The industry is responding with the development of voluntary guidelines for television and other consumer ads. Although voluntary, they set a significant standard of behavior that companies should follow carefully or else suffer the wrath of public and regulatory actions.

Risk management

Goals are to Maximize Benefit and Minimize Risk

Today's health care products are developed and used within a complex system involving a number of key participants. As illustrated in this figure for risk mamagement. Fig. 11.23 participants include (1) manufacturers who develop and test products and submit applications for their approval to the FDA; (2) the FDA, which has an extensive premarketing review and approval process and uses a series of postmarketing surveillance programs to gather data on and assess risks; (3) the health care delivery system, including its many providers and elements; and (4) patients, who rely on the ability of this complex system to provide them with needed interventions while protecting them from injury. In many cases, the roles of the participants in this system evolved independently, and in some cases, the roles are not clearly defined. Collectively, risk management is the phrase used to capture the people, processes, information, and outcomes dealing with the risks in using pharmaceuticals with patients in the health care system [65].

The choice to use a drug, biological product, or device involves balancing the benefits to be gained with the potential risks of using a product. As illustrated in Fig. 11.23, an elaborate system has developed in the United States with the goals of maximizing the benefits and minimizing the risks associated with using medical products. Under this system, medical products must undergo FDA approval before marketing. FDA's premarketing review involves (1) developing criteria for the evidence of product safety and effectiveness that manufacturers must submit to FDA, and (2) evaluating the data manufacturers submit to see if the product meets the statutory standard for market approval. Briefly, the system works as follows. After a systematic development process that includes clinical trials, the manufacturer submits an application to the FDA for approval. After a thorough review of the data, FDA makes a decision to approve or not approve a product to treat a specific condition, based on a benefit-risk analysis for the intended population and use. Although medical products are required to be safe, safety does not mean zero risk, as all medical products are associated with risks. A *safe medical product* is one that has reasonable risks, given the magnitude of the benefit expected and the alternatives available. One result of

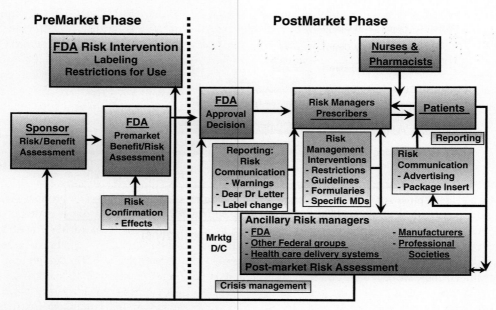

Fig. 11.23. MC: Risk Management Pre - & Postapproval

FDA's premarketing evaluation of a new product is the approval of its labeling. The labeling must indicate which patients are appropriate for treatment, identify the product's potential adverse side effects, and explain how the product should be used to maximize benefits and minimize adverse side effects. Risks are categorized into adverse events, precautions and warnings in product labeling.

Once approved, products move swiftly into the marketplace for use by prescribers and patients in much larger numbers and much more variable situations than during its clinical investigation. As shown in Fig. 11.23 on balancing risks and benefits, after FDA evaluates the risks and benefits for the population, the prescriber then is central to managing risks and benefits for the individual. In addition, patients make decisions about treatment choices based on their personal valuation of benefits and risks. In the context of an individual treatment decision, FDA's role in reducing risk involves ensuring that accurate, substantiated, and balanced information about a product is available to the prescriber and the patient. This system, when functioning well, succeeds in managing a balance between benefit and risk. But FDA's mission to ensure the safety of medical products cannot be accomplished without effective partnerships with health care practitioners, the companies, and the public.

FDA also operates postmarketing surveillance programs intended to identify unexpected risks of approved products. When new risks are identified in a medical product, the manufacturer adds them to the labeling, or, if serious enough, they may trigger an agency reevaluation of the approval decision. Recent concerns about the safety of medical products have focused on several types of risks. For newly approved products, concerns have centered on unanticipated side effects that emerge after a product is on the market. In addition, concerns have been raised about FDA's ability to ensure the appropriate use of regulated products in medical practice. For example, how far should the agency intrude into traditional areas of medical practice when the safe use of a product requires practitioner training, or frequent patient blood testing? Is the agency responsible when a medical product is used beyond the parameters of the approved labeling? Some reports have focused on the human and economic costs of medication errors, while others are concerned about serious adverse events that have occurred even when a medical product has been used appropriately. Because each of these types of risks has a different source, effective management of each is likely to be different. To understand the complexity of managing the risks associated with using medical products, it is important to understand the different types of risks and their sources.

- Product defects
- Known side effects:
  - Avoidable
  - Unavoidable
- Unexpected side effects
  - Long-term side effects
  - Off-label effects
  - Effects in unstudied populations
- Medication/Device errors
- Remaining uncertainties

Fig. 11.24. MC: Adverse Events, Types of Product Risks

What are the risks involved with using medical products? In general, the sources of medical product risks can be thought of as falling into the following five categories: (1) product defects, (2) known side effects, both avoidable and unavoidable, (3) unexpected side effects (4) medication or device errors, and (5) remaining uncertainties. When using a medical product results in a patient's serious injury or death, the patient is said to experience a serious adverse event (Fig. 11.24).

Product Defects

Historically, product defects have been an important source of medical product–associated injuries. A significant portion of FDA's resources are currently devoted to regulating product quality. Although additional resources are needed to maintain and enhance current oversight activities, the risks associated with defective medical products are relatively well managed. FDA research, surveillance, and inspections form the cornerstone of FDA efforts to keep product defects to a minimum. The risks associated with poor product quality are not the subject of this report.

Known Side Effects

When using a drug or other medical product, a patient runs the risk of experiencing reactions resulting from the product's interaction with the body. For pharmaceuticals, these reactions are commonly termed "side effects". They usually have been identified and are indicated as possible risks in a product's labeling. Known side effects are the source of the majority of injuries and deaths resulting from product use. During product development and the premarketing review process, manufacturers and the FDA focus on identifying and understanding this very large category of risks. The risks must be identified, described, and measured before a sound overall risk–benefit decision can be made on the product's approval. After approval, product labels describe how to select patients, how to select and modify the dose schedule, how to avoid interacting treatments, how to monitor for toxicity, and what measures to use to avoid or mitigate toxicity. If additional side effects are identified during the postmarketing phase, the manufacturer changes the product's labeling information to reflect these possible side effects.

Avoidable Side Effects

Some known side effects are predictable and avoidable. To avoid them, the health care practitioner must select the best treatment and plan for the appropriate measures to manage the risks, for example, patient hydration for products that are toxic to the kidneys, or dose adjustments for patients with impaired kidney function. A medical practitioner can choose the wrong therapy for a specific condition (e.g., using antibiotics for viral infections). Alternatively, a practitioner may prescribe the appropriate therapy but fail to individualize the therapy or monitor the patient for signs of toxicity. Examples of avoidable side effects include the consequences of known drug–drug interactions or side effects caused by prescribing an inappropriate dosage in the elderly. Communicating the potential for these types of risks to health care practitioners and explaining how to minimize them are major goals of product labeling. Occasionally, to further reduce such risks, additional restrictions are placed on the use of a product, its availability, or its promotion. But, generally, existing regulatory controls are intended to provide the necessary information to the product users who rely on them to use the product safely.

Problems resulting from poor product selection by a practitioner can be reduced by interventions, such as targeted medical education and/or national or institutional guidelines for disease management or treatment algorithms, but are largely not amenable to FDA action. Reducing the risks related to poor product use requires collaboration by the manufacturer, the FDA, health care professionals, the various components of the health care delivery system, and patients.

Unavoidable Side Effects

In many cases, known side effects are unavoidable, even with the best medical practice because they can occur even when a product is used appropriately. Although estimates vary, the overall human and economic costs of unavoidable side effects are high. The risk of experiencing such side effects is the inevitable price for gaining the benefits of treatment. Superinfection after antimicrobial chemotherapy, fatigue and depression from interferon use, and bone marrow suppression from chemotherapy are common, predictable, and usually unavoidable side effects. Successfully managing these risks centers on ensuring that both the practitioner and patient are fully aware of the risks involved in treatment and that the patient is carefully monitored.

Medication or Device Errors

Medication or device errors involve the incorrect administration of the prescribed product or incorrect operation or placement of a medical device. Errors also can involve the unintended substitution of the wrong product for the prescribed product. Errors arise, for example, when a confusing product name results in the wrong product being dispensed, or when inattention results in an overdose of an intended drug. Substantial numbers of injuries and deaths occur annually from medication or device errors. In general, these errors are believed by experts to result from systemic problems, rather than from a single individual's mistake. Such errors are not totally preventable but can be minimized through interventions to the system.

Many outside organizations are involved with identifying and reducing medication errors. In its final report, The

President's Advisory Commission on Consumer Protection and Quality in the Health Care Industry called on interested parties to jointly develop a health care error reporting system to identify errors and prevent their recurrence. As a result, the Quality Interagency Coordination (QuIC) committee was formed on March 13, 1998. In addition, The Institute of Medicine within the National Academy of Sciences issued a report in 1999 on preventing medication errors.

In light of a series of highly publicized major adverse events in hospitals during 1995, the Joint Commission on Accreditation of Healthcare Organizations (JCAHO) reexamined existing processes for evaluating and monitoring such events and established a more consistent approach to reducing the likelihood of medication errors in all types of health care organizations. By 1996, JCAHO had established a sentinel event reporting policy that would provide for (1) a safe harbor context to encourage the self-reporting of sentinel events, (2) the establishment of a database of such serious events to determine their demographics and epidemiology, (3) the sharing of this aggregate information among health care organizations, (4) the continuous development and dissemination of information about the sentinel event causality through root cause analysis, and (5) an emphasis on the concept of prospective systems analysis to minimize errors and protect against the effects of errors through improved design and redesign of health care processes and systems.

In 1995, a National Coordinating Council for Medication Error Reporting and Prevention (NCC MERP) was formed. The NCC MERP is a collaborative effort to (1) increase awareness of medication errors and methods of prevention; (2) stimulate reporting to a national system for review, analysis, and development of recommendations to reduce and prevent medication errors; (3) stimulate the development and use of a medication error reporting and evaluation system; (4) examine and evaluate the causes of medication errors; and (5) develop strategies relative to system modifications.

In 1997, the American Medical Association (AMA) announced the formation of the National Patient Safety Foundation (NPSF). The NPSF is a collaborative effort in pursuit of three goals: (1) serve as an educational forum for building awareness among providers and the public about patient safety, errors in health care, and preventive strategies; (2) support new research designed to analyze risk factors in health care to develop practical tools and solutions; and (3) serve as a clearinghouse for research information, best practices protocols, and preventive tools regarding patient safety risk factors.

In December 1997, the Health Care Financing Administration (HCFA) proposed "Conditions of Participation in Medicare and Medicaid" that would require hospitals to routinely monitor for adverse drug events and medication errors.

The Institute for Safe Medication Practices (ISMP) and the United States Pharmacopeia (USP) operate a voluntary medication error reporting system (MERS). In addition, the USP has recently introduced MedMARx, an Internet-accessible database software program designed to anonymously report, track, and benchmark medication error data in a standardized format for hospitals nationwide.

During the premarketing review process, FDA works to reduce the risk of medication and device errors by evaluating product design and packaging, reviewing product names, and reviewing product labeling, dose, and dose modification instructions. In the postmarketing period, FDA is taking a more active role in attempting to identify common use errors and in developing strategies to reduce those errors. Examples of these efforts include the publication of safety alerts, public health advisories, guidances, brochures, and other educational information.

Unexpected Side Effects

Unexpected side effects are those that were not identified as potential risks prior to product marketing. The contribution of serious adverse events resulting from unexpected side effects to the overall rate of serious adverse events is relatively small. Working together, manufacturers, clinicians, and the FDA have created an elaborate product development and premarketing review system to identify risks prior to marketing and thus minimize the occurrence of unexpected side effects. This system enables most of these types of risks to be identified.

There are risks, however, that are difficult to identify before a product goes on the market. Some very rare, serious, or life-threatening side effects may be recognized only after marketing. These rare side effects are not usually identified during medical product development because they happen so infrequently. As is the case with a medication or device error that results in injury or death, serious adverse events resulting from unknown side effects often gain widespread media attention because they are less acceptable to the public than injury resulting from known side effects. When these kinds of serious adverse events happen, they lead to questions about the quality of FDA's premarketing review process.

Long-term Effects

Another type of uncertainty relates to the long-term outcomes of many medical interventions, including pharmaceutical or device interventions. Because long-term studies to assess these types of risks usually are not required prior to product approval or for continued marketing, considerable uncertainty exists about long-term side effects (particularly in the chronic disease setting). Pharmaceuticals, in particular, may provide short-term benefits but may be associated with increased mortality or other serious long-term injuries.

Effects of Off-label Uses

Marketed products frequently are used to treat conditions that were not studied during clinical development (i.e., off-label uses). When products are used *off-label*, there is usually greater uncertainty about both the benefits and risks because less information on safety and effectiveness is available. Unexpected adverse events may occur in this context.

Effects in Populations Not Studied

Some groups (children, pregnant women) may not be studied before marketing. Additional uncertainties about risks (and benefits) occur with use in unstudied populations (see http://www.fda.gov/oc/tfrm/Part1.html).

Remaining Uncertainties

Given current scientific and medical knowledge, it is not possible to learn everything about the effects of a medical product during its research and approval. For example, new information about long-marketed products (e.g., digoxin) often becomes available as a result of further scientific study or new technologies. Therefore, a degree of *uncertainty* always exists about both benefits and risks from medical products. Several types of uncertainties exist.

Outcome and Pharmacoeconomics Research

Pharmacoeconomic trials are not required for the drug approval process by regulatory agencies. However, these types of economically focused studies are paramount to discussions with purchasers and government agencies, given that in 2003, U.S. health expenditures equaled $1.5 trillion, about 14% of gross domestic product. Pharmacoeconomics is the study of net economic impact of pharmaceutical selection and use on total cost of delivering health care. It is a more global perspective (balanced the cost and consequences of treatment) and actually involves health care utilization (efficiency) versus efficacy and safety. The focus is on value (a multivariate concept), that is, costs and consequences, plus clinical, economic, and humanistic (QOL) dimensions.

A variety of pharmacoeconomic studies may be developed to evaluate the health *and* financial implications of drug therapy (or lack thereof). These studies are generated from an applied science necessitating different and broader clinical and cost data versus typical clinical trials. For example, some new biotech agents clinically advance therapy for untreatable diseases but without generating significant economic offsets. So what is the incentive to add these types of drugs on a hospital formulary? Often, pharmaceutical advances offer improvements in humanistic and qualitative measures of life, such as a significant impact on patient quality-of-life or daily functional status. These types of soft measures often offer economic impacts to a society as a whole and on occasion offer consideration for health policy debate. Pharmacoeconomics studies offer another perspective to the value of health care, and often much of this work occurs after product approval [1, 66–72].

Goal: Establish more global **value** of products to health care system, as well as patient care

- Product decision-makers AND Policy-makers
- Health care organizations
- Disease management companies
- Pharmacy benefit management companies
- Federal health care programs (including Tricare and Department of Veterans Affairs)
- CMS, especially Medicare, Part D
- Compendia
- P & T Committees

FIG. 11.25. Outcomes Research - Customers

Outcomes research as suggested in Fig. 11.25 is targeted for a variety of different audiences than a typical clinical trial. The latter is focused primarily on the provider and regulatory agency, as well as compendia and the committees that determine formulary status. Therefore, the focus is on efficacy, dosing, and safety in the medically appropriate patient group. Outcomes and pharmacoeconomic studies also are of interest to providers, but also a much more broad group of people and health care entities, as listed in this slide, both product and policy decision-makers desire these data. They strive to understand the full impact of a new product on health care including costs of care. In government, there are policymakers responsible for product access decisions, such as Medicare and Medicaid in the U.S. and in most European and Asian countries, a drug pricing body. Among payors, we have insurance companies and self-insured companies. In health care entities, we have various gatekeepers for health care, such as pharmacy benefit managers, managed care organizations, group purchase organizations, and pharmacy and therapeutics committees, making decisions about product availability. Each of these constituencies possesses a uniquely different perspective with different desires, the patient versus provider versus payer, and inpatient versus clinic/physician office versus managed care settings.

The variety of needs for various types of pharmacoeconomic (PE) studies, the unique design requirements, the added cost to drug development, and their possible effect on product approvals impact the timing for conduct of these types of *PE studies* (Fig. 11.26). At the time of approval, some key data demonstrating health care impacts generally is needed in working with the various gatekeepers for product access as suggested above. However, the many settings and perspectives for pharmacoeconomic research creates a need for several different studies possibly to be done, which likely would occur in the postmarketing period and would be done often by the medical affairs group with expertise in pharmacoeconomics.

As explained in the chapter on study types, cost studies may be cost-minimization, cost-effectiveness, cost–utility, or

- Timing – pre- & post-drug approval
- Types of data and studies:
  - Pharmacoeconomic: Cost - Benefit, - Utility, - Effectiveness, - Minimization
  - Productivity, Days of work lost
  - Quality-of-Life/Outcomes
  - Patient preferences, Willingness-to-pay
  - Drug (DUE) or Morbidity (MUE) Utilization Evaluation
- Investigator Initiated trials
- Limits of PE in R&D:
  - Data availability
  - Impact on registrational trials
  - Applicability (extrapolation) of results
  - Cost
- Benefits – Real world data; demanded now by payers

FIG. 11.26. Types of Outcome/PE Research in Med. Aff.

cost–benefit studies, each with design and data requirements, again distinct from clinical trials. Specific trials may be appropriate for perhaps productivity parameters, patient or provider preferences, willing-to-pay feedback, and quality of life parameters. Quality of life studies focus on humanistic issues versus standard clinical parameters. The study design demands for PE significantly differs from the standard RDBPC clinical trials for approval; for example, the typical patients in a managed care organization, flexibility in dosing, different end points, and cost data for the components of care for drugs and many ancillary services. Most of these data are not part of the registrational trials, such that these data will be need to be collected later most likely in the phase 4 period so that product registration and approval are not held up. Also, the phase 2 or 3 patients are not representative of settings for these gatekeepers (e.g., managed care or clinic environments). A question of applicability of data is important. In working with health care entities, some will desire medication utilization review or morbidity utilization review studies to be done as part of their acceptance to use the products in their institutions. The cost of all these trials is substantial, adding to product development costs, such that a revenue stream from product sales helps justify more PE studies done later in the product life cycle. As can be readily observed from these types of studies, a different research discipline exists requiring its own research expertise and staff for a company. The benefits of the PE studies is collection of real-world data that is demanded by payors, followed by successfully obtaining reimbursement for the product or achieving access for the product with various populations of patients.

Figure 11.27 demonstrates the factors in the PE area that are of interest to payors and gatekeepers that will impact *drug coverage* decisions. With this four-point scale, seven factors achieve a 75% level of desire for such information, including drug cost, the condition being treated, contracting issues regarding product purchase, comparative drug information, physician demand, short-term medical savings, and patient demands. The other three factors are desired as well but just not as much: long-term medical savings, health-related quality of life, and nonmedical costs [73].

One example is provided for an outcomes-type *pharmacoeconomic study* (Fig. 11.28). This study focuses on sumatriptan for migraine headaches and its occupational impact on work. It is a cost-effectiveness study from both the employer

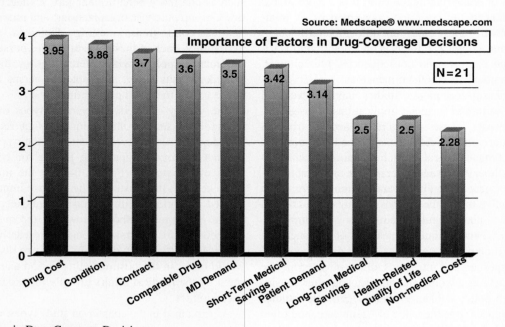

FIG. 11.27. Factors in Drug Coverage Decisions

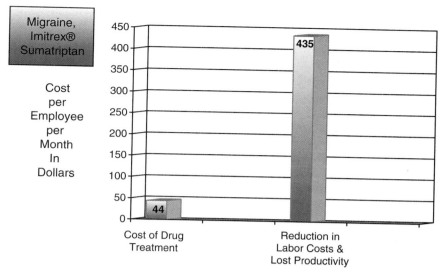

FIG. 11.28. PE Research Study Example in Migraine
Source: Legg RF et al. Cost Benefit of Sumatriptan to an Employer. J Occupation Environmen Med, 1997;39(7).

and patient perspectives. Labor costs and the cost of lost productivity were assessed, as well as drug product costs. The cost to treat migraine headaches was $44 per employee per month. Product-related benefits were reduction of labor costs and a decrease in lost productivity, saving $435 per employee per month for the employer [74].

Late-Phase Clinical Trials

"Late-phase" pharmaceutical clinical trials can be bucketed into studies designed to support current product labeling *(phase 4)* or clinical trials intended to explore additional product indications *(phase 2)* for a marketed product. Phase 4 studies are valuable postmarketing clinical trials designed to better understand a targeted disease marketplace, compare safety and efficacy between patient subgroups, and/or capture adverse events across a broader commercially available patient population. In general, the studies are intended to expand the knowledge about the product in more practical settings, expand the variety of investigators, and be published. Some studies will be required during the postmarketing period by the regulatory authority as part of the approval, usually addressing adverse event issues, or, in cases of expedited applications, continuing key studies to elaborate upon efficacy, especially if surrogate markers were used initially, and safety [75].

The scope, number, and types of studies in the medical affairs' *phase 4 study plans* have a panoply of factors that need to be considered (Fig. 11.29). Regulatory input includes the negotiated final labeling, giving the boundaries of the work regarding indications and dosing, and any study prerequisites for approval, voluntary by the company or dictated by the regulatory authority. Company plans are manifold that influence phase 4 plans; clinical study plan, life cycle plan, marketing plan (target audiences), and publication plans. The product's profile includes uses, doses, adverse events, formulation, dosage and administration, which will have advantages or limitations to explore. KOLs need to be brought into the loop, as they will generate study ideas, study design issues, possible investigators, practical health care issues, and bring an external perspective and prioritization. Further patient care needs, that is, unmet medical needs, will be identified by the company or advisors. Competitive products for the same indication will offer issues for comparison, or their labeling will suggest topics to study. Payors and health care systems will request information that may yet not be available and necessitate

Late Phase Development (Phase 3B & Post-Marketing)

- Scope of proposed labeling of new product:
  - Indications & side effects especially
- Commitment of company
- Overall R&D research plans and Life cycle plan
- Marketing plans (KOLS, Publication plans)
- FDA approval requirements
- Product profile needs
- Investigator interests and needs:
  - Advisory boards of KOLs
- Competitive situation
- Pharmacoeconomic needs
- Budget & staffing limits

FIG. 11.29. Phase 4 Studies in MA: Factors Influencing Planning pharmacoeconomic or quality-of-life studies, which will encourage formulary consideration. Existing treatment guidelines may necessitate specific studies to help in the product's integration into the choices in guidelines. The company needs expanded experience with the product in real-life settings to understand how the product will be used [75].

To better understand usage issues, formulation, or product safety, phase 4 trials often extend for years after a drug or biological product is formally FDA approved [54]. Often, a potential investigator at a medical center will have a specific research idea that will be proposed to a company for a research grant, called a "investigator-initiated" trial. *Types of phase 4 research* areas are quite extensive, creating many opportunities for follow-on research and important publishable studies to enhance product use and safety (Fig. 11.30).

- Patient subgroups for efficacy and safety, based on age, sex, disease subtype, disease severity, disease duration, concurrent disease, or pharmacogenomics;
- Special populations; pregnancy, pediatrics, geriatrics;
- Dosing (induction vs. maintenance; varied dose levels; divided doses; longer duration); administration; stability;
- Pharmacokinetics (ADME in special situations, e.g., renal failure; patient subgroups, impact of food/meals);
- Pharmacoeconomics in new health care settings (hospitals, clinics, managed care), with different perspectives (payor or employer or provider or patient), with study designs that mimic normal practice situations;
- Concurrent drugs—added efficacy or safety, or drug interactions;
- Adverse events for specific issues and postmarketing surveillance of drug during its use in typical clinical situations;
- Patient registries, wherein data is collected on patients with a disease and their treatments and demographics or patients receiving a specific treatment to understand the types of patients;
- Usage trials (real-life settings) that document the drug's typical usage in dosing (amount, schedule, duration), patient description, patient groups, and/or concurrent drugs.

Phase 2 studies are intended to expand upon existing approved product insert labeling. Such trials are exploratory in nature and can be company sponsored or initiated by a potential investigator. The studies tend to be small and focused on a specific issue; if interesting findings occur, they may be expanded. These types of trials are traditionally conducted according to the rigors of good clinical practice guidelines.

Planning and implementing the phase 4 research program from an operational perspective should be done well in advance of the product approval, such that the program will be ready to be up and running at launch. About 2 years before the expected approval, the plans, organization, staffing, budgeting, and processes are prepared, which often will take a year to complete (Fig. 11.31). The rough plan is created as early as the end of a favorable phase 2 program and is edited at the end of phase 3 as more data is provided and opportunities are better understood. At NDA filing, the program is put into place such that staffing can be done, procedures written, advisory boards convened, study ideas entertained, protocols written, investigators identified, and contracts agreed to. Hence, launch day for the commercialization of the product is the official launch of the phase 4 program as well.

Internal organization must be worked out with R&D, regulatory and legal, human resources, and marketing. People must be recruited, hired, and trained. A variety of key operational processes need to be developed for investigator and site assessments, protocol reviews, study contact, monitoring and reporting, study budgeting, program budgeting, IRB approvals, adverse event collection and analysis, field and internal communications, drug/product provision and tracking, record keeping, and study abstract and report generation.

- Investigator initiated vs Company sponsored
- Types of research areas: (Phase 4, within labeling):
  - Patient subgroups based on age, sex, disease type, disease severity, disease duration, & concurrent disease
  - Dosing (induction vs maintenance, varied dose levels, divided doses, & longer duration); Administration; Stability
  - Pharmacokinetics; Pharmacoeconomics; Pharmacogenomics
  - Concurrent drugs – added efficacy, or drug interactions
  - Adverse events & Postmarketing surveillance
  - Patient registries
  - Usage trials (Real-life settings for dosing, safety)
- Exploratory phase 2 trials:
  - Outside labeling
  - FDA approvals, added reporting (especially safety)

FIG. 11.30. Types of Medical Affairs Research

- Timing of P.4 planning: @ P.2, P.3, NDA, annually
- Internal organization, planning, andi mplementation processes:
  - Review process: participants, decision-makers, integration with corporate process, and budgets
  - Implementation: Timing, staffing, budgeting, communications (PIs)
- Staffing needs (at company):
  - Number & complexity of trials, budget size, Co. sponsored & PI-initiated, monitoring, data collection & analysis needs, report writing, and publications support
- Departmental budgeting criteria:
  - # trials, # PIs, # patients, geography, and staffing
  - Monitoring complexity, data analysis, and importance of studies

FIG. 11.31. Planning & Implementing MA Clinical Trials
Source: Evens RP et al. Drug Info J. 1996;30(2):583

All operations need either integration with or separation from existing internal procedures, as necessary. Systems will need to be developed or integrated with existing systems for patient selection and accruals, data collection, and study monitoring. Staffing at the professional and support levels must be worked out, including training on company's products and processes. Levels and types of people are predicated on the program: size of program and individual studies, study indications, company sponsorship with protocol writing and investigator initiated, monitoring needs, statistical needs, audit needs, and report writing. The department budget is created based on staff, studies (number, size, complexity), scope of work (company sponsored, investigator sponsored, pharmacoeconomics), statistical requirements, systems and procedures, and geography to be covered [75, 76].

Investigator-initiated (I-I) studies need to be addressed after product approval, as any potential investigator can study any issue with a company's product by using the commercially available product in their patients (Fig. 11.32). The advantages for a potential investigator are funding for their research and possibly drug being provided, as well as access to company expertise. Many unique ideas for research topics will occur outside the company. It behooves the company to participate as much as possible in such trials for several reasons: creative research ideas, company responsibility to meet unmet needs, company responsibility to collect adverse experiences, especially in new situations, the product need for a continual flow of new data and publications, the expertise at the company in its staff and in-house data that can benefit outside investigators, and opportunity to identify new markets for the product or new data to create product advantages.

Research proposals for the I-I studies involve four components. For a potential investigator, the first step to obtain funding is to create a concept sheet for the study before writing a full protocol. This concept will be a one-page synopsis of the research idea and protocol, which will be reviewed and hopefully approved by the company. The full protocol will then be submitted for approval and funding consideration, including all the standard elements of a protocol (e.g., rationale and objectives, patient selection, dosage and administration of products, monitoring parameters, etc.). The budget will be submitted along with a discussion of the capabilities of the investigators and the site for the study. Budgets are generally moderate in size for I-I studies, as little as $5,000 but up to as much as $150,000. The criteria for budgeting of the study is the number of patients, complexity of the protocol, duration of the study, importance of the study idea, and opportunity for the company's product. Overhead to the investigative site through the investigator usually is limited by the company to about 25%. The internal processes for the company include acknowledgments of study ideas, protocol review committees, budget review committees, protocol design negotiations, monitoring visits and audits (depending on the importance of the study, adverse event potential, need for assistance by investigators), contracting between the site and company, legal review for patent issues, IRB documentation, interim report requirements, and preparation and review of abstracts and manuscripts. Companies generally require the opportunity to review and comment on any publication in abstract form or full paper [54].

Postmarketing commitments for clinical studies are an agreement between a company and the regulatory authority, in

- Study purposes:
  - Significant medical issue for the disease and drug
  - Opportunity for company's drug to have possible benefit
  - Publishable research effort

- Content of research proposals:
  - Ist step is concept sheet (1 pg.), followed by a protocol if favorably reviewed
  - Protocol elements: background, objectives, endpoints, design, statistics plan
  - Investigator credentials; Patients available; Institution's capability
  - Budget proposal (staff, monitoring, labs, drugs, stat analysis, travel, overhead)

- Study process (at company):
  - Acknowledge receipt - Review & approval - Changes in protocol negotiated - Contract agreement (size of study, budget, timeframe, responsibilities) - IRB agreement - Interim reports - Abstract or manuscript preparation & review

- Grants:
  - Budgets of modest size ($10,000-150,000) based on importance of idea, patient number, complexity of study, duration of study, & upside for product
  - Overhead, usually limited for phase 4 to about 25% or less

FIG. 11.32. Clinical Grants, Investigator Initiated

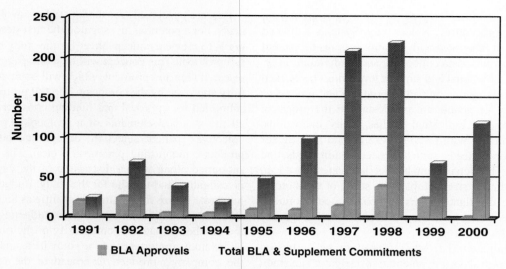

FIG. 11.33. Post-Marketing Study Commitments (BLA)

this case the FDA. Often, they are part of the approval process, and they may be voluntary by the company or required. The figure in Fig. 11.33 presents a bar graph of the number of *post-marketing commitments* associated with Biologic License Applications (BLAs) approved between 1991 and 2000. During this 10-year period, 163 BLAs were approved, 79 of which had at least one postmarketing commitment. These BLAs, combined with supplemental applications received during this period, generated a total of 927 postmarketing commitments. Of this number, 193 commitments addressed clinical safety and efficacy studies and would be subject to status reporting under FDAMA 130 implementation. As can be observed in the graph, the requirements have substantially increased over the second half of the decade. For biologics (BLA), only 44 of 301 commitments have been categorized as completed [77].

The figure in Fig. 11.34 presents a bar graph of the number of *postmarketing commitments* associated with New Drug Applications (NDAs) approved between 1991 and 2000. During this period, 1,090 NDAs were approved. These NDAs plus supplemental applications generated a total of 2,328 postmarketing commitments. The numbers of commitments include those associated with changes to approved applications (supplements), as well as original applications (NDA). Of these, 1,737 commitments addressed clinical safety and efficacy and would be subjected to status reporting under FDAMA 130 implementation. Out of 2,400 commitments, 882 have been completed [77].

At a start-up company that has existed to perform research, they will have their first lead product nearing the health care marketplace at about 8–10 years after their beginning (Fig. 11.35). These companies often have only 50–250 employees who are expert in research and development of a molecule; they often have limited financial capital or experienced people or even leadership to move to this next stage of evolution, that is, a marketing company with a product for patient use. The most common situation is a partnership with a full-fledged company with experience in the particular therapeutic area and the resources (budget and staff) to launch a new product. Also, commonly the start-up company desires to gain the experience and eventually the resources to launch and carry forward with the marketing, as well as follow-on research with this product and others independently. Hence, they will co-market and co-develop the molecule. A good example would be in 2005 when Amylin Pharmaceuticals received approval for exenatide (Byetta™) for type 2 diabetes mellitus, their first product, and partnered with Lilly. This figure lists the challenges, programs, process, and resources for the start-up company [78].

The limits and challenges for medical affairs (MA) for a start-up company include a low budget and few staff, necessitating a virtual company situation using consultants, vendors (outsourced services), the partner company, and some company staff. However, even before staffing, the planning phase for MA needs to encompass strategies for the corporation, R&D, S&M, and the product. Here expert consultants in key areas can assist the company in planning and early implementation. Product needs will be based on the product profile, marketing's plans, the competitive situation, and availability and capabilities of the partner. Many government regulations and laws exist, as discussed earlier, that

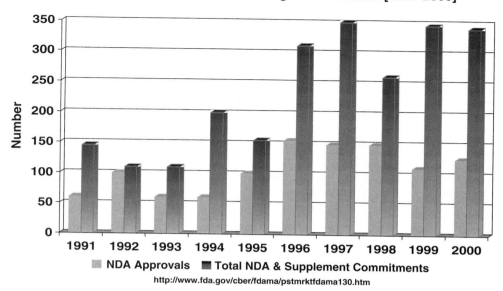

Fig. 11.34. Post-Marketing Study Commitments (NDA)

- Limits & Challenges:
  - Strategies and Needs: Corporate, R&D, S&M, Product
  - Customer expectations
  - Requirements (Government)
  - Resources
- Scope of Programs:
  - Core 1st: Phase 4 Research, Information, Safety, & Medical Education
  - Added 2nd: Formulary, MSLs, Marketing aide, Opinion-leaders, Publications
- Process:
  - A-S-P-I-R-E process
  - Analyses: Gaps for product, company, staff, competition
  - Operating guidelines
  - Timeframe
- Resources:
  - Staff, budget, equipment, & systems
  - Outsourcing

Fig. 11.35. Medical Affairs at Start-Up Companies

must be understood and integrated into company practices and operating guidelines. Safety is an exceptionally labor intensive and highly procedural and regulatory challenge, such that it could be done by the experienced and well-financed partner company. Medical information support is a customer expectation and could be done by a vendor who is integrated intimately with the company. Phase 4 studies are a high priority and necessity for key opinion leader development, future publications, and added information for safer and the best possible product usage. Perhaps the company may want to focus their new resources in MA here, given its priority and need for control and learning by the company. Alternatively, vendors can have almost turnkey operations for phase 4 study programs, which however will need oversight and integration into other company research efforts.

The scope of programs and services must be decided, based on strategies, needs, strengths, and weaknesses that are identified, in the three areas of information, research, and education. We would suggest that a minimum for MA at a start-up would include safety monitoring, medical information, some medical education, and phase 4 studies.

A common plan is also to employ their own medical science liaisons on a relatively small scale; they can be an excellent outreach to key researchers, clinicians, and customers for working relationships and mutually beneficial outcomes in education and follow-on research.

The processes to make these decisions include several steps in the planning and implementation phases for a new organization. First, an A-S-P-I-R-E® process can be used; that is, assess/analyze, strategize, plan, and implement, followed later by review and edit. Analyses are primarily gap analyses regarding resources, staff, product, studies, and competition. Information needs to be collected from R&D, S&M, and L&R, as well as ensuring their input in MA planning. Resource assessment and resource planning encompass the staff (number, backgrounds in types and experiences, training, internal vs. partner vs. outsource vendor), equipment and related systems, and then operating budget (staff compensation for salary, bonus, and benefits, operations, overhead, training, grants to customers for research and educational programs, equipment [especially for communication and information processing and delivery, such as computers, and cell phones], and systems to collect and disseminate data and information). Operating guidelines need to be written that will meet regulations and fulfill services and programs objectives. Vendor assessment, audit, and selection need to be done for outsourcing programs or services. Partner assessment needs to be done, as well as plans for coordination of efforts. These steps need to start ideally 2 years before product approval and will continue with refinements and changes over the 2 years postlaunch [78].

References

1. Health Care Industry Market Update. Available at www.cms.gov/reports/hcimu/hcimu_01102003.pdf
2. Lamberti MJ (editor). An industry in evolution, 4th ed. Thomson CenterWatch, Boston, MA, 2003.
3. CenterWatch. 2002 in review: An industry in the balance. CenterWatch 2003;10(1):1-10.
4. Drug Topics Redbook 2005, 109th ed. Medical Economics Company, 2004.
5. Magazzini L, Pammolli F, Riccaboni M. Dynamic competition in pharmaceuticals. Patent expiry, generic penetration, and industry structure. Eur J Health Econ 2004 Jun;5(2):175-82.
6. Suh DC, Manning WG Jr, Schondelmeyer S, Hadsall RS. Effect of multiple-source entry on price competition after patent expiration in the pharmaceutical industry. Health Serv Res 2000 Jun;35(2):529-47.
7. Kirking DM, Ascione FJ, Gaither CA, et al. Economics and structure of the generic pharmaceutical industry. J Am Pharm Assoc (Wash) (United States), Jul-Aug 2001;41(4):578-84.
8. U.S. FDA. Current Good Manufacturing Practice for Active Pharmaceutical Ingredients: Guidance for Industry—Manufacturing, Processing, or Holding Active Pharmaceutical Ingredients. Washington, DC: U.S. Food and Drug Administration, 1998.
9. Bender A, Shannon N. The firewall mandate. Pharmaceutical Executive 2003 (October supplement):20-26.
10. PhRMA Report, Prescription Medicines 24 Years ago and Today: Changing Trends, Enduring Needs, 2003.
11. Peddicord TE, Baker M, Oki J, Mouser JF, Hooks MA, Korth-Bradley JA. Casting call from industry: Reel in and retain appropriate information, release the rest. Pharmacotherapy 2002;22(7):934-8.
12. FDA Modernization Act of 1997 (FDAMA), November 21, 1997. Available at www.fda.gov/oc/fdama/default.htm.
13. Health Insurance Portability and Accountability Act of 1996 (HIPPA). Code of Federal Regulations 45.164.502(b) and 164.514(d). Public Law 104-191.
14. Department of Health and Human Services. Standards for Privacy of Individually Identifiable Health Information. Billing Code 4150-04M. Federal Register, Dec 28, 2000;82461–829 (45 CFR Parts 160–164).
15. Shannon ME, Malecha SE, Cha AJ, Moody ML. Evaluation and critical appraisal of a random sample of drug information practice in United States academic and industry medical information centers. Drug Info J 2000;34(4):1133-8.
16. Doyle RI, Song KH, Baker RP. An industry-wide evaluation of drug information services. Drug Info J 2000;34(4):1139-48.
17. Gundaker LF. Developing an automated telephone triage system for a new product launch. Drug Info J 2000;34(4):1035-39.
18. Code of Federal Regulations 21.202.1(I) (2). Drugs. Prescription drug advertising.
19. Food, Drug & Cosmetic Act Section 201 (m) Drugs. Labeling.
20. Basara LR, Juergens JP. Patient package insert readability and design. Am Pharm 1994 Aug;NS34(8):48-53.
21. Williams-Deane M, Potter LS. Current oral contraceptive use instructions: an analysis of patent package inserts. Fam Plann Perspect 1992 May-Jun;24(3):111-5.
22. Rowe HM. Patient package inserts: the proper prescription? Food Drug Law J 1995;50(1):95-124.
23. Code of Federal Regulations 21.310.501. Patient Package Insert, Oral Contraceptives.
24. Code of Federal Regulations 21.310.515. Patient Package Insert, Estrogens.
25. Code of Federal Regulations 21.310.516. Patient Package Insert. Progestational products.
26. Frequently asked questions about CDER. Available at http://www.fda.gov/cder/about/faq/default.htm7.
27. Code of Federal Regulations 21.99. Dissemination of Information on Unapproved / New Uses for marketed drugs, biologics and devices.
28. Wolgemuth RL. Realizing the promise of the US Food and Drug Administration Modernization Act. Clin Ther 1998;20(Suppl C):C26-31.
29. Wager E, Field EA, Grossman L. Good publication practice for pharmaceutical companies. Curr Med Res Opin 2003;19:149-54. Available at: http://www.gpp-guidelines.org/.
30. Pilnick A, Dingwall R, Starkey K. Disease management: definitions, difficulties and future directions. Bull World Health Organ 2001;79(8):755-63.
31. Ara S. A literature review of cardiovascular disease management programs in managed care populations. J Manag Care Pharm 2004 Jul-Aug;10(4):326-44.
32. Pelletier KR. A review and analysis of the clinical- and cost-effectiveness studies of comprehensive health promotion and

disease management programs at the worksite: 1998-2000 update. Am J Health Promot 2001 Nov-Dec;16(2):107-16.
33. Posey LM. Pfizer cuts disease management deal with Florida Medicaid but nurses may provide services. Pharmacy Today 2001;7(8):1-7
34. Wall Street Journal, July 9, 2001. Pfizer makes a deal with Florida Medicaid.
35. Duke KS, Raube K, Lipton HL. Patient-assistance programs: Assessment of and use by safety-net clinics. Am J Health-Syst Pharm 2005;62(7):726-31.
36. Kennedy DL, Uhl K, Kweder SL. Pregnancy exposure registries. Drug Saf 2004;27(4):215-28.
37. Gare BA. Registers in the study of pediatric rheumatic diseases—limitations and potential. J Rheumatol 1996;23(11):1834-7.
38. Goldstein SL, Somers MJ, Brophy PD, et al. The Prospective Pediatric Continuous Renal Replacement Therapy (ppCRRT) Registry: Design, development and data assessed. Int J Artif Organs 2004;27(1):9-14.
39. Code of Federal Regulations 21.310.305. Food and Drugs. New Drugs. Records and reports concerning adverse drug experiences on marketed prescription drugs for human use without approved new drug applications.
40. Code of Federal Regulations 21.314.80. Drugs. Post-marketing reporting of adverse drug experiences.
41. Code of Federal Regulations 21.600.80. Biological products. Post-marketing reporting of adverse experiences.
42. FDA Report, May 1999. Managing the Risks from Medical Product Use: Creating a Risk Management Framework. Available at www.fda.gov/oc/tfrm/riskmanagement.html.
43. Office of Drug Safety, Food and Drug Administration. Available at www.fda.gov/cder/ODS/default.htm.
44. MedWatch form 3500A. Available at http://www.fda.gov/medwatch/report/hcp.htm.
45. Medical product safety information. MedWatch. The FDA safety information and Adverse event reporting program. Available at http://www.fda.gov/medwatch/safety.htm.
46. Chin J. Biotechnology's special forces: Field-based medical science liaisons. J Commercial Biotechnology 2004;10(4):312-18.
47. Malecha S, Spears JB, Sylvestri MF. Managing medical liaisons. Strategies in an evolving environment. DIA Forum 2003;39(3):18-21.
48. Morgan DK, Domann DE, Collins GE, Massey KL, Moss RJ. History and evolution of field-based medical programs. Drug Info J 2000;34(4):1049-52.
49. Schneider BA. Maximizing your investment—The MSL force. DIA Forum 2004;40(4):28-30.
50. PhRMA Code Calls for Changes in Drug Marketing: A Newsmaker Interview with John T. Kelly, MD, PhD, Aug. 9, 2002. Medscape Medical News 2002.
51. OIG Compliance Program Guidance for Pharmaceutical Manufacturers. Office of Inspector General, Department of Health and Human Services. Fed Reg 2003;68(86):23731-43.
52. Higgins D, Goize S. OIG Draft Compliance for Pharmaceutical Companies. Health Law Reporter 2002;11(41):1-7 (BNA Professional Information Center Health Care).
53. Wolin MJ, Ayers PM, Chan EK. The emerging role of medical affairs within the modern pharmaceutical company. Drug Info J 2001;35(2):547-55.
54. Salter FJ, Kramer PF, Palmer-Shevlin NL. Pharmaceutical industry medical communications departments: regulatory and legal perils and pitfalls. Drug Info J 2000;34(4):1009-15.
55. Werner AL, Murray KM. Preparing for a product launch in a medical communications department. Drug Info J 2000;34(4):1021-33.
56. Werner AL, Poe TE, Graham JA. Expanding medical services to internal customers. Drug Info J 2000;34(4):1053-61.
57. Begg C, Cho M, Eastwood S, Hood R, Moher D, Olkin J. Improving the quality and reporting of randomized trials. JAMA 1996;276:637-9.
58. Anonymous. Thousands served. Pharmaceutical Executive Medical Education Supplement, August 2004:42-43.
59. Herman R, Schonbachler D. PhRMA Code: new rules, same game. Pharmaceutical Executive Medical Education Supplement, August 2004:10.
60. Code of Federal Regulations 24.1001.952(d). The personal services and management contracts safe harbor.
61. Raineri BD. Focus on feedback. Pharma can still have advisory meetings—in some ways, even better than before. Pharmaceutical Executive Medical Education Supplement, August 2004:27-34.
62. Optrompke J. Navigating US conflict-of-interest rules when commercializing research. Nature Biotechnology 2004;22(7):921-3.
63. Anonymous. The AMCP format for formulary submissions, version 2.1. J Managed Care Pharm 2005;11(5 Supplement):1-29.
64. Anonymous. A winded FDA races to keep up with drug ads that go too far. USA Today 2005;May 31:1A, 4A
65. Morris L. The risk management mandate. Pharmaceutical Executive 2004;24(5):98-110.
66. Ruchlin HS. New directions in pharmacoeconomic research: The next step. Drug Info J 2002;36(4):909-17.
67. Kim J, Morris CB, Schulman KA. The role of the Food and Drug Administration in pharmacoeconomic evaluation during the drug development process. Drug Info J 2000;34(4):1207-13.
68. Bootman JL, Townsend RJ, McGhan WF. Principles of pharmacoeconomics. Harvey Whitney Books Co., Cincinnati, OH, 1991.
69. Santell JP. Projecting future drug expenditures—1996. Am J Health Syst Pharm 1996;53(2):139-50.
70. Neumann PJ, Claxton K, Weinstein MC. The FDA's regulation of health economic information. Health Aff (Millwood) 2000;19(5):129-37.
71. Revicki DA, Osoba D, Fairclough D, Barofsky I, Berzon R, Leidy NK, Rothman M. Recommendations on health-related quality of life research to support labeling and promotional claims in the United States. Qual Life Res 2000;9(8):887-900.
72. Marra CA, Levine M, McKerrow R, Carleton BC. Overview of health-related quality-of-life measures for pediatric patients: application in the assessment of pharmacotherapeutic and pharmacoeconomic outcomes. Pharmacotherapy 1996;16(5):879-88.
73. Armstrong EP, Abarca J, Grizzle AJ. The role of pharmacoeconomic information from the pharmaceutical industry perspective. Drug Benefit Trends 2001;13(3):39-45.
74. Legg RF, Sclar DA, Nemec NL, et al. Cost benefit of sumatriptan to an employer. J Occup Environ Med 1997;39(7):652-7.

75. Evens RP, Flynn J, Mapes D. Preventing the pitfalls in planning phase IV clinical trials: a biotechnology experience. Drug Info J 1996;30(2):583-91.
76. PhRMA. Principles on conduct of clinical trials and communication of clinical trial results. 2003. Available at http://www.phrma.org/publications/publications//2004-06-30.1035.pdf
77. Report to congress. Reports on postmarketing studies [FDAMA 130]. Available at http://www.fda.gov/cber/fdama/pstmrktfdama130.htm.
78. Evens RP, Sylvestri M, Boone S. Medical affairs for biotechnology/pharmaceutical start-ups: strategies for launch (research and medical communication). R&D Strategy Session, Drug Information Association 2005 Annual Meeting, June 27, 2005, Washington, DC.

12
Special Considerations in Research

1
Cardiovascular Drug Development

C. Michael White, Jessica Song, and Jeffrey Kluger

State of Cardiovascular Research	275
New Modalities	275
Cardiovascular Drug Development	278
Standard of Care	279
Clinical Trial Types	281
Acute and Chronic Therapy Trials	281
Special Populations Studies	284
Conclusions	286
References	290
	290

Cardiovascular research is addressed in this chapter as a special therapeutics area, especially given its different and unique study needs and designs, the many indications for use of the products, the extensive pharmaceutical markets in patient numbers and sales potential, and the many products that exist and are investigational agents.

The cardiovascular market for drug therapy is huge with many products and many pharmacologic categories to treat each disease. Research continues to improve the care of patients; these challenges unique to cardiovascular products are addressed in this section of the book. Given the crowded list of products for each disease, the issues of "me-too" drugs will be discussed, both successful add-ons and products with no advances. Cardiovascular treatment guidelines are ubiquitous in medical practice, and their impact on research will be reviewed. Five types of study designs common to cardiovascular research will be addressed: superiority (drug and placebo), superiority (drug + drug vs. drug), noninferiority, acute trials, and chronic trials. Special populations are important in cardiovascular disease, such that four areas are covered: obesity, ethnicity, polymorphisms, and drug interactions.

State of Cardiovascular Research

Cardiovascular disease is the number one cause of mortality in the United States, almost twice that of any other cause, leading to substantial *opportunity for product development* (Fig. 12.1.1). Approximately 12 million to 16 million people currently have coronary artery disease, and the incidence is expected to double within 50 years. An additional 5 million currently have congestive heart failure, and the incidence is continuing to rise as well. The top two selling drugs in the United States are both cardiovascular drugs. The two cholesterol- lowering drugs together, atorvastatin (Lipitor®) and simvastatin account for some $11.2 billion dollars. As such, the cardiovascular drug market is immense and aimed at the treatment of cardiovascular disorders such as hypertension, hyperlipidemia, ischemic heart disease, congestive heart failure, arterial/venous diseases and arrhythmias [1, 2].

In most cases, utilization of drugs, devices, or procedures is based on strong evidence from clinical trials (Fig. 12.1.2). In many cases, standard treatments have already been shown to reduce either morbidity or mortality or both. For every one of

- Huge market opportunities:
  - 16.5 million patients with coronary artery disease, 5 million patients with congestive heart failure
  - Drugs targeted against acute manifestation of disorder or chronic therapy to slow progression or prevent recurrence
  - Undertreated groups, e.g., blacks & hispanics
- 4 1* Areas: Myocardial infarction, congestive heart failure, stable angina, arrhythmias:
  - Evidence based approach extensively utilized:
    - Multiple drugs already proven to reduce events in the marketplace
    - Major clinical trials conducted show benefits of therapy
- Major diseases covered by national consensus treatment guidelines (impact study designs)

FIG. 12.1.1. State of Cardiology Research I
Source: www.americanheart.org/statistics/index.html; American Heart Association., 2003 ACC/AHA Guidelines for CHF.http:www.acc.org/clinical/guidelines.

- Strong baseline basic science and epidemiologic research in cardiac disease
- Extensive government, foundation, and pharmaceutical company expenditures on basic and clinical research
- Successful drug classes have multiple "me-too" drugs within them. New therapeutic targets for drug discovery are important but need to be incorporated into standard of practice.

FIG. 12.1.2. State of Cardiology Research II

these aforementioned disorders, there are national expert panels convened to derive consensus guidelines for their treatment using an evidence-based medicine approach with clinical trials, best practices, and expert opinions.

For each cardiovascular disorder, there are numerous pharmacologic and nonpharmacologic options for therapy, and several drugs are usually employed for each disorder. For example, the average hypertensive patient with diabetes mellitus will need 2–3 antihypertensives for proper blood pressure control. A patient with a myocardial infarction will need 3–5 drugs acutely in the peri-infarction period and a minimum of four drugs in the postinfarction period in order to achieve optimal therapy. Heart failure patients chronically take 3–4 drugs to treat their disorder, and during acute events several other short-acting products are used as well. Using a larger number of drugs for these cardiovascular disorders is likely to occur in the future. In 2001, the worldwide market for cardiovascular drugs was the largest of all drugs at $85 billion (out of about $405 billion) with a 13% compound annual growth rate.

The pathophysiology of cardiovascular disorders is well understood as is the pharmacology of many standard treatments. Given the large health impact of cardiovascular disease, the Heart Lung and Blood Institute of the National Institutes of Health, American Heart Association, private foundations, and the pharmaceutical industry have invested hundreds of millions of dollars annually into basic science, epidemiologic, and clinical research. The number of active INDs in the cardio-renal category with the FDA was 556 in 2001 (one drug can have multiple INDs). In 2000, cardiovascular investigators in the United States received clinical grants totaling $876 million (all investigator grants equaled about $6 billion), separate from the company's staff and operations and CRO costs of about $24 billion). The median development time for a drug for cardiovascular illnesses is 7.1 years for pharmaceutical companies [3].

Successful drug classes (e.g., ACE inhibitors) will frequently have multiple drugs within each class (e.g., enalapril, captopril, monopril, lisinopril, etc.). New therapeutic targets and the drugs that modulate them need to be understood in light of the current standards of care. There are clearly unique advantages and challenges when performing drug development in the cardiovascular realm.

In cardiovascular research, the pharmaceutical industry and academia have had a symbiotic relationship regarding disease mechanisms and the *mechanisms of action* of drugs (Fig. 12.1.3). This symbiotic research agenda and often shared discoveries have led to major advances in disease understanding and better drugs with more therapeutic impact, not just improving the clinical signs of disease but long-term outcomes. A better understanding of pathophysiologic and physiologic cardiovascular disease mechanisms have led to the development of new drug targets and subsequently new drug categories. In cardiology, discovery of the L-type calcium channel led to numerous dihydropyridine and non-dihydropyridine calcium channel blockers, which are used for angina and hypertension. Similarly, the discovery of the renin–angiotensin–aldosterone system has led to the development of angiotensin converting enzyme (ACE) inhibitors and angiotensin II receptor blockers (ARBs) for hypertension, postmyocardial infarction, and

Breakthroughs in Research in Cardiology via Collaborations of Industry & Academia

- New MOAs & Drug Categories (over 25 years):
  - Calcium channnel blockade (hypertension)
  - ACE, Angiotensin converting enzyme inhibition (hypertention & heart failure)
  - Angiotensin II antagonists (hypertension)
  - Thrombolytic enzymes (AMI, coronary syndrome, stroke)
  - Clotting IIB/IIIA and Xa antagonists (AMI, stroke)
- Outcomes:
  - Disease mechanism elucidation
  - Second indications for new diseases
  - New treatment options
  - Disease control and survival

FIG. 12.1.3. State of Cardiology Research III

12.1. Cardiovascular Drug Development

heart failure. The discovery of the enzyme HMG CoA reductase in the cholesterol synthesis cascade led to selective inhibitors called HMG-CoA reductase inhibitors, which have significantly greater LDL-cholesterol lowering potency with fewer adverse effects than traditional therapy. These drugs have revolutionized the treatment of hyperlipidemia and coronary artery disease patients. Research led to the knowledge that acute myocardial infarctions and thrombotic strokes resulted primarily from thrombus formation in the coronary or carotid arteries rather than from vasospasm. Once this was known, the development of drugs that could enhance plasminogen activity was accomplished (i.e., thrombolytic drugs dramatically improved outcomes for patients with acute myocardial infarctions and thrombotic stroke). Aspirin has been used as an antiplatelet drug for decades but it has limited antiplatelet efficacy. The discovery that the glycoprotein IIb IIIa receptor was the final common pathway in platelet aggregation led to glycoprotein IIb IIIa receptor drugs, which provided short-term intense antiplatelet effects and improved the outcomes of patients undergoing acute coronary syndromes or coronary angioplasty. Although heparin had been used for many decades, increased knowledge of which clotting factors were being inhibited when different parts of the heparin molecule were being used led to the development of low-molecular-weight heparins and factor Xa antagonists. These anticoagulants have enhanced efficacy and improved pharmacokinetics and safety versus regular heparin in patients with acute myocardial infarctions, stroke, and joint replacement.

As described above, emerging scientific knowledge leads to new targets for disease. In the case of the aforementioned drug classes, new drug classes were developed or drugs already used for conditions such as hypertension were subsequently broadened in their use for diseases such as myocardial infarction and congestive heart failure. Agents such as ACE inhibitors and ARBs were found to provide additional cardiovascular benefits in patients with myocardial infarctions and heart failure due to their impact on left ventricular remodeling and other effects. Ultimately, these therapies enhanced patient survival in the diseases providing benefit far in advance of the antihypertensive effects for which the drugs were originally developed.

A major advance in cardiology is the focus on *terminal end points* (outcomes) (Fig. 12.1.4). Traditionally, epidemiologic studies show relationships or correlations between a surrogate marker and a disease. Normal ranges are constructed for the surrogate marker, such as diastolic blood pressure. Then studies are conducted to determine if a drug regimen can correct abnormalities outside these normal ranges. The studies were then evaluated by regulatory bodies for approval of these indications. The benefits of surrogate makers are that they allow efficient evaluation of drugs (lower number of subjects, shorter follow-up time, lower expense). However, surrogate markers can lead clinicians to incorrect conclusions in some cases. Immediate-release nifedipine lowered blood pressure as well as traditional antihypertensives but increased the risk of people having a heart attack rather than reducing it. Oral milrinone dramatically improved patient symptoms in heart failure but increased mortality by generating ventricular arrhythmias. Beta-blockers were contraindicated in heart failure patients for decades because they increased patient symptoms and reduced cardiac output. However, beta-blockers have long-term survival benefits in heart failure making them a drug of choice for the disorder.

Evolution of Cardiovascular Endpoints
- Surrogate markers (correlations) to Hard endpoints
- Arrythmias:
  - ECGs & PVCs to Survival
- Hypertension:
  - Bpd change of 3 mmHg to Organ function
- Coronary artery disease:
  - Symptoms & ST segment depression to Survival
- Impacts:
  - Complexity of research
  - Size of trials
  - Time and cost of research
  - More meaningful outcomes in health care

Fig. 12.1.4. State of Cardiology Research IV

So cardiology has moved from relying on surrogate end points to an evaluation of outcomes. Instead of just looking at electrophysiologic studies or changes in the electrocardiogram (ECG) such as ST segment change, number of days of event-free survival and overall mortality are evaluated in sentinel arrhythmia studies. In hypertension studies, the occurrence of end organ damage (stroke, myocardial infarction, renal protection) is now evaluated rather than simply looking at blood pressure. In coronary artery disease, overall survival is evaluated in sentinel trials rather than simply evaluating symptoms or ECG changes.

While this focus on evidence-based medicine has led to new concepts and understanding of disease pharmacotherapy, it has also increased the complexity of the trials that are needed. The new tests are more complicated and sophisticated requiring special equipment and staff. Multicenter trials with thousands of patients followed up over several years are now commonly needed to generate the statistical power to demonstrate statistically significant and clinically meaningful changes in these terminal end points. These trials are very expensive to conduct related to the size of patient samples and test sophistication. In addition, they extend the product development program time period considerably thus losing patent life, which can be hundreds of millions of dollars of lost revenue per year.

There is an abundance of *diagnostic and treatment devices* employed in cardiology (Fig. 12.1.5). For example, we now use 12-lead electrocardiograms, ambulatory Holter monitors, electrophysiologic studies, and implantable defibrillators (to detect arrhythmias in the outpatient setting and shock the patient out of the arrhythmias), and ablation technology to cure some

Equipment Use Issues in Cardiology
- Types:
  - e.g., Electrocardiograms (in-clinic vs holter monitors), Echocardiograms, Plethysmography, Stress tests - ergometers and tread mills with ECGs
- Variables:
  - Type of equipment
  - Equipment calibration
  - Technician (training, experience, & use of equipment)
  - Timing (related to disease course, time of day)
  - Concurrent drug use (CV and non-CV)
  - Cardiologist readings
  - Cost

FIG. 12.1.5. State of Cardiology Research V

Possible Research Questions
- Target event reduction (AMI) vs disease parameter improvement (LDL)
- New MOA as new and additional disease mechanism
- Efficacy superiority
- Additive drugs – more efficacy
- Adverse event reduction
- Drug interactions advantage
- Metabolism advantage
- Special populations
- Pharmacogenomics
- Improved dosing schema

Need to improve over manifold drugs already available

FIG. 12.1.6. State of Cardiology Research VI

forms of arrhythmias. Echocardiograms are used to determine heart function and pumping ability or to evaluate endothelial function of arteries. Blood pressure and blood flow can be taken while seated via a plethysmograph or can be taken while ambulating. Newer technology allows the determination of the components of blood pressure (cardiac output and systemic vascular resistance). MRIs and electron bean tomography are used to evaluate heart structure, function, and calcification. Cardiac stress tests can use treadmills, bicycle ergometers, or other machines. Nuclear stress tests using exercise or drugs evaluate areas of coronary blockages and determine if the myocardium fed by that stenosed segment is viable or not.

When these devices are used in research, many factors need to be considered. What type of equipment and what technique is being used? When ECGs are being read, is it by a single cardiologist or multiple cardiologists at multiple sites? Are they using a 12-lead ECG or just a single lead? Do they measure the "u" wave when measuring QT duration? Are they averaging the cardiac cycle they measure or using a single cycle? It was thought that the HMG CoA reductase inhibitor atorvastatin (Lipitor®) increased fibrinogen levels. However, the test they used was an immunonephalometric test, which can have trouble detecting fibrinogen in hypertriglyceridemic samples. As it turns out, atorvastatin reduced triglycerides making it easier to detect the fibrinogen. Fibrinogen testing using the Claus method where the fibrinogen in the samples were clotted out and measured found no impact of atorvastatin on fibrinogen. If laboratory samples are being run, what is the coefficient of variation for the standard? Can the technicians really run the tests with acuity? Are the samples/tests being run at the correct time? It you give a drug that takes 30 minutes to be absorbed and sample for results 10 minutes after ingestion, a lack of effect doesn't preclude a future effect. Even seated blood pressure can be confounded if the technician doesn't allow the patients to rest before taking the reading or doesn't have the arm positioned correctly. Even the timing of testing may impact a study result based on how busy the test site, the stage of disease, patient postiton (standing, sitting, ambulatory), and circadium rhythms for hormone and catecholamine release in patients. Therefore, the use of equipment in studies will require a variety of steps to be elucidated in the study protocol; for example, standardization of testing equipment across study sites, calibration of equipment periodically to ensure operational efficiency and consistency, test procedures spelled out for all technicians to follow, test readings (e.g., ECGs) often to be done at a central site with the same cardiologists for all patients, training of all study monitors to ensure consistency as much as possible across sites, and in the final end-of-study analyses an intersite comparison of data to help be sure consistency was actually achieved.

There are many *questions* that can be asked which can provide data for *drug development* (Fig. 12.1.6). Several of the questions are framed in the figure that can help to focus a research group, company, or individual in on a new area of work, be it in discovery or development. The types of research questions are quite varied beyond efficacy and safety, which certainly are the focus of a NDA/CTD in product development, encompassing mechanisms of disease and drugs, pharmacokinetics and metabolism, dosing, drug interactions, and more. These points will be developed further along in the presentation.

New Modalities

In cardiology, new drugs from new classes are commonly used in place of, or adjunctively with, other drugs currently being employed (Fig. 12.1.7). As such, an evaluation of the potential advantages and disadvantages of these drugs are important. Does this new therapeutic drug class have advantages over older drugs? Advantages could be pharmacologic, pharmacokinetic, or clinical in nature [8–12].

ACE inhibitors block not just the afterload and preload enhancing effects of angiotensin II but actually directly reduce angiotensin II levels. This gave them pharmacologic advantage over the combination of hydralazine and isosorbide dinitrate in heart failure with better patient tolerability.

- Does drug in new class have the potential to replace another drug class?
  - Does it have the potential for fewer side effects, more convenient dosing, or better efficacy than established drug classes:
    - Enalapril (Vasotec®) vs hydralazine/isosorbide dinitrate
    - Low molecular weight heparin vs unfractionated heparin
    - Ximelagatran (Exanta®)
- Will new drug use be adjunctive with another drug?
  - Does basic science or epidemiological studies suggest that both processes are activated in disease:
    - Eplerenone (Inspira®) plus ACE inhibitor post-MI
    - Carvedilol (Coreg®) plus ACE inhibitor in CHF
    - Ximelagatran (Exanta®) plus aspirin vs aspirin alone

Cohn J. *N Engl J Med* 1991;325:303-10.; Antman EM. *Circulation* 1999;100:1593-1601.; Pitt B. *N Engl J Med* 2003;348:1309-21.; Colucci WS. *Circulation* 1996;94:2800-6.; Sander S, White CM. Formulary 2004, In press. Wallentin S. Lancet 2003;362:789-97.

FIG. 12.1.7. New Therapeutic Modalities

Low-molecular-weight heparins had lower binding to proteins within the body and therefore had much more predictable pharmacokinetic effects. This allowed for standard subcutaneous dosing without blood monitoring for the level on anticoagulation as compared with weight-based intravenous dosing of heparin with periodic monitoring.

After 51 years, there is a potential oral anticoagulant other than warfarin. Ximelagatran does not have drug–vitamin K interactions, drug–drug interactions with CYP2C9 inhibitors, and can be used in standard doses rather than doses that are altered periodically based on a laboratory test. However, it does have an Achilles' heel because it raises liver function tests and in a small subset of patients can elevate bilirubin levels as well. Additional safety studies are needed before this drug can be approved.

Drugs can be used adjunctively with other drugs. Sometimes a new drug blocks a system or subsystem within the body which causes pathological damage but for which current drugs are not adequate to treat. ACE inhibitors initially reduce aldosterone concentrations but over time, the concentrations of aldosterone rise to pretreatment values. ACE inhibitors still prevent angiotensin II–induced pathological cardiovascular remodeling but the benefits of aldosterone suppression are lost. Eplerenone is an aldosterone antagonist that has been studied with ACE inhibitors and found to provide additional benefits among patients with a myocardial infarction or diabetes mellitus and concurrent left ventricular dysfunction.

Similarly, carvedilol has been shown to work adjunctively with ACE inhibitors in patients with heart failure to further reduce the risk of mortality. Carvedilol is a multicomponent adrenergic neurohormonal antagonist. It was known for some time that higher concentrations of circulating catecholamines were related to the risk of death in heart failure, but ACE inhibitors were not capable of interfering with this system, setting the stage for adjunctive therapy.

Finally, aspirin is an antiplatelet drug used after a myocardial infarction to prevent recurrent unstable coronary syndromes. No anticoagulants are currently being employed for chronic management. In the recent ESTEEM clinical trial, the oral direct thrombin inhibitor ximelagatran was shown to provide additional benefits over and above aspirin alone. Because this trial was a dose-ranging study for ximelagatran, a follow-up study will be needed before it can be routinely implemented [8–14].

Cardiovascular Drug Development

As described in (Fig. 12.1.8), many cardiovascular drug classes have multiple drugs in them. As an example, there are currently 6 HMG CoA reductase inhibitors, 11 ACE inhibitors, and 6 angiotensin receptor blockers. It is hard to find a cardiovascular drug class without at least two approved drugs and/or several investigational drugs on the way. This is mostly due to the lucrative cardiovascular market available for those classes. I have termed these drugs in crowded drug classes as *"me-too"* drugs.

Health care practitioners have an inherent bias against "me-toos" and do not like to learn new dosing. Once they have memorized a drug name and dosing regimen, know how to titrate therapy, and know some of the intricacies of the drug, they are less likely to abandon the drug for a "me-too." As such, there is an inherent value to being the first drug on the market in the class. However, a "me-too" can be successful if it can show important advantages to health practitioners or the health care system. Although pricing and contracting are important considerations in drug placement and marketing, it is outside the focus of this chapter. However, being the preferred drug on a formulary can compel a physician to use the drug entity for their patients.

Questions that the manufacturer needs to address when deciding to develop a "me-too" include (1) does the new drug have potential advantages versus other drugs in the class, and (2) does it have an apparent Achilles' heel? Advantages could

- Join the crowd?
  - Six HMG CoA RIs, 11 ACE inhibitors, 6 ARBs
- Doctors hate learning new dosing regimens
- Does the drug have potential advantages in the marketplace?
  - Renally eliminated when all others are CYP metabolized
  - Longer duration, tighter binding to enzyme
  - Less potential for side effects
- Does it have an apparent Achilles Heel?
  - QTc interval prolongation, neutropenia, drug interactions, anal leakage, hepatotoxicity

FIG. 12.1.8. "Me-Too" Drug Development

include greater potency than other drugs, a different mechanism of metabolism/elimination, longer or shorter half-life (depending on the intended indication), or lower potential for side effects. These advantages will provide health practitioners an incentive to learn the new drug, dosing, and titration schedules because of the advantage to their patients. An apparent Achilles' heel includes excessive QTc prolongation or modest QTc prolongation in a class where this has not been identified with other drugs. Other Achilles' heels include neutropenia, severe drug interaction profile, or an unpalatable adverse effect such as hirsutism, anal leakage, or hepatotoxicity.

Ticlopidine, the first ADP inhibitor on the market, was used primarily for aspirin failures, in cases where aspirin was contraindicated, or adjunctively with aspirin in patients who had percutaneous coronary intervention. However, it required periodic blood monitoring to prevent the development of neutropenia. Neutropenia enhances the risk of life-threatening infections. Clopidogrel, the second ADP inhibitor, showed similar efficacy to ticlopidine without requiring blood monitoring, *successful me-too* (Fig. 12.1.9). Shortly after its release, clopidogrel utilization virtually replaced ticlopidine [3].

Clopidogrel also embarked on an aggressive clinical trial program. It was compared directly against aspirin in the CAPRIE trial for patients with arteriosclerotic conditions such as peripheral vascular disease, coronary artery disease, and cerebrovascular disease. Although the results of this trial were favorable versus aspirin, the pricing of clopidogrel limited the growth in market share for initial therapy. Clopidogrel was then studied in the CURE trial as adjunctive therapy with aspirin for unstable angina or non-ST segment elevation myocardial infarction patients. With additional benefits versus aspirin alone, they have expanded their potential market.

Low-density lipoprotein (LDL) cholesterol treatment goals are established by the National Cholesterol Education program. In their Adult Treatment Panel II guidelines, they had set aggressive goals for LDL that most patients were not obtaining with standard therapies. These goals were communicated effectively to practitioners and they believed in the target LDLs established, but only a minority of patients were achieving the goals. Atorvastatin was the sixth HMG CoA reductase inhibitor on the market. It had CYP3A4 interactions like simvastatin and lovastatin. It didn't have large multicenter trials showing reductions in cardiovascular events for primary or secondary prevention like pravastatin or simvastatin. However, it was by far the most effective LDL-cholesterol reducing drug on the market. Atorvastatin improved LDL goal attainment and was relatively easy to use (Fig. 12.1.9). Just a 10 mg starting dose of atorvastatin could reduce LDL by as much as the maximum or near maximum doses of fluvastatin, lovastatin, or pravastatin [4].

Atorvastatin has also embarked on a clinical trials program that has found cardiac event reductions in primary prevention (ASCOT-LLA), like pravastatin and lovastatin, but also investigated the impact of more aggressive lowering (MIRACL) of cholesterol and found additional benefits. This trial and the PROVE IT trial both show that even the new NCEP ATP III guidelines that have come out after atorvastatin's release are probably not low enough. If NCEP ATP IV guidelines have even lower LDL goals, it will further reduce the chances that older competitors like lovastatin, pravastatin, and fluvastatin will be able to bring patients to goal. This is very important for market share maintenance in a drug class where one drug has already gone generic and two others will move to generic status over the next several years.

There are many examples of drugs that have come on the market as later *me-too drugs* and sell poorly (Fig. 12.1.10). Three examples come to mind: telmisartan, moexipril, and dofetilide. Telmisartan is an angiotensin II type 1 receptor blocker. Even though it has potent blockade of angiotensin II, it is not commonly used. This is because other agents (irbesartan and candesartan) came out before it, are equally as effective for blood pressure reduction, and have a lower risk of drug interactions. Telmisartan is a P-glycoprotein inhibitor and raises the concentrations of digoxin. Although only a minority of patients eligible to receive telmisartan for hyper-

- **Clopidogrel (Plavix®):**
  - ADP inhibitor for antiplatelet effects
  - Proven to work as well as ticlopidine with less GI problems and neutropenia risk (no WBC monitoring)
- **Atorvastatin (Lipitor®):**
  - HMG CoA Reductase inhibitor to reduce cholesterol
  - Tighter binding to HMG CoA Reductase than other statins
  - Guidelines and Marketplace moving towards lowering LDL-C goals
  - Came to market with FDA dose which reduced LDL-C more than other agents
  - Less lipophilicity and lower potential for CYP3A4 drug interactions
  - No new adverse effects

FIG. 12.1.9. Successful "Me-Too" Drugs
Source: Patrons C. Platelet-active drugs. *Chest* 1998;114;470s-8s.; White CM.. Dyslipidemias. In: PSAP, 5th Edition. ACCP, Inc, Kansas City, MO.. 2004: pg 16590.

- Inconveniences or problems established drugs don't have
- **Telmisartan (Micardis®, angiotensin receptor blocker):**
  - P-glycoprotein inhibitor with digoxin interaction
  - No efficacy advantages over irbesartan and candesartan
- **Moexipril (Univasc®, ACE inhibitor):**
  - Drug food interaction
  - Ideally dosed twice a day rather than once daily
  - Six other true once a day ACE inhibitors without interactions
- **Dofetilide (Tykosyn®, Class III antiarrhythmic):**
  - Serious drug interactions (cation tubular secretion inhibitors)
  - Sotalol (Betapace®) not have same interaction potential

FIG. 12.1.10. Unsuccessful "Me-Too" Drugs
Source: Song JC, White CM. *Pharmacotherapy* 2000;20:130-9.; Song JS, White CM. *Clin Pharmacokinet* 2002;41:207-24.; White CM. *Pharmacother* 1998;18:588-99.

tension would be on concurrent digoxin, it is something that practitioners need to be aware of [5–7].

Four of the 11 ACE inhibitors have very low utilization. One of these is moexipril. Moexipril has a drug–food interaction, trough to peak ratio of less than 50%, and no clinical trials evaluating event reductions. Agents with a trough to peak ratio of less than 50% should optimally be dosed twice a day. Only one other ACE inhibitor has a drug–food interaction, several other ACE inhibitors are true once-a-day drugs, and many ACE inhibitors have clinical trials proving event reductions (renal protection in diabetes mellitus patients, mortality reductions in heart failure, or mortality reductions after a myocardial infarction).

Dofetilide is a class III antiarrhythmic drug. It can be used in conversion and maintenance of sinus rhythm in patients with atrial fibrillation. Its closest competitor, sotalol, can only be used in sinus rhythm maintenance. It does not have beta-blocking effects like sotalol, which would be an advantage for dofetilide over sotalol among patients with Raynaud phenomenon, heart failure, or moderate to severe persistent asthma. However, it has a multitude of problematic drug interactions. Dofetilide is a substrate for CYP3A4 and therefore has numerous well-known drug interactions. It is also contraindicated with several cation tubular secretion inhibitors. Health care practitioners are not familiar with this mechanism of interaction and would therefore need to invest substantial time in order to be able learn the drugs that should be avoided. Given the narrow therapeutic window for class III antiarrhythmics and the risk of severe polymorphic ventricular arrhythmias resulting from enhanced dofetilide blood concentrations if an interaction occurred, physicians have largely opted to use alternative therapy for atrial fibrillation.

Standard of Care

Previously, it was elucidated that treatment guidelines and *standards* are developed for all major cardiovascular diseases (Fig. 12.1.11). Having aggressive LDL-lowering goals in the NCEP ATP II and III guidelines for cholesterol management led to the high-potency agents atorvastatin and simvastatin dominating the HMG CoA reductase inhibitor market [14].

However, there are also important research and drug development implications to these guidelines. Large studies initially found ACE inhibitors to be superior to placebo in congestive heart failure. However, it would be unethical to evaluate a new ACE inhibitor versus placebo in a large clinical trial because it would be denying an effective therapy to patients in need. Beta-blockers and carvedilol were subsequently shown to provide mortality benefit in addition to ACE inhibitors. As such, it would be difficult to do a large placebo-controlled study among subjects not receiving ACE inhibitors and beta-blockers as well.

This provides only three main options that would routinely be acceptable to institutional review boards (a group of people

- Based mostly on large clinical trials:
  - Most guidelines published jointly by American Heart Association and American College of Cardiology
- NYHA class II or III heart failure:
  - Large studies show improved survival if patients take beta-blockers and ACE inhibitors
  - New potential heart failure drugs cannot routinely be studied in people denied these therapies
  - You can study:
    - New drug in people unable to take one or both of these drugs versus placebo
    - New drug in addition to standard of care
    - New drug versus standard of care
    - Special populations, e.g., with concurrent diseases

FIG. 12.1.11. Standard of Care in Cardiology
Source: http:www.acc.org/clinical/guidelines.

who review potential projects and ensure that patients are being protected, in and as much as possible, from harm). First, you can use the new drug among patients who inherently cannot receive the standard-of-care drugs. This would include patients with contraindications to the standard-of-care drug, people who previously failed therapy with the standard-of-care drug, or people who did not tolerate the previous drug therapy. Second, you could use your drug in addition to the standard-of-care drug and look for additional benefits. Third, you could compile several small trials showing similar or superior effects in surrogate end points germane to the disease state covered and then do a head-to-head comparison. Such surrogate end point studies would have to be compelling enough to get practitioners and patients be willing to accept the risk of losing an accepted therapy in exchange for the experimental drug [15].

Clinical Trial Types

Over the past few decades, the outcomes of major clinical trials have profoundly influenced clinical practice along with the development and modification of clinical guidelines such as the American Heart Association's Advanced Cardiac Life Support (ACLS), the National Cholesterol Education Program (NCEP) Adult Treatment Panel (ATP) III, and the seventh report of the Joint National Committee (JNC VII) on Prevention, Detection, Evaluation, and Treatment of High Blood Pressure. Moreover, *clinical trial* have resulted in modification of the FDA-approved prescribing information for various cardiovascular drugs such as candesartan (Figs. 12.1.12 and 12.1.13).

At present, the majority of standard-of-care drug therapy recommendations developed for cardiovascular disease management by expert panels are based on evidence derived from large superiority trials, including placebo-controlled or active-controlled clinical trials. Superiority trials are conducted under

- Disprove the null hypothesis by showing that the groups are different
- Placebo control – comparison to determine if something is better than nothing:
  - Post-cardiothoracic surgery atrial fibrillation prophylaxis:
    - AFIST and AFIST II trials: Amiodarone vs placebo
- Active control (mano y mano) – comparison to show one therapy is better than another:
  - Hypertension:
  - Irbesartan (Avapro®) vs losartan (Cozzar®) for blood pressure lowering
  - LIFE trial: losartan vs atenolol for event reduction in HTN + LVH
  - IDNT trial: Irbesartan vs control and amlodipine (Norvasc®) therapy for renal protection in type II d.m.

FIG. 12.1.12. Clinical Trial Types I: Superiority Study
Source: White CM. *Ann Thorac Surg* 2002;74:69-74.; White CM. *Circulation* 2003;108[Suppl ll];200-6.; Dahlof B. *Lancet* 2002; 359:995-1003.; Lewis EJ et al. *N Engl J Med*. 2001;345;851-860.

- Adjunctive therapy vs active control with one drug:
  - To see if new therapy + standard of care is better than standard of care alone
  - ARREST Trial: Amiodarone + standard advanced cardiac life support (ACLS) vs standard ACLS alone:
    - Study led to addition of amiodarone to ACLS guidelines creating a new standard of care

FIG. 12.1.13. Clinical Trials II: Superiority Study
Source: Kudenchuk PJ. *N Engl J Med* 1999;341;871-8; American Heart Association, *Circulation* 2000;102(suppl l):1-384.

the assumption that any differences among treatments or between some treatment and a control are purely due to chance. This hypothesis of "no effect" due specifically to an intervention is known as the null hypothesis. The goal of active-controlled and placebo-controlled superiority trials is to reject the null hypothesis, thereby demonstrating that a significant difference between two different treatments or between placebo and some treatment exists, respectively [15–18].

The primary aim of a placebo-controlled study is to determine if some form of treatment is superior to no intervention in regard to an outcome. Because bias can potentially emerge when a favorable response to placebo occurs as a result of expectation of an effect, double-blinding so that neither patients nor study investigators know which subjects are in the treatment or placebo groups is usually implemented in the study design. Recently, two randomized, double-blinded, placebo-controlled studies, the AFIST (Atrial Fibrillation Suppression Trial) and AFIST II trials, compared the efficacy of amiodarone with that of placebo in reducing the risk of atrial fibrillation among elderly open heart surgery (OHS) patients receiving β-blockade.

The AFIST trial randomized 220 OHS patients (>87% received β-blocker) to receive placebo or amiodarone: fast load (6 g orally more than 6 days, starting 1 day before OHS) or slow load (7 g orally more than 10 days, starting 5 days before OHS). The primary end point was the development of any type of atrial fibrillation (AF) detected by continuous electrocardiogram monitoring, including symptomatic, post-operative, and recurrent AF. When compared with the placebo group, patients receiving amiodarone had a significantly (41%) lower risk of atrial fibrillation ($p = 0.01$). In addition, amiodarone-treated subjects demonstrated a significant (77%) reduction in the risk of symptomatic atrial fibrillation ($p = 0.001$) and cerebrovascular accidents ($p = 0.04$). Of note, previous amiodarone studies included patients with limited background β-blockade, a treatment modality that has been shown to reduce postoperative OHS AF. The AFIST II trial randomized 160 cardiothoracic surgery (CTS) patients to receive placebo or amiodarone (hybrid intravenous and oral regimen delivering the equivalent of 6.9 g oral amiodarone) and then to atrial septal pacing or no pacing using a 2×2 factorial design. The primary end point was the occurrence of atrial fibrillation within 30 days of CTS. Patients receiving amiodarone demonstrated a significantly (43%) lower risk of atrial fibrillation ($p = 0.037$) versus placebo. No additional clinical benefit was associated with adjunctive pacing, as there was no significant difference in AF incidence between the pacing and no-pacing groups.

The principal goal of active-controlled trials is to determine whether one therapeutic option is superior to other options. Studies utilizing this type of design have clearly made an impact on consensus reports such as JNC VII and NCEP ATP III. The Antihypertensive and Lipid Lowering Treatment to Prevent Heart Attack Trial (ALLHAT) was a randomized, double-blind, clinical trial that included high-risk patients with hypertension. The trial enrolled 33,357 patients (after doxazosin arm was removed) aged 55 years and older, representing a diverse population including large numbers of African Americans and Hispanics, women, and diabetic patients. ALLHAT was designed to determine the combined incidence of nonfatal myocardial infarction (MI) and fatal coronary heart disease (CHD) during initial treatment with chlorthalidone, amlodipine, and lisinopril. If blood pressure goal was not achieved on the maximum tolerated doses of these agents, open-label medication (atenolol, reserpine, clonidine, hydralazine) was added at the physician's discretion. The primary end point was combined nonfatal MI and CHD deaths. No difference between the amlodipine and chlorthalidone groups for the primary outcome of CHD deaths and nonfatal MI was observed (relative risk of amlodipine, 0.98, 95% CI, 0.90 to 1.07). Similarly, there was no difference between the lisinopril and chlorthalidone groups for the combined incidence of CHD deaths and nonfatal MI (relative risk of lisinopril, 0.99, 95% CI, 0.91 to 1.08). The outcome of the ALLHAT study provided the rationale for ACE inhibitor, calcium channel blocker, and diuretic use in high CAD risk patients and in diabetic patients, as recommended by the JNC VII report.

The recently updated NCEP ATP III guidelines calls for more intensive cholesterol-lowering treatment, especially in patients at high risk for CHD. Recent trial evidence suggests that a LDL goal of less than 70 mg/dL may be of benefit in very high-risk patients. The PROVE IT (Pravastatin or Atorvastatin Evaluation and Infection Therapy) trial evaluated 4,162 patients, hospitalized for an acute coronary syndrome (ACS) within the preceding 10 days. Patients were randomized to receive 40 mg of pravastatin daily or 80 mg of atorvastatin daily and followed for 18 to 36 months (mean, 24). The primary end point was a combination of all-cause mortality, MI, unstable angina requiring rehospitalization, revascularization (performed at least 30 days after randomization), and stroke. When compared with the pravastatin group, the event rate of the primary end point was significantly lower for patients receiving maximum-dose atorvastatin (hazard ratio reduction: 16%, 95% CI, 5 to 26%, p = 0.005). Moreover, the median LDL-cholesterol levels achieved by the pravastatin- and atorvastatin-treated patients were 95 mg/dL and 62 mg/dL, respectively (p < 0.001). Shortly after publication of the PROVE IT trial results, the expert panel of the updated NCEP ATP III guidelines proposed an optional LDL-cholesterol goal of less than 70 mg/dL for ACS patients with established cardiovascular disease.

In addition to influencing clinical guidelines, the outcomes of active-controlled trials can lead to modification of FDA-approved prescribing information for cardiovascular drugs. A recent randomized, double-blind, parallel group, forced titration study (n = 611) compared the antihypertensive effects of once-daily treatment with candesartan 32 mg and losartan 100 mg on 24-hour (trough) and 48-hour post-dose blood pressure. After 8 weeks of therapy, candesartan reduced trough and 48-hour post-dose blood pressure by 3.3/1.4 mm Hg and 4.6/2.9 mm Hg more than losartan, respectively (p < 0.05). Other smaller comparative trials yielded similar outcomes, thereby resulting in the addition of the statement, "In a total of 1268 patients with mild to moderate hypertension who were not receiving other antihypertensive therapy, candesartan cilexetil 32 mg lowers systolic and diastolic blood pressure by 2 to 3 mm Hg on average more than losartan potassium 100 mg" to the prescribing information for candesartan [15–26].

Placebo-controlled trials are inappropriate for evaluating the efficacy of therapeutic regimens for certain conditions, such as cardiac arrest due to ventricular fibrillation, as providing no active treatment for this type of situation could certainly lead to excessive deaths in the placebo group. A more suitable and ethical study would compare the efficacy of standard-of-care with that of new therapy added to standard-of-care. Shortly before the publication of the new and revised 2000 guidelines for ACLS by the American Heart Association, a randomized, double-blinded study assessed the efficacy of amiodarone added to standard-of-care treatment compared with standard-of-care treatment of out-of-hospital cardiac arrest due to ventricular fibrillation (n = 504). Patients were randomized to receive 300 mg of intravenous amiodarone or placebo after unsuccessful resuscitation (received three or more shocks from an external defibrillator). Use of additional antiarrhythmic agents was permitted for both groups of patients. The primary end point was hospital admission with a spontaneously perfusing rhythm. Amiodarone-treated patients were more likely to achieve the primary end point than were recipients of placebo (44% and 34%, respectively; p = 0.03). The favorable outcome of this study led to support (in the 2000 ACLS guideline) of the use of amiodarone for the treatment of ventricular fibrillation/pulseless ventricular tachycardia [27, 28].

There are some situations where clinicians are interested in demonstrating the equivalence of two treatment options, vis-à-vis *noninferiority* (Fig. 12.1.14). An equivalence trial can be of value when the standard active comparator has been shown to be beneficial for a serious medical condition and the new treatment offers advantages in safety, cost, or convenience. In an equivalence trial, the null hypothesis is that a minimum difference exists between the treatment groups. The goal of an equivalence trial is to reject the null hypothesis, thereby demonstrating that no difference exists between the treatment groups. One common failing of equivalence trials is the assumption that equivalence of two treatments validates the efficacy of both treatments. In order to confirm the efficacy of the two treatment groups, inclusion and exclusion criteria should mirror the ones utilized in previous trials of the standard active comparator. In addition, the rates of achieving the primary end point should be similar to that observed in previous trials of the active comparator. Other study design considerations include the dosing regimen of the standard comparator, the use of concomitant medications/interventions, the primary end point, the schedule of primary outcome measurements, and the inclusion of intention to treat and per protocol analyses.

Fibrinolytic treatment of MI patients within 6 hours of symptom onset has been shown to reduce mortality by 23%. Moreover, prior to the publication of the ASSENT-2 (Assessment of the Safety and Efficacy of a New Thrombolytic outcomes), rapid infusion (90 minutes) of the

- Proving the null hypothesis correct:
  - Different than not achieving significance in a superiority trial:
    - You can only say that one drug is not better than the other
  - Trying to show that a new drug is similarly effective to a standard drug
  - ASSENT II Trial:
    - Showed that tenectaplase (TNKase®) was not inferior to alteplase (Activase®) in acute myocardial infarction
  - VALIANT Trial:
    - Showed that valsartan was not inferior to captopril in post myocardial infarction

FIG. 12.1.14. Clinical Trials III: Non-Inferiority Study
Source: ASSENT II Investigators. *Lancet* 1999;554:716-22.; Pfeffer MA. *N Engl J Med* 2003;349:1893-906.

tissue-plasminogen activator alteplase represented the standard-of-care treatment option for pharmacological reperfusion in acute MI. Recently, a new fibrinolytic agent, tenecteplase, was added to the armamentarium of pharmacological reperfusion agents. This drug offers a more convenient dosing schedule (bolus dose given over 5–10 seconds) and has better fibrin specificity than alteplase.

ASSENT-2 was a randomized, double-blinded clinical trial that included early onset MI (symptoms within 6 hours before randomization) patients. The trial enrolled 16,949 patients aged 18 years and older, of which 80% and more than 50% were on concomitant β-blockers and ACE inhibitors, respectively. ASSENT-2 was designed to demonstrate the equivalence of single-bolus tenecteplase compared with front-loaded alteplase. The primary end point was death due to any cause at 30 days. The null hypothesis presented in the ASSENT-2 trial stated that the 30-day mortality rate seen after tenecteplase treatment would be at least 1% higher than the 30-day mortality rate reported after alteplase use. Alternatively, the null hypothesis stated that the relative risk in 30-day mortality would be at least 14% higher with tenecteplase treatment in comparison with alteplase treatment. The 30-day mortality rates for the tenecteplase- and alteplase-treated patients (6.18% and 6.15%, respectively) were similar to the rate seen with alteplase in the GUSTO-1 trial (6.3%). The prespecified criteria of equivalence was fulfilled, as the 95% one-sided upper limits of the absolute and relative differences in 30-day mortality were 0.61% and 10%, respectively. The study investigators concluded that tenecteplase matched the efficacy of alteplase, but its ease of administration may facilitate the institution of early reperfusion therapy in acute MI patients.

Several large, randomized, placebo-controlled studies have assessed the role of ACE inhibitors for post acute MI patients with left ventricular dysfunction or clinical signs of heart failure. In SAVE (Survival and Ventricular Enlargement), AIRE (Acute Infarction Ramipril Efficacy), and TRACE (Trandolapril in Patients with Reduced Left-Ventricular Function After Acute Myocardial Infarction), a total of 5,966 patients with evidence of clinical heart failure or evidence of left-ventricular dysfunction were randomized to receive ACE inhibitor (captopril, ramipril, and trandolapril, respectively) or placebo between 3 and 16 days after acute MI. After a median treatment duration of 31 months, mortality rates were 23.4% and 29.1% in the ACE inhibitor and placebo groups, respectively (odds ratio 0.74, 95% CI, 0.66 to 0.83, $p < 0.0001$).

While ACE inhibitors are considered first-line agents for the treatment of post-MI patients with signs of heart failure or evidence of left-ventricular dysfunction, nearly 5% to 10% of patients who receive these drugs develop a dry cough. Furthermore, angiotensin II can be generated by non-ACE-dependent pathways catalyzed by other enzymes, including cathepsin G, elastase, tissue plasminogen activator, chymostatin-sensitive angiotensin II generator enzyme, and chymase. Through antagonism of the angiotensin II type I receptor, angiotensin II receptor blockers (ARBs) induce absolute inhibition of angiotensin II activity and thus may offer clinical benefits beyond those achieved with ACE inhibitors.

The VALIANT study was a randomized, double-blinded, clinical trial that included acute MI (between 0.5 to 10 days previously) patients with signs of heart failure or evidence of left-ventricular systolic dysfunction. The trial enrolled 14,808 patients aged 18 years and older, of which 70% were on concomitant β-blockers. VALIANT was originally designed to demonstrate the superiority of valsartan alone (target dose: 160 mg twice daily) or in combination with captopril (target doses: 160 mg/day and 150 mg/day, respectively) compared with captopril monotherapy (target dose: 150 mg/day) in reducing all-cause mortality. However, in the event that valsartan did not prove to be superior to captopril, the noninferiority of valsartan relative to captopril was to be assessed. The primary end point was all-cause mortality. The null hypothesis presented in the VALIANT trial stated that the relative risk in all-cause mortality would be at least 13% higher with valsartan treatment in comparison with captopril treatment. The all-cause mortality rates for the valsartan-, combination-, and captopril-treated patients (19.9%, 19.3%, and 19.5%, respectively) were similar to the rate seen with captopril in the SAVE trial (20%). The hazard ratio for all-cause mortality in the valsartan group as compared with the captopril group was 1.00 (97.5% CI, 0.90 to 1.11, $p = 0.98$), and the hazard ratio for all-cause mortality in the combined treatment group as compared with the captopril group was 0.98 (97.5% CI, 0.89 to 1.09, $p = 0.73$). The prespecified criteria of equivalence was fulfilled, as the 97.5% one-sided upper limit of the relative difference in all-cause mortality was 11%. The study investigators concluded that valsartan matched the efficacy of captopril in reducing all-cause mortality rates in acute MI patients with signs of heart failure or evidence of left-ventricular dysfunction [7, 29–36].

Acute and Chronic Therapy Trials

The therapeutic regimens utilized in clinical trials are usually indicated for acute or chronic medical conditions. *Acute medical conditions* warranting hospitalization or occurring during hospitalization for a different condition may include acute myocardial infarction, acute decompensated heart failure, hypertensive crisis, postoperative atrial fibrillation, and pulmonary embolism (Fig. 12.1.15). In general, because therapy is targeted toward alleviating a life-threatening condition such as an occluded artery or toward improving a seriously compromised parameter such as blood pressure or pulmonary capillary wedge pressure, short-term therapy (hours to days) is the standard in acute therapy trials. Outcomes evaluated in these trials can range from changes in hemodynamic parameters to all-cause mortality. Because the measured outcomes have been shown to occur soon after the acute event, follow-up times usually do not exceed 30 to 60 days.

- Disorders like acute myocardial infarction, acute decompensated heart failure, postoperative atrial fibrillation, hypertensive crises, and pulmonary embolism
- Can be treated with short term therapy
- can be conducted with short term follow-up (usually up to 30 days):
  - VMAC trial:
    - Nesiritide vs standard of care in acute decompensated heart failure

FIG. 12.1.15. Acute Therapy Trials
Source: McBride BF, White CM. *Pharmacother* 2003;23:997-1020.; White CM. *Ann Pharmacother* 1999;33:1063-72.; White CM. *Circulation* 2003;108[Suppl II]:200-6

At present, there is unequivocal evidence that the earlier (within 12 hours) thrombolysis is administered to an acute MI patient, the more favorable the outcome. In the GUSTO-I trial of 41,021 patients, accelerated tissue plasminogen activator (t-PA; 15 mg bolus dose, 0.75 mg/kg infused over 30 minutes, and 0.50 mg/kg infused over 60 minutes) was compared with three other intravenous thrombolytic regimens (streptokinase monotherapy or in combination with t-PA). The primary end point in this randomized, double-blinded trial was all-cause mortality at 30 days of follow-up. The GUSTO-1 (Global Utilization of Streptokinase and Tissue Plasminogen Activator for Occluded Arteries) trial found a 30-day mortality rate of 6.3% for the front-loaded t-PA regimen, which was significantly lower than the 7.2% mortality with streptokinase and subcutaneous heparin ($p = 0.001$) and less than the 7.4% mortality with streptokinase and intravenous heparin ($p = 0.001$). The improvement in mortality was already apparent after only 24 hours, with t-PA–treated patients having a significantly reduced mortality rate. Of note, the Kaplan–Meier mortality curves for all treatment groups arrived at a plateau within 3 weeks, well before the 30-day follow-up. Several thrombolytic trials following GUSTO-I utilized short-term treatment (seconds to hours) and 30- to 35-day follow-up times for mortality.

To date, the majority of decompensated heart failure trials have evaluated short-term end points such as hemodynamic and symptomatic changes. The Vasodilation in the Management of Acute CHF (VMAC) study enrolled 489 hospital patients with decompensated heart failure and dyspnea at rest and randomized them to receive intravenous nesiritide, intravenous nitroglycerin, or placebo, in addition to standard therapy. The primary end points were patient self-evaluation of dyspnea at 3 hours and change in pulmonary capillary wedge pressure (PCWP) at 3 hours among patients who had undergone pulmonary artery catheterization. At 3 hours, the mean reduction in PCWP was greater with nesiritide (−5.8 mm Hg) compared with nitroglycerin (−3.8 mm Hg; $p = 0.03$) and placebo (−2 mm Hg; $p < 0.001$). The investigators found that nesiritide was more effective at reducing the symptoms of dyspnea versus placebo ($p = 0.03$), but the difference between nesiritide and nitroglycerin was not significant on this measure.

The OPTIME-CHF (Outcomes of a Prospective Trial of Intravenous Milrinone for Exacerbations of Chronic Heart Failure) assessed the in-hospital management of 951 patients with acute NHYA class III or IV heart failure exacerbation, but not in cardiogenic shock. In addition to standard diuretic and ACE inhibitor therapy, patients were randomized to a 48-hour infusion of either milrinone or placebo. The primary end point was the total number of days hospitalized for cardiovascular causes within 60 days after randomization. The investigators found no difference in the number of days hospitalized for cardiovascular causes from the time of randomization to day 60, whether patients received milrinone or not. Despite the neutral results, this study was noteworthy for examining an intermediate-term end point rather than a short-term end point.

Hypertensive crisis represents a medical emergency that must be dealt with immediately in order to prevent complications such as hemorrhagic stroke, renal failure, MI, or pulmonary edema. Patients presenting with hypertensive emergency are usually started promptly on parenteral antihypertensive therapy in an intensive care unit. Blood pressure reductions approaching 25% should occur over 2 to 3 hours, with parenteral therapy continuing for an additional 6 to 12 hours before instituting oral antihypertensive therapy. Some clinical studies evaluating the efficacy of antihypertensive agents in hypertensive crisis have used diastolic blood pressure reduction after a few hours of treatment as a primary outcome measure. Follow-up times up to 48 hours are commonly used to monitor for adverse events, death, and laboratory parameters such as blood urea nitrogen and creatinine.

Postoperative atrial fibrillation occurs in 27% to 40% of patients undergoing cardiac surgery, with a mean onset time of 2 to 3 days after surgery. Kaplan–Meier analyses for 30 to 50 days atrial fibrillation–free survival from various clinical studies assessing the efficacy of amiodarone as prophylaxis against postoperative atrial fibrillation have demonstrated a plateau occurring within 10 days. The duration of amiodarone prophylactic treatment ranged from 5 to 13 days in published clinical trials, with follow-up times of 30 days.

Patients with acute massive pulmonary embolism can experience profound hemodynamic compromise, as evidenced by diminished cardiac output and elevations in pulmonary artery pressure. Thrombolytic therapy is widely accepted as the treatment of choice for acute massive pulmonary embolism, with alteplase and streptokinase representing the most commonly used agents. At present, with the exception of two recent trials, most trials have not assessed the potential long-term benefits of thrombolytic therapy in acute massive pulmonary embolism. Short-term treatment regimens of 2 hours were used and the predominant primary end point was immediate hemodynamic improvement. Follow-up times of 18 to 28 hours have been reported with some clinical trials [16, 33, 37–51].

In contrast with acute therapy trials, *chronic therapy trials* evaluate the effects of interventions on disease states requiring

- Most cardiac disorders require chronic management:
  - Factors such as hyperlipidemia and hypertension are generally not curable:
    - Need chronic suppressive therapy to attain and maintain normal levels
    - Studies evaluating antihyperlipidemic efficacy require several weeks (4-6) of treatment
    - Studies evaluating impact of modifying risk factors on actual cardiac events require large populations and a long evaluation time

FIG. 12.1.16. Chronic Therapy Trials
Source: White CM.. Dyslipidemias. In: PSAP, 5th Edition. ACCP, Inc, Kansas City, MO.. 2004: pg 16590.; Jones PH. *Am J Cardiol* 2003;93:152-160.; SOLVD Investigators. *NEJM* 1991;325:293-302.

continuous therapy, as diseases such as hyperlipidemia and hypertension are generally incurable (Fig. 12.1.16). Patients with hyperlipidemia and hypertension require chronic suppressive therapy to attain and maintain normal serum cholesterol levels and blood pressure, respectively. Clinical trials have demonstrated that within 4 weeks of discontinuing cholesterol-lowering and blood-pressure lowering medications, patients' serum cholesterol levels and blood pressure revert back to pretreatment levels.

Hydroxy-3-methylglutaryl coenzyme A reductase inhibitors (HMG CoA RIs) represent the most powerful class of drugs for lowering low-density lipoprotein cholesterol (LDL-C). Since their introduction in the 1980s, there have been at least seven HMG CoA RIs approved by the Food and Drug Administration (FDA) for clinical use, with a total of six currently available for use in the United States. The time to achieve peak reductions in LDL-C has been shown to approach 4 to 6 weeks for fluvastatin, lovastatin, pravastatin, and simvastatin, whereas atorvastatin and rosuvastatin require 2 to 4 weeks to exert maximal LDL-C–lowering effects. Consequently, forced-titration studies assessing the dose-response effects of HMG CoA RIs have used 6-week intervals between dosage adjustments. Moreover, the duration of the active treatment phase of HMG CoA RI efficacy studies has often been 6 weeks.

Studies evaluating the impact of LDL-C reduction on major coronary events, including mortality, require large populations and a long evaluation time. The Heart Protection Study (HPS) examined the efficacy of simvastatin treatment in 20,536 adults (aged 40 to 80 years) who were at high risk for a cardiovascular disease event. Patients were randomized to simvastatin (40 mg/day) or placebo. The primary end points included a composite of fatal and nonfatal vascular events for subcategory analysis and total mortality for overall analysis. The HPS investigators estimated that among 20,000 high-risk patients followed for an average of 5 years, there might be 1,500 coronary deaths, plus similar numbers of non-fatal MIs. They also estimated that if cholesterol-lowering therapy resulted in a 25% reduction in 5-year CHD mortality and a 15% reduction in all-cause mortality, a study of that magnitude would likely demonstrate such effects at sufficient levels of statistical significance (i.e., >90% power to achieve $p < 0.01$) [4, 52–67].

Special Populations Studies

Until recently, heart failure trials included very few African Americans, a *special population* that is characterized by having higher mortality rates due to this condition than non-black patients (Fig. 12.1.17). Moreover, before the publication of the results from A-HeFT (The African-American Heart-Failure Trial), no prospective data on drug therapy for African Americans with left-ventricular dysfunction was available to guide therapy in this population. Retrospective data from major clinical trials, including V-HeFT (The Veterans Administration Cooperative Vasodilator-Heart Failure Trial) I, V-HeFT 2, and SOLVD (Studies of Left Ventricular Dysfunction) suggested that African Americans did not derive benefit from ACE inhibitor therapy compared with hydralazine/nitrate treatment and that better outcomes were realized with hydralazine/nitrate treatment compared with placebo. In contrast, non-black patients have been shown to have better mortality outcomes with ACE inhibitor (enalapril) therapy compared with hydralazine/nitrate. Furthermore, data from the BEST (The Beta-Blocker Evaluation of Survival Trial) study hinted that race may have played a role in the lack of benefit seen with bucindolol (beta-blocker) therapy compared with placebo, as African Americans comprised 23% of the study population.

The post hoc analysis of V-HeFT included 180 African-American subjects and 450 non-black patients. During a follow-up time of 2 years, hydralazine/nitrate treatment resulted in a 44% lower mortality rate in black patients compared with placebo ($p = 0.04$) and a 10% lower mortality rate in non-black patients compared with placebo (p-value non-significant). Outcomes from A-HeFT have further established the role of vasodilators in the management of heart failure in African Americans. A-HeFT randomized 1,050 self-classified African Americans with NYHA class III–IV heart failure and reduced left-ventricular function to receive a fixed-dose combination of isosorbide dinitrate (20–40 mg t.i.d.)–hydralazine

- Ethnic Differences:
  - Differences in blood pressure response between Caucasian and African American patients
  - Etiologies for certain diseases:
    - African-Americans have hypertension as #1 reason for CHF vs Caucasians with coronary disease as #1 reason
  - Differences in innate metabolic ability of drugs:
    - Due to genetic polymorphisms

FIG. 12.1.17. Special Population Studies I
Source: JS, Nappi JM. *Ann Pharmacother* 2002;36:471-8.; Exner DV. *NEJM* 2001;344:1351-7.; Chow MSS. *J Clin Pharmacol* 2001; 41:92-6.

(37.5–75 mg p.o. t.i.d.) or placebo. Nonischemic cardiomyopathy was the cause of heart failure in 77% of the patients. Conversely, ischemic cardiac disease has been shown to be the most likely etiology of heart failure in trials enrolling non-black patients. At randomization, 84% of the patients were taking ACE inhibitors, 71% beta-blockers, and 88% diuretics. The primary end point was a composite of all-cause mortality, first heart failure hospitalization, and change in quality-of-life score at 6 months. The trial was stopped early after a significant benefit emerged among patients receiving vasodilator therapy at 10 months (mean) follow-up. When compared with placebo, the addition of fixed-dose isosorbide dinitrate–hydralazine to standard therapy improved survival by 43% ($p = 0.02$), decreased heart failure hospitalization by 33% ($p = 0.001$), and improved quality of life ($p = 0.02$).

Exner and associates recently reanalyzed pooled data from the SOLVD prevention and treatment trials. A total of 800 African-American subjects were matched with 1,196 white patients from the two studies, with left-ventricular ejection fraction, gender, randomly assigned therapy (enalapril or placebo), and age comprising the matching criteria. However, there were some differences between the black and matched white patients in regard to medical history, as whites were more likely to have prior ischemic heart disease and blacks were more likely to have a history of hypertension. All-cause mortality and heart failure hospitalizations were prospectively recorded throughout mean follow-up times of 35 months in the prevention trial and 33 months in the treatment trial. At 1 year, nonsignificant reductions in systolic blood pressure (SBP) and diastolic blood pressure (DBP) from baseline occurred in blacks receiving enalapril, whereas significant reductions in SBP and DBP of 5.0 mm Hg ($p < 0.001$) and 3.6 mm Hg ($p < 0.001$), respectively, were noted in white patients receiving enalapril. Enalapril therapy did not change the risk of death from any cause among either the African American or the matched white patients. However, enalapril treatment was associated with a 40% reduction (95% CI, 32 to 47%, $p < 0.01$) in the risk of heart failure hospitalization among white patients compared with similar black patients.

The BEST trial randomized 2,708 patients with NYHA class III (>90%) or IV heart failure and reduced left-ventricular ejection fraction (≤35%) to bucindolol, a nonselective beta-blocker (target dose: 100–200 mg/day) or placebo. The primary end point was death due to any cause. After 2 years of follow-up, the data and safety monitoring board halted the study, as the primary end point was not significantly different between the two treatment groups. However, a prespecified subgroup analysis showed that there was a significant reduction in mortality in the white population (hazard ratio 0.82, 95% CI 0.7 to 0.96, $p = 0.01$), but no effect in the African-American population (hazard ratio 1.17, 95% CI 0.89 to 1.53, $p = 0.27$). Of note, previous heart failure studies, specifically the CIBIS-II and the MERIT-HF trials, showed positive benefits associated with bisoprolol and metoprolol, respectively. However, unlike the BEST trial, which included a large proportion of African-American patients (23%), the CIBIS-II and MERIT-HF trials were carried out mostly in Europe, with a predominantly non-black population. Interestingly, the US Carvedilol Heart Failure Trials Program (217 black, 877 non-black patients) showed a similar level of benefit with carvedilol in both black and non-black patients.

Many factors may have a potential role in accounting for racial differences in response to vasodilators, ACE inhibitors, and beta-blockers in heart failure patients. The AHeFT and V-HeFT investigators proposed that the bioavailability of nitric oxide might be lower in blacks than in whites. Because organic nitrates induce vasodilation by functioning as nitric oxide donors, the superior efficacy of the hydralazine–nitrate combination observed in AHeFT and V-HeFT might be explained by enhanced nitric oxide availability in black patients. The decreased bioavailability of nitric oxide observed in black patients might also account for the lesser response to ACE inhibitors observed in black patients enrolled in the SOLVD trial. ACE inhibition leads to enhanced nitric oxide release through increased kinin activity. Because non-black patients have higher bioactivity of endogenous nitric oxide than blacks, they are more likely to derive benefit from ACE inhibitor therapy. A potential explanation for racial differences in response to bucindolol and carvedilol is that hypertension in blacks is more responsive to a beta-blocker when it is combined with an alpha-antagonist. Carvedilol, unlike bucindolol, acts as a nonselective beta-blocker and as an alpha-antagonist [68–74].

In the past decade, the application of *pharmacogenomic research* to elucidate gene alterations or deletions that impact drug absorption and metabolism along with the patients' response to therapy has been used to explain some of the variability in these aspects of drug therapy (Fig. 12.1.18). In particular, much attention has been directed toward evaluating the variability in the expression of the cytochrome P-450 (CYP) family of enzymes, the role of P-glycoprotein gene alteration in drug disposition, and the effect of aberrant genes encoding the myocardial ion channels on QT-prolongation.

- Genetic polymorphisms:
  - Gene alterations or deletions that alter biologic functioning or changes in drug metabolism:
    - CYP2D6 metabolizer status:
      - 7% of Caucasian & African American population are poor metabolizers for CYP2D6 drugs
      - 1% of Chinese
      - Alters blood levels of CYP2D6 substrates such as propafenone
    - P-glycoprotein gene alteration (MDR-1 gene defect):
      - Diminishes functioning of this efflux pump
      - Elevates concentrations of substrates such as digoxin
    - Genetic long QT syndrome:
      - Genetic polymorphism of myocardial K and Na ion channels
      - Increased baseline QTc interval enhancing risk of Torsade de Pointes arrhythmia

FIG. 12.1.18. Special Population Studies II
Source: Abernethy DR. *Circulation* 2000;101:1749-53.; Chow MSS. *J Clin Pharmacol* 2001;41:92-6.; Tanigawara Y. *Ther Drug Monitor* 2000;22:137-40. Roden D. *N Engl J Med* 2004;350:1013-22.

Genetic polymorphisms of the CYP2C subfamily (CYP2C8, CYP2C9, CYP2C18, CYP2C19) have been shown to increase the risk of drug toxicity in susceptible individuals exposed to drugs metabolized by CYP2C isoenzymes. The CYP2C9 isoenzyme is responsible for the biotransformation of one of the most commonly used cardiovascular drugs, warfarin. The clinical consequence of polymorphisms of the CYP2C9 isoenzyme can be profound, as the dose required for adequate anticoagulation was shown to be markedly lower for patients with impaired CYP2C9 activity in some reports. Approximately 2% to 6% of Caucasians exhibit poor metabolizer (PM) phenotype of mephenytoin oxidation (metabolized by CYP2C19), whereas PMs represent 19% to 23% of the Japanese population. Polymorphisms of CYP2C9 arise from two alleles, CYP2C9*2 (Arg144Cys substitution) and CYP2C9*3 (Ile359Leu mutation), with similar or lower frequencies seen in Asian populations compared with Caucasians. Less is known about the interethnic differences in the incidence of CYP2C9/CYP2C19 PM phenotypes among other Asian populations. One study showed that no significant differences in CYP2C19 phenotype were found among Japanese, Chinese, Thai, and Vietnamese populations.

Another CYP isoform that is known to be important for cardiovascular drug biotransformation is CYP2D6, the isoenzyme responsible for the metabolism of propafenone to its major metabolite, 5-hydroxypropafenone. Propafenone is a sodium channel blocker that possesses weak beta-blocking activity in addition to its antiarrhythmic properties. Individuals with a lower level of CYP2D6 activity are theoretically at increased risk of central nervous system side effects, as they will exhibit markedly greater beta-adrenoceptor blockade with propafenone. Studies have shown that approximately 7% of African Americans and Caucasians are poor metabolizers of the CYP2D6 probe dextomethorphan, whereas poor metabolizers represent 1% of the Chinese population.

The P-glycoprotein (Pgp) membrane transporter is encoded by the human multidrug-resistance 1 (MDR1) gene and serves as a renal drug transporter, as an efflux pump from the capillary endothelial cells of the brain, and as an impediment to drug absorption in the intestinal wall. Pgp has a critical role in the absorption and excretion of digoxin, a drug commonly used by patients with heart failure or atrial fibrillation. Intestinal Pgp limits the bioavailability of substrate drugs such as digoxin, by pumping them from the enterocytes back into the intestinal lumen. Single nucleotide polymorphisms (SNPs) in the MDR1 gene can potentially lower intestinal Pgp levels, thereby increasing digoxin plasma concentrations in individuals homozygous mutant for the MDR1 SNPs.

In recent years, the most common cause of the withdrawal or use of marketed drugs has been attributed to excessive QT_c prolongation associated with polymorphic ventricular tachycardia, or torsades de pointes (TdP). After the removal of grepafloxacin from the marketplace due to QT_c interval prolongation and associated arrhythmias (7 deaths), the U.S. Food and Drug Administration requested that manufacturers of recently developed fluoroquinolones document the effects of their products on QT_c interval duration. QT_c interval prolongation is a surrogate marker of cardiotoxicity, as the risk of developing TdP is proportional to the magnitude of QT_c prolongation. The Committee for Proprietary Medicinal Products stated that prolongation of the QT_c interval by more than 30 seconds and in excess of 60 ms should be classified as a potential adverse effect and a definite adverse effect, respectively. Factors that predispose to QT_c prolongation and higher risk of TdP include electrolyte disturbances, cardiac disease, hypertension, diabetes mellitus, hypothyroidism, obesity, alcohol/cocaine abuse, increased age, female gender, and congenital long QT_c interval syndrome. Genetic studies have identified six separate genes that, if altered, can cause the congenital long QT syndrome. The effects of human ether-a-go-go gene (HERG) inhibition on QT_c prolongation have been particularly well studied in recent years. Mutation of the HERG-encoded rapidly activating delayed-rectifier potassium channel (Ikr) is a significant contributor to QT_c prolongation, as repolarization delay results from potassium accumulation within the myocyte. The likelihood of inducing TdP increases with the generation of repetitive early after depolarizations, a phenomenon that can result from delayed repolarization [70, 75–87].

Drug interactions are an important consideration in the treatment of patients with cardiovascular disease, as many of the drugs used by these patients are associated with serious adverse effects. Drug interactions can be classified as either pharmacokinetic or pharmacodynamic in nature (Fig. 12.1.19). Pharmacokinetic interactions modify the absorption, metabolism, or excretion of a drug, whereas pharmacodynamic interactions change the pharmacologic response to a drug. Commonly reported pharmacokinetic interactions may involve alterations in drug biotransformation mediated by the cytochrome P-450 system (CYP), P-glycoprotein modulation (Pgp), and inhibition of cation tubular secretion. Pharmacodynamic interactions can result in synergistic or antagonistic responses to drug therapy.

- Pharmacokinetic drug interactions:
  - P-glycoprotein interactions
  - Cytochrome P450 interactions
  - Cation tubular secretion inhibitors
- Pharmacodynamic drug interactions:
  - Synergy:
    - ARB + ACE I for renal protection in diabetics
  - Antagonism:
    - Magnesium attenuating ibutilide QTc interval prolongation

FIG. 12.1.19. Special Population Studies III
Source: White CM. *Formulary* 2002;27:588-93;. Finch C. *Arch Intern Med* 2002;162:985-92.; Jacobsen P. *Kidney Int* 2003;63:1874-80.; Caron MF. *Pharmacotherapy* 2003;296-300. Tanigawara Y. *Ther Drug Monitor* 2000;22:137-40.; Abernethy DR. *Circulation* 2000;101:1749-53.

Over the past few decades, numerous studies documented an interaction between rifampin and drugs that are substrates for CYP3A4. Rifampin is a potent inducer of CYP3A4, and recent evidence suggests that it is an inducer of Pgp. Greiner and associates found, in 8 healthy men, that when digoxin 1 mg (orally or intravenously) was administered after a 14-day course of rifampin 600 mg/day, the AUC (area under the plasma concentration versus time curve) and C_{max} (maximum plasma concentration) of digoxin were decreased by 43% and 58%, respectively. Intestinal Pgp levels were increased nearly four fold with rifampin therapy. Because it has a narrow therapeutic range (0.8–2 µg/L), even minor reductions in digoxin plasma concentration can result in therapeutic failure, especially among atrial fibrillation patients who require higher levels for rate control.

The 3-hydroxy-3-methylglutaryl coenzyme A reductase inhibitors (statins) are commonly used antihyperlipidemic agents that are well tolerated and relatively safe. The most common adverse effects are headache and gastrointestinal-related, but myopathy has also been of some concern. In August 2001, cerivastatin (Baycol®; Bayer Corporation) was removed from the market after causing 31 deaths associated with rhabdomyolysis in the United States and 52 deaths worldwide. Statin-induced myotoxicities are dose-related and related to the lipophilicity of the drug. Other drug-related properties that may increase the risk of myopathy are high systemic exposure, high bioavailability, limited protein binding, and potential for drug–drug interactions metabolized by CYP pathways. Three of the six statins on the U.S. market, atorvastatin, lovastatin, and simvastatin, are CYP3A4 substrates. A review by Omar and Wilson of all reports of statin-associated rhabdomyolysis reported to the FDA (between November 1997 and March 2001) showed that in ~35% of all cases (n = 601), CYP3A4 inhibitors were used concomitantly. The importance of avoiding concomitant administration of CYP3A4 inhibitors is highlighted in the most recently updated monograph of simvastatin, as concurrent use of itraconazole, ketoconazole, erythromycin, clarithromycin, protease inhibitors, large quantities of grapefruit juice (>1 quart/day), or nefazodone is contraindicated.

Dofetilide is the only class III antiarrhythmic agent indicated for both acute cardioversion of atrial fibrillations/atrial flutter and maintenance of normal sinus rhythm. It is primarily eliminated renally (60%) through glomerular filtration and cationic tubular secretion. Cimetidine, ketoconazole, prochlorperazine, megestrol, and trimethoprim (including in combination with sulfamethoxazole) inhibit tubular secretion of dofetilide. Consequently, simultaneous administration of these agents will result in elevated dofetilide plasma concentrations. Because the incidence of torsades de pointes (TdP) increases with elevated dofetilide plasma concentrations, its use is contraindicated in patients receiving these agents.

At present, results from studies of type 1 diabetes patients with nephropathy support the important role of the renin–angiotensin–aldosterone system (RAAS) in progression of diabetic renal disease, as angiotensin-converting enzyme inhibition has been shown to reduce albuminuria. It has been suggested that monotherapy with ACE inhibitors is insufficient for complete inhibition of the RAAS, as demonstrated by the generation of angiotensin II by ACE-independent pathways such as chymase. Angiotensin receptor blocker (ARB) therapy will circumvent the effects of angiotensin II generated by ACE-independent pathways because it prevents the emergence of unfavorable effects secondary to the action of angiotensin II through the angiotensin II type 1 receptor. However, unlike ACE inhibitors, the degradation of bradykinin, a powerful vasodilator, is not prevented by treatment with ARBs. A recent study conducted by Jacobsen et al. lends support to increased benefit realized with dual blockade of the RAAS with respect to diabetic renal disease progression. Jacobsen and associates enrolled 24 type 1 diabetes patients with nephropathy in a randomized, double-blind cross-over study to compare the effects of combining maximum-dose ACE inhibitor with an ARB with that of monotherapy with maximum-dose ACE inhibitor on urinary albumin excretion rate. After an 8-week course of treatment with placebo or irbesartan, 300 mg/day, added on top of enalapril, 40 mg/day, albuminuria was shown to be reduced by 25% with combination therapy (95% CI, 15 to 34, p < 0.001).

Ibutilide is a class III antiarrhythmic agent indicated for the rapid conversion of recent-onset atrial fibrillation or flutter. Ibutilide has been shown to be an effective pharmacologic agent for conversion to normal sinus rhythm, with efficacy rates of up to 50% and 70%, respectively, among patients with atrial fibrillation or flutter. However, widespread use of this agent has been limited by the risk of TdP, with an estimated occurrence rate of 4%. Recently, Caron and associates demonstrated the antagonistic effect of prophylactic administration of intravenous magesium sulfate on the QT_c-prolonging effect of ibutilide. Caron et al. enrolled 20 patients with atrial fibrillation or flutter in a prospective, randomized, double-blind, placebo-controlled trial to evaluate the effect of intravenous magnesium sulfate on the QT_c intervals of patients undergoing chemical cardioversion with ibutilide. The QT_c interval in magnesium sulfate–treated patients was not significantly altered from baseline at 30 minutes after the final dose of ibutilide, whereas the QT_c interval at the same time point in placebo-treated patients was 18% higher than that at baseline (p = 0.01) [75, 76, 88–99].

To date, data evaluating the safety of using weight-based dosing of cardiovascular drugs in *obese patients* are limited (Fig. 12.1.20). Use of low-molecular-weight heparins (LMWHs) is widespread, as current indications for use of these agents include treatment and prophylaxis of deep-vein thrombosis and management of acute coronary syndromes. Three LMWH products are currently available in the United States and include enoxaparin (Lovenox®), dalteparin (Fragmin®), and tinzaparin (Innohep®). Because clinical trials have included only limited numbers of obese patients, the optimal dosage of LMWHs has not been established for this patient population.

LMWHs are dosed according to body weight, but because intravascular volume does not have a linear relationship with

- Obese or emaciated subjects:
  - Weight or body surface area may not be adequate in calculating drug clearance since neither reflects the size or function of the liver or kidney
  - Do you use a standard dose, dose by ideal weight, or dose by actual weight?
    - Need to study to find out
- What about age?

FIG. 12.1.20. Special Population Studies IV-Obesity
Source: Winter ME (Ed). Basic Clinical Pharmacokinetics. Applied Therapeutics, Inc 1994.

total body weight, weight-adjusted dosing in obese patients could lead to excessive anticoagulation. Findings from recent studies suggest that LMWH pharmacodynamics are independent of body mass index (BMI) and body weight and are comparable with those in healthy normal-weight volunteers. Clinical studies evaluating the anticoagulant response to enoxaparin, tinzaparin, and dalteparin have included subjects weighing up to 144 kg, 165 kg, and 190 kg, respectively. However, because these studies included few patients with a BMI of >50 kg/m^2 or a TBW >150 kg, the American College of Chest Physicians recommends monitoring of anti-Xa activity in these patients. Peak activity levels should be obtained 4 hours after a subcutaneous dose of a LMWH [89, 100–104].

Conclusions

In *conclusion*, the prevalence of cardiovascular disease, large number of cardiac drugs on the market, ability to niche new drugs for subpopulations, and extensive therapeutic targets for new drugs make this area of drug discovery and development very exciting (Fig. 12.1.21). Treatment guidelines and established standards of care ensure that evidence-based drugs are used more extensively and impact study design as we have seen. New drugs against new targets need to deliver better results than older, more established drugs or be better tolerated or safer. New drugs in the same class as older, more established

- Cardiovascular disease is prevalent and a major market
- Many drugs already available and well studied for the treatment of most diseases
- Treatment guidelines and standards of care are common, impacting study design
- New drugs need to have significant pharmacologic, pharmacokinetic, or therapeutics benefits
- Drug can be either a replacement or adjunct for standard therapies
- The potential for the drug in the marketplace determines which types of studies to employ to either alter current practice or niche the drug

FIG. 12.1.21. Conclusions

drugs need to do the same. These benefits can be pharmacologic, pharmacokinetic, therapeutic, or pharmacoeconomic in nature. By fully assessing the potential benefits, different study types can be designed to allow for substantiation of these potential benefits in order to impact the marketplace.

References

1. American Heart Association. Heart disease and stroke statistics. www.americanheart.org/statistics/index.html
2. American Heart Association, 2003 ACC/AHA Guidelines for CHF. Available at http:www.acc.org/clinical/guidelines.
3. Lamberti MJ (ed). State of the Clinical Trials Industry. A Sourcebook of charts and statistics. Thomson CenterWatch, Boston, MA, 2005.
4. Patrons C. Platelet-active drugs. Chest 1998;114:470s-8s.
5. White CM.. Dyslipidemias. In: PSAP, 5th Edition. ACCP, Kansas City, MO, 2004:16590.
6. Song JC, White CM. Pharmacologic, pharmacokinetic, and therapeutic differences among angiotensin II receptor antagonists. Pharmacotherapy. 2000 Feb;20(2):130-9. Review.
7. Song JC, White CM. Clinical pharmacokinetics and selective pharmacodynamics of new angiotensin converting enzyme inhibitors: an update. Clin Pharmacokinet. 2002;41(3):207-24. Review.
8. White CM. Pharmacologic, pharmacokinetic, and therapeutic differences among ACE inhibitors. Pharmacotherapy 1998;18:588-99.
9. Cohn JN, Johnson G, Ziesche S, Cobb F, Francis G, Tristani F, Smith R, Dunkman WB, Loeb H, Wong M, et al. A comparison of enalapril with hydralazine-isosorbide dinitrate in the treatment of chronic congestive heart failure. N Engl J Med. 1991 Aug 1; 325(5):303-10.
10. Antman EM, McCabe CH, Gurfinkel EP, Turpie AG, Bernink PJ, Salein D, Bayes De Luna A, Fox K, Lablanche JM, Radley D, Premmereur J, Braunwald E. Enoxaparin prevents death and cardiac ischemic events in unstable angina/non-Q-wave myocardial infarction. Results of the thrombolysis in myocardial infarction (TIMI) 11B trial. Circulation. 1999 Oct 12;100(15):1593-601.
11. Pitt B, Remme W, Zannad F, Neaton J, Martinez F, Roniker B, Bittman R, Hurley S, Kleiman J, Gatlin M. Eplerenone Post-Acute Myocardial Infarction Heart Failure Efficacy and Survival Study Investigators. Eplerenone, a selective aldosterone blocker, in patients with left ventricular dysfunction after myocardial infarction. N Engl J Med. 2003 Apr 3;348(14):1309-21. Epub 2003 Mar 31. Erratum in: N Engl J Med. 2003 May 29;348(22):2271.
12. Colucci WS, Packer M, Bristow MR, Gilbert EM, Cohn JN, Fowler MB, Krueger SK, Hershberger R, Uretsky BF, Bowers JA, Sackner-Bernstein JD, Young ST, Holcslaw TL, Lukas MA. Carvedilol inhibits clinical progression in patients with mild symptoms of heart failure. US Carvedilol Heart Failure Study Group. Circulation. 1996 Dec 1;94(11):2800-6.
13. Sander S, White CM. Ximelagatran: a new oral anticoagulant. Formulary 2004;39:398-404.
14. Wallentin L, Wilcox RG, Weaver WD, Emanuelsson H, Goodvin A, Nystrom P, Bylock A; ESTEEM Investigators. Oral ximelagatran for secondary prophylaxis after myocardial infarction: the ESTEEM randomised controlled trial. Lancet. 2003 Sep 6; 362(9386):789-97.

15. Chronic heart failure in the adult. ACC/AMA 2005 guideline update. Available at www.acc.org/quality and science/clinical/topic.htm#guidelines
16. White CM, Giri S, Tsikouris JP, Dunn A, Felton K, Reddy P, Kluger J. A comparison of two individual amiodarone regimens to placebo in open heart surgery patients. Ann Thorac Surg. 2002 Jul;74(1):69-74.
17. White CM, Caron MF, Kalus JS, et al. Intravenous plus oral amiodarone, atrial pacing, or both strategies to prevent post-cardiothoracic surgery atrial fibrillation: the Atrial Fibrillation Suppression trial II (AFIST II). Circulation 2003;108(suppl. II):II-200-6.
18. Dahlof B, Devereux RB, Kjeldsen SE, Julius S, Beevers G, de Faire U, Fyhrquist F, Ibsen H, Kristiansson K, Lederballe-Pedersen O, Lindholm LH, Nieminen MS, Omvik P, Oparil S, Wedel H; LIFE Study Group. Cardiovascular morbidity and mortality in the Losartan Intervention For Endpoint reduction in hypertension study (LIFE): a randomised trial against atenolol. Lancet. 2002 Mar 23;359(9311):995-1003.
19. Lewis EJ, Hunsicker LG, Clarke WR, Berl T, Pohl MA, Lewis JB, Ritz E, Atkins RC, Rohde R, Raz I; Collaborative Study Group. Renoprotective effect of the angiotensin-receptor antagonist irbesartan in patients with nephropathy due to type 2 diabetes. N Engl J Med. 2001 Sep 20;345(12):851-60.
20. Guidelines 2000 for Cardiopulmonary Resuscitation and Emergency Cardiovascular Care. Part 6: advanced cardiovascular life support: 7C: a guide to the International ACLS algorithms. The American Heart Association in collaboration with the International Liaison Committee on Resuscitation. Circulation 2000;102:II42-57.
21. Grundy SM, Cleeman JI, Merz CNB, et al., for the Coordinating Committee of the National Cholesterol Education Program. Implications of recent clinical trials for the National Cholesterol Education Program Adult Treatment Panel III guidelines. Circulation 2004;110:227-39.
22. Chobanian AV, Bakris GL, Black HR, et al. Joint National Committee on Prevention, Detection, Evaluation, and Treatment of High Blood Pressure. National Heart, Lung, and Blood Institute; National High Blood Pressure Education Program Coordinating Committee. Seventh Report of the Joint National Committee on Prevention, Detection, Evaluation, and Treatment of High Blood Pressure. Hypertension 2003;42:1206-52.
23. Candesartan (Atacand®) prescribing information. AstraZeneca LP, Wilmington, DE, 2004.
24. Giri S, White CM, Dunn AB, et al. Oral amiodarone for prevention of atrial fibrillation after open heart surgery, the Atrial Fibrillation Suppression Trial (AFIST): a randomised placebo-controlled trial. Lancet 2001;357:830-6.
25. The ALLHAT Officers and Coordinators for the ALLHAT Collaborative Research Group. Major outcomes in high-risk hypertensive patients randomized to angiotensin-converting enzyme inhibitor or calcium channel blocker vs diuretic: the Antihypertensive and Lipid-Lowering Treatment to Prevent Heart Attack Trial (ALLHAT). JAMA 2002;288:2981-97.
26. Cannon CP, Braunwald E, McCabe CH, et al., for the Pravastatin or Atorvastatin Evaluation and Infection Therapy-Thrombolysis in Myocardial Infarction 22 Investigators. Intensive versus moderate lipid lowering with statins after acute coronary syndromes. N Engl J Med 2004;350:1495-504.
27. Vidt DG, White WB, Ridley E, et al. CLAIM Study Investigators. A forced titration study of antihypertensive efficacy of candesartan cilexetil in comparison to losartan: CLAIM Study. J Hum Hypertens 2001;15:475-80.
28. Kudenchuk PJ, Cobb LA, Copass MK, et al. Amiodarone for resuscitation after out-of-hospital cardiac arrest due to ventricular fibrillation. N Engl J Med 1999;341:871-8.
29. Guidelines 2000 for Cardiopulmonary Resuscitation and Emergency Cardiovascular Care. The American Heart Association in Collaboration with the International Liaison Committee on Resuscitation. Circulation 2000;102 (Suppl 1):1-384.
30. Assessment of the Safety and Efficacy of a New Thrombolytic (ASSENT-2) Investigators; Van De Werf F, Adgey J, Ardissino D, Armstrong PW, Aylward P, Barbash G, Betriu A, Binbrek AS, Califf R, Diaz R, Fanebust R, Fox K, Granger C, Heikkila J, Husted S, Jansky P, Langer A, Lupi E, Maseri A, Meyer J, Mlczoch J, Mocceti D, Myburgh D, Oto A, Paolasso E, Pehrsson K, Seabra-Gomes R, Soares-Piegas L, Sugrue D, Tendera M, Topol E, Toutouzas P, Vahanian A, Verheugt F, Wallentin L, White H. Single-bolus tenecteplase compared with front-loaded alteplase in acute myocardial infarction: the ASSENT-2 double-blind randomised trial. Lancet. 1999 Aug 28;354(9180):716-22.
31. Pfeffer MA, McMurray JJV, Velazquez EJ, et al., for the Valsartan in Acute Myocardial Infarction Trial Investigators. Valsartan, captopril, or both in myocardial infarction complicated by heart failure, left ventricular dysfunction, or both. N Engl J Med 2003;349:1893-906.
32. Jones B, Jarvis P, Lewis JA, Ebbutt AF. Trials to assess equivalence: the importance of rigorous methods. BMJ 1996;313:36-9.
33. Fibrinolytic Therapy Trialists' (FTT) Collaborative Group. Indications for fibrinolytic therapy in suspected acute myocardial infarction: collaborative overview of early mortality and major morbidity results from all randomized trials of more than 1000 patients. Lancet 1994;343:311-22.
34. Assessment of the Safety and Efficacy of a New Thrombolytic (ASSENT-2) Investigators. Single-bolus tenecteplase compared with front-loaded alteplase in acute myocardial infarction: the ASSENT-2 double-blind randomised trial. Lancet 1999;354:716-22.
35. Kober L, Torp-Pedersen C, Carlsen JE, et al., for the Trandolapril Cardiac Evaluation (TRACE) Study Group. A clinical trial of the angiotensin–converting-enzyme inhibitor trandolapril in patients with left ventricular dysfunction after myocardial infarction. N Engl J Med 1995;333:1670-6.
36. Pfeffer MA, Braunwald E, Moye LA, et al. on Behalf of the SAVE Investigators. Effect of captopril on mortality and morbidity in patients with left ventricular dysfunction after myocardial infarction: results of the survival and ventricular enlargement trial. N Engl J Med 1992;327:669-77.
37. The Acute Infarction Ramipril Efficacy (AIRE) Study Investigators. Effect of ramipril on mortality and morbidity of survivors of acute myocardial infarction with clinical evidence of heart failure. Lancet 1993;342:821-8.
38. McBride BF, White CM. Acute decompensated heart failure: a contemporary approach to pharmacotherapeutic management. Pharmacotherapy 2003;23:997-1020.
39. White CM. Prevention of suboptimal beta-blocker treatment in patients with myocardial infarction. Ann Pharmacother. 1999 Oct;33(10):1063-72. Review.

40. The GUSTO investigators. An international randomized trial comparing four thrombolytic strategies for acute myocardial infarction. N Engl J Med 1993;329:673-82.
41. International Joint Efficacy Comparison of Thrombolytics. Randomised, double-blind comparison of reteplase double-bolus administration with streptokinase in acute myocardial infarction (INJECT): trial to investigate equivalence. Lancet 1995;346:329-36.
42. Publication Committee for the VMAC (Vasodilatation in the Management of Acute CHF). Intravenous nesiritide vs nitroglycerin for treatment of decompensated congestive heart failure: a randomized controlled trial. JAMA 2002;287:1531-40.
43. Cuffe MS, Califf RM, Adams KF, et al. Outcomes of a Prospective Trial of Intravenous Milrinone for Exacerbations of Chronic Heart Failure (OPTIME-CHF) Investigators. Short-term intravenous milrinone for acute exacerbation of chronic heart failure: a randomized controlled trial. JAMA 2002;287:1541-7.
44. Black HR, Bakris GL, Elliott WJ. Hypertension: epidemiology, pathophysiology, diagnosis and treatment. In: Hurst's: The Heart, 10th ed. McGraw-Hill, New York, 2001:1553-1604.
45. Reisin E, Huth MM, Nguyen BP, Weed SG, Gonzalez FM. Intravenous fenoldopam versus sodium nitroprusside in patients with severe hypertension. Hypertension 1990;15(suppl 1):I-59-I-62.
46. Tumlin JA, Dunbar LM, Oparil S, et al. Fenoldopam, a dopamine agonist, for hypertensive emergency: a multicenter randomized trial. Fenoldopam Study Group. Acad Emerg Med 2000;7:653-62.
47. Hirschl MM, Binder M, Bur A, Herkner H, Mullner M, Woisetschlager L. Safety and efficacy of urapidil and sodium nitroprusside in the treatment of hypertensive emergencies. Intensive Care Med 1997;23:885-8.
48. Daoud EG, Strickberger SA, Man CK, et al. Preoperative amiodarone as prophylaxis against atrial fibrillation after heart surgery. N Engl J Med 1997;337:1785-91.
49. Meneveau N, Schiele F, Metz D, et al. Comparative efficacy of a two-hour regimen of streptokinase versus alteplase in acute massive pulmonary embolism: immediate clinical and hemodynamic outcome and one-year follow-up. J Am Coll Cardiol 1998;31: 1057-63.
50. Jerjes-Sanchez C, Ramirez-Rivera A, Arriaga-Nava R, et al. High dose and short-term streptokinase infusion in patients with pulmonary embolism: prospective with seven-year follow-up trial. J Thromb Thrombolysis 2001;12:237-47.
51. Sors H, Pacouret G, Azarian R, Meyer G, Charbonnier B, Simonneau G. Hemodynamic effects of bolus vs 2-h infusion of alteplase in acute massive pulmonary embolism. A randomized controlled multicenter trial. Chest 1994;106:712-7.
52. Meyer G, Sors H, Charbonnier B, et al. Effects of intravenous urokinase versus alteplase on total pulmonary resistance in acute massive pulmonary embolism: a European multicenter double-blind trial. The European Cooperative Study Group for Pulmonary Embolism. J Am Coll Cardiol 1992;19:239-45.
53. Jones PH, Davidson MH, Stein EA, et al. Comparison of the efficacy and safety of rosuvastatin versus atorvastatin, simvastatin, and pravastatin across doses (Stellar* Trial). Am J Cardiol 2003;89:152-60.
54. The SOLVD Investigators. Effect of enalapril on survival in patients with reduced left ventricular ejection fractions and congestive heart failure. N Engl J Med. 1991 Aug 1;325(5): 293-302.
55. Product information. Pravachol. Bristol-Myers Squibb, Princeton, NJ, 2003.
56. Product information. Crestor. AstraZeneca Pharmaceuticals LP., Wilmington, DE, 2003.
57. Product information. Lescol. Novartis Pharmaceuticals, East Hanover, NJ, 2002.
58. Product information. Lipitor. Parke-Davis (Pfizer), Parsippany, NJ, 2004.
59. Product information. Zocor. Merck & Co., White House Station, NJ, 2004.
60. Product information. Mevacor. Merck & Co., White House Station, NJ, 2002.
61. Jones P, Kafonek S, Laurora I, Hunninghake D. Comparative dose efficacy study of atorvastatin versus simvastatin, pravastatin, lovastatin, and fluvastatin in patients with hypercholesterolemia (the CURVES study) Am J Cardiol. 1998 Mar 1;81(5):582-7. Erratum in: Am J Cardiol 1998 Jul 1;82(1):128.
62. Karalis DG, Ross AM, Vacari RM, et al. Comparison of efficacy and safety of atorvastatin and simvastatin in patients with dyslipidemia with and without coronary heart disease. Am J Cardiol 2002; 89: 667-71.
63. Van Dam MJ, Penn HJAM, Hartog FR, et al. A comparison of the efficacy and tolerability of titrate-to-goal regimens of simvastatin and fluvastatin: a randomized, double-blind study in adult patients at moderate to high risk for cardiovascular disease. Clin Ther 2001;23:467-78.
64. Insull W, Kafonek S, Goldner D, et al. Comparison of efficacy and safety of atorvastatin (10 mg) with simvastatin (10 mg) at six weeks. Am J Cardiol 2001;87:554-9.
65. Farnier M, Portal JJ, Pascal M. Efficacy of atorvastatin compared with simvastatin in patients with hypercholesterolemia. J Cardiovasc Pharmacol Ther 2000;5:27-32.
66. Ballantyne CM, McKenney J, Trippe BS. Efficacy and safety of an extended-release formulation of fluvastatin for once-daily treatment of primary hypercholesterolemia. Am J Cardiol 2000;86:759-63.
67. Mogensen CE, Neldam S, Tikkanen I, et al. for the CALM study group. Randomised controlled trial of dual blockade of renin-angiotensin system in patients with hypertension, microalbuminuria, and non-insulin dependent diabetes: the candesartan and lisinopril microalbuminuria (CALM) study. BMJ 2000;321: 1440-4.
68. Heart Protection Study Collaborative Group. MRC/BHF Heart Protection Study of cholesterol lowering with simvastatin in 20,536 high-risk individuals: a randomised placebo-controlled trial. Lancet 2002;360:7-22.
69. Kalus JS, Nappi JM. Role of race in the pharmacotherapy of heart failure. Ann Pharmacother 2002;36:471-8.
70. Exner DV, Dries DL, Domanski MJ, Cohn JN. Lesser response to angiotensin-converting-enzyme inhibitor therapy in black as compared with white patients with left ventricular dysfunction. N Engl J Med 2001;344:1351-7.
71. Chow MSS, White CM, Lau CP, Fan C, Tang MO. Evaluation of CYP2D6 oxidation of dextromethorphan and propafenone in a Chinese population with atrial fibrillation. J Clin Pharmacol 2001;41:92-96.
72. Taylor AL, Ziesche S, Yancy C, Carson P, D'Agostino R, Ferdinand H, et al.; African-American Heart Failure Trial Investigators. Combination of isosorbide dinitrate and

hydralazine in blacks with heart failure. N Engl J Med 2004;351:2049-5.
73. Carson P, Ziesche S, Johnson G, Cohn JN. Racial differences in response to therapy for heart failure: analysis of the Vasodilator-Heart Failure Trials. J Card Fail 1999;5:178-87.
74. The Beta-Blocker Evaluation of Survival Trial Investigators. A trial of the beta-blocker bucindolol in patients with advanced heart failure. N Engl J Med 2001;344:1659-67.
75. Hare JM. Nitroso-redox balance in the cardiovascular system. N Engl J Med 2004;351:2112-2114.
76. Abernethy DR, Flockhart DA. Molecular basis of cardiovascular drug metabolism: implications for predicting clinically important drug interactions. Circulation 2000;101:1749-53.
77. Tanigawara Y. Role of p-glycoprotein in drug disposition. Ther Drug Monitor 2000;22:137-40.
78. Roden DM. Drug-induced prolongation of the QT interval. N Engl J Med 2004;350:1013-22.
79. Kanazawa H, Okada A, Higaki M, Yokota H, Mashige F, Nakahara K. Stereospecific analysis of omeprazole in human plasma as a probe for CYP2C19 phenotype. J Pharm Biomed Anal 2003;30:1817-24.
80. Goldstein JA. Clinical relevance of genetic polymorphisms in the human CYP2C subfamily. Br J Clin Pharmacol 2001;52:349-55.
81. Yamada S, Onda M, Kato S, et al. Genetic differences in CYP2C19 single nucleotide polymorphisms among four Asian populations. J Gastroenterol 2001;36:669-72.
82. Parker RB, Yates CR, Soberman JE, Laizure SC. Effects of grapefruit juice on intestinal p-glycoprotein: evaluation using digoxin in humans. Pharmacotherapy 2003;23:979-87.
83. De Ponti F, Poluzzi E, Cavalli A, Recanatini M, Montanaro N. Safety of non-antiarrhythmic drugs that prolong the QT interval or induce torsades de pointes: an overview. Drug Safety 2002;25:263-86.
84. Owens RC. Risk assessment for antimicrobial agent-induced QT interval prolongation and torsades de pointes. Pharmacotherapy 2001;21:301-19.
85. Malik M, Camm AJ. Evaluation of drug-induced QT interval prolongation: implications for drug approval and labelling. Drug Safety 2001;24:323-51.
86. Kang J, Wang L, Chen X-L, Triggle DJ, Rampe D. Interactions of a series of fluoroquinolone antibacterial drugs with human cardiac K^+ channel HERG. Mol Pharmacol 2001;59:122-6.
87. Bischoff U, Schmidt C, Netzer R, Pongs O. Effects of fluoroquinolones on HERG currents. Eur J Pharmacol 2000;406:341-43.
88. Anderson ME, Mazur A, Yang T, Roden DM. Potassium current antagonist properties and proarrhythmic consequences of quinolone antibiotics. J Pharmacol Exp Ther 2001;296:806-10.
89. White CM. HMG CoA reductase inhibitor-induced muscle toxicity: risks, monitoring, and management. Formulary 2002;37:588-93.
90. Finch C. Rifampin and rifabutin drug interactions: an update. Arch Intern Med 2002;162:985-92.
91. Jacobsen P, Andersen S, Rossing K, Jensen BR, Parving HH. Dual blockade of the renin-angiotensin system versus maximal recommended dose of ACE inhibition in diabetic nephropathy. Kidney Int 2003;63:1874-80.
92. Caron MF, Kluger J, Tsikouris JP, Ritvo A, Kalus JS, White CM. Effects of intravenous magnesium sulfate on the QT interval in patients receiving ibutilide. Pharmacotherapy 2003;23:296-300.
93. Winter ME. Clearance. In: Koda-Kimble MA, ed. Basic Clinical Pharmacokinetics, 3rd ed. Applied Therapeutics, Inc. Vancouver, 1994:26-36.
94. Newman CB, Palmer G, Silbershatz H, Szarek M. Safety of atorvastatin derived from analysis of 44 completed trials in 9416 patients. Am J Cardiol 2003;92:670-6.
95. Rosenson RS. Current overview of statin-induced myopathy. Am J Med 2004;116:408-16.
96. Jamal SM, Eisenberg MJ, Christopoulos S. Rhabdomyolysis associated with hydroxymethylglutaryl-coenzyme A reductase inhibitors. Am Heart J 2004;147:956-65.
97. Thompson PD, Clarkson P, Karas RH. Statin-associated myopathy. JAMA 2003;289:1681-90.
98. Product information. Zocor. Merck & Co., White House Station, NJ, 2004.
99. Kalus JS, Mauro VF. Dofetilide: a class III-specific antiarrhythmic agent. Ann Pharmacother 2000;34:44-56.
100. Product information. Tikosyn. Pfizer Pharmaceuticals, New York, 2004.
101. Hirsh J, Raschke R. Heparin and low-molecular-weight heparin. The Seventh ACCP Conference on Antithrombotic and Thrombolytic Therapy. Chest 2004;126:188S-203S.
102. Product information. Lovenox. Aventis Pharmaceuticals Inc., Bridgewater, NJ, 2004.
103. Product information. Fragmin. Pharmacia & Upjohn Company, Kalamazoo, MI, 2004.
104. Product information. Innohep. Pharmion Corp, Boulder, CO, 2003.

2
Infectious Diseases

Ralph H. Raasch

"Bad Bugs, NO Drugs"—The Diminishing Antibiotic Pipeline	294
Problem of Antimicrobial Resistance	297
Rational Antibacterial Drug Development	300
Proposed Approaches to Address "Bad Bugs, NO Drugs"	306
References	307

In 2005, antimicrobial drug discovery, development, and research are at a critical stage of evolution. The first several decades of the antibiotic era, since the mid-1940s, were marked with huge successes in terms of the discovery and clinical use of a multitude of new innovative agents. Even whole new categories of products were developed by the pharmaceutical industry, such as quinolones and nucleoside transcriptase inhibitors. These new drugs improved the outcomes of patients with infections that in the pre-antibiotic era were routinely fatal.

The last decade of the 20th century has been characterized by infections more refractory to therapy for a variety of reasons, usually related to the marked immunosuppression of infected hosts (less natural resistance and eradication of infections) and the inherent or newly developing resistance of the infecting organism to antimicrobial drugs. Immunosuppression is an important therapeutic tool in many medical fields. In oncology, therapeutic improvements have led to long-term treatments with cancer becoming a chronic disease in many cases. Arthritis and bowel diseases are treated with anti-inflammatory and immunosuppressive biological products. Corticosteroids are used chronically in a host of diseases. Interferon indications have expanded. Also, inappropriate excessive prescribing of antibiotics for respiratory infections, which are often the common cold and the flu, exacerbates the resistance development problem. This situation has reached the point that The Infectious Diseases Society of America (IDSA) has generated the warning phrase "Bad Bugs, No Drugs – As Antibiotic Discovery Stagnates . . . A Public Health Crisis Brews" (www.idsociety.org).

This chapter will discuss why this warning has been generated and will cover a variety of important issues in current antibiotic development. These issues include the continuing problem of antimicrobial resistance, the ongoing challenges of discovering new drugs with novel (new) mechanisms of action coupled with the search for new antimicrobial drug targets, the emergence (and demand) to use pharmacodynamic data in the determination of dose selection, and the design of trials to show any clinical differences between antimicrobial drugs. The focus of the discussion herein will be with antibacterial drugs, although where relevant, remarks will also be made in the context of antifungal and antiviral agents. Antiparasitic drugs will not be given consideration in this chapter.

"Bad Bugs, NO Drugs"—The Diminishing Antibiotic Pipeline

Figure 12.2.1, *The Diminishing Antibiotic Pipeline*, documents the marked reduction in the number of FDA-approved antibacterial drugs over the past 20 years [1]. At the same time, note

12.2. Infectious Diseases

New Agent Approvals

Year	Antibacterial	Antifungal	Antiviral (HIV)
1983-1987	16		1
1988-1992	14		2
1993-1997	10		6
1998-2002	7	2	9
2003-2004	9		3

FIG. 12.2.1. The Diminishing Antibiotic Pipeline

Type of Compound	Number
Anti-HIV	12 (6 with novel mechanisms)
Antiviral	5
Antibacterial	5
Antiparasitic	5
Antifungal	3

*15 largest companies for 2003

FIG. 12.2.3. New Antimicrobial Compounds in R&D

that the development and approval of antiviral agents, particularly antiretroviral drugs, has been a tremendously successful endeavor. The data in this and the next several figures comes from the analysis of FDA internal and online databases from 1980 through 2003 and does not include topical antimicrobials, antibodies, immunomodulators, and vaccines. In addition, the development programs of the 15 largest pharmaceutical companies (i.e., Merck, GSK, BMS, Pfizer, Abbott, Wyeth, etc.) and seven major biotechnology companies (i.e., Amgen, Genentech, Chiron, etc.) were examined. Internet listings and drug development descriptions were reviewed for an understanding of potential new drugs under development and whether or not any new drugs come from a new pharmacological class. It is easy to see that the number of newly approved antibacterial drugs has decreased markedly in the 20 years between 1985 and 2005. It is also remarkable to note that of the 225 new molecular entities (NMEs) approved by the FDA during the 5 years January 1998 to December 2002, only seven (3%) were for new antibacterial drugs. No new antibacterial drugs were approved during 2002 [2].

Figure 12.2.2 lists the year of FDA approval of these seven *new agents* from 1998 to 2002, as well as two new agents approved in 2003. In 2004, only one new antibacterial drug from the 15 largest companies was approved, telithromycin (Ketec®, Sanofi-Aventis). Taking these 10 new agents into account, it is disappointing to see that only two drugs (linezolid and daptomycin) have been developed with a novel or new mechanism of action. The other new drugs have generally been additions to the quinolone and beta-lactam classes of antibiotics. Linezolid (Zyvox®, Pharmacia) is unique in that it acts at a binding site on the bacterial 23S ribosomal RNA of the 50S subunit, thereby inhibiting protein synthesis [3]. Macrolides (azithromycin, clarithromycin, etc.) work slightly differently; they bind to the P site of the 50S ribosomal subunit, provoking dissociation of t-RNA from ribosomes and blocking protein synthesis. Daptomycin (Cubicin®, Cubist) binds to bacterial cell membranes, provoking a change in membrane potential by rapid depolarization. The depolarization causes an inhibition of DNA, RNA, and protein synthesis [4].

Figure 12.2.3 shows the *new antimicrobial agents* in *research and development* in 2003 from the 15 largest pharmaceutical manufacturers. You should note, in comparison, the larger number of antiviral agents in development, particularly for HIV indication, as well as relatively equal development numbers of antifungal and antiparasitic drugs. Within the subsequent 24 months, only telithromycin has been approved by the FDA (April 1, 2004). It is structurally related to macrolide antibiotics. Tigecycline (Tigacyl®, Wyeth) was approved in 2005. The five *antibacterial agents* in *R&D in 2003* are summarized in Fig. 12.2.4.

Antibacterial Agents Approved Since 1998

Year	Drug	Novel Mechanism
1998	Rifapentine	No
1999	Quinupristin/dalfopristin	No*
	Moxifloxacin	No
	Gatifloxacin	No
2000	Linezolid	Yes
2001	Cefditoren pivoxil	No
	Ertapenem	No
2003	Gemifloxacin	No
	Daptomycin	Yes
2004	Telithromycin	No

*similar mechanism to macrolides

FIG. 12.2.2. "New" Antibacterial Agents

Drug (Manufacturer)	Novel Mechanism
Gerenoxacin (BMS)	No (fluoroquinolone)
Telithromycin (Aventis)	No (ketolide)** - approved
ABT-773 (Abbott)	No (ketolide)
BAL5788 (LaRoche)	No (cephalosporin)
Tigecycline (Wyeth)	No (glycylcyline)*** - approved

* 15 largest companies
** mechanism similar to macrolide
*** mechanism similar to tetracycline

FIG. 12.2.4. New Antibacterial Agents in R&D, 2003*

Biotechnology companies to date have not been in the business of antibacterial drug development. Surveys of the development programs of the seven largest biotech companies show that only one new antibacterial agent is in development with a novel mechanism of action (Fig. 12.2.4) [1]. And finally, the reduction in antibacterial development does not appear to be a result of a global cutback in drug development research. Taking data from the annual reports of 10 large companies (Merck, Pfizer, BMS, Abbott, Eli Lilly, etc.), this documented reduction in the development and marketing of new antibacterial drugs has occurred despite a 30% increase in overall research and development expenditures between 1998 and 2002 [1].

It should be appreciated that the development of new drugs within an antibiotic class may obviously be advantageous. New but mechanistically similar agents may have improved safety profiles, or have pharmacokinetic characteristics that allow for new routes of administration (oral versus intravenous), or have more convenient dosing regimens, or be evaluated for indications not previously approved. There may also be expansion of the spectrum of antimicrobial activity as agents within the same class are developed (i.e., first- vs. second- vs. third- vs. fourth-generation cephalosporins). In our current era of emerging antibacterial drug resistance, only new drugs with different mechanisms of action are likely to be capable of slowing the problem of significant infections caused by multiple-resistant bacteria. At a time when new, effective drugs are increasingly needed, the antibacterial drug pipeline is down to only a trickle of new drugs [1].

Now that these data on new antibacterial and antiviral drug development are appreciated, two logical questions arise. First, why has HIV drug development flourished while antibacterial drug development declined? Figure 12.2.5 addresses the reasons for the continued *successful development* of *anti-HIV* or antiretroviral drugs. First, it should be observed that the important economic barriers that have slowed antibacterial drug development have not yet significantly inhibited research and development of antiretroviral agents. Current trends in the research directions of many companies are to the development of agents for long-standing, chronic diseases, such as diabetes mellitus, heart disease, and neurological disorders. Because of many scientific advances in cell and cancer function with improved outcomes from treatment, cancer has become a more chronic disease and the largest research area. Once a patient with a chronic disease is started and stabilized on a medication, they are likely to stay on that medication for a long time, assuming no side effects arise. On the other hand, most bacterial infections are obviously not lifelong, and the antibacterial drug treatment is only short-term. HIV is a lifelong infection, and antiretroviral drugs are generally continued indefinitely. HIV fits into the research and development trend of chronic diseases. Second, there is less competition within the formulary of approximately 20 antiretroviral agents than there is for the more than 90 antibacterial drugs. That is not to say there is no competition within the anti-HIV marketplace, but it is likely that there will be more profitable sales initially from a new antiretroviral medication than a new antibacterial drug (unless this agent is remarkably unique), simply because there are less HIV medications from which to choose [1]. Third, FDA approval of antiretroviral agents often is quicker and easier than the approval process traditionally applied to antibacterial drugs. HIV drugs generally are given fast-track accelerated status by regulatory authorities, related to the routinely life-threatening nature of HIV infections, ensuring a 6-month review and approval period or less, instead of 1 year or more. Instead of clinical end points that have been used in the pivotal trials of antibacterial drugs (clinical and microbiologic cure rates), many antiretroviral drugs have been approved on the basis of an evaluation of the changes in surrogate markers, usually a reduction in HIV-viral load and/or an increase in the CD4-lymphocyte count. The surrogate markers are very objective and less open to interpretation than many clinical end points used in infection-treatment trials, as well as directly related to disease morbidity. And finally, HIV/AIDS lobbying groups and patient support groups have campaigned aggressively and successfully in promoting the development and more rapid approval of antiretroviral agents, leading to the current accelerated approval timeline used by the FDA. To summarize, HIV is a chronic disease, with high patient need and demand, measurable surrogate markers, and a rapid approval mechanism in place—it makes more sense from a business perspective for pharmaceutical companies to develop antiretroviral drugs.

To close and summarize this section of the chapter, the next question that follows the above discussion (and has been somewhat addressed already) is why is industry leaving antibacterial drug development (Fig. 12.2.6)? Surprisingly, FDA statistics show that the anti-infective/immunologic area is the fourth largest research area for active INDs (in 2001, 1,258 vs. 1,610 for the top category neuropharmacological) [5]. However, many INDs are for new infection sites or new organisms for the same and existing antibiotics. Simply stated, many companies believe "there are better ways to invest research dollars" [6] regarding patient care opportunities, research innovations, and potential financial success. This realization has prompted Aventis, BMS, Eli Lilly, GSK,

- HIV is a life-long infection - it fits into R&D trend for chronic diseases
- Less Competition:
  - About 20 HIV drugs
  - Over 90 antibacterial agents
- Use of surrogate markers (rather that clinical endpoints) in clinical trials
- Public demand high with informed patient support groups
- FDA Regulatory reform – Accelerated approval timeline (6 months or less) for life-threatening disease

FIG. 12.2.5. Successful Development for HIV

12.2. Infectious Diseases

- Development dilemma - "There are better ways to invest research dollars."
- Antibacterial marketplace is no longer attractive.
- Chronic diseases are a more cost effective approach to drug development.
- New drug discovery has been difficult and not particularly productive.
- Drug approval has become increasingly complex.

FIG. 12.2.6. Industry Leaving Antibacterial Development

Roche, and Wyeth to spin off, greatly curtail, or eliminate their efforts in antibacterial research because the marketplace for antibacterial drugs is no longer attractive versus the costs of development and the high development challenges. As discussed above, chronic diseases are a more cost-effective approach to drug development, as well as the basic science of disease mechanisms has not advanced.

The discovery of new agents with novel mechanisms of action has been *a dilemma* and not very productive (Fig. 12.2.2: 10 new drugs, two new mechanisms of action). The development of multiple new agents on the basis of recent scientific advances has yet to occur regarding the acquisition of genomic data of bacteria. And antibacterial drug approval has become more, rather than less, complex in number of procedures required per patient, the number of trials for an NDA, the number of patients, and more time being consumed of the patent life. The cost of drug development by the industry has skyrocketed to more than $800 million, raising the bar for products to return the sales figures to pay for past and fuel future research. Yet, officials at the FDA are questioning whether the current approval process is as rigorous as it should be in the assessment of the effectiveness and safety of new drugs. The marketing and then retraction from the market of several quinolone antibiotics (i.e., trovafloxacin, sparfloxacin) only supports this contention.

The emergence and spread of multiple-antibiotic-resistant bacteria is the main scientific rationale for the development of new agents. However, what happens when one of these agents is approved and reaches the marketplace (i.e., linezolid or daptomycin) Fig. 12.2.7? In most hospital settings with a formulary system, these new drugs are reserved or restricted for use only against bacteria that have resistance to multiple other drugs, or where there is documented intolerance to other agents. Using linezolid and daptomycin as examples, many formulary-based hospitals restrict these drugs for use to treat vancomycin-resistant enterococcal (VRE) infections or for oxacillin-sensitive staphylococcal infections where the patient has a true vancomycin allergy. The rationale for the restriction is that uncontrolled use, over time, would promote further resistance in bacteria for which there is already resistance to multiple other drugs. And then if further resistance occurs, there would be no drugs to which the multiple-drug-resistant bacteria are sensitive. But this restriction, or required approval process by an infectious disease expert, greatly diminishes commercial viability and economic return (Fig. 12.2.7). Thus, efforts to preserve new agents may restrict their clinical availability and commercial viability. Second, even if there is no drug resistance, efforts to preserve new drugs would not be appropriate if the newer agent was clinically superior to previously available drugs. However, it has been difficult to show clinical superiority with new drugs, particularly when any head-to-head comparison is designed to show noninferiority and not necessarily to document superiority of one agent over another. Thus, in the absence of clear superiority, new drugs remain restricted because of the concern that uncontrolled use will provoke eventual emergence of resistance in already multiply-resistant bacteria, and we are back to the situation of "Bad Bugs, NO Drugs." Thus, under these circumstances of conservatism and restriction, the economic question arises as to whether there is an adequate ability to recoup research and development investments by new drug sales? Many companies who have had long-standing antibacterial drug development programs are obviously saying no [6–8]. But that is not to say that all companies are abandoning all development. Proposed solutions to the current antibiotic drug development dilemma are being discussed. But before these possible solutions are described, a brief review of antimicrobial resistance and the process of traditional antibacterial drug development will be reviewed.

- New drugs (where there is not yet resistance) are reserved or restricted; diminishes commercial viability
- Efforts to "preserve" new superior agents may restrict their clinical availability
- Under these circumstances of conservatism and restriction, is there adequate ability to recoup R&D investments by drug sales?

FIG. 12.2.7. Antibacterial Drug Development Dilemma

Problem of Antimicrobial Resistance

It is clear that the use of antimicrobial drugs for treatment or prophylaxis of infection in humans, and for agricultural and veterinary purposes, provides the selective pressure for the emergence and dissemination of drug-resistant bacteria. Higher resistance rates are usually noted first in clinical isolates from hospitalized patients because of the more intensive antibiotic use in that environment, particular the intensive care units. But, in general, given time, resistance rates from community isolates begin to approach that of isolates from patients within the hospital. Bacteria possess mechanisms to create resistance to antimicrobial agents (e.g., capsules to shield themselves, plasmid transfer between organisms, genes

- Multiple resistance examples:
  - Staphylococcus aureus (oxacillin)
  - Enterococcus sp. (vancomycin)
  - Streptococcus pneumoniae (penicillin G)
  - E. coli and Klebsiella sp. (extended spectrum beta-lactamase producers)
  - Pseudomonas aeruginosa (ciprofloxacin and imipenem-cilastatin)
- Bacterial mechanisms for resistance (capsules, plasmids)
- Dichotomy regarding recommendations for reduction in antibiotic usage vs. Encouragement for development of new agents

FIG. 12.2.8. Problem of Antimicrobial Resistance

responsible for extrusion of drugs from a cell). Multiple examples of *antimicrobial resistant* exist (Fig. 12.2.8), including Gram-positive cocci (*Staphylococcus aureus*, *Streptococcus pneumoniae*, *Enterococcus*), and Gram-negative rods (*Escherichia coli*, *Klebsiella*, and *Pseudomonas aeruginosa*).

A discussion of the epidemiology of the emergence of resistance for each of these organisms is beyond the scope of this chapter, but a simple review of the timeline of availability of certain agents and the emergence of resistance is instructional. However, before doing so, a certain dichotomy of argument needs to be appreciated. Basic to the discussion of maneuvers to reduce the emergence of resistance is the recommendation to reduce the overall usage of antibacterial drugs. And there is some documentation that resistance rates decrease with a reduction in the use of certain antibiotics. For example, Seppala et al. showed in Finland that the widespread decrease in availability of macrolide antibiotics resulted in an increase in sensitivity rates of group A streptococci to erythromycin [9]. So the dichotomy that exists is the overall recommendation to curb the use of antibacterial drugs (or to consider the restricted use of these new agents) versus the encouragement for the development of new drugs. As has been suggested above, why should new drugs be developed if there is a prevailing attitude to restrict or even diminish their use in an effort to avoid further resistance [10]?

Figures 12.2.9, 12.2.10, and 12.2.11, respectively, show the emergence of *resistance* of *S. aureus* to *penicillin G* and oxacillin and *Enterococcus* to vancomycin, respectively. As shown in Fig. 12.2.9, penicillin became available in large quantities for commercial use in the mid-1940s. Penicillin resistance was rare, but it did not take too long (3–4 years) before more than 50% of the clinical isolates of *S. aureus* became penicillin resistant secondary to the ability to generate an enzyme, penicillinase, that renders penicillin G inactive [11]. By 1980, more than 90% of *S. aureus* isolates were penicillin resistant. In order to treat penicillin-resistant isolates, chemical modifications of the penicillin molecule were made, resulting in the development of methicillin, oxacillin, and other penicillinase-resistance penicillins in the 1970s.

Figure 12.2.10 shows the rare frequency of *methicillin-resistant S. aureus* (MRSA) in 1975, but after a decade or so of use, the frequency of MRSA clinical isolates began to significantly emerge [12, 13]. The rate of MRSA in most large hospital systems is more than 50%, and there are now reports of further emergence of MRSA clinical isolates from community-acquired staphylococcal infections. As MRSA became more frequently responsible for staphylococcal

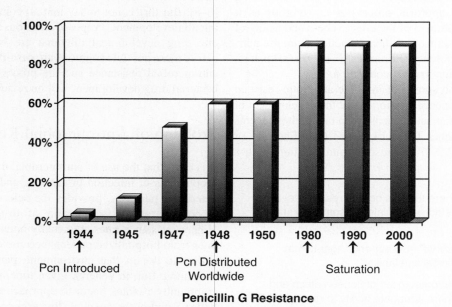

FIG. 12.2.9. Penicillinase Production in S. Aureus (Reprinted from International Journal of Antimicrobial Agents, Vol. 16 (suppl. 1), Livermore, D.M., *Antibiotic resistance in* staphylococci pages 3-10, Copyright 2000, with permission from Elsevier BV and the International Society of Chemotherapy.)

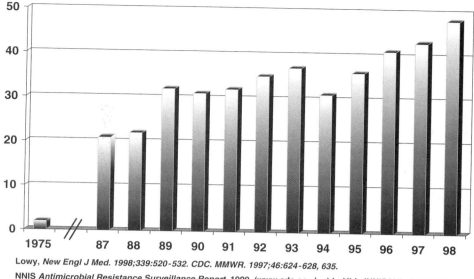

Fig. 12.2.10. Methicillin Resistance: Staph. aureus

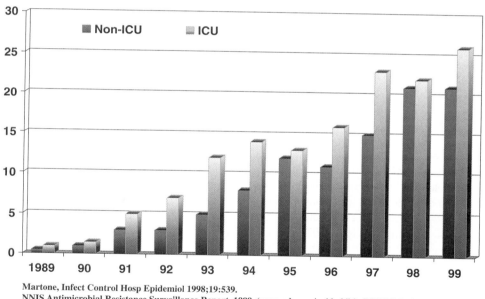

Fig. 12.2.11. Vancomycin Resistant Enterococci

infections during the 1980s and 1990s, the increased use of vancomycin was necessary to adequately treat these infections.

As a result of increased vancomycin use, the prevalence of *vancomycin-resistant enterococci* (VRE) also increased markedly during the 1990s, as shown in Fig. 12.2.11. Note, initially, the frequency of VRE infections was higher in intensive care units (ICU), where MRSA infections and vancomycin use is particularly common, but, given time, resistance rates for non-ICU patients are just as high as for patients in intensive care. These figures simply and nicely document the impact of antibiotic use on antibiotic resistance. And this phenomenon is no different from other bug–drug combinations. The principles of evolution and selective pressure dictate that, given time, bacteria exposed to an antibiotic will generate resistance mechanisms that promote its survival. Given this record of decreased drug effectiveness due to increased drug resistance, what should be the current approach to the development of antibacterial drugs? These issues will be discussed next.

Rational Antibacterial Drug Development

The previous discussion has introduced reasons for the continued justification for antibacterial drug development. Clearly, there is a medical need. The emergence of MRSA and VRE infections was the unmet medical circumstance that led to the development of linezolid and daptomycin. Second, there has been the discovery of "new" bacterial diseases, such as Legionnaire disease (*Legionella pneumophila*), cat scratch disease (*Bartonella henselae*), and even peptic ulcer disease (*Helicobacter pylori*) that may provide particular bacterial targets for new drugs [7]. This section of the chapter will describe the steps involved in the development of new antimicrobial drugs (again, focusing on antibacterial agents), taking into account "classic" antibacterial drug development combined with new technology primarily derived from the emergence of genomic science and research.

Figure 12.2.12 summarizes the possible multiple *goals* for the development of new antibiotic drugs, and the current approach to antibiotic development is intended to satisfy one or more of these goals [7, 8]. As has been described above, antibacterial drug development has sought to discover drugs with a unique mechanism of action so that the agent is the first one within a new product category. Despite this effort, the recent development of new agents with new mechanisms has been relatively disappointing, as we have examined. Nevertheless, new agents within a class of drugs may have a therapeutic advantage against older drugs by having increased activity (or having lower minimum inhibitory concentrations; MICs) against a range of susceptible bacteria. The increased activity of later generation drugs within a class may allow for the continued use of the newer agent—but this continued use of new drugs within a class may only last for a while, as more intensive resistance mechanisms continue to develop so that previously susceptible bacteria are now resistant to all drugs within the same class. An example of this phenomenon can be demonstrated by evaluating the development of extended-spectrum beta-lactamase (ESBL) production in *E. coli* or *Klebsiella pneumoniae*. As new third-generation cephalosporin antibiotics became available, more isolates of *E. coli* were susceptible to the third-generation in comparison with the older first-generation drugs, such as cefazolin. However, given time, and the always expanding ability of bacteria to develop and then translocate resistance mechanisms, there are now strains of *E. coli* able to generate an ESBL that renders them resistant to all cephalosporins.

Developing drugs with no (very optimistic) or less resistance (more realistic) than currently available drugs against bacteria continues to be an important goal in antibiotic development. As has been noted, the problem of VRE infections drove the development of linezolid, quinupristin/dalfopristin (Synercid®, Monarch), and daptomycin. The appearance of penicillin-resistant *Streptococcus pneumoniae* has driven the development of multiple quinolone antibiotics active against this pathogen. This type of discovery may make a "blockbuster" type of drug available—but as we have already discussed, restriction and approval policies in health systems can limit the very broad, and perhaps uncontrolled, use of new agents. The development of new drugs may allow for an investigation of effectiveness in multiple types of infection, depending upon the types of pivotal protocols put into place. That is, multi-organ system efficacy can be evaluated for multiple infections (such as pneumonia, or urinary tract infection, or septicemia) that may not have undergone evaluation for earlier drugs within the same antibiotic class. And, finally, new drug development can proceed with the goal to market an agent with more compliance-friendly dosage or administration requirements. Even in the context of intravenous agents, the development of ertapenem (Invanz®, Merck), the newest carbapenem, with a once-a-day dosing regimen, is an example. Even though ertapenem has a narrower Gram-negative spectrum than other carbapenem agents, its daily dosing schedule makes it a useful drug, particularly for patients requiring home antibiotics. Companies also need to develop both parenteral and oral forms of the same product for more serious hospital-based infection and outpatient situations, respectively. Added clinical trials will be needed.

Despite the "doomsday" implication of the phrase "Bad Bug, NO Drugs," it should be appreciated that certain relatively unique *opportunities* or circumstances pertinent to antibiotics have made antibiotic development very successful in the past and are still relevant today [15–17]. These opportunities are summarized in Fig. 12.2.13. First, drug targets are usually defined, that is, sites of action and the affiliated biochemistry are appreciated, such as cell wall synthesis or integrity, protein synthesis, or nucleic acid synthesis, so that compounds that inhibit a step (or steps) in a process can be identified and tested. *In vitro* studies, such as susceptibility testing, synergy studies, and killing curve evaluations, are

- Unique mechanism of action for new product category
- Higher sensitivities of target microorganisms
- Broader spectrum of action
- Less resistance with resistant organisms than marketed products
- Multi-organ system efficacy (multiple diseases, e.g., pneumonia, urinary tract, and septicemia)
- Better dosage and administration (daily vs. multidose per day, oral vs. intravenous)

Fig. 12.2.12. Research Goals in Antibiotic Development

- Drug targets are usually defined.
- *In vitro* studies are standardized and document antimicrobial effects.
- *In vivo* (animal) models established - reassuring that clinical trials in humans will be successful.
- Patients in clinical trials easily monitored by defined biomarkers.
- Pharmacodynamics is used in dose selection.
- Safety issues are "acceptable" because of often short term (not chronic) treatment.
- Mean total development time is less vs. other therapies.
- Cost per patient for clinical trials on average less.

FIG. 12.2.13. Opportunities Antibiotic Development Attractive

- Preclinical Testing Stages:
  - New Drug Discovery
  - Establishment of structure activity relationships (SAR)
  - Animal Models: pharmacokinetics and efficacy
  - Animal Toxicology
- Clinical Testing Phases:
  I. Healthy volunteers: safety and pharmacokinetics
  II. Efficacy for a defined therapeutic indication (pathogen)
  III. Efficacy against a pathogen for multiple indications
- Differences with antimicrobials:
  - Animal models of infection
  - HIV phase I/II trials
  - Phase III trials, multiple indications

FIG. 12.2.14. "Classic" Antimicrobial Drug Development

standardized and are used both in research and in the clinic to document antimicrobial effects. Most infections have an established *in vivo* animal model that has been shown to correlate reasonably well with outcome results in human trials. Meningitis, endocarditis, sepsis, and peritonitis models are examples of animal infections that are frequently used in the preclinical evaluation of a new agent to begin the process of determining the types of indications that will be studied in human infections. Clinical trial monitoring of patients is relatively simple, using routine vital signs (i.e., temperature, pulse rate), laboratory testing (complete blood counts and differential), and symptom resolution to help judge whether or not a clinical cure is achieved. Microbiologic cures are also relatively simple to determine and are based on the findings of a negative repeat culture from a specimen from an accessible site of infection (i.e., blood, sputum, cerebrospinal fluid, peritoneal fluid) that was previously positive.

Advances in the science of pharmacokinetics and pharmacodynamics have prompted a more rigorous determination of appropriate dose selection for use in clinical trials. Certainly, a drug's serum half-life is an important determinant for frequency of dosing, but other more recent developments in antibiotic pharmacodynamics (such as intracellular elimination characteristics, and concentration vs. time-dependent killing) are other factors that are taken into account to attempt to determine an optimal antibiotic dose. And finally, because most antibacterial drug trials are short-term, and not for chronic treatment, safety issues are less problematic, at least during the initial clinical trials phase (phases II and III) of drug development. Rarer adverse events (as is true for most agents) are typically discovered with longer term use, or when the population that receives the drug expands markedly after a product's approval. Hepatotoxicity from trovafloxacin and thrombocytopenia secondary to linezolid are two examples where significant side effects were discovered with larger exposed populations and longer therapy than in the shorter term approval trials. All of the above constitute reasons why many companies have traditionally been active in antibacterial drug development and for the successful marketing of more than 90 antibacterial drugs. Mean total development time and the cost per patient in a clinical trial have traditionally been less than for other therapies (e.g., 5.3 years vs. 7.1 for cardiovascular and 10.4 for central nervous system). But, despite this successful track record, as we have already examined, the antibacterial pipeline is less robust. Perhaps it would be useful to look at the various aspects of "classic" antimicrobial drug development next and try to understand if any modifications within this traditional approach can be made to improve the apparent future of antibacterial drug development.

The framework of *antimicrobial drug development* is really no different from the development process of any other class of drugs, as shown in Fig. 12.2.14 [18, 20]. The preclinical testing stages attempt to identify a new drug entity, understand how it works (including toxicological effects), structure–activity relationships, and how it is absorbed, distributed, metabolized, and eliminated in animal species. The animal models of infections, as noted previously, are excellent predictors of human infections, superior to other disease areas in product development. After successful studies in animal models of infection, the drug is given to humans in the typical phase I through phase III sequence of studies. A notable exception to this ordered sequence of events occurs with antiretroviral drugs, where the phases of studies, particularly phase I and II studies, become blurry, as HIV-positive patients are phase I study subjects to evaluate pharmacokinetics, initial activity, and initial safety. Antibiotic studies also differ from typical drug development in regard to one study possibly including multiple indications, based on multiple organisms and multiple sites of infection.

New antibiotic discovery is influenced by factors associated with microorganism characteristics and is accomplished with a variety of approaches. How does a company decide to pursue the development of a new agent from among all of the infectious diseases and causative microorganisms possible? Important *considerations with microorganisms* are summarized

- For a particular infectious disease, organism is of sufficient prevalence.
- Organism causes severe or serious disease.
- Available therapies are limited due to multidrug resistance.
- Clinical correlation of *in-vitro* resistance is associated with poorer clinical outcome

FIG. 12.2.15. Organisms & Antimicrobial Development

in Fig. 12.2.15 [7]. The most viable development programs for a new agent active against a particular organism, as well as for a specific infectious disease indication, focus their efforts on those organisms that have a high prevalence of causing the infection. This is obvious—higher prevalence organisms means more patients with a particular infectious disease (for example, *Streptococcus pneumoniae* and community-acquired pneumonia) will be candidates for the new drug, which then could translate into more patient use and new drug sales. Other promoting characteristics include organisms that cause severe or serious disease, and if reasonably prevalent, should provoke the scientific community to try and do something about it. The HIV epidemic and the relatively rapid emergence of multiple antiretroviral drugs is the best example of this concept. If there are limited available therapies due to multiple drug resistance (as has been discussed), then a "window of opportunity" (unmet medical need) is present to consider the development of an agent that treats an infection with very limited options. And last, if there are correlations with *in vitro* resistance for a particular bug–drug combination with poorer clinical outcomes, then alternative options can be approached rather than waste time, money, and effort in clinical research that is likely not to show effectiveness of the drug.

New antibiotic discovery has traditionally been based on *screening of molecules* from natural and synthetic compounds (Fig. 12.2.16). This tedious process has been successful, leading to the development of many currently used and important antibacterial agents, such as penicillins, cephalosporins, vancomycin, streptomycin, erythromycin, chloramphenicol, and tetracycline. The earliest screening involved fermentation broths prepared from soil samples and the detection of antibacterial activity on agar plates seeded with different bacteria. If a broth showed antibacterial effects, then the next huge step was to attempt to isolate an antibacterial agent though multiple steps of chromatographic separation and biologic assays. Then, a lead (or predominant) structure could be characterized. Once characterization (or the specific structure) was understood, then *in vitro* chemical modification could be possible that might allow for changes in potency, antimicrobial spectrum, and pharmacokinetics (Fig. 12.2.16).

Recombinant DNA technology can be used to promote the synthesis of new products by isolating genes involved in antibiotic production by bacteria. And natural synthetic antibiotic pathways can be "diverted" within bacteria by introducing substrate precursors that may then yield new compounds with antibacterial activity. But, in general, these refinements have produced agents with minor changes in chemical structure to a lead compound, or do not yield enough product to make commercial production realistic [18].

Lead compound isolation, purification, and determination of structure has promoted a successful antibiotic development industry related to *chemical modifications* of the structure to expand antimicrobial activity (Fig. 12.2.17). Multiple derivatives of an existing antibiotic can be synthesized to understand the structure–antibacterial activity relationships between different compounds. Structural changes have led to many successful new drugs, starting (as was discussed earlier) with the penicillinase-resistance penicillins, and includes the anti-pseudomonal penicillins and carbapenems (meropenem, ertapenem), macrolides (clarithromycin, azithromycin), and quinolones (moxifloxacin, gatifloxacin). As these examples above illustrate, new structures can have the advantage of a new spectrum in comparison with older drugs but may also offer advantages in potency, oral bioavailability, and half-life.

The sequence of many bacterial genomes is now understood, opening up the possibility that new drug entities could be found by *genome hunting* that inhibit crucial bacterial

- Classic screening of large number of natural and synthetic compounds
- Direct isolation from soil/marine microorganisms
- Examples: penicillins, cephalosporins, vancomycin, streptomycin, erythromycin, chloramphenicol & tetracycline
- Characterization of lead structure followed by chemical modification for enhanced activity or kinetics
- Genetic modification of microorganisms to produce novel antimicrobial products
- Diversion of natural pathways by introduction of substitute precursors

FIG. 12.2.16. Antibiotic Discovery - Screening

- Synthesis of multiple derivatives of existing antibiotics (SAR)
- Examples: anti-pseudomonal penicillins, macrolides, carbapenems, quinolones
- Possible advantages of new structures:
  - Spectrum
  - Potency
  - Oral bioavailability
  - Half-life
  - Adverse events

FIG. 12.2.17. Antibiotic Discovery - Structural Changes

12.2. Infectious Diseases

- Identification of target gene in microorganisms
- Clone gene; express; crystallize recombinant protein
- Determine protein structure
- Screen small molecules computationally-looking for a protein inhibitor
- Test in whole microorganisms
- Successful process in HIV development; not really developed yet for antibacterial drugs

FIG. 12.2.18. Antibiotic Discovery - Genome Hunting

- First administration to rodents (IV, SC, PO) to determine pharmacokinetics and initial safety
- Additional pharmacokinetics for Cmax, half-life, AUC in beagle dogs and monkeys
- Metabolism (renal, hepatic, & other) of new agents initially defined
- Evaluation of effectiveness possible in experimental infections:
  - Models available for pneumonia, endocarditis, osteomyelitis, abscess formation, and meningitis in rodents and rabbits
  - Infection models in immunosuppressed animals may include measurements on microbial clearance from tissues

FIG. 12.2.20. Animal Models in Antibiotic Development

enzymes (new drug targets), particularly enzymes not inhibited by current drugs (Fig. 12.2.18). A target gene is identified; the gene is cloned and expressed, yielding a recombinant protein product that can be purified and crystallized; the structure and active site of the protein (enzyme) can be discovered; via computational chemistry and high-throughput screening methods, thousands of small molecules can be screened for activity against the active site and optimized for binding; and then candidate compounds can be tested in whole bacteria [18]. Only in HIV infections has this genomics process show some success.

The goals of *bacterial genomics* are to find new targets in bacteria that are critical to their survival, such as structural proteins and enzymes (Fig. 12.2.19). Also, the target enzymes hopefully will allow new products to be developed that are inhibitors different from existing products, which also may avoid existing microbial resistance mechanisms. However, no product with these properties has been discovered by 2005. This process using genomics, which is very complicated, is still maturing, and new products derived to date have been few. The field is currently "target rich, but compound poor."

Once a lead compound is developed and its antimicrobial spectrum understood on the basis of standardized *in vitro* susceptibility testing, then the compound must proceed to testing in standard animal models. An assessment of safety, pharmacokinetics, and efficacy in experimental *animal* infection *models* are the goals of this phase of drug development (Fig. 12.2.20). The simplest way to understand preliminary pharmacokinetics and safety of a new compound is to administer escalating doses intravenously, subcutaneously, and orally to rodents (usually rats). Blood samples are obtained from each animal at set times, and antibiotic concentrations are assessed either by biologic or chromatographic assay, allowing for the evaluation of standard pharmacokinetic parameters (C_{max}, T_{max}, half-life, AUC, total body clearance, etc.). Additional studies must be conducted in additional, larger animal species, such as beagle dogs and monkeys, for further evaluation of pharmacokinetic parameters, and metabolic routes of elimination, such as the kidney or biliary tract.

Various mathematical models have been derived so that the pharmacokinetic data collected for animal species can be used to predict the various pharmacokinetic parameters in humans. Initial infection studies in animals (mice, rats, or rabbits) may be conducted to evaluate whether a drug under development can be protective of an animal given a lethal peritoneal injection of bacteria. The infection models used in animals can determine whether a new drug is effective for a specific indication or under circumstances of immunosuppression. If a new antibiotic is shown to be protective for 50% or more of the tested animals, and it protects at a relatively low dose, this data usually correlates with testing that shows the agent has potent *in vitro* antibacterial activity.

Safety studies of antibiotics in animals are very straightforward and differ little from preclinical toxicology studies for other classes of drugs (Fig. 12.2.21). Single or multiple doses of drug considerably above the dose anticipated for humans

- Identification of new targets (proteins, enzymes) critical for organism survival
- Development of enzyme inhibitors for enzymes not affected by current drugs
- However, new products (drugs) developed as above have not been productively derived
- "Target rich, but compound poor."

FIG. 12.2.19. Antibiotic Discovery - Bacterial Genomics

- Acute and chronic dosing studies
- Tests for assessment of genetic safety
- Tests for adverse events in reproductive capability
- Intended to help identify a toxic dose for each new drug
- Assessments: hematology, chemistry, renal & liver function, & histology (gross and microscopic)

FIG. 12.2.21. Safety Studies in Animals

are given over 24 hours in the acute toxicology studies. These protocols are intended to find the maximum tolerated dose (MTD) of the antibiotic. Chronic safety studies can use a variety of dosing regimens, but the duration of exposure is usually for 30 days or longer. And then for antibiotic candidates that have proceeded to phase II human trials, 6-month animal safety studies must be completed in two species. Multiple assessments are conducted throughout the dosing in the toxicology studies, with blood sampling for hematology, chemistry, and renal and liver function. At the end of dosing, animals are sacrificed and multiple tissues are collected for gross and microscopic histological evaluation [20]. Once these toxicology studies are finished, all preclinical data is forwarded to the FDA as an Investigational New Drug (IND) application for their evaluation. The FDA has 30 days to review the data and must approve the IND before clinical trials can proceed in humans.

Antibiotics are studied in phase I, II, and III protocols as proposed by the pharmaceutical company and agreed to by the FDA. The goals for the *phase I clinical trials* are summarized in Fig. 12.2.22. Previous animal studies usually provide reasonable starting points for dosing regimens in humans, but one of the major goals in phase I studies is to verify human pharmacokinetic parameters that were projected from the animal data. Human pharmacokinetic parameters, blood levels, tissue levels (especially useful for extravascular infection sites), and routes of elimination are defined, but in a population, typically, of young, healthy subjects, an ideal population. Also, oral bioavailability needs to be determined for oral product approval. Dose escalation studies may also be done for toxicity assessment and determination of the maximum tolerated dose (MTD). And, as has been mentioned previously, antiretroviral agents are tested under most circumstances just in HIV-positive research volunteers, which allow combining goals of phase I and some phase II goals, such as safety in target disease patients.

The FDA has published its perspective with regard to *dose selection for efficacy* trials on the basis of the pharmacokinetic-pharmacodynamic (PK-PD) data generated in phase I investigations (Fig. 12.2.23). Their overriding consideration in generating this perspective is that wise use of PK-PD data may help "shrink the size of clinical trials" by studying fewer

- Dose selection should be based upon a detailed development plan, including PK-PD relationships
- Data and protocols necessary:
  - Microbiologic data (microbial sensitivities)
  - *In vitro* and animal model data
  - Phase I PK data
  - Design of Phase II studies

FIG. 12.2.23. Dose Selection for Efficacy Trials & FDA

necessary patients and expediting product approvals through more flexible regulations. PK-PD can be useful in overall drug development because proper dose selection will decrease the size of the study database (by reducing treatment failure rates and frequency of adverse events) and hence decrease the time and expense to complete the development program. Therefore, the guidance from the FDA is that dose selection should be based on very detailed development plans, taking into account a rigorous understanding of PK-PD relationships. The types of data and protocols necessary to generate the dose selection plan include basic microbiologic data (MIC, antibiotic killing characteristics), *in vitro* and animal model data, and the phase I human PK data, which can then be incorporated into the dosing regimens proposed within the design of the phase II clinical trials [20].

Appropriately designed *phase II trials* assess the new drug's efficacy in patients with a specific, diagnosed infection, which includes the organism (genus and species) and the site (organ or tissue) of infection Fig. 12.2.24. The existing resistance status of the organism to existing antimicrobial agents is documented, especially if it is to be part of the indication for the new product. A direct comparison with another agent with a known efficacy record for the same infection indication must be made, and this design is a significant difference from most drug classes that can compare products to placebo, which is not ethical or desirable in infectious disease. Most early clinical studies are conducted in patients with non-life-threatening infections, such as urinary tract infections (UTIs) or skin and soft tissue infections (SSTIs), assuming the antibiotic has activity against common UTI and

- Necessary to verify human pharmacokinetic parameters as projected from animal experiments
- Blood levels, elimination half-life, urinary recovery, oral bioavailability usually determined
- Indicates metabolism and distribution of a new agent under optimum conditions (young, healthy subjects)
- Dose escalation studies to assess toxicity or to find maximum tolerated dose (MTD)
- HIV drugs usually tested in HIV-positive subjects

FIG. 12.2.22. Phase I Clinical Studies

- Assessment of efficacy in patients with diagnosed infection
- Direct comparison with another agent with a known efficacy record
- Initial studies usually run in patients with non-life-threatening infections, i.e., UTI, SSTI, assuming pertinent bacteria is a pathogen at the site of infection
- Assessment of clinical and microbiologic cures
- Duration of follow-up, variable but very important

FIG. 12.2.24. Phase II Clinical Trials

SSTI-causing bacteria. The overall end points in these trials are both clinical and microbiologic cure rates. Clinical cure happens when the relief of symptoms and the resolution of signs associated with the infection resolve. For example, for pneumococcal pneumonia, clinical cure would occur when the patient's fever and chills go away, the patient is documented to be afebrile and not producing sputum, and their chest x-ray improves and eventually normalizes. X-ray readings should be done by third-party providers who are also blinded to the study drug being used. Inter-rater variability for this parameter needs to be considered in the later multicenter trials. Bacteriological cure occurs when there is documentation of eradication of the causative organism from a specimen at the site of infection, after treatment, and as appropriate, by follow-up culture several weeks after therapy. Specimen collection and storage and culturing processes are important procedures to be standardized especially with the later multicenter trials that are usually done. Again, for pneumococcal pneumonia, a previously positive sputum culture for *Streptococcus pneumoniae* at the start of therapy that is rendered negative at the end of treatment would be a microbiologic cure. These clinical trials should clearly define the period of follow-up for evaluation of clinical and microbiologic cures. For example, for more insidious infections, such as osteomyelitis, a long-term follow-up (1–2 years) is appropriate to understand cure rates, whereas for UTIs, follow-up evaluations can occur much more quickly (days to weeks).

Phase III clinical trials are usually designed to study efficacy of a new drug for multiple infectious disease indications (Fig. 12.2.25). But efficacy must be demonstrated for a specific organism at each site of infection, or the indication will not be approved. In order to achieve sufficient sample size to document efficacy against multiple pathogens at multiple sites, a large overall sample size (several thousand patients) may need to be compiled. For example, if a new agent for treatment of *E. coli* is undergoing evaluation, sufficient patients must be studied where the organism is present in urine, blood, lung, and peritoneum to get an indication for UTIs, bloodstream infection, pneumona, and peritonitis, respectively. Both a microbiological and clinical cure again need to be established for each organism and each site of infection in phase III. Obviously, if the indication is to extend to similar infections, but caused by *Klebsiella pneumoniae*, additional patients with infection with this bacteria must be recruited and studied. These types of studies, especially those with broad indications, require years to complete. Also, these studies involve many investigative sites, medical centers or clinics, each with their own resistance patterns, sepsis care guidelines, and existing drugs of choice, all of which must be taken into consideration in the protocols and data analysis. Intersite standardization must be set up and variability needs to be assessed for any tests with subjective variability, such as x-rays, specimen collection, and culture techniques. Completion of the protocol and subsequent data analysis of phase III trials requires the greatest amount of time and expense from the sponsor. The clinical development of an antibiotic averages 5.3 years [5].

The most meaningful data regarding relative efficacy and toxicity of a new agent are generated in prospective, randomized, double-blind, *comparative trials* (Fig. 12.2.26). However, in addition to efficacy and toxicity, these trials can also document improvement (or not) in side-effect profile, dosing frequency, duration of treatment, or cost of therapy. The development of broad-spectrum antibiotics with reduced dosing frequency supports the evaluation new parameters (single-agent therapy, given less frequently, and at possibly lower overall cost) in newer comparative trials than the traditional evaluation of efficacy alone. Duration of therapy is an important efficacy criterion, especially for fungal (topical and gynecologic) infections that may require 1 to 4 weeks of treatment, which has been used successfully in product development.

For some infections, usually at extravascular sites of infection such as osteomyelitis and endocarditis, long-term therapy is required to eradicate the organism, and, along with long-term follow-up, is necessary for efficacy determinations. Combination therapy may be the indication for a new product, which will require microbial studies of each agent separately in "*in vitro*" and animal studies for the organisms, before combination therapy in clinical trials will be done. Dosing will be a bigger challenge, as well as pharmacokinetics, in order to obtain the most effective combination therapy.

- They are usually designed to show efficacy and safety for multiple indications in same clinical trial.
- Efficacy is demonstrated for each specific organism (usually bacteria) at each infection site.
- Cure rates are clinical and microbiological.
- Example: a new agent active against E. coli–sufficient patients must be studied where organism present in urine, blood, lung, peritoneum, or other site wherever approval is sought.
- Multi-center trials and inter-site variability is checked for resistance patterns and practices, standardization and/or analysis.

FIG. 12.2.25. Phase III Clinical Trials

- Most meaningful data obtained in prospective, randomized, double-blind, comparative trials
- Improvement (or not) in side effect profile, dosing frequency, duration of treatment, and cost of therapy
- Duration of follow-up to detect possible relapses of infection – especially important for endocarditis, osteomyelitis, & cellulitis
- Combination therapy, more preclinical work, and dosing challenges

FIG. 12.2.26. Considerations in comparative Clinical Trials

After completion of the phase III studies, all data is compiled, evaluated, summarized, and presented to the FDA in the form of a New Drug Application (NDA). If the NDA is approved, which may take 12 to 24 months for a new antibacterial drug (usually less time for an antiretroviral agent), then the agent can be made commercially available [20].

Proposed Approaches to Address "Bad Bugs, NO Drugs"

If the previous narrative can be significantly distilled down to important summary statements, the following remarks would begin that *summary* (Fig. 12.2.27). Current antimicrobial development (1990s and early 2000-plus) is effectively producing new antiviral (particularly HIV) and antifungal compounds (azoles and echinocandins). On the other hand, antibacterial drug development is in decline and has been so since the mid-1990s. The search for new antibacterial targets has not been especially fruitful to date, thereby increasing the difficulty in identifying new compounds with different mechanisms of action. Many large pharmaceutical companies have reduced or abandoned antimicrobial development programs. Smaller biotechnology companies appear to be at the forefront of new drug discovery and approval, exemplified by Cubist (daptomycin) and perhaps by Intermune and Vicuron if their drugs, oritavicin and dalbavancin, respectively, receive FDA approval. Genomics technology holds promise for unique mechanisms of action for products, but none to date have progressed significantly.

Further *barriers* have been identified that slow the development of new drugs active against resistant bacteria (Fig. 12.2.28) [15–17]. The cost and timeline for development of such an agent are not attractive—scientific challenges and business considerations suggest that research and development funds can be more effectively spent in other areas of therapeutics. There has been a demand for more and "higher quality" data, and the necessity for comparative trials complicates the development process necessary for marketing approval (more tests and more studies and more cost and less use with restrictive guidelines). Furthermore, there is the claim that the lack of clear regulatory guidance promotes uncertainty

- Effectively producing new antiviral (particularly HIV) and antifungal compounds (azoles, echinocandins)
- Antibacterial development is in decline, and has been so throughout the 1990's to now.
- Smaller, biotechnology companies are largely at the forefront of new drug discovery.
- For example: Cubist (daptomycin, approved 2003), InterMune (oritavicin), Vicuron (dalbavancin)

FIG. 12.2.27. Summary-Current Antimicrobial Development

- Cost and timeline of development
- Stringency for active comparison, noninferiority trials necessary for registration (approval)
- Demand for more and "higher quality" data
- Lack of clear regulatory guidance on what it would take to bring a novel agent to market
- Restrictions in usage for new products

FIG. 12.2.28. Barriers to Development

(and then reasoned hesitance) on what it takes to bring a novel agent to market.

The Infectious Diseases Society of America (IDSA) is leading an effort to create appropriate guidelines and incentives to attract industry back to antibacterial drug development [21]. These *R&D proposals* are summarized in Fig. 12.2.29. The previous set of antibiotic development guidelines generated by the FDA was published almost 10 years ago in 1997. A new set of guidelines should be generated to help pharmaceutical sponsors in their efforts to produce rigorous and robust data with the best possible study designs to enhance the approval process of new antibacterial agents. The use of surrogate markers of efficacy should be used and applied wherever possible—this had been a clear advantage in the development of antiretroviral drugs. Appropriate surrogate markers, however, for non-HIV infections need to be more specifically defined and ensure applicability and predictability of clinical success. In order to expand on financial return, certain drugs should be specified as "orphan drugs," especially if the medical community will significantly restrict its usage to resistant patients only, which will prolong their period of patent exclusivity and return on the research investment. As was discussed earlier, expanded PK-PD data sets, and use of animal model data, should be allowed to design clinical trials with a smaller number of subjects. And finally, with the use of tighter control populations, a reduction in the number of efficacy trials should be required for a new indication. Of course, these latter proposals, with a smaller number of study subjects and reduced number of trials, comes

- Generation of new guidelines for antibiotic development (previous guidelines dated 1997)
- Develop a process for clinical trials that target specific pathogens (i.e., MRSA) & specific diseases (sepsis)
- Use and apply specific surrogate markers of efficacy
- Designate specific drugs as "orphan drugs" for market exclusivity
- Use valid PK-PD data and animal model data to allow for smaller clinical trials
- With tighter controls, reduce number of efficacy trials required for a new indication

FIG. 12.2.29. Proposals to Spur Antibiotic Development

- Gootz TD. Discovery and development of new antimicrobial agents, Clin Microbiol Rev 1990;3:13.
- Spellberg B et al. Trends in antimicrobial drug development: implications for the future. Clin Infect Dis 2004;38:1279.
- Coates A et al. The future challenges facing the development of new antimicrobial drugs. Nature Rev 2002;1:895.
- Wenzel RP. The antibiotic pipeline – challenges, costs, and value. N Engl J Med 2004;351:523.
- Projan SJ. Why is big pharma getting out of antibacterial drug discovery? Curr Opin Microbiol 2003;6:427.

FIG. 12.2.30. Selected References

at a time when many question the ability of the drug development community (industry, academia, FDA) to recognize the risks of new drugs. Overall regulatory requirements may get more complicated in this environment rather than more simple or less costly regulations. Nevertheless, it is crystal clear that an urgent need is apparent for finding ways to promote the research and development of new antibacterial drugs effective against multidrug-resistant bacteria [15–17]. Suggested readings (references) are provided in Figure 12.2.30.

References

1. Spellberg, B.; Powers, J.H.; Brass, E.P.; Miller, L.G.; Edwards, Jr., J.E. Trends in Antimicrobial Drug Development: Implications for the Future. Clin. Infect. Dis. 2004;38(9):1279-1286.
2. CDER Drug and Biologic Approved Reports. U.S. Food and Drug Administration. Dec. 2004. Available at www.fda.gov.
3. Linezolid, Zyvox® Product Information. Pfizer, 2004.
4. Daptomycin, Cubicin® Product Information. Cubist Pharmaceuticals, 2003.
5. Lamberti MJ. An Industry in Evolution, 4th ed. Thomson CenterWatch, Boston, MA, 2003.
6. Projan, S.J. Why Is Big Pharma Getting Out of Antibacterial Drug Discovery? Curr. Opin. Microbiol. 2003;6:427-430.
7. Thompson, C.J.; Power, E.; Ruebsamen-Waigmann, H.; Labischinski, H. Antibacterial Research and Development in the 21st Century—An Industry Perspective of the Challenges. Curr. Opin. Microbiol. 2004;7:445-450.
8. Wenzel, R.P.; The Antibiotic Pipeline—Challenges, Costs and Values. N. Engl. J. Med. 2004;351(6):523-526.
9. Seppela, H.; Klaukka, T.; Vuopio-Varkila, J. et al. The Effect of Changes in the Consumption of Macrolide Antibiotics on Erythromycin Resistance in Group A Streptococci in Finland. N. Engl. J. Med. 1997;337:441-446.
10. Bax, R.; Mullan, N.; Verhoef, J. The Millenium Bugs—The Need for and Development of New Antibacterials. Int. J. Antimicrob. Agents 2000;16:51-59.
11. Livermore, D.M. Antibiotic resistance in staphylococci. Int. J. Antimicrob. Agents 2000;16:3-10.
12. Lowy, F.D. Staphylococcus aureus infections. N. Engl. J. Med. 1998;339:520-532.
    CDC. MMWR. 1997;46:624-628, 635.
13. NNIS Antimicrobial Resistance Surveillance Report. 1999. Available at www.cdc.gov/ncidod/hip/NNIS/AR_Surv1198.htm.
14. Martone, W.J. Spread of vancomycin-resistant enterococci; why did it happen in the United States. Infect. Control Hosp. Epidemol. 1998;19(8):539-545.
15. Bush, K. Antibacterial Drug Discovery in the 21st Century. Clin. Microbiol. Infect. 2004;10(Suppl. 4):10-17.
16. Powers, J.H. Antimicrobial Drug Development—The Past, the Present, and the Future. Clin. Microbiol. Infect. 2004;10(Suppl. 4): 23-31.
17. Projan, S.J.; Shlaes, D.M. Antibacterial Drug Discovery: Is It All Downhill From Here? Clin. Microbiol. Infect. 2004;10(Suppl. 4): 18-22.
18. Gootz, T.D. Discovery and Development of New Antimicrobial Agents. Clin. Microbiol. Rev. 1990;3(1):13-31.
19. Coates, A.; Hu, Y.; Bax, R.; Page, C. The Future Challenges Facing the Development of New Antimicrobial Drugs. Nat. Rev. 2002;1:895-910.
20. Office of Drug Evaluation IV, Center for Drug Evaluation and Research, FDA/IDSA/ISAP Workshop, April 16, 2004.
21. Available at www.idsociety.org.

3
Oncology

Suzanne F. Jones and Howard A. Burris III

Background Issues 309
Phase 1 Clinical Trials 310
Phase 2 Clinical Trials 313
Other Design Issues in Oncology 316
Era of Targeted Therapy 317
References 319

The drug development process in the area of oncology is unique due to the nature of the diseases, often resulting in serious acute morbidity and mortality, and the cytotoxic nature of many of the agents that are in practice and being evaluated. The overall goal of palliation versus cures generally is different in oncology. As some cancers have become better controlled with drug therapies than in the past (e.g., non-Hodgkin lymphoma, acute lymphoblastic leukemia, and testicular cancer), some cancers in certain patients are becoming chronic diseases.

Oncology products very commonly produce substantial toxicity on multiple organ systems simultaneously that can be quite debilitating and even life-threatening to the patients even at normal routine doses. Preliminary dose finding, tolerability, and pharmacology studies cannot be conducted in normal healthy volunteers, as is traditionally done in other disease states, due to the toxicity profiles of the drugs. As a result, these trials must be conducted in patients with cancer, many of whom have advanced stage disease and multiple other concomitant underlying medical conditions. The investigational drugs in early trials (phases 1 and 2) often are studied in patients who already have failed on standard therapies, creating a higher hurdle for efficacy than most drug categories. The standard of care in almost all of oncology is to employ structured treatment protocols involving product combinations, similar to research trials, except that they are standardized and widely accepted usually in some national treatment guideline. The training of oncology fellows follows this research focus, and they most always participate in research protocols at their tertiary care institutions.

This chapter will first discuss general issues about the opportunity for research in the oncology area, which has become a major focus of both the biotechnology and pharmaceutical companies with the private, academic, and government oncology research community. Phase 1 trials in oncology are unique in the patients, end points, and dosing schema, which are covered next. Design features of phases 2 and 3 trials are the subsequent topics including alternative end points. Special design issues are addressed for elderly, supportive care products, and adverse event work. Targeted therapy is becoming the standard in oncology product development, which is covered in the final section, which offers more individualized treatment.

12.3. Oncology

Background Issues

In the 1960s to 1980s, basically only a handful of companies were involved in the science and business of oncology (e.g., Bristol-Myers, Burroughs-Wellcome, Lederle, and Pharmacia). The need has always been great in oncology given its significant morbidity, disability, and mortality. With the phenomenal advance of science in the past 10–20 years, many secrets of cell function and their relationships to abnormal cell growth have been discovered leading to many new targets and therapeutic opportunities; for example, tyrosine kinases, metalloproteinases, ribozymes, apoptosis capase enzymes, cell growth factors (e.g., vascular endothelial GF), oncogenes (e.g., Her2neu, Bcrl), proteasomes, and JAK/STAT proteins. Oncology research for new products exploded in the 1990s and to the present with many new drugs and better outcomes. Many pharmaceutical companies have followed the lead of the biotechnology industry to focus on oncology product development and then the marketing of the novel products. Now, most pharmaceutical and many biotech companies focus in oncology; for example, Roche, Genentech, Novartis, Pfizer, GlaxoSmithKline, AstraZeneca, Sanofi-Aventis, Bristol-Myers Squib, Biogen-Idec, and Amgen with a 2003 total market of $29.97 billion and a projected 2008 *market* of $59.27 billion (Fig. 12.3.1). Supportive care products comprise the largest area for hematologic support, pain relief, and nausea/vomiting prevention and treatment. The biggest indications as far as markets in decreasing order are the cancers of the breast, gastrointestinal (colorectal/stomach), lung, lymphoma, prostate, leukemia, and ovarian. There are now 20 blockbuster products approved worldwide for use in oncology (sales over $500 million in 2004); that is, Taxotere® for breast, lung, and prostate cancers, Gleevec® for leukemia, Eloxatin® and Avastin™ for colorectal cancer, Gemzar® for breast, lung, and pancreatic cancers, Arimidex® and Herceptin® for breast cancer, Casodex®, Leuplin®, and Zoladex® for prostate cancer, and Rituxan®/MabThera® for lymphoma, plus supportive care products, Procrit®/Eprex®, Aranesp®, and NeoRecormon®/Epogin® for anemia, Neulasta™ and Neupogen® for myelosuppression and infection prevention, Durasegic® for pain, Zometa® for hypercalcemia, and Zofran® for nausea and vomiting. The pipeline for oncology is the largest in the pharmaceutical and biotechnology industries, numbering about 500–600 in clinical trials, including about 300 biotechnology products [1–3].

Cancers are common diseases that increase in frequency with age. New cases for 2002 from the National Cancer Institute's (NCI) Surveillance, Epidemiology and End Results (SEER) database are shown in Fig. 12.3.2. Breast cancer in women and prostate cancer in men are the most predominant. Cancer is the second most common cause of death in the United States. Deaths occur in 194 of every 100,000 in the population in the United States. Figure 12.3.2 also shows the death rates for the various cancers listed per 100,000 of the population. Over the past decade (available SEER statistics, 1992 to 2002), the death rates have been modestly decreasing about 9% overall and ranging from 4.7% to 28.3%. The reasons are manifold, including more public awareness, improved and earlier diagnosis, new cancer drugs and biologicals, and better supportive care therapies. Even though improvements have occurred, given these statistics of disease prevalence, much work needs to be done in reducing the disease burden and creating more cures of cancer. The pace of research has increased dramatically and new discoveries are being made in disease mechanisms and new drugs, which fosters hope of even more significant advances [4].

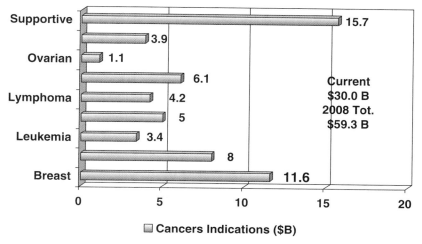

FIG. 12.3.1. Indications and Markets (Projected 2008)

Cancers	Cases 2002	Deaths* 2002	%Decr. 1992-2002
Breast	205,000	25.5	19.3
Prostate	189,000	28.1	28.3
Lung	169,400	54.9	6.8
Colorectal	148,300	19.6	16.9
Lymphoma	60,900	7.6	7.3
Leukemia	30,800	11.7	9.3
Ovarian	23,300	9.0	4.7
Overall		193.5	9.4

* Deaths per 100,000

FIG. 12.3.2. Common Cancers

Another unique feature in cancer research among diseases is the extensive formal *research network* that exists in the United States and around the world (Fig. 12.3.3). Individual practice groups of oncologists, such as the Sarah Cannon Center, are also major research centers in one city/state area (Nashville, Tennessee) that can perform all stages of drug research. Consortia of health care institutions that specialize in oncology practice and research are created (e.g., National Cancer Center Network [NCCN] for 28 major university-based medical centers), in order to foster research and improve the development of new therapies. The government through the National Cancer Institute has major commitment to perform basic research at their laboratories and sponsor basic and clinical research through grants. Special cancer research networks, "oncology groups" such as the Southwest Oncology Group (SWOG) or the National Surgical Adjuvent Breast and Bowel Project (NSABP), were formed to create collaboration among clinicians, academicians, and government and also standardization of research protocols, including coordinating design, conduct and data analyses, across the United States in specific cancers or broad areas. Professional societies provide the forums to present and share the data, and also create practice guidelines, among the oncology practice and research communities, including quality control mechanisms in their review processes for participation. Finally, a key driving force for product development is the biotechnology and pharmaceutical companies, who have a large percentage of the scientific advances within their umbrellas, the new molecules under development, the financial backing, and the drug development expertise to work with this network of oncology researchers to bring products to market.

Phase 1 Clinical Trials

Phase 1 clinical trials are the first trials conducted in humans after the completion of preclinical and animal toxicology testing (Fig. 12.3.4). In other disease states, these trials are typically conducted in normal healthy volunteers. However, due to the cytotoxic nature of the drugs that are being tested and the attendant serious potential toxicities, phase 1 oncology trials must be conducted in cancer patients. With traditional cytotoxic agents, multiple phase 1 trials exploring a variety of dosing schedules (i.e., once every three weeks dosing, weekly dosing, prolonged infusions, etc.) are conducted in order to determine the optimal schedule for subsequent trials. The first dose is very low and very unlikely to produce any toxicity, that is, a single dose equal to 1/10 of the lowest dose to show activity in the most sensitive animal species. The primary goals of these studies are to identify the dose-limiting toxicities (DLTs), to determine the maximum tolerated dose (MTD) and a recommended phase 2 dose, and also to assess the pharmacokinetic profile of the drug (blood and tissues levels, absorption, distribution, metabolism, and elimination).

Traditionally, a modified *Fibonacci dose escalation* has been used to determine the MTD and DLT for a drug

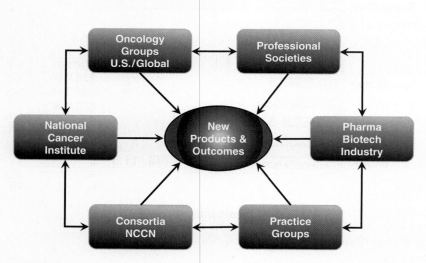

FIG. 12.3.3. Oncology Research Network

- First trials conducted in humans:
  - Oncology patients vs normals
  - Multiple trials often conducted to determine optimal schedule (q3week, weekly, prolonged infusion)
- Dosing Schemas (5 approaches)
- Dosing typically initiated with single dose at 1/10th the LD10 in the most sensitive animal species
- "Traditional" trial goals:
  - Identify dose-limiting toxicities (DLT)
  - Identify the maximum tolerated dose (MTD) & recommended phase II dose
  - Assess pharmacokinetics [PKs] (metabolism & clearance)

FIG. 12.3.4. Phase I Clinical Trials

"Traditional" Modified Fibonacci Dose Escalation

- Starting dose = LD10; subsequent dose increments: 100%, 67%, 50%, 40%, 33%, 33%, 33%...
- 3 patients treated and observed per dose level prior to escalation
- MTD = highest dose studied for which incidence of DLT was < 33%
- Disadvantages: number of patients, time, & patients treated at subtherapeutic doses

FIG. 12.3.5. Phase I Trials Dose Escalation Schemas – 1

Accelerated Titration Designs (multiple designs)

- Initial dose levels enroll one patient per cohort
- Dose escalate in 40 or 100% increments until DLT observed in 1 patient or grade 2 toxicities in 2 patients during cycle 1
- Subsequent cohorts expanded to 3 patients with dose escalations of 40% (± intrapatient escalation)
- Advantages: fewer patients/dose levels; fewer patients treated at subtherapeutic doses
- Disadvantages: interpatient variability in toxicity and pharmacokinetics

FIG. 12.3.6. Phase I Trials Dose Escalation Schemas – 2

(Fig. 12.3.5) [5]. In order to maximize patient safety, the first phase 1 trial is typically initiated with a single dose schedule at 1/10 of the dose at which 10% of the most sensitive animal species died, or lethal dose$_{10}$ (LD$_{10}$). Three patients are enrolled per dosing cohort (either at staggered intervals or simultaneously) and are observed for treatment-related toxicities for an entire treatment cycle before dose escalation can occur. Subsequent doses are escalated from the starting dose in increments of 100%, 67%, 50%, 40%, and ≤33% for all subsequent doses. The MTD is defined as the highest dose studied for which the incidence of DLT is <33%. The inherent disadvantages for this type of phase 1 dose escalation study include the number of patients enrolled and longer time required to proceed through the various dose levels, as well as the number of patients who are treated at subtherapeutic dose levels.

Typically, the starting dose for phase 1 trials is significantly less than the dose recommended for subsequent phase 2 efficacy trials. As a result, various alternative dose escalation schemas have been devised to speed patient accrual and trial completion, as well as minimize the number of patients who are exposed to subtherapeutic doses (Fig. 12.3.6). *Accelerated titration designs* are one alternative for phase 1 dose escalation trials [6]. In these trials, the initial dose levels enroll only one patient per cohort. Dose escalation proceeds in 40–100% increments until DLT is observed during the first treatment cycle in a single patient or grade 2 toxicities are observed in two patients. Subsequent dosing cohorts are expanded to three patients, with dose escalations of 40% between dose levels. Intrapatient dose escalation may also be incorporated into the trial to further decrease the number of required patients, but the majority of trials do not allow for intrapatient dose escalation. Accelerated titration designs allow fewer patients to be enrolled at subtherapeutic dose levels but may also provide limited toxicity and pharmacokinetic data due to the small number of patients enrolled at the early dose levels.

Another alternative phase 1 trial design is to use *pharmacologically guided dose escalation* (PGDE) (Fig. 12.3.7) [7, 8]. In these trials, the dose is rapidly escalated (doubled) until a target area under the curve (AUC) that was previously derived from animal pharmacokinetic data is obtained. Although PGDE does potentially save dose levels when compared with the modified Fibonacci schema, it does have some disadvantages. Because subsequent dosing is based on AUC, "real-time" pharmacokinetics are required throughout the trial, adding cost, sophistication, and patient blood sticks for drug levels. Trial results may also be affected by the sensitivity of the drug assay, interspecies pharmacology differences, interpatient pharmacology variability, the presence of an active drug metabolite, and the correlation between extracellular drug concentrations and intracellular tumor concentrations.

Pharmacologically Guided Dose Escalation (PGDE)

- Rapid dose escalation (double dose) until target AUC reached (derived from animal pharmacokinetic data)
- Disadvantages:
  - Requires "real-time" PKs; assay sensitivity
  - Interspecies pharmacology differences
  - Extrapolating preclinical data
  - Interpatient variablility
  - Drug metabolites
  - Extracellular drug concentrations versus intracellular tumor concentrations

FIG. 12.3.7. Phase I Trials Dose Escalation Schemas – 3

Escalation with Overdose Control (EWOC)

- Bayesian model that adjusts dose escalation based on occurrence or absence of toxicity from accumulated trial data
- Advantages:
  - Provides confidence interval for MTD at end of trial
  - Controls probability that patient will receive an overdose
- Disadvantages:
  - Requires pretreatment assessment of covariate;
  - Escalation curves must be re-generated as data accumulates

FIG. 12.3.8. Phase I Trials Dose Escalation Schemas – 4

A third dose escalation schema is a Bayesian model that adjusts dose escalation based on the occurrence or absence of toxicity from accumulated trial data while controlling for the probability of overdosing (Fig. 12.3.8). This method is known as *escalation with overdose control* (EWOC) [9, 10]. The use of EWOC allows the dose to be adjusted according to a pretreatment assessment of covariate while incorporating patient safety into the dosing algorithm. Optimally, EWOC would minimize the number of patients treated at subtherapeutic or severely toxic dose levels and maximize the number of patients treated at optimal dose levels. The disadvantages of EWOC are the requirement for a pretreatment assessment of covariate and the need to regenerate dose escalation curves as data accumulates.

The *modified continual reassessment method* (mCRM) is a similar Bayesian method that constantly modifies the dose–toxicity curve based on toxicity experience, resulting in updated predictions of MTD (Fig. 12.3.9) [11, 12]. Theoretically, the use of the mCRM would minimize the number of patients treated at subtherapeutic dose levels and optimize the dose levels that are likely to be clinically relevant. Again, the problems with using this Bayesian method are that the MTD must be estimated before the initiation of the trial based on preclinical data, and the dose escalation curves must constantly be regenerated based on real-time data from limited numbers of patients. As a result, the recommended MTD may be influenced by both interspecies and interpatient variability. Dose-related pharmacokinetic data is also limited with the mCRM due to the limited numbers of patients enrolled at each dose level.

The establishment of the MTD, or the dose at which ≥33% of patients experience DLT, is the *traditional end point* that has been used for phase 1 trials of antineoplastic agents (Fig. 12.3.10). Dose-limiting toxicities are predefined for each trial and typically consist of treatment-related hematologic or non-hematologic toxicities that are considered dose-limiting based on their severity or duration (e.g., treatment-related febrile neutropenia). The dose that is recommended for subsequent phase 2 trials may be the actual MTD or 1 dose level below the MTD. In order to obtain additional pharmacokinetic and toxicity data necessary for subsequent trials, the phase 1 trial may also be expanded at the recommended phase 2 dose level to enroll 10–20 additional patients. Attempting to include patients similar to those who will be enrolled on subsequent trials in the expansion of the recommended phase 2 dose may further establish the appropriateness of the dose for a particular patient population. Investigators may also choose to determine separate MTDs in heavily and minimally pretreated patients, as patients who have received numerous prior chemotherapy regimens may experience more toxicities due to prior treatments. However, the definitions that are employed for "heavily" and "minimally" pretreated patients are not well-defined and will vary based on individual experiences.

There are also several *alternative phase 1 clinical trial end points* that may be used, if appropriate (Fig. 12.3.11). Preclinical

"Traditional" Endpoint = MTD (dose at which ≥ 33% of patients experience DLT)

- DLT: hematologic or non-hematologic toxicities considered dose-limiting based on their severity or duration; predefined for each trial
- Recommended phase II dose may be MTD or 1 dose level below MTD
- Expand recommended phase II dose (10-20 patients) to obtain additional PK and toxicity data prior to starting subsequent trials
- May determine separate MTDs in heavily and minimally pretreated patients

FIG. 12.3.10. Phase I Clinical Trial Endpoints – 1

Modified Continual Reassessment Method (mCRM)

- Bayesian method which constantly modifies dose-toxicity curve based on toxicity experience resulting in updated predictions of MTD
- Advantages: fewer patients than Fibonacci method
- Disadvantages:
  - Must estimate MTD based on preclinical data before trial
  - Interspecies variability
  - Interpatient/variability
  - Limited PK data (1 patient/dose level)
  - Requires re-analysis at each dose level based on current toxicity data

FIG. 12.3.9. Phase I Trials Dose Escalation Schemas – 5

Alternative Trial Endpoints:

- Target plasma concentration: defined as "effective concentration" in *in vitro* or animal studies
- Biochemical endpoint: activation or inhibition of critical pathway enzymes
- Blood flow/perfusion: evaluated by magnetic resonance imaging or positron emission tomography

FIG. 12.3.11. Phase I Clinical Trial Endpoints – 2

12.3. Oncology

- **Patient population:**
  - Typically patients who have failed standard therapy or for whom no standard therapy exists
  - Measurable or evaluable disease
- **Adequate organ function required:**
  - I.e., kidney, liver, bone marrow
- **Number of prior chemotherapy regimens may/may not be limited**
- **Labor/time intensive for patient and staff due to multiple required tests:**
  - I.e., PK samples, PD samples, pharmacogenomic samples, skin/tumor biopsies, MUGAs, EKGs, specialized lab tests

FIG. 12.3.12. Phase I Clinical Trials

- **Single-agent trials to determine efficacy in specified disease setting**
- **"Traditional" endpoint: objective response rate (CR + PR)**
- **Disease settings:**
  - E.g., tumor, cell type, stage, pretreatment status, & performance status
- **Early stopping rules:**
  - I.e., Gehan, Fleming, Simon, and multinomial
  - Limit patient exposure to ineffective treatments
  - Maximize speed of development
- **Sometimes incorporated into phase I or phase III trials to condense development timeline**
- **Alternative endpoints possible for cytostatic agents**
- **FDA may allow drug approval with Phase II trials**

FIG. 12.3.13. Phase II Clinical Trials

in vitro studies or animal studies may be able to predict an "effective plasma concentration" that may be used as a target concentration to be obtained in the phase 1 human trials. The availability of nontraditional anticancer agents, such as biologic or targeted therapies, has also resulted in the establishment of alternative phase 1 clinical trial end points such as biochemical end points (i.e., activation or inhibition of critical pathway enzymes that can be measured in blood or tumor samples) or perfusion end points that can be assessed by positron emission tomography (PET) or some other imaging technique. With some agents, it may be appropriate to employ an alternative trial end point in combination with the traditional MTD end point in order to establish the most appropriate dose for subsequent trials.

Patients included in *phase 1 clinical trials* have typically failed standard therapy or no standard therapy exists for their disease (Fig. 12.3.12). Because safety is the primary end point of the trial, and response is generally a secondary end point, patients are usually allowed to have either measurable or evaluable disease, and the number of prior chemotherapy regimens may or may not be limited. It is important for patients to have adequate kidney, liver, and bone marrow function at baseline, even if heavily pretreated or late-stage disease, in order to appropriately determine the safety profile of the drug. Phase 1 trials are critical for determining the preliminary safety profile and recommended doses for subsequent trials. As a result, these trials may be labor and time intensive for both the patients and health care personnel due to multiple tests that may be necessary to establish safety, such as pharmacokinetic, pharmacodynamic, or pharmacogenetic measurements from blood, skin, or tumor biopsies; specialized laboratory tests, procedures, or scans.

Phase 2 Clinical Trials

Phase 2 clinical trials are typically single-agent trials conducted to determine the efficacy of a particular compound in a specified cancer setting (Fig. 12.3.13). The "traditional" end point for these trials is objective response rate (complete [CR] + partial responses [PR] combined) based on changes observed in the dimensions of measurable tumor lesions at predetermined time points. Patient selection is influenced substantially by the cancer setting; it includes a complex set of parameters. Cancer is a set of diseases involving specific and different tissue types, the diseases progress over time in stages each with a different prognosis, cell types may differ within the same stage, the cancer may metastasize to other organs as a complication, cancer may debilitate patients (variable performance status) changing the amount of chemotherapy possible or the side effect potential of a new drug, and the amount of pretreatment can impact subsequent drug activity, all of which needs to be taken into consideration in patient selection for both phases 2 and 3 trials. The treatments in cancer settings also differ from other disease in that the patients will normally receive combination treatment in a series of prescribed cycles of therapy (often six or more cycles) that often can require about 6 months or more, with rest periods in between for the patients' system to recover from the side effects of treatment.

Although the data generated from phase 2 studies is important, it typically is not adequate alone for submission to regulatory agencies for drug approval. However, if the drug demonstrates significant antitumor activity in a disease setting where no standard therapy is currently available, the FDA will allow product approval at this early stage of research to make available major therapeutic advances, but phase 3 type confirmatory trials still will have to be done after marketing. Because product approvals are rare with phase 2 data, many pharmaceutical companies may try to condense the drug development timeline by incorporating a "mini" phase 2 trial at the end of a phase 1 trial or the beginning of a phase 3 trial. Alternatively, the use of early stopping rules (such as Gehan, Fleming, Simon, and the multinomial rule) before full patient enrollment in a study protocol may also help speed the phase 2 trials [13–16]. Specific stopping criteria that incorporate response and/or disease progression data are established prior to trial initiation. If the trial does not meet the specified criteria at the end of the initial enrollment stage (i.e., 14–20 patients),

- Disease evaluated approximately every 8 weeks with same method of assessment
- Measurable lesions: ≥ 10 mm with spiral CT
- Responses categorized as CR, PR, SD, or PD, based on measurements
- Bidimensional measurements historically used
- RECIST criteria: new system evaluating single tumor dimension; measurable and non-measurable lesions included in response
- Documented responses (CR or PR) confirmed no less than 4 weeks after initial scans
- New drug response often required by FDA in patients that failed on standard treatments

FIG. 12.3.14. Determination of Disease Response

- Large, multicenter, randomized trials:
  - Comparing new agent to "standard therapy" in a specific disease setting
  - Cancer setting & patient selection challenges
  - Cornerstone of submissions for FDA approval
- Patients may be stratified by prognostic factors
- "Traditional" endpoint:
  - Overall survival (statistically powered for proving superiority or equivalence)
  - Multiple alternative endpoints are also currently being evaluated.
- Interim data analysis by a Data Safety Monitoring Board (DSMB) to assure best interest of patients

FIG. 12.3.15. Phase III Clinical Trials

the trial is terminated. If sufficient activity is observed, patient accrual continues until trial completion (i.e., 40–60 patients). By employing the early stopping rule criteria, the number of patients exposed to ineffective therapy may be minimized. Furthermore, this design also minimizes the probability of a false-negative result (type II error), so drugs that may actually have activity are not rejected prematurely.

In most oncology trials, the *disease response* is evaluated approximately every 8 weeks with the same method of assessment at each evaluation (Fig. 12.3.14). For solid tumors, measurable lesions are defined as lesions outside of any prior radiation fields that are ≥10 mm by spiral computed tomography (CT). Responses are then categorized as a complete (CR) response, partial response (PR), stable disease (SD), or progressive disease (PD), based on measurements obtained from disease assessments. Historically, bidimensional measurements were used to calculate response [17, 18]. A CR generally means that no measurable lesion exists for a predetermined specified period of time. A PR can be a specified measurable tumor reduction (e.g., 50%, over the same time period). SD occurs when the disease has not worsened during the evaluation period. More recently, a new tumor evaluation system defined as "RECIST" (Response Evaluation Criteria in Solid Tumors) has been proposed and is gradually being adopted as standard of care [19]. In contrast with the previous criteria, the RECIST criteria utilize unidimensional tumor measurements and incorporate both measurable (target) and nonmeasurable (nontarget) lesions in the response criteria. Regardless of the response assessment method, all documented responses (CR or PR) should be confirmed with repeat scans no less than 4 weeks after the scans documenting the initial response. These responses for phase 2 trials are assessed often in patients who have failed on prior chemotherapy, which creates a higher hurdle of efficacy for oncology than most other therapeutic areas.

Phase 3 trials are large, multicenter, randomized trials comparing the new agent (alone or in combination with other approved drugs) to what is considered "standard therapy" in a specific disease setting (Fig. 12.3.15). Phase 3 trials can be done through the oncology networks who would have access to the sufficiently large number of patients needed and also a collection of practice groups and university centers. These trials are typically the cornerstone of a New Drug Application submission package for approval by regulatory authorities. Because patients are randomized between treatment regimens, they may be prospectively stratified by a particular prognostic factor (such as disease stage or performance status). The benefits of stratification are to help ensure that the results in one treatment arm are not biased by the prognostic factor, and to find a major subgroup who may respond better to new treatment than the whole population. The "traditional" end point for these trials has been overall survival, and enrollment numbers are determined prospectively based on statistical power for proving either superiority or equivalence. The extent of improvement in survival desired will impact sample size; for cancers with higher 5-year mortality rates, an improvement of 10%, that is, from 10% to 20% will be desirable, but with cancers with better survival, a change from 50% to 60% may not be a sufficiently desirable goal. The survival end point requires longer period of monitoring patients for this determination versus response rates. The statistical analysis may also include an interim analysis, which is typically performed and reviewed by an independent (from the sponsors and investigators) Data Safety and Monitoring Board (DSMB). It is the responsibility of the DSMB to ensure that it is in the best interest of all patients to continue the trial to completion or not, based on both toxicity and efficacy issues. Phase 3 trials are typically very large, so an interim analysis by a DSMB may prevent patients from receiving an inferior treatment regimen.

There are also multiple *alternative trial end points* that may be used as secondary end points (Fig. 12.3.16). Time to disease progression is an alternative end point that occurs earlier than overall survival and is not influenced by subsequent or crossover therapy that patients may receive. Time to disease progression is also sensitive to prolonged periods of stable disease, so this may be an appropriate end point to choose for

- Time to progression:
  - Occurs earlier than survival
  - Not obscured by subsequent or crossover therapy
  - Measured in all patients
  - Disease assessment schedule can create bias
  - Sensitive to stable disease/cytostasis
- Fractional survival at a designated time point:
  - Percentage of patients surviving at a designated time interval from the start of the study (i.e., 1 year)
  - May identify regimens that provide clinical benefit for patients in the absence of a tumor response

FIG. 12.3.16. Phase II/III Trials Alternative Endpoints – 1

a drug that may be associated with cytostasis rather than cytotoxicity. A second alternative end point is fractional survival at a designated timepoint. This end point is defined as a percentage of patients surviving at a designated time interval from the start of the study (i.e., 1 year). This alternative end point would be used most appropriately in a setting where the treatment regimen may provide clinical benefit in the absence of a measurable tumor response, such as pancreatic cancer or non-small cell lung cancer (particularly patients receiving treatment in the second or third line setting).

Clinical benefit or symptom improvement is another *alternative clinical trial end point* (Fig. 12.3.17). In order for this end point to be accepted as credible, the end point criteria must be clearly defined prior to the initiation of the trial. Because the end point relies on improvement in disease-related symptoms, patient accrual may be somewhat difficult as patients with disease-related symptoms often have advanced disease and a poor performance status. However, in diseases where visual imaging of the tumor is difficult, such as pancreatic and prostate cancer, improvements in "clinical benefit" may be very beneficial. Gemcitabine (pancreatic cancer) and mitoxantrone (prostate cancer) are two drugs that were approved by the FDA based on clinical benefit response [20, 21].

Health-related quality of life may be a component used to determine clinical benefit. Quality of life is a subjective assessment of the psychological, emotional, and physical functioning of a person that may be affected by the underlying disease and any treatment administered for the disease [22]. Multiple questionnaires and scales have been developed over the past three decades to determine quality of life in patients participating in clinical trials. Quality of life data becomes increasingly important in large, randomized controlled trials for which there is no significant difference in the primary end point of the study (i.e., overall survival) between the comparative treatment regimens. In this setting, significant differences in quality of life may cause physicians to favor one regimen over another. Differences in quality of life data may also be extremely important in clinical trials evaluating the benefits of supportive care agents for the prevention or management of treatment-related side effects. However, problems surrounding methodologic issues and the interpretation of quality of life data have limited the use of these clinical data. Some of the limitations of quality of life data that have been defined include patient compliance, analysis and reporting of missing data, prospectively stating a clear hypothesis, and limited reporting of detailed results [23].

Tumor marker–based outcomes is another *alternative end point* that can be used in phases 2 or 3 clinical trials (Fig. 12.3.18). This end point may be particularly useful in diseases, such as prostate, pancreatic, and ovarian cancer, where it is difficult to assess measurable disease by traditional methods. By measuring changes in serum biomarkers such as PSA and CA-125, one may be able to determine response to chemotherapy in the absence of measurable disease as well as early disease progression in the setting of small-volume disease [24, 25]. However, specific response criteria based on significant, reproducible changes in the serum biomarker must be established in order for changes in serum tumor marker levels to be accepted as a surrogate clinical trial end point.

The development of novel imaging technologies, including PET and dynamic contrast-enhanced magnetic resonance imaging (DCE-MRI), may also result in alternative clinical trial end points [26]. These new imaging techniques alone or

- Clinical benefit/symptom improvement:
  - Presence of disease-related symptoms may be associated with advanced disease or poor performance status so accrual may be difficult
  - May be beneficial in tumors that are difficult to measure via imaging (i.e., pancreatic and prostate cancer)
  - Endpoints must be clearly defined
- Health-related quality of life:
  - Multiple validated scales to assess physical, emotional, social, and functional well-being
  - May include disease-specific subscale; influenced by patient compliance
  - Endpoints must be defined.

FIG. 12.3.17. Phase II/III Trials Alternative Endpoints – 2

- Tumor marker based outcomes:
  - Used to measure both response and relapse/progression
  - May also be used for stratification; specific criteria needed
- Novel imaging technologies:
  - Imaging techniques may be used to assess cellular metabolism, cell proliferation and apoptosis, resistance to chemotherapy, and angiogenesis/perfusion
- Biomarker or targeted endpoint:
  - Target must be measurable
  - Could potentially increase efficiency and accelerate drug development process

FIG. 12.3.18. Phase II/III Trials Alternative Endpoints – 3

in combination with standard imaging tests, such as computed tomography, may allow researchers to assess a drug's effects on cellular metabolism, cellular proliferation, apoptosis, and angiogenesis or perfusion.

The identification of biomarkers relevant to the mechanism of action of the anticancer drug is a final potential alternative end point for oncology clinical trials. If a critical, measurable biomarker target (such as an intracellular enzyme pathway) can be established for an agent, then activation or inhibition of this biomarker could potentially be used as an alternative clinical trial end point.

Other Design Issues in Oncology

Due to improvements in health care, the life expectancy of persons living in the United States has increased over the years. Furthermore, the incidence of cancer increases with increasing age. As a result, the number of *elderly patients* with cancer is expected to increase significantly over the next 30 years (Fig. 12.3.19). Traditionally, "elderly" has been defined as patients >65 years of age. Based on this age definition, more than half of newly diagnosed cancer cases occur in elderly patients, but elderly patients comprise only 25% of the clinical trial population [27]. In fact, when cancer drug registration trials were analyzed for elderly patient participation, the discrepancy between cancer incidence and trial participation actually increased with age [28]. Although the incidence of cancer increases in patients over the age of 70 years, clinical trial participation in this age group actually decreased. Multiple barriers for the recruitment of elderly patients into clinical trials have been identified [28, 29]. One reason that elderly patients may be underrepresented in cancer clinical trials is disqualification due to eligibility criteria. Age-associated declines in performance status and organ function may prohibit elderly patients from enrolling in clinical trials. The presence of excluded multiple comorbid conditions or concomitant medications may also prevent patient enrollment. Many physicians may also have preconceived ideas about treatment tolerability that may prevent them from enrolling elderly patients on clinical trials or treating elderly patients with chemotherapy

- "Elderly" empirically are defined as >65 years of age.
- More than half of the new cancer cases occur in elderly patients, but they comprise only 25% of the clinical trial population.
- Eligibility criteria may disqualify older patients from trial participation.
- Preconceived ideas about treatment tolerability may prevent trial participation; elderly often under treated in practice.
- Frequency of clinic visits and treatment-related expenses may also inhibit trial participation.

FIG. 12.3.19. Elderly Patients in Clinical Trials

- Common Signs/Symptoms:
  - Fatigue/Anemia, Mucositis, Nausea/Vomiting, Neuropathy, Neutropenia/Infections, Pain
  - Often debilitating to patients
  - Usually related to disease and chemotherapies
- Design Issues:
  - Possible cause of dropouts
  - Common limits to achieving full dosing and full responses to chemotherapy
  - Manifestations need to be treated during drug studies (uniformity required)
- Assessments:
  - Usually quality-of-life or humanistic parameters
  - Symptom specific vs. general tools, e.g. SF-36
  - Acceptance by clinicians and regulators needed

FIG. 12.3.20. Supportive Care Product Development

in clinical practice. Elderly patients are often undertreated for this reason with lower doses and less toxic combinations of drugs. Finally, the frequency of clinic visits and treatment-related expenses may inhibit trial participation of elderly patients. In the future, it will be critical to obtain clinical trial data regarding dosing and tolerability of new agents in elderly patients so that this data can be extrapolated to the ever-increasing population of elderly patients with cancer.

Another area of oncology drug development that has grown tremendously over the past two decades is the arena of *supportive care* (Fig. 12.3.20). Patients with cancer experience numerous underlying disease or treatment-related signs and symptoms, such as fatigue with anemia, mucositis, nausea and vomiting, neuropathy, neutropenia and infection, and pain. Any or all of these symptoms can potentially limit the chemotherapy doses patients are able to receive and can debilitate patients, resulting in decreased quality of life. Clinical trials evaluating the effectiveness of supportive care agents are difficult to conduct because patients are typically receiving therapy simultaneously for the treatment of their underlying malignancy. Therefore, it is critical to be able to determine what effects may be attributed to the supportive care agent versus any disease treatment. Because many of the improvements in the supportive care arena are subjective in nature (decreased pain, nausea, fatigue, etc.), quality of life questionnaires and scales are critical components of these clinical trials. Often, the generic questionnaires for quality of life, such as the often used and validated SF-36 Scale, do not alone sufficiently demonstrate the true benefits of a supportive product. Therefore, over the past 10–15 years, disease specific, that is, symptom specific, scales needed to be created and validated (and agreed upon by the regulatory authorities), which was done for example for nausea and vomiting for Zofran® and anemia for Procrit® and Aranesp®. Plus, these specific parameters were developed simultaneously with the clinical development process for product approval. Acceptance of such nontypical and new disease parameters

- Determine if treatment-related toxicities outweigh benefits
- Acute hematologic or non-hematologic toxicities:
  - Very common with these drugs
  - Primarily reported of serious nature
- Tolerance of toxicity is high in oncology.
- Common grading systems:
  - WHO and NCI/CTC
  - Grading of severity (1 to 4 levels)
- Criteria for patient screening, data collection, data analysis, & data display/publication need to be established
- Late toxicities are frequently not evaluated; most recent CTCAE (version 3.0) includes late effects.
- Adverse event reporting needs to be standardized across clinical trials so data may be compared.

FIG. 12.3.21. Adverse Event Reporting

was needed also by the oncology research and practice community, which was both another research and educational challenge for the pharmaceutical industry during product development [30–32].

The primary goal of chemotherapy treatment in most patients is palliation rather than cure. As a result, it is critical to be able to determine whether treatment-related toxicities outweigh the benefits of treatment. The toxicities that are primarily reported in clinical trials are the acute hematologic or non-hematologic *adverse events* of a serious nature (Fig. 12.3.21). In general, toxicities of cancer products (cytostatic agents especially) are much more common and severe than most drugs. The level of acceptance of toxicity in oncology is much higher than other therapeutic areas given the frequently mortal outcome of the disease, the potential survival benefit of the new product, and the shorter duration of therapy (weeks to months). As stated earlier, adverse effects are the primary end points of phase 1 clinical trials in oncology. Also, the therapeutic index of cancer drugs is small, and cancer drugs are dosed at fairly toxic levels commonly to achieve the desired efficacy.

The two most common grading systems for treatment-related toxicities are the World Health Organization criteria (WHO) and the National Cancer Institute Common Toxicity Criteria (NCI CTC) [33]. Both of these systems grade individual toxicities based on the severity of the symptom. For example, adverse effects in oncology are reported at a frequency level at each grade level of severity from 1 mildest to 4 most severe. However, currently there is no one standard guideline for the evaluation or reporting of adverse events experienced in clinical trials. Criteria for patient assessment, data collection, data analysis, data display, and data publication of treatment-related adverse events need to be established and incorporated into oncology clinical trials. In drug development and application for approval, the adverse event monitoring and reporting are key agreements between the regulatory authorities and the pharmaceutical company for the protocols. Because clinical trials are conducted over a finite period of time, potential long-term treatment-related adverse events may not be evaluated. Furthermore, many of the late effects of chemotherapy are not included in the grading systems for treatment-related toxicities. In an effort to improve reporting of long-term toxicities, the most recent version of the NCI CTC (version 3.0) does include grading for late effects of cancer treatments. As more cancer treatment options become available, it will become critical to have standardization of adverse event assessment and reporting so that treatments can be compared across clinical trials and in the clinic setting.

Era of Targeted Therapy

Over the past decade, a tremendous amount of progress has been made in understanding the molecular basis of cancer, and this has led to the development of *targeted therapies* (Fig. 12.3.22). The discovery of mechanism-based biomarkers of a biological aspect of cancer (e.g., EGFR [epidermal growth factor receptor], VEGF [vascular endothelial growth factor], mTOR [mammalian target of rapamycin]) has led to the development of numerous drugs that either stimulate or inhibit these targets [34–36]. In the best case scenario, the target would preferentially be expressed in tumor tissue and not in normal tissue. However, in many of the compounds developed to date, this is not the case. In order to speed the development and increase the specificity of these compounds for cancer cells, researchers have developed novel molecular and cellular technologies such as immunological and PCR-based assays, genomic and proteomic analyses, and gene expression profiling with DNA microarrays to assess the effectiveness of target inhibition [37–39].

Targeted therapy is the new mantra of oncology product development (Fig. 12.3.23). Many of the targeted compounds in development inhibit cell growth, resulting in cytostasis, rather than causing cell death or cytotoxicity. The toxicity profiles of the targeted therapies also tend to be different from the profiles of traditional chemotherapy agents. As a result, alternative drug development strategies from those that have

- Progress in understanding molecular basis of cancer has led to development of targeted therapies (e.g., EGFR, VEGF, mTOR)
- Biomarker:
  - Mechanism based measurable/testable endpoints of biological aspects of cancer
  - May be used to assess effectiveness of target inhibition
- Molecular and cellular technologies are used to develop biomarker-based targeted therapies:
  - Immunological and PCR-based assays
  - Genomic and proteomic analyses
  - Gene expression profiling (DNA microarrays)
  - Detection of circulating cancer cells

FIG. 12.3.22. Cancer Drugs in Era of Targeted Therapy – 1

- Targeted therapy is burgeoning area of oncology research that may warrant alternative development strategies for drug approval.
- Most compounds inhibit cell growth and may not cause tumor regression (cytostatic vs cytotoxic).
- Target may not be unique to tumor cells (may be expressed in normal tissues as well).
- Numerous differences between targeted therapies and cytotoxic compounds; as result "traditional" drug development process may not be applicable.

FIG. 12.3.23. Cancer Drugs in Era of Targeted Therapy – 2

historically been utilized for traditional chemotherapy agents may be warranted.

Preclinical studies of targeted therapies are critical in order to develop and validate a biomarker of target inhibition. The primary goal of therapy is to inhibit the specified target in tumor tissue. As a result, the only reliable means of determining target inhibition is to perform serial tumor biopsies. This may be possible in small subsets of patients in early clinical trials, but it is certainly not feasible in large clinical trials. Surrogate markers of tumor inhibition need to be established early in development so that they can be validated throughout the clinical development process. These markers could potentially be measured in blood, serum, or normal skin tissue. Any of these samples would certainly be more readily available than tumor biopsy specimens on a large-scale basis. Preclinical studies with targeted therapies may be difficult due to interspecies differences that may prevent data extrapolation from animals to humans. Furthermore, additional preclinical toxicology testing may be necessary to support the prolonged dosing in humans that may be warranted due to the cytostatic nature of the drugs.

Preclinical studies are an important part of oncology product development to help screen for product activity and toxicity (Fig. 12.3.24). Biomarkers for cancer disease exist and more can be developed and validated in animals. A limit to animal studies is interspecies variation possibly limiting extrapolation of animal data to humans. As patients are living longer with cancer related to earlier and better diagnosis and improved therapy, treatments are more chronic in nature and may require added longer term animal studies for toxicity.

The targeted compounds are typically less toxic than traditional chemotherapy drugs. As a result, numerous dose levels may be required to establish the MTD, and in some cases DLT may not be encountered (Fig. 12.3.24). Because many targeted agents are not cytotoxic in nature, the initial phase 1 work may be conducted in normal healthy volunteers. The use of normal healthy volunteers for initial drug testing could potentially speed the completion of phase 1 trials in cancer patients, as fewer dose levels will need to be assessed. The use of accelerated titration designs with aggressive dose escalation schemas may also help speed the phase 1 trials. On the other hand, there are several things that may actually slow the completion of phase 1 trials. Depending on the specificity of the targeted therapy, patients may need to be screened at baseline, with only those patients who express or overexpress the drug target being enrolled in the trial. This "patient selection" process could slow patient accrual if the target is not common in the majority of patients. Phase 1 trials of targeted therapies may also be somewhat cumbersome as drug developers view these trials, because additional blood samples, skin biopsies, or tumor biopsies may be incorporated into the trials to define the pharmacokinetics and pharmacodynamics of the compound and validate successful target inhibition. Noninvasive imaging techniques, such as DCE-MRI, color Doppler ultrasounds, and PET scans, may also be incorporated to determine their utility as surrogate markers for drug activity.

Historically, *phase 2 trials* have been used to determine the preliminary antitumor activity of chemotherapy agents in various disease settings in order to make a decision about the feasibility of subsequent *phase 3 trials* in a specific disease setting (Fig. 12.3.25). With cytotoxic compounds, antitumor activity has been determined with standard tumor response assessment criteria. However, due to the cytostatic nature of the targeted therapies, the traditional measurements of

- Preclinical studies:
  - Optimal to develop and validate biomarker of target inhibition
  - Interspecies differences may prevent data extrapolation from animals
  - Prolonged dosing may warrant more toxicology data
- Phase I studies:
  - DLT may not be optimal endpoint.
  - Consider patient "selection" based on target expression or dependence
  - Larger, multi-faceted trials may be needed to address PK, PD, and safety issues before proceeding to later trials.
  - Early trials may be done in normal volunteers.

FIG. 12.3.24. Cancer Drugs in Era of Targeted Therapy – 3

- Phase II trials:
  - Patients may need to be "selected" based on target expression or tumor types known to be dependent on target for tumor proliferation
  - Questionable feasibility with standard disease response criteria
  - May be omitted (phase I/III expanded)
  - May use alternative endpoint for proof of concept
- Phase III trials:
  - Must select appropriate endpoints
  - May need to be conducted in earlier stages of disease, so endpoints may take longer to attain
  - Rational combinations with chemotherapy, radiation, or other targeted therapies warranted

FIG. 12.3.25. Cancer Drugs in Era of Targeted Therapy – 4

antitumor activity may not be applicable. As a result, it may be difficult to establish specific end points for phase 2 trials. As a result, some pharmaceutical companies have opted to proceed directly from phase 1 trials to phase 3 trials with targeted agents. Other companies have chosen to redesign the concept of phase 2 clinical trials in order to obtain additional data about the clinical activities of their compounds. Instead of measuring antitumor activity by RECIST or other traditional criteria, alternative trial end points such as prolonged disease stabilization, time to tumor progression, or prolonged survival may be used. These trials might also incorporate assessment of the molecular target assays in an attempt to determine factors that may predict efficacy of the targeted therapy or randomize patients to a range of dose levels to help determine the most appropriate dose for subsequent testing. Unfortunately, the use of alternative end points may warrant an increase in the number of patients enrolled on the phase 2 trials to as high as 100–300 patients. Phase 2 trials of traditional cytotoxic agents utilizing tumor measurement criteria typically require 40–50 patients. Although activity in phase 2 trials does not guarantee successful phase 3 trials, larger phase 2 trials with additional supportive data regarding target specificity may help prevent drugs with borderline activity and target nonspecificity from progressing into large randomized phase 3 trials that consume both patients and resources.

Study end point selection is critical for the success of phase 3 trials. Again, because targeted therapies are cytostatic in nature rather than cytotoxic, alternative primary and secondary end points may be warranted. Some of the end points that can be considered are described in the previous sections of this chapter. The success of phase 3 trials may also depend on patient selection. It is crucial to determine whether patients need to be prospectively screened for the presence or overexpression of the drug target in early clinical trials. Lack of patient screening could result in negative phase 3 trial results; whereas screening for an appropriate "patient subset" could produce positive results.

Finally, the incorporation of targeted agents into standard treatment regimens containing cytotoxic chemotherapy or radiation therapy has been quite challenging. Although the small molecule EGFR tyrosine kinase inhibitors demonstrated antitumor activity as single agents in patients with refractory non-small cell lung cancer, the addition of either gefitinib or erlotinib to standard chemotherapy did not improve disease response or survival in four large randomized trials [40–43]. These data were quite surprising and disappointing for the researchers conducting the clinical trials and has led to renewed investigation into the optimal dose and schedule of the anti-EGFR agents, particularly when combined with conventional therapies. The identification of factors predictive of response and resistance is also an area of intense investigation.

The development of molecularly targeted therapies is a rapidly burgeoning area of oncology clinical research. It is hoped that the availability of specifically targeted therapies will result in increased antitumor activity and improved toxicity profiles of anticancer agents. However, because of numerous differences between molecularly targeted compounds and traditional cytotoxic chemotherapy drugs, radical changes may be warranted in all facets of the oncology drug development process.

References

1. Anonymous. Oncology, market indicators. Nat Rev Drug Discov 2005 (May Suppl):S19.
2. Humphreys A, Mayer R. 11th Annual Report: World's best selling medicines. Med Ad News 2005;24(5):1, 24-37.
3. King J. Top 10 areas of research. 5th annual report. R&D Directions 2004;10(9):36-43.
4. SEER Surveillance, Epidemiology and End Results. Stat database: incidences and mortality in U.S. (1992, 2002), National Cancer Institute, 2005.
5. Eisenhauer EA, O'Dwyer PJ, Christian M, et al. Phase I clinical trial design in cancer drug development. J Clin Oncol 2000,18: 684-692.
6. Simon R, Freidlin B, Rubinstein L, et al. Accelerated titration designs for phase I clinical trials in oncology. J Natl Cancer Inst 1997;89:1138-1147.
7. Collins JM, Grieshaber CK, Chabner BA. Pharmacologically guided phase I trials based upon preclinical development. J Natl Cancer Inst 1990;82:1321-1326.
8. Dees EC, Whitfield LR, Grove WR, et al. A phase I and pharmacologic evaluation of the DNA intercalator CI-958 in patients with advanced solid tumors. Clin Cancer Res 2000;6: 3885-3894.
9. Babb J, Rogatko A, Zacks S. Cancer phase I clinical trials: efficient dose escalation with overdose control. Stat Med 1998;30: 1103-1120.
10. Cheng JD, Babb JS, Langer C, et al. Individualized patient dosing in phase I clinical trials: the role of escalation with overdose control in PNU-214936. J Clin Oncol 2004;22:602-609.
11. O'Quigley J. Another look at two phase I clinical trial designs. Stat Med 1999;18:2683-2690.
12. Rinaldi DA, Burris HA, Dorr FA, et al. Initial phase I evaluation of the novel thymidylate synthase inhibitor, LY231514, using the modified continual reassessment method for dose escalation. J Clin Oncol 1995;13:2842-2850.
13. Gehan EA. The determination of the number of patients required in a preliminary and a follow-up trial of a new chemotherapeutic agent. J Chron Dis 1961;13:346-353.
14. Fleming TR. One-sample multiple testing procedure for phase II clinical trials. Biometrics 1982;38:143-151.
15. Simon R. Optimal two-stage designs for phase II clinical trials. Control Clin Trials 1989;10:1-10.
16. Dent S, Zee B, Dancey J, et al. Application of a new multinomial phase II stopping rule using response and early progression. J Clin Oncol 2001;19:785-791.
17. Green S, Weiss GR. Southwest Oncology group standard response criteria, endpoint definitions, and toxicity criteria. Invest New Drugs 1992;10:239-253.
18. Miller AB, Hoogstraten B, Staquet M, Winkler A. Reporting results of cancer treatment. Cancer 1981;47:207-214.

19. Therasse P, Arbuck SG, Eisenhauer EA, et al. New guidelines to evaluate the response to treatment in solid tumors. J Natl Cancer Inst 2000;92:205-216.
20. Burris HA III, Moore MJ, Andersen J, et al. Improvements in survival and clinical benefit with gemcitabine first-line therapy for patients with advanced pancreas cancer: a randomized trial. J Clin Oncol 1997;15:2403-2413.
21. Johnson JR, Williams G, Pazdur R. End points and United States Food and Drug Administration approval of oncology drugs. J Clin Oncol 2003;21:1404-1411.
22. Patrick-Miller LJ. Is there a role for the assessment of health-related quality of life in the clinical evaluation of novel cytostatic agents? Clin Cancer Res 1990;9:1990-1994.
23. Bottomley A, Efficace F, Tjomas R, et al. Health-related quality of life in non-small-cell lung cancer: methodologic issues in randomized controlled trials. J Clin Oncol 2003;21:2982-2992.
24. Rustin GJS, Bast RC Jr., Kelloff GJ, et al. Use of CA-125 in clinical trial evaluation of new therapeutic drugs for ovarian cancer. Clin Cancer Res 2004;10:3919-3926.
25. Kelloff GJ, Coffey DS, Chabner BA, et al. Prostate-specific antigen doubling time as a surrogate marker for evaluation of oncologic drugs to treat prostate cancer. Clin Cancer Res 2004;10:3927-3933.
26. Park JW, Kerbel RS, Kelloff GJ, et al. Rationale for biomarkers and surrogate end points in mechanism-driven oncology drug development. Clin Cancer Res 2004;10:3885-3896.
27. Lewis JH, Kilgore ML, Goldman DP, et al. Participation of patients 65 years of age or older in cancer clinical trials. J Clin Oncol 2003;21:1383-1389.
28. Talarico L, Chen G, Pazdur R. Enrollment of elderly patients in clinical trials for cancer drug registration: a 7-year experience by the US Food and Drug Administration. J Clin Oncol 2004;22:4626-4631.
29. Townsley CA, Selby R, Siu LL. Systematic review of barriers to the recruitment of older patients with cancer onto clinical trials. J Clin Oncol 2005;23:3112-3124.
30. Boccia R. Darbepoetin alfa in anemia management. A vital role in cancer care. Supportive Oncol 2005;3(2 Suppl. 1):3-38.
31. Dale DC. Advances in the use of colony-stimulating factors for chemotherapy-induced neutropenia. Supportive Care 2005;3(2 Suppl. 1):39-72.
32. Stiff PJ. Oral mucositis therapy comes of age. Supportive Care 2005;3(2 Suppl. 1):73-83.
33. Trotti A, Bentzen SM. The need for adverse effects reporting standards in oncology clinical trials. J Clin Oncol 2004;22:19-22.
34. Arteaga CL, Baselga J. Clinical trial design and end points for epidermal growth factor receptor-targeted therapies: implications for drug development and practice. Clin Cancer Res 2003;9:1579-1389.
35. Hicklin DJ, Ellis LM. Role of the vascular endothelial growth factor pathway in tumor growth and angiogenesis. J Clin Oncol 2005;23:1011-1027.
36. Thompson JE, Thompson CB. Putting the rap on akt. J Clin Oncol 2004;22:4217-4226.
37. Vande Woude GF, Kelloff GJ, Ruddon RW, et al. Reanalysis of cancer drugs: old drugs, new tricks. Clin Cancer Res 2004;10:3897-3907.
38. Tyers M, Mann M. From genomics to proteomics. Nature 2003;422:193-197.
39. Mohr S, Leikauf GD, Keith G, et al. Microarrays as cancer keys: an array of possibilities. J Clin Oncol 2002;20:3165-3175.
40. Giaccone G, Herbst RS, Manegold C, et al. Gefitinib in combination with gemcitabine and cisplatin in advanced non-small cell lung cancer: A phase III trial-INTACT 1. J Clin Oncol 2004;22:777-784.
41. Herbst RS, Giaccone G, Schiller JH, et al. Gefitinib in combination with paclitaxel and carboplatin in advanced non-small cell lung cancer: a phase III trial-INTACT 2. J Clin Oncol 2004;22:785-794.
42. Gatzemeier U, Pluzanska A, Szczesna A, et al. Results of a phase III trial of erlotinib HCL (OSI-774) combined with cisplatin and gemcitabine (GC) chemotherapy in advanced non-small cell lung cancer (NSCLC). Proc Am Soc Clin Oncol 2004;22:617 (abstract 7010).
43. Herbst RS, Prager D, Hermann R, et al. TRIBUTE- A phase III trial of erlotinib HCL (OSI-774) combined with carboplatin and paclitaxel (CP) chemotherapy in advanced non-small cell lung cancer (NSCLC). Proc Am Soc Clin Oncol 2004;22:617 (abstract 7011).

4
Pediatrics

Philip D. Walson

The Case for Pediatric Research with Drugs	322
American Academy of Pediatrics and Research	325
Legislative and Government Initiatives in Pediatric Research	327
The Science in Pediatric and Clinical Studies	333
References	337

Pediatric drug development has been slow and difficult for all the reasons that "Children are not just little adults." There are real as well as merely perceived problems in doing trials in children. However, recent advances in research, education, and legislation have all begun to improve the chances that children will benefit as much as their parents and grandparents from modern pharmaceutical advances. This chapter will cover the current problems with the use of medications in children, and the justification for doing research in this vulnerable population, as well as the actual design and conduct of clinical trials in children.

Actual use data illustrate the current problems with the use of medications in children. These problems are based on the many knowledge gaps concerning pediatric use. Most (70–80%) medications are used in children without adequate safety, efficacy, or even dosing information in children; no pediatric labeling in the package insert existed up until recently. Formulations are created extemporaneously from adult's dosage forms for children without bioavailability and stability, as well as safety, efficacy, and dosing information. This is despite the fact that medications have the greatest potential to improve health and survival in children as opposed to elderly adults; where most health care spending is focused.

The history of medication use in children is replete with examples of how dangerous it can be to use medications in children based solely on adult data. Medications used in children must be studied in children to be certain that the risk to benefit ratio is maximized. In fact, the powers of the FDA are in large part the result of medical disasters in children that provided impetus for the FDA to demand adequate evidence of safety and efficacy before the sale or promotion of products. Unfortunately, the history of drug regulation is full of examples of legislation that came from children damaged or killed by the lack of appropriate studies yet resulted in safer medicines for adults but little benefit to children. It is not too large an overstatement to say that the drug development system in place today was built on the bodies of dead children. The Federal Food, Drug, and Cosmetic Act in 1938 that created FDA regulations of drugs for safety followed more than 100 deaths from a sulfanilamide liquid; the 1962 Kefauver–Harris amendments creating effectiveness and safety requirements and an active FDA approval process followed thalidomide and chloramphenicol tragedies. Only recently has legislation been designed to guarantee that drugs given to children will be adequately studied in children.

In part, past resistance to doing adequate studies in children was based on well-meaning concerns about exploitation of children or exposure to excessive risks. While legitimate concerns, the decision whether to test or not test medications in children must be based on a risk/benefit analysis of both testing and the alternative. Allowing medications that have not been adequately tested to be given to children "off-label" is the alternative. Such use is essentially unregulated experimentation without consent or assent and does not produce scientifically or clinically acceptable knowledge. When pediatric use prior to studies was compared with what was found after testing, it was found that only two-thirds of the time was "best practice" of pediatric "experts" correct, according to comments from Diane Murphy of the U.S. FDA [1, 2]. Clearly what pediatricians, specialists or not, believe is true is not always correct. In fact, drugs are tested in adults because it is known that "opinions," no matter how "expert," are simply no substitute for adequate controlled clinical trials. Children as well as adults should not be subjected to the uncontrolled, unregulated, unscientific, useless "experiment" of unstudied, off-labeled use.

This chapter will discuss some of the ethical, scientific, and practical issues involved in pediatric clinical trials, as well as the legislation that has encouraged or allowed pediatric studies. Finally, some of the findings and benefits of studying medication in children will be summarized.

The Case for Pediatric Research with Drugs

Because drugs cannot be labeled or promoted for use without adequate information from clinical trials, some 70–75% of medications are not adequately labeled for use in children. Practitioners are therefore forced to either avoid using potentially effective medications in children or use them "off-label." Such off-label use of medications in children is common because many children receive medications. In the Canadian Health System, examination of *prescribing in children* (more than 1 million claims) showed up to 76% are treated with antibiotics (some studied and labeled for use in children, at least older children), but also 18% with respiratory diseases, 7% for acne, 3% with attention deficit hyperactivity disorders (ADHD), 2% with depression, as listed on Fig. 12.4.1 [3, 4]. Considering the millions of children treated, this amounts to a huge unregulated experiment, without consent/assent and without the opportunity to provide useful scientific labeling.

♦ Antibiotics	76%
♦ Respiratory/Asthma	18%
♦ Acne	7%
♦ Contraception	4%
♦ ADHD	3%
♦ Depression	2%
♦ Gastrointestinal	2%
♦ Epilepsy	0.6%
♦ Diabetes	0.4%

Canadian Health system
- 1.03 million claims
- Children 12-18% of Canadian population

MacLoad SM. DIA Annual Meeting, June 17, 2005; Pediatrics and Child Health 2003; 8 Suppl.A:

FIG. 12.4.1. Prescription Drug Use in Children

- Limited populations
- Limited marketing potential
- Difficulty in conducting trials in children
- Ethical concerns
- Lack of endpoints
- Belief that dosing is just by weight
- Inexperienced investigators

FIG. 12.4.2. Knowledge Gap in Pediatrics

There are many reasons for the *lack of studies/knowledge* concerning drug use in children (Fig. 12.4.2). There are limited numbers of children with specific conditions whose parents (or themselves) are willing to be in clinical trials. The relatively small pediatric markets limit the financial incentives for pharmaceutical companies to perform pediatric studies. A pharmacokinetic trial was estimated to cost $250,000 to $750,000 by the U.S. General Accounting Office in 2001; an efficacy trial costs about $1 million to $7.5 million each. A pediatric indication will require a battery of such studies [5]. Pediatric trials are more difficult to conduct because of limited numbers of investigators willing to do studies, as well as few parents and subjects willing to be in studies. Also, drug responses can vary substantially (more or less activity or even opposite effects) in children versus adults or the animal models for diseases. Dosing and formulation challenges exist for children for study designs. Placebo comparisons create added risks. There are numerous ethical concerns about doing studies in any vulnerable population, including children, for example, patient/parental consent, placebo trials, compensation. Therapeutic and toxic end points are more difficult to assess in children. Many practitioners feel that it is sufficient to merely "scale down" pediatric doses by weight. Additional difficulties include a lack of investigators who are well trained to do pediatric clinical trials, related to some extent to the facts that they take longer, are more difficult to design and get approved, that they are more expensive (per subject) to do than are adult studies, and have limited academic, professional, or financial benefit to investigators.

The *lack of pediatric data* is being increasingly recognized *globally* (Fig. 12.4.3). Press reports, statements by international groups such as the International Union of Pharmacology (IUPHAR), the European Federation of Pharmaceutical Industries and Associations EFPIA, the British Forum for the Use of Medicines in Children, the U.S. Food and Drug Association (FDA), the International Congress on Harmonization (ICH), and the American Academy of Pediatrics (AAP) have all released statements on the need for pediatric studies. These organizations, particularly the AAP, have been responsible for the documentation from experts of the extent and gravity of the problem and creating the groundswell of public outcry and stimulus to

12.4. Pediatrics

- Growing awareness of deficiencies in therapeutic information for children
  - Press reports
  - International Union of Pharmacology
  - European Federation of Pharmaceutical Industries and Associations (EFPIA)
  - British Forum for the Use of Medicines in Children
  - FDA & ICH
  - AAP
- Eyes are on U.S. policy

FIG. 12.4.3. Global Advocacy

legislative bodies to pass legislation that allows and requires pediatric drug trials. The United States has led the world in writing new drug regulations and passing legislation to protect children, foster research, create new incentives for industry, and create new information for more safe and effective drug use. The European Union is in process of changing their laws and regulations (2004–2006) to improve incentives for pediatric studies and labeling, while ensuring protection of this vulnerable population.

Legislation both encouraging and insisting on pediatric trials is based on a *historical review* of the problem and a risk/benefit analysis of drug testing, as well as the alternative continuation of the limited testing paradigm of the past (Fig. 12.4.4). The most compelling reasons that drugs used in children must be studied in children is because results of studies in adults simply can't always be used to make decisions about drug use in children and because such testing is safer and more ethical than the alternative continued off-label use. The ethical stand that children must be "protected" from experimentation ignores the fact of off-label use without data. It can be argued that the widespread, off-label use of medications in children without adequate data is essentially unregulated experimentation without the possibility of benefiting either society, other children, or perhaps even the patient who receives such treatment. Children are not just little adults for drug use, as well as in other aspects of medicine and life. The paradigm of pediatric drug testing has been so completely altered that now there are actually drugs tested, labeled, and used in children where in fact adult use is "off-label," as indicated by this recent warning on a pediatric cancer drug that warns that "There are insufficient data to determine whether geriatric or adults in general respond differently from pediatric patients." The tables have definitely begun to turn for children; and a review of the history of drug use indicates that it has taken too long.

The FDA's drug approval and labeling system is arguably part of the best drug development system in the world. The legislative authority upon which it is based until recently was a process that benefited adults almost exclusively, despite the fact that the FDA's authority came from events that killed mostly children as the *history* display in Figure 12.4.5. In 1902, the first pure drug act was a result of an episode when children died from diphtheria antitoxin taken from animals infected with tetanus. The 1938 revision of the Food Drug, and Cosmetic Act, which required proof of safety and created a new federal approval process, came as a result of the deaths of more than 100 children poisoned by ethylene glycol used for a new formulation of a sulfanilamide "elixir" to treat children. Finally, the 1962 revision of the act (the Kefauver–Harris amendments), which required safety and efficacy to be proved with active FDA approval required, was prompted by the thalidomide disaster, which harmed untold numbers of unborn children, and the choramphenicol disaster in neonates and toddlers. These examples all illustrate the folly of using adult medicines in children, of reformulating products for

- "Pediatrics does not deal with miniature men and women, with reduced doses and the same class of diseases in smaller bodies, but….it has its own independent range and horizon…"
  Dr. Abraham Jacobi, 1889

- "Altered Reality"
  "There is insufficient data to determine whether geriatric or adults in general respond differently from pediatric patients".

FIG. 12.4.4. Why Pediatric Trials? – Historical View

- Adult Process (FDA) largely through child deaths
- 1902 – Diphtheria antitoxin - horse had tetanus
- 1938 – Sulfanilamide - ethylene glycol
- 1962 – Thalidomide – teratogen &
    Chloramphenicol – neonatal circulatory collapse
- New laws, no new pediatric data
  - No incentives
  - Fear of exploitation of children

FIG. 12.4.5. Historical Perspective

pediatric use without adequate studies, and for exposing developing children to drugs not tested in developing humans.

However, the legislative authority these events gave to the FDA did not produce much pediatric data. In part this was because of a lack of adequate incentives to industry to do necessary and rather expensive studies rather than merely warn against pediatric use. Additionally, it was the result of ethical concerns about "exploiting" children who would be in studies and the ethical problems of consent/assent in children. For studies to be done, children, parents, pediatricians, other physicians, and legislators, as well as national and international organizations needed to be convinced of the need to do studies.

Beyond the many tragedies and toxicity issues that justify the study of drugs in children, many *needs* for information in pediatrics exist (Fig. 12.4.6). The argument favoring such studies is based on a risk/benefit analysis of pediatric studies. Early childhood health is an important determinant of intellectual achievement and later adult health. Improving child health pays much larger dividends that attempting to reverse or ameliorate the effects of diseases in adults; especially at the extremes of age. Healthy children make healthier, happier, more productive adults, and many of the things that children need to grow up healthy are related to drugs and therapeutics. In fact, making drug therapies available to all children may be one of the most cost-effective measures available to improve future health and well being. This is clearly true of immunizations but is also true of other therapies as well. Chronic diseases such as diabetes, asthma, infections, arthritis, and so forth, are both common in children and can be effectively treated. Clearly, there are benefits to appropriate therapy of children. However, there are risks associated with treating children based on adult data. There are common pharmacokinetic (PK) and pharmacodynamic (PD) differences in children of all ages, as a result of developmental changes in physiology and behavior. It is not always safe to select a drug dose, route, formulation, duration, or indication in neonates, infants, children, or even adolescents based on adult data, weight, and "expert" opinion. Dr. Diane Murphy of the FDA recently reviewed information that came from pediatric labeling studies done under the FDA modernization act's exclusivity provisions. She reported that as many as one-third of off-label pediatric use was flawed. Yet millions of children continue to routinely be given medications for which adequate pediatric studies have not been done. This unregulated investigation without consent is routine. The risks of this "experimentation without consent" must be compared with the regulated, supervised, limited study of drugs in children. In addition, such off-label use does not produce scientifically valid information that can be used to help other children.

Use of medications without adequate labeling is considered "*off-label*" use (Fig. 12.4.7). This is extremely common in pediatric practice, because of the lack of research, publications, and labeling for pediatric patients. Between 1973 and 1997, reviews indicate that as many as 71–81% of medications used in children did not have adequate labeling for children [6, 7]. Despite numerous attempts by the FDA to encourage and even facilitate pediatric studies, and despite numerous "promises" by pharmaceutical companies to do postapproval pediatric studies, in fact only 9 of 27 approved drugs in 1997 had any pediatric labeling. Studies indicate that only 32% of 243 products in Europe had pediatric labeling, as well as 46% to

1. Early childhood health is an important determinant of intellectual achievement and later adult health
2. Many needs of children worldwide are related to drugs and therapeutics
3. Availability of drug therapies may be the most cost effective measure available to improve future health and well-being
4. Chronic diseases are common in children and need drug treatment, e.g., diabetes, asthma, infections, arthritis
5. PK / PD differences are common in children, differing even between neonates - infants - children - adolescents
6. Unregulated investigation without consent occurs routinely in pediatric practice
7. Millions of children taking medications without labeling information

FIG. 12.4.6. Why pediatric research? – Needs

- Absent pediatric labeling
  - E.g., 71% - 81% of approved drugs in 1973 - 1997
  - E.g., In 1997, only 9 of 27 NMEs with some labeling
  - E.g., in Europe, 32% of 243 products with labeling
- Off - Label prescribing
  - E.g., 2,262 hospital Rx in 5 European countries, 46%
  - E.g., Netherlands hospital, 66% (76% in neonates)
  - E.g., Outpatient Rx (1,925), 15.5%
- Off - label formulations
- Disease impacts off - label use
  - 13.2% off - label use in all outpatients
  - 55% of cardiovascular drugs
  - 45% of anti - inflammatory drugs
  - 33 - 40% of antidepressants

FIG. 12.4.7. Off-Label Product Use in Children
Source: Scheiner MS. Nature Reviews Drug Discovery 2003; 2(12): 949-61.

76% of hospital prescriptions for children. Also, up to 15.5% of outpatient medications had inadequate pediatric labeling, which predominantly involves quite common diseases in pediatric patients wherein the drugs mostly have labeled indications. In addition, there are many oral, IV, and other medications that lack appropriate pediatric formulations, necessitating extemporaneous compounding of doses for children, all of which is without labeling. The extent of the problem is universal although there are disease-specific differences in percentages of off-label use. While off-label use of general pediatric outpatient drugs (e.g., antibiotics and antipyretics) may be lower, off-label use is much higher in certain diseases, including cardiovascular, anti-inflammatory, and psychiatric diseases.

American Academy of Pediatrics and Research

While there is clearly a need for pediatric studies, concerns about "exploitation" of children are legitimate. In response to these concerns, the *AAP published guidelines* for doing studies in children (Fig. 12.4.8) [8, 9]. These guidelines are available on the AAP Web page. In addition, the AAP helped the FDA draft their guidelines for pediatric studies, as well as the International Congress on Harmonization guidelines; both of which are covered later in this chapter. One of the ethical concerns about doing pediatric studies is the issue of compensation of investigators and pediatric subjects. In general, the guidelines admonish against "undue" rewards to health care providers or "undue" incentives for coercing patients to participate in studies. The institutional review board or ethics committee is charged with determining what is "undue." It is unfortunate that not all IRBs/ECs have, nor do FDA guidelines require, adequate pediatric representation (parental or community), because the decision as to what is "undue" requires knowledge of such things as childhood development, disease variation in pediatrics, standard of care, and community realities. As with all ethical issues, there are many aspects to the decision as to whether a given payment for participation is "ethical" or not. The AAP acknowledged that payment is "in accord with the traditions and ethics of society to pay people who participate and cooperate in activities that benefit others." However, they point out that problems in pediatrics involve the "serious ethical questions" that arise when payments are to adults acting on behalf of minors. For these reasons, the AAP recommended that remuneration not be beyond a token gesture for participation, best if not discussed before a decision is made to participate, and that the IRB must review (and approve) any proposed remuneration. Payment for participation is only one of the aspects of pediatric trials that requires special consideration compared with adult studies. However, the ethics of "exploitation" are not limited to pediatrics. More thought should be given to the ethics of paying adults as well. An 18- or even 21-year-old "adult" college student may receive much more to be in some studies than they can earn working without risking IRB concerns. It is certainly possible to coerce young or even older adults unethically to participate.

The AAP guidelines allow for the provision of funds and facilities to reimburse children and their families for participation as well as for direct and indirect costs (e.g., parking, meals, baby-sitter, etc.) incurred because of participation in a study but state that such payment should be "fair and not an inducement" to enroll in a study in which they otherwise would refuse to participate.

Conflicts of interest and payments are required to be revealed mutually by sponsors of the research as well as the institution and its investigators (Fig. 12.4.9). However, knowledge of certain conflicts of interest (such as the possibility of financial gain if the research is successful or completed quickly) could clearly impact the willingness of subjects to enroll in studies and yet are not yet required to be revealed to subjects and their guardians. Details of compensation paid to the institution or participating investigators or referring physicians for enrolling or completing subjects are also not yet required. Clearly, details of any remuneration, financial or

- **Payment of Providers**
  "....avoid undue rewards to health care providers.....undue incentive for coercing patients to participate in a study."
- **Payment for Participation**
  - It is in accord with the traditions and ethics of society to pay people who participate and cooperate in activities that benefit others."
  - Serious ethical questions…when payment…to adults acting on behalf of minors in return for allowing minors to participate as research subjects
  - Remuneration… not be beyond a token gesture….for participation…
  - Best if not discussed before the study's completion
  - Waiver of medical costs…. may be permitted.
  - The IRB should review any proposed remuneration."

FIG. 12.4.8. American Academy of Pediatrics
Source: 1995 Guidelines for the Ethical Conduct of Studies to Evaluate Drugs in Pediatric Populations (RE9503) *www.aap.org*

- Payment for Participation (Compensation)
  - Investigator may make funds and facilities available to reimburse the child (or the family)
  - For any direct or indirect costs incurred because of the child's involvement in the study
  - Must be fair and not an inducement."
- Conflict of interest and Commitment
  - Required for sponsor and institution
  - But not yet required for subjects/guardians
- Investigator Compensation Disclosure
  - Required for Sponsor and Institution
  - Not yet required to subject/guardian but good idea."

FIG. 12.4.9. American Academy of Pediatrics – Payments/Conflicts

FIG. 12.4.10. Inducements Parent & Child Compensation

- Magnitude - How Much?
- Basis - Time, Discomfort, Complexity
- Type - Money, Toys, Care, Attention
- Focus - Subjects vs. Guardians
- Realistic, Ethically Acceptable Payment for Time & Inconvenience
- Finders Fees - Identify Potential Subjects Without Influence
- Actual & Implied Guidelines Needed
- Treating Physician's Influence
- Implied/Actual Preferential Care

FIG. 12.4.11. Participants - Inducements

otherwise, for participation should be disclosed even if not required. Publications, presentations, grants, and academic promotion could constitute "remuneration" to academic investigators just as raises, bonuses, higher stock prices, or stock options are "remuneration" to sponsor representatives. The possibility of benefiting from insider trading is both unethical and illegal and yet has occurred.

While direct payment of subjects is preferred by most for adult studies, there are a number of "*inducements*" that may be more acceptable to IRBs, guardians, and children (Fig. 12.4.10). Such "remuneration" includes toys, saving accounts or savings bonds payable only to the child, gift certificates to toy stores, movie passes, Internet access, and educational gifts. Again, the type and amount of compensation for participation, as well as when in the study process these should be revealed to the guardians or child participants, must be approved by the local IRB and should not be so great as to make a child or guardian enroll in a study that they otherwise would not agree to. Determining what is appropriate is not simple, however. What is "excessive" to one guardian/child might not be adequate for another. The IRB must weigh the risks and benefits of the study, as well as the consent process and the subject population when deciding on what is "undue" inducement rather than merely adequate "compensation." Studies have shown that many children (as well as their guardians) enroll in studies to help other children, rather than to be "paid," and most (>90%) children who are subjects in a study when asked indicate they would do so again. Therefore, IRBs must at least consider the possible benefits associated with participation in addition to those associated with the actual treatments studied in their risk/benefit analyses.

Adult studies can use increasing inducements, often financial, to increase enrollment. Adults can be given an enrollment fee and fees for completion of a study, which can be several hundred dollars, and all compensation requires IRB approval because it impacts reasonability and ethics of patient consent. In general, these practices are not allowed for vulnerable populations including children. National and international guidelines allow subjects to be compensated for their participation, but this must be "*reasonable inducement*" and not be "coercive"; it must not induce a child or their guardian to participate in a study that they otherwise would not agree to (Fig. 12.4.11). However, deciding what is "reasonable" can be difficult and is the purview of the local IRB based on local realities as well as the particular study. The IRB must consider how much is being offered as well as what the amount is based on. Compensation can be based on a number of things including time spent, discomfort, and complexity of the study. It can be in the form of money, toys, special care, or attention. It can focus on the subject or on their guardian(s) or both. Regardless of what or how much, it must be judged by the IRB to be realistic and ethically acceptable payment for the time and inconvenience that the study requires.

Adult studies in the past have also used "finder's fees" to reward referrals of subjects. This practice is generally considered unethical for any study including pediatric studies. However, paying for the cost of looking for and explaining studies to potential subjects can be ethically acceptable. What must be avoided is paying only for subjects who enroll so as to avoid incentives that induce referral sources to increase enrollment by subjects under their care. Unfortunately, there are no national or international guidelines on what constitutes "unreasonable" inducements. Such guidelines are needed but the decision about what is "reasonable" or not will still remain a local IRB decision that must be based on local realities, as well as the risks and benefits of the particular study proposed. Any inducement must avoid inducing treating physicians to exert undue influence on their patients and must avoid implying or providing preferential care for subjects compared with usual patients. Finally, the AMA has clearly stated that it feels that treating physicians must not be involved in the actual consent process. Treating physicians can explain studies to potential subjects but they must remove themselves from the actual consent process to allow patients to say no and not worry about how refusal will effect their future care or relationship with their treating physicians.

The *AAP*, after reviewing the alternatives to doing studies in children, made the unequivocal *statement* that it is unethical not to do such studies (Fig. 12.4.12). However, such a

12.4. Pediatrics

It is unethical to NOT do studies in children
Ethics

- Relative Risks/Benefits ("minimal risk")
- Equipoise
- Informed Consent
  - Child "assent"
  - Responsible adult "consent"

FIG. 12.4.12. AAP Statement

statement requires clarification. The relative benefits of the particular study proposed must clearly outweigh the risks of the study being done. Therefore, there are different criteria for studies that offer little or no benefit than for studies that offer potential life-saving treatment. There are also different criteria for studies that have little or no risk (e.g., simple questionnaires or drawing of a small amount of extra blood during a medically indicated venipuncture) than for studies that involve major risks (e.g., surgical procedures or toxic chemotherapy). Studies are therefore evaluated by IRBs according to a "minimal risk" assessment [9]:

(a) No more than minimal risk with possibility of direct benefit to the subjects.
(b) No more than minimal risk without the possibility of direct benefit to subjects.
(c) More than minimal risk with possibility of direct benefit to the subjects.
(d) More than minimal risk with no possibility of direct benefit but the possibility of benefit to society or others.

Minimum risk is characterized as follows:

1. Problem is encountered ordinarily in daily life not related to any specific population.
2. Focus is on equivalence to the daily lives of average, healthy, normal children.
3. Risk is considered in relation to the ages of the children to be studied.
4. Risk assessment includes the duration, as well as the probability and magnitude of the potential harm or discomfort.
5. Minor increase over minimal would, for example, include urine collection via catheter, lumbar puncture, skin punch biopsy with topical anesthesia, or bone marrow aspirate with topical anesthesia

In addition, in order for any study to be justified, there must be "equipoise"; that is, lack of known superior benefits of a given therapy. For example, pediatric studies using a placebo generally require that there not be an accepted therapy that can be compared.

The consent process is perhaps the major difference between pediatric and other clinical trials. Children are "vulnerable" subjects. Ethical studies require both the reasonable "assent" of pediatric subjects as well as the "consent" (actually "permission") of those responsible for them to be obtained. The consent process is also difficult in "vulnerable" adults such as prisoners or the handicapped, but pediatric consent and assent requires more education, time, and skill than does any other adult "consent" process. Difficulty in obtaining consent/assent is the most common reason for the delay or even failure of pediatric studies to meet study timelines and recruitment goals.

Legislative and Government Initiatives in Pediatric Research

Most of the early legislation (1902–1968) that established the role of the *FDA* could be considered *"pediatric initiatives"* because it was passed after events that harmed children [5, 6, 9]. In addition, only fairly recently (1990s) the FDA has written several guidelines and "Pediatric Rules" that attempted to simplify and encourage pediatric drug testing. The last major FDA attempt was the 1994 Pediatric Rule. This "lowered the bar" for pediatric trials. In essence, the FDA under then-commissioner and pediatrician Dr. Kessler stated that for diseases that are essentially the same in pediatric and adult patients, the FDA would waive its requirement for "two scientifically valid" safety and efficacy trials. Instead, it would require only two pharmacokinetic studies designed to identify doses based on age-related differences in drug handling. The FDA also allowed sponsors to submit published or unpublished studies in pediatric subjects that had not been FDA approved prior to completion. Finally, the FDA allowed sponsors to delay completion of proposed pediatric studies until after adult marketing was approved. Unfortunately, none of these efforts was very successful in promoting pediatric trials.

Fortunately for children, the FDA Modernization Act (FDAMA), which was passed in 1997, contained a provision that proved critical for pediatric drug development (Fig. 12.4.13). The pediatric provision of FDAMA gave sponsors a 6-month extension of marketing exclusivity or patent extension in exchange for doing requested pediatric studies of an already marketed drug. The provision was voluntary; it did not "force" companies to do the studies the FDA thought were required. However, the financial incentives associated with

- FDAMA (Section 111) 1997
  - 6 mo. marketing exclusivity extension
  - Completing FDA requested ped. studies
  - Voluntary - Sunsets January 1, 2002
- Final Rule 1998 (Dec.2000)
  - Pediatric studies required for new & marketed drugs and biological products
  - for which there may be a "meaningful therapeutic benefit" or
  - "substantial number of patients"
  - Court struck down
  - Pediatric Research Equity Act

FIG. 12.4.13. FDA Pediatric Initiatives

the exclusivity or patent extension were adequate to produce a surge of pediatric studies (lists of those requested and done available at the FDA Web site, www.fda.gov/pediatric). This provision of FDAMA created a change in the paradigm for pediatric drug development. Suddenly, the infrastructure was developed to design, approve, and conduct pediatric studies. Pediatric studies went from being rare to common. Companies, FDA staff, and investigators found that it was possible to do pediatric studies and get products labeled for use in children. There was concern, however, that the incentive to do pediatric studies was excessive and that the program was voluntary. Also, there was no way for the FDA to demand that necessary studies could be mandated. In addition, FDAMA was to be in effect only until January 1, 2002, when it "sunset," or ceased to exist. Fortunately, FDAMA was replaced by improved legislation called the Best Pharmaceuticals for Children Act (BPCA) that will be discussed later.

Perhaps because FDAMA's pediatric provisions were temporary and voluntary, and because the 1994 Pediatric Rule had not produced much new pediatric labeling, the FDA issued the "Final" Pediatric Rule in 1998 (Fig. 12.4.13). This "rule" stated that for any new drug being developed that was likely to be used in children and for which such use offered the possibility of "meaningful therapeutic benefit" for "a substantial number of patients," the FDA would "demand" pediatric studies be done. If the sponsor refused to do so, they would risk having their adult indications denied. This "rule" was struck down by the court. While the judge acknowledged that the rule "might make good legislation," he ruled in favor of the plaintiff that the FDA did not have the authority to make such a rule, that is, force a company to do studies for an added indication (pediatrics or otherwise). Fortunately, Congress stepped in and passed the Pediatric Research Equity Act (PREA) in 2004, which gives the FDA the regulatory authority to do what the FDA had proposed in its 1998 Final Pediatric Rule.

Though very effective, FDAMA had some problems. There was no funding given to FDA to support the increased numbers of pediatric studies. In fact, the FDA was even specifically forbidden to create any additional advisory or regular FDA panels. There was also no way to deal with off-patent or orphan drugs. The *Best Pharmaceuticals for Children Act* (BPCA) both continued the pharmaceutical aspects of the FDAMA legislation but also improved on the original version (Fig. 12.4.14) [8, 10].

BPCA also gives 6 months of exclusivity/patent protection. It also has provisions for promoting labeling of off-patent medication. It created and funded a new FDA Division of Pediatrics (changed to Pediatrics and Bioterrorism). It created an industry-supported charitable fund for study of off-patent drugs (to this date rather sparsely supported), and mandated (but did not provide funds for) spending $200 million on pediatric studies of off-patent drugs.

Off patent drugs as well as on-patent drugs qualify, but the benefits for off-patent products are limited by generic avail-

- General elements
  - Exclusivity (6 months)
  - Off-patent & On-patent drugs
  - FDA – NIH collaboration
  - Office of Pediatric Therapeutics
  - Adverse events all reported for 1 year
- On-patent
  - Industry (PPSR) or FDA initiation
  - Written Requests
  - Company or NIH study (referral)
  - NIH studies - buspirone, morphine, zonisamide
- Off-Patent Drugs & List of Drugs
  - 2003 – Written requests #20 (Iterative process & annual)
  - e.g., lorazepam, nitroprusside, baclofen, azithromycin, rifampin
  - Inputs from FDA, Adv. Comm., NIH, AAP, USP

Fig. 12.4.14. Pediatric Legislation - BPCA

ability. On-patent drugs (or drugs with remaining exclusivity) are handled as they were under FDAMA. Letters describing the specific studies requested can be drafted by industry, proposed pediatric study requests (PPSR), or can come directly from the FDA. These result in a formal "written request" (WR) drafted by the FDA. The FDA either can modify the WR or refuse to grant a WR proposed by a company PPRS because of failure to make a case for reasonable health benefit to children. Although the act is very specific in describing what needs to be done to qualify for patent or exclusivity extension, protocols are not required, and the studies may not need to demonstrate efficacy or safety. Granting of exclusivity is based solely on whether exactly what was proposed in the "written request" was in fact performed or not. Label changes, including results of negative trials (another one of the improvements of BPCA over FDAMA), are based on study results, however. Only new data can be used to satisfy a WR, including ongoing studies yet to be submitted to the FDA. The FDA web site contains lists of hundreds of both completed and ongoing trials that have resulted from this process.

Off-patent drugs for which the FDA determines pediatric studies are needed can be studied by the originator, but if the originator declines to do the studies requested, then the FDA in collaboration with the NIH can do the studies through the traditional NIH (request for application) RFA process. There have been a limited number of off-patent study written requests to date (also available at the FDA web page). Selection of drugs for a written request is based on a complicated collaboration between FDA, NIH, and a number of other organizations.

These studies are all coordinated by the newly created Office of Pediatric Therapeutics at FDA. The BPCA also mandates that all adverse events be reported for at least 1 year. Unfortunately, as did their predecessors FDAMA, both the BPCA and the Pediatric Research Equity Act (PREA) will both sunset after a limited period (in 2007). This

12.4. Pediatrics

- Written Request (from FDA)
  - Letter to sponsors requesting pediatric studies
  - Specifics – Indication, population, types of studies, safety, follow-up, timeframe for response
  - Requirements for drugs: prevalence – 50,000 patients annually, meaningful therapeutic benefit, adequate safety in adults and animals
- Adult data extrapolation
  - Adequate and well-controlled
  - Disease pathogenesis known well
  - Supplemental information may be necessary (safety, PK, PD)

FIG. 12.4.15. Regulatory Pediatric Initiatives

is to allow evaluation of the efficacy as well as costs (human and others) associated with the legislation. Depending on which side of the issue you reside, this will either allow Congress to identify "excessive" profit-taking by industry or allow observers to verify that FDA isn't "abusing" its authority and demanding too many or unrealistic pediatric trials.

Under the BPCA, the *written request* (WR) is key to the pediatric extension program (Fig. 12.4.15). The WR is a letter that comes from the FDA to a sponsor that describes exactly what must be done to qualify for an extension. The WR includes the indication, population, types of studies, the number of subjects, specific characteristics of subjects (e.g., age, gender, weight, and condition(s); that is, enrollment criteria), duration of the study, study end points, safety assessments, follow-up, time frames, and analytical work to be done. In order to qualify for the program, there must be "at least 50,000 pediatric patients" treated annually with the medication, the medication must provide a potential "meaningful therapeutic benefit," and there must be acceptable safety data in adults as well as animals. The *adult data* from which pediatric studies are based must have been adequate and well controlled. The pathogenesis of the disease being treated must be well characterized. The written request may also request supplemental information, such as on efficacy, safety, pharmacokinetics, or pharmacodynamics of the drug.

The *Pediatric Research Equity Act* (PREA) was passed by Congress in response to the court decision that struck down enforcement of the 1998 Final Pediatric Rule (Fig. 12.4.16)

- Requires pediatric studies for all new drugs
- Labeling required for new indications, dosage forms, dosing regimens, route of administration, ingredient
- Pediatric Advisory Committee
- Mimics Pediatric Rule
- Waivers
- Retroactive to 4-1-1999
- Orphan drugs exempt
- Sunset 10-1-2007

FIG. 12.4.16. PREA (Pediatric Research Equity Act) 2003

[1, 5]. This act empowers the FDA to require sponsors to do pediatric studies for all new drugs for which the FDA decides could provide a significant therapeutic advantage for a significant number (50,000/year) of pediatric patients and for which appropriate pediatric formulations can be developed. The studies focus on safety, efficacy, and dosage. PREA also requires labels to indicate all that is known about the new product in terms of any new indications, dosage forms, dosing regimens, route of administration, negative trials, adverse effects, and ingredients. This act also established an FDA Pediatric Advisory Committee to advise the FDA on pediatric studies and labeling. In essence, this act codified the 1998 Final Pediatric Rule; it gives the FDA the authority to mandate pediatric studies of new drugs whether or not the sponsor of the new drug wants to or would seek pediatric labeling. There are provisions to grant waivers of pediatric studies as well as exemptions, when in the opinion of the FDA pediatric studies are unnecessary, impractical, or can be delayed until adult studies are further along or even completed. PREA also allows for exemptions from pediatric studies based on clinical, scientific, or even practical problems, such as the inability after "good faith efforts" to develop appropriate pediatric formulations, when the condition does not occur in pediatrics, or strong evidence exists that it is likely treatment would not be effective. It was made retroactive to April 1, 1999 (the date of the negative court decision on the Final Rule), so that there would not be a "window" of opportunity for sponsors to get drugs through without FDA review of the need for pediatric studies. Orphan drugs are exempt because they do not meet the population requirements. This act "sunsets" at the same time as the BPCA on October 1, 2007. The Congress will review the outcome of both BPCA and PREA and decide on whether the acts should be renewed, altered, or allowed to "sunset" without renewal. Hopefully for children, this act will be renewed on or before this date. For marketed drugs and biological products, pediatric studies and labeling can be required under PREA to be done in following circumstances: a large number of children could use the product, lack of labeling poses significant risks to children, and a meaningful therapeutic benefit likely exists compared with existing therapies.

The *ICH* set up a series of *guidelines* on drug development designed to standardize the process throughout at least the developing world (Fig. 12.4.17) [11]. There are signatory countries, observers, and uninvolved countries, but this group has produced documents that guide how each step of the drug development process is conducted; including pediatric studies.

ICH documents cover recruitment, payments, general principles, bridging studies, and pediatric studies. Both the ICH documents and the AAP guidelines state that inducements used to recruit subjects should not be "inappropriate" or "coercive" and that payments for participation in a pediatric study can cover reimbursement and subsistence costs and that any compensation should be reviewed by the IRB/IEC. The ICH general principles state that (a) medicines should be studied in the population for which the medication is intended

- **Recruitment:**
  - A manner free from inappropriate inducements either to the parent(s) or legal guardian or the study participant.
- **Payments:**
  - Reimbursement and subsistence costs covered in the context of a pediatric clinical study.
  - Any compensation should be reviewed by IRB/IEC.
- **General Principles:**
  - Medicines have proper evaluation in intended population
  - Product development should include pediatric studies
  - Pediatric should not delay adult availability
  - Responsibility shared by company, regulators, providers

FIG. 12.4.17. International Conference on Harmonization

- Variability in drug activity
- NIH Differences
- FDA Status for trials
- Industry Clinical Trials
- Present Studies
  - Knowledge
  - Science
  - Participants

FIG. 12.4.19. Pediatric Studies

to be used (including pediatric patients), (b) that drug development should include pediatric studies but (c) that pediatric studies should not delay approval for use in adults. The ICH states that responsibility for pediatric drug studies should be shared by sponsors, regulators, and providers.

Bridging studies are done to investigate how experience in one population applies to another (Fig. 12.4.18) [12]. In a sense, most pediatric studies are bridging studies (from adults to children). These bridging studies are designed to investigate how intrinsic factors (e.g., genetics and physiology), environment, and disease impact drug efficacy or toxicity. Development in children is well-known with some drugs to produce physiologic and pharmacologic differences (e.g., stimulants such as methylphenidate causing a sedative effect in children with attention deficit disorders). Genetic differences in some children can have dramatic impact on drug action. Pharmacokinetic differences can be observed in children especially very the young (neonates and infants), who do not have fully developed excretory organ function in the liver and kidney. Yet slightly older children (approximately 4–6 years of age) may be in fact clear drugs more rapidly than adults, because their renal and hepatic function is better that it is even in healthy adults. Extrinsic factors (e.g., environment, diet, comedication, and procedures) also need to be addressed in bridging studies of possible differential effects in children versus adults.

Both FDA and ICH guidelines are designed to facilitate proper *pediatric clinical trials* (Fig. 12.4.19). Pediatric clinical trials are designed to identify age and developmentally caused differences in drug action or dosage requirements. Pharmaceutical trials are designed to decide how to best market, label, and promote drug products. The NIH has traditionally been involved in the study of the pathophysiology of disease, as well as studies of effective prevention and treatments. The NIH has conducted many pure clinical trials with either its own funds or in collaboration with industry, but in general the NIH has concentrated on funding ideas rather than deliverable contracts for clinical trials. This focus is quite different than that of industry. Both NIH and FDA are under the same branch of government (Heath and Human Services), but they have many differences with respect to clinical trials. Despite recent attempts to harmonize the two, there are still great differences in how they design, approve, fund, monitor, and review the outcome of trials and their focus (NIH on science and FDA on regulation). One outcome of the BPCA legislation is the collaboration between FDA and NIH to design, conduct, and review the results of studies of off-patent medications. This collaboration will require both groups to understand and adapt to the "culture" of the other. However, there is already a large amount of collaboration between the two agencies and as a result pediatric trials are generating both basic and clinical knowledge and science. This benefits current and future patients. In fact, perhaps the greatest benefactors have been the participants themselves. Pediatric trial subjects (and their guardians) overwhelmingly (>90%) agree to be in another study after their first. They also report that the single most common reason (37% of all subjects) given for agreeing to participate is "the ability to help others." This altruism is only one of the reasons that pediatric trial subjects are so enjoyable to work with. This subject of "perceived" benefit must also be considered into the risk/benefit of study participation. Consideration should be given to whether disallowing children to participate doesn't deprive them of a satisfying and personally rewarding experience.

- New Scenario, extrapolation from old to new for safety & efficacy
- Intrinsic factors
  - Genetic
  - Physiological
  - Pharmacokinetic
- Extrinsic
  - Environmental
  - Medical practice

FIG. 12.4.18. ICH – Bridging Studies (Variable Activity in Children)

12.4. Pediatrics

- HHS - NIH and FDA
- COI's - who/what?
- Budgeting - grants/contracts
- Compliance/Monitoring
  - Individual vs. Institutional
- Motives/Aims
  - Science vs. Commercial
- Intellectual
- Labeling
- Publication

FIG. 12.4.20. FDA/NIH Differences – 1

- Children mandated
- Ideas vs Deliverables
- GCP - ICH
- SOPs - Standards
- Design
- Monitoring
- Risks
- Record Retention

FIG. 12.4.21. NIH Differences – 2

As mentioned previously, despite both being under the same governmental branch of Health and Human Services, there are many differences in how the *FDA and NIH* view pediatric clinical trials (Fig. 12.4.20). For example, conflict of interest (COI) review differs between them. Even the amount of financial involvement considered *de minimus* differs between these agencies. In addition, the kinds of COI can also differ. Budgeting processes for industry and NIH studies also differ greatly. Companies compensate investigators more strictly based on per patient costs and overhead for enrollment of patients, whereas NIH uses a more global budgeting that includes study operational costs, costs of physical plant to some extent, data analysis, travel, and more overhead expense. In addition to differing with respect to the "deliverables," one concentrates on scientific productivity (e.g., publications), whereas the other (companies and the FDA) concentrates on subjects (enrolled and completed), safety and efficacy, and study completion culminating in a new drug application. NIH exists to create science, while the FDA regulates the industry to protect the public, ensure accurate and fair labeling of drug products, and foster new drug availability through new science. The two agencies also differ greatly in the type and amount of compliance monitoring that occurs, as well as the individual versus institutional responsibilities. Companies and FDA have extensive quality assurance programs for studies, especially case report forms, medical documentation, monitoring parameters, adverse events, and dispensing and administration of drugs, all of which is audited by both the company and the regulatory authorities. They also have differing "motives" for doing studies. Although both focus on bringing new drugs and new indications to patients and providers, one focuses more on science (NIH) in a non-profit world, whereas the other focuses on commercialization meeting needs of the public for new drugs and investors, too, who demand a financial return. There are differing levels of intellectual involvement, and until recently the importance of presentation and publication of results differed greatly. For companies, success is measured by study completion and successful new drug applications, more than publications, while NIH focuses on publication of their work.

Other differences exist between NIH and FDA (Fig. 12.4.21). For example, while pediatric trials have increased since FDAMA, BPCA, and PREA, pediatric studies are still not mandated for all drugs and in fact there are many "loopholes" that prevent some pediatric studies from ever being conducted. One example is the waiver that is possible if no "suitable" pediatric formulation can be produced after "good faith" attempts. The NIH however has simply the mandate for the inclusion of children (unless the guardian can give good reasons not to include children). In fact, the NIH has also mandated inclusion of women in trials.

Funding is based on ideas and science for NIH versus deliverables for FDA. The lack of the equivalent of good clinical practices (GCP) or ICH guidelines (the major industry and FDA operational requirements in conducting trials) is a major problem with NIH trials (no GCPs) that is slowly being addressed. The lack of clear standard operating procedures (SOPs) even in general clinical research centers is a difference in how performance sites view regulation. The design of studies can differ greatly for an NIH study versus an FDA-approved protocol. The number and type of monitoring visits differs greatly between NIH and FDA, that is, major requirement for FDA (often 3–5 visits per study of 1 year in duration). The FDA does thousands of monitoring visits per year as opposed to the rare monitoring visits done by NIH. There may also be major differences in the allowable risks of study participation, as well as records' design, data recording, data analysis, and retention or data after study completion.

An industry publication (CenterWatch) listed the number of pediatric studies being done by the industry in 2003 and the types of study (Fig. 12.4.22) [7]. In 2003, there were 190 studies listed, of which 32 were for cancer, 25 for vaccines, 17 infectious diseases, 16 for cystic fibrosis, and 16 for cardiovascular drugs. This number and distribution should be compared with updated FDA data; especially as these data are from before enactment of PREA, which gave the FDA the authority to mandate clinical trials of new drugs developed largely for adults.

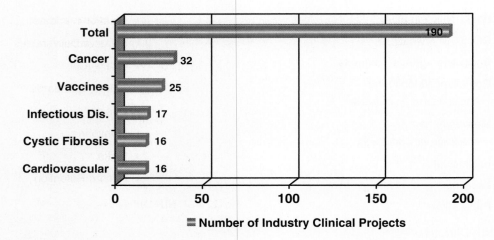

FIG. 12.4.22. Drugs in Pediatric Research (2003)
Source: Korieth K. State of the Clinical Trials Industry. Thomson Center Watch, 2005

The FDA continuously updates data on the *status of pediatric studies* being done for exclusivity as well as post talks on various pediatric initiatives (Fig. 12.4.23) [13]. These can be accessed at www.fda.gov/pediatrics. This report illustrates the number and type of studies that have been or are being performed in response to a written letter from the FDA as part of either the original FDAMA or current BPCA. By early 2005, almost 700 studies of more than 120 products were being conducted or planned. These studies would recruit more than 43,000 pediatric subjects in studies of either efficacy and safety (35%), PK and safety (30%), safety alone (15%), PK/PD (9%), or other (11%) pediatric topics. There are additional drugs for which the FDA feels that pediatric studies are needed, but for which studies have not yet been designed or conducted. These include drugs for infectious diseases, cardiovascular drugs, central nervous system drugs, and oncology drugs.

The FDA also breaks down these studies into categories based on the type of product for which studies are being done (Fig. 12.4.24) [14]. The *exclusivity initiative* is responsible for the largest number of studies. Of the 692 written requests issued, 374 came from industry for 300 products. Of these industry-proposed studies, 108 products had studies permitted by the FDA, and 112 products had exclusivity granted. In addition, 90 of these resulted in label changes. These 90 label changes included 89 new pediatric sections with 17 dosing changes, 22 new safety sections, and 6 pediatric formulation sections. Importantly, approximately one-third of the drugs studied to date have resulted in important dosing, safety, or indication information that differed from what was believed to be "best practice" by pediatric subspecialists and generalists. Clearly, the process has proved again that children are not little adults and that "evidence based" pediatric therapeutics is flawed approximately one-third of the time.

- Studies - 694
- Products – over 120
- Number of patients over 43,400
- Study Categories
  - Efficacy & Safety   35%
  - PK & Safety   30%
  - Safety   15%
  - PK/PD   9%
  - Other   11%
- Drugs with Studies Needed: ID-10, CV-9, CNS-6, Onc-2

FIG. 12.4.23. FDA Status for Post-Exclusivity (2005)
Source: Murphy S. Div. Ped. Drug Dev., CDER, FDA, DIA Annual Meeting, June 18, 2005

- **Exclusivity Initiative**
  - Industry proposals   374
  - Products with WRs   300
  - Products with studies submitted 108
  - Exclusivity granted   112
  - New Labels   90
- **Studies Requested**
  - WRs   692
- **Off-patent Drugs**
  - WRs   7
  - RFPs from NIH   3

- **Exclusivity – On-Patent Findings (new labels-2005)**
  - Dosing change   17
  - Safety info   22
  - Formulation (peds)   6
  - New labels for pediatrics   89

FIG. 12.4.24. FDA Results for Pediatrics
Source: Murphy S. NICHD Symposium, May 2005

12.4. Pediatrics

FDAMA had no provisions to deal with sponsors who were not willing or interested in conducting trials that the FDA felt were required. The BPCA, however, provided a mechanism to study such drugs whether on or off patent drugs. For on-label drugs, the FDA can demand studies be done. For off-patent drugs, if "innovator" industry sponsor is unwilling to conduct the studies requested by the FDA, then the National Institute for Children's Health and Disease (NICHD) in collaboration with the FDA has the ability to contract; and in fact is conducting at least four studies to date. There have been seven written requests issued for off-patent drugs which resulted in four Requests for Proposals by the NIH (NICHD).

The Science in Pediatric and Clinical Studies

As mentioned, the studies being done voluntarily under FDAMA and BPCA, as well as because of FDA-mandated studies under PREA, have provided extremely useful *information for pediatric patients* and practitioners (Fig. 12.4.25). They have emphasized and clarified developmental differences in PK/PD behavior of drugs in children and that these changes are continuously occurring during development. They have also both shown how disease type and state can alter drug handling and effects but also how different diseases (diagnosis, course, presentation, and management) can be in pediatric patients of different ages. The studies have illustrated the problems in administering drugs to children; only in part because of the scarcity of pediatric appropriate formulations. We have learned and relearned the problems with using patient or parental reports for efficacy and toxicity. We have learned that doses that were commonly used, even if recommended by the best subspecialists, were not always correct. Both under- and overdosing occur because of a lack of data. We have discovered both unusual toxicity and different propensity to toxicity in different age groups. We have discovered important drug–drug and drug–diet interactions unique to treating children. We have learned that the indications for use of drugs can be quite different in children. In summary, we have discovered that one-third of the time, "expert" opinions about pediatric dosages are incorrect,

- PK/PD differences
- Constant changes
- Disease differences
- Administration problems
- Formulations
- Self report
- Wrong doses
- Unusual Toxicity
- Interactions
- Indications
- Expert opinions

FIG. 12.4.25. Information in Pediatrics

- Midazolam - IV
- Neurontin®
- Propofol
- Prozac®

FIG. 12.4.26. Knowledge – Examples

"eminence" based prejudices that are disproved when appropriate trials are conducted.

There are many *examples* of what studies of drugs given to children have revealed (Fig. 12.4.26). The 89 label changes listed on the FDA web site should be reviewed for a more complete list as well as the presentations of this issue by Dr. Diane Murphy also listed on the web site. Some examples, however, include the fact that giving the IV formulation of midazolam (Versed®) orally to children was not as effective as an oral midazolam preparation developed for this use. We learned that the pediatric mg/kg dose of gabapentin (Neurontin®) needed for seizure control was in fact higher than what is needed in adults. It was learned that ICU sedation of children with a combination of propofol and midazolam produced more, not less, toxicity than did older, traditional, less expensive alternatives with equal efficacy.

There are a number of *age-related pharmacokinetic (PK)* differences that must be considered in pediatric trials (Fig. 12.4.27). For example, the PK behavior of a drug can change even during a single observation period or hospitalization. This is especially true in neonates where both weight and PK processes can change significantly during a single hospitalization. One of the causes of these changes is the rapid changes in drug metabolism that change with ontogeny during the neonatal and infant periods. PK studies in children must consider these changes and the environmental and genetic factors that affect them.

A number of new methods have been developed for dosing pediatric PK studies. These are necessary because traditional adult, dense sampling PK studies are difficult or impossible to do in pediatrics. Methods include optimal sampling designs where very few ("sparse") samples are collected during the absorption, distribution, and elimination phases, and then mathematical (largely Bayesian) techniques are used to define PK behavior. In addition, rather than collect multiple samples at fixed, or selected times after dosing, population PK models have been used to use very few samples, even single samples, collected at multiple times after dosing in a population of children rather than multiple samples collected in fewer children. A particular advantage of such studies is that they can be done in children already receiving medication "off-label," which increases enrollment success.

Pharmacogenetic studies have shown that it may be possible to predict both individual susceptibility to drug toxicity, as well as to predict the most effective dosing regimens based on pharmacogenetic testing. Such testing can be done to

- Ontogeny
- Optimal Sampling
- Population PK
- Pharmacogenetics
- Experience

FIG. 12.4.27. Science -Pharmacokinetic Issues in Pediatric Clinical Trials

identify individual drug metabolizing enzyme capability, as well as individual drug transporter and/or drug target (e.g., receptor number, type or affinity) differences. Another potentially important development is the use of so-called phase 0 testing or "mimi" dosing, where drug doses, not expected to produce any pharmacologic or toxic effect, are given to characterize drug metabolism pathways, as well as individual differences in drug handling. Although not yet accepted by the U.S. FDA, such testing holds particular promise for children.

These developments have been made possible by the experiences gained in doing pediatric PK studies. Perhaps one of the least appreciated aspects of the recent trends in pediatric studies is that they have produced the infrastructure, personnel, and experience necessary to develop safer, higher recruiting, less invasive, more informative drug studies that can be done in children. This infrastructure necessary for pediatric trials is rapidly appearing in industry, academia, and the private sector, as well as at the FDA.

Optimal sampling techniques have been developed to allow PK studies in pediatric populations in whom multiple sampling is problematic (Fig. 12.4.28). These designs minimize sample numbers and total volume collected. Properly done studies are extremely useful in pediatrics. However, proper design requires *a priori* knowledge of drug disposition, as well as an appreciation for the principles of the technique. For example, samples must be drawn during the "data rich" portions of the concentration versus time profile; in general during the absorption, distribution, and elimination phases. For drugs that obey multiple compartment kinetics, this may require sampling during each elimination phase. Finally, studies must be designed to avoid circular mathematical reasoning where sampling and analysis simply confirm preexisting, incorrect assumptions about drug handling. Optimal sampling techniques are a good example of what can be done; both well or badly to simplify PK studies in children.

The need to develop new paradigms for studies in children are related to the fact that children particularly represent a "vulnerable" population. As such, many adult *trial designs* acceptable to subjects and IRBs are problematic in children (Fig. 12.4.29). Use of placebos rather than active comparators is a particular concern; both because of the ethical problems of giving children known inactive comparators (i.e., placebo) and because the lack of prior studies makes it difficult to know which if any "active comparator" is really effective in children. The "standard of care" may be demanded by an IRB but in fact cannot be accepted as such without adequate testing in children. This leads to confusion or rejection of protocols by IRBs if members are personally convinced as to the efficacy of a given comparator when the investigator, sponsor, or FDA remain unconvinced. Other design issues involve how to treat subjects who withdraw from studies or demand to be put in treatment groups that involve unproved therapy.

Population PK methods have been discussed briefly (Fig. 12.4.30). These methods use sparse samples collected in a population with specific characteristics (age, gender, size, disease, etc.) to predict using Bayesian techniques what the PK behavior of a drug would be in a population or member of the population. The techniques are based on explicit assumptions about the population as a whole, as well as about the specific sample points. These techniques can be used to model PK, pharmacodynamic (PD), or PK/PD relationships. Of the techniques, there are two common methods. One is the two-stage method where data from a population is used to predict average

- Minimize sample number & blood volume
- Depends upon *a priori* knowledge of disposition
- Mathematically discern "data rich" portion of concentration v. time profile
- Must prove predictability of relationship without the use of circular mathematical reasoning

FIG. 12.4.28. Science – Optimal Sampling

- Control Groups – Active comparator
- Placebos
- Withdrawals / Treatment groups

FIG. 12.4.29. Science – Design Issues

12.4. Pediatrics

- Based on explicit assumptions
- Can use for PK, PD or PK/PD
- Established methods
  - Two-stage
  - Nonlinear mixed-effects
- Validity of analyses depends upon validity of assumptions

FIG. 12.4.30. Science – Population Pharmacokinetics

behavior and then each individual data point (sample concentration) is used to refine the model. The second model is a nonlinear mixed-effects model where individual patient characteristics (e.g., weight, renal or hepatic function, etc.) are used to predict which individual characteristics best predict an individual's behavior in relation to the entire population (e.g., whether an individual's elevated creatinine predicts higher concentrations or slower elimination than expected in an entire population). The validity of the methods to "model" (predict) population or individual PK or PD behavior depends on the validity of the assumptions made about the population, the individual, and the sample (collection time, accuracy of the assay, time after dosing, absorption characteristics, etc.).

While recent developments have shown that it is possible to do good pediatric studies, there are a number of *investigator deficiencies* that limit pediatric studies (Fig. 12.4.31). First, there are a limited number of trained investigators, and they are not distributed evenly either geographically or by practice site, specialty, or type. Second, at the same time that pediatric studies are becoming more common, pediatricians are disproportionately leaving the ranks of active investigators. A growing shortage of trained, experienced pediatric investigators is predicted in the near future. Third, of the thousands of investigators doing pediatric studies, they differ widely in their experience in conducting clinical trials (adult or pediatric), and very few are actually certified clinical investigators. Fourth, many investigators lack experience or knowledge of the drug development process. Finally, there is a wide range of experience in terms of dealing with the pharmaceutical industry research and FDA. It is hoped that as principal investigator (PI) training and certification programs become more available, these deficits will be corrected and the quality of pediatric studies will increase even as the number and type of studies expand.

- Number and Distribution
- Pediatricians leaving research
- Experience base in research
- Pediatricians having research

FIG. 12.4.31. Science – Investigator Deficiencies

- Benefits from regulated studies, trained investigators/staff, and monitoring
- Subjects and society benefit from knowledge gained
- Improved outcomes over routine care, even if placebo
- Personal benefits/willingness to do again

FIG. 12.4.32. Participants – Benefits

As pediatric studies improve and expand, pediatric trial *participants* could realize more *benefits* (Fig. 12.4.32). Participants can benefit from having their own therapies investigated in well designed, regulated studies done by trained investigators and staff, with ongoing review of the quality and safety of the studies. This stands in stark contrast with prior events where children were placed on untested therapies without consent (or even knowledge), without careful scientific design, without ongoing monitoring or eventual publication or dissemination of results. Subjects in studies themselves, as well as patients, with similar diseases and society in general also can benefit from appropriate pediatric trials. Support for these claims include a recent review where 23 of 24 studies showed that children in clinical trials who received placebo had better outcomes than did children who had received "standard" care. In addition, more than 90% of children who are in one trial state they are willing to be in other trials.

One very successful, generally accepted way to increase both quality and participation in pediatric studies is the *use of networks* (Fig. 12.4.33). The Children's Oncology Group (the combination of the prior Southwest Oncology Group and the Children's Cancer Study Group) is perhaps the best model of the power of pediatric studies to advance knowledge, as well as to generate excellent recruiting success. Pediatric cancer care is systematically studied in the vast majority (>90%) of children with cancer. This incredible enrollment success is both the result of, and the cause of, the fact that pediatric cancer survival has shown tremendous improvement; especially when compared with the rather modest improvement in adult cancer survival. Adult cancer patients in contrast are seldom enrolled in studies. The success of COG in enrolling subjects can be directly tied to their success in improving survival, as well as their decades of experience designing and conducting

- COG Children's Oncology Group
  - SWOG + CCSG
  - "Model" Network
- PPRU-Pediatric Pharmacology Research Unit
  - 13 Pediatric Centers
- RUPP-Research Units for Pediatric Psychopharmacology
  - Centers for diagnoses
    (eg. Autism, ADHD, Mood disorders, etc.)

FIG. 12.4.33. Participants – Use of Networks

ethical, practical, yet scientifically valid studies. Much can be learned about successful recruitment from COG.

Other more recent pediatric networks exist such as the Pediatric Pharmacology Research Network (a group of 13 large pediatric centers with both experience in doing pediatric trials and large numbers of trained, experienced pediatric investigators with infrastructure support necessary to run pediatric trials). The Research Units for Pediatric Psychopharmacology are yet another example of such a network. Other networks involve neonatal medicine, pulmonary diseases, rheumatology, and so forth.

Although it is clearly possible to do good efficient pediatric clinical trials, there are a number of *challenges* to doing so (Fig. 12.4.34) [5, 7]. Pediatric trials are expensive, and more expensive than adult trials. Age-appropriate formulations are not always available and can be expensive or even impossible to produce. It has been claimed that because pediatric studies are often slow to complete, this could cause delay in getting products approved for adult use. In fact, the FDA can allow adult indications to be approved pending completion of pediatric studies. There is a danger in this approach because such "promises" to complete pediatric studies after adult approval were seldom kept by the pharmaceutical industry. Whereas theoretically possible with PREA, the FDA would have difficulty removing an effective adult medication if promised pediatric studies were never started or never completed.

Pediatric studies can require more complex product labeling, but this is not a legitimate reason to not do the studies. Unfortunately, certain important aspects of the labels, even those required, are not always complied with. For example, all labels are required to have information on breast milk excretion, yet this information is not always included even when known. With time it is hoped that the FDA and industry will develop more effective, efficient, and readable labels. Lack of FDA, practitioner, and industry experience designing and conducting pediatric studies is still a concern despite the great improvements since adoption of FDAMA, BPCA, and PREA. With time and reapproval of legislation, this problem may cease to be a problem, but there are still issues with study design, placement, contracting, conduct, and even reporting. Unfortunately, there is also a shortage of adequately trained and experienced investigators to conduct pediatric studies. And paradoxically, while the number of pediatric studies is increasing rapidly, the number of qualified pediatric investigators is disproportionately decreasing. Perhaps the greatest problem created by a lack of adequate experience in industry, FDA, and investigators is pediatric-unfriendly protocol design. Many FDA study guidelines must be rewritten for pediatrics. For example, the antihypertension guidelines require stopping effective medication for 2 weeks prior to starting randomized placebo-controlled treatment. This design is simply unethical and unnecessary. Placebos are a special problem in pediatrics; both because the latest version of the Declaration of Helsinki demands that they not be used if there are effective alternatives available, and because the use of placebos greatly increases problems with recruitment. Also, studies and even formulations that are appropriate for one pediatric population are not appropriate for others. Finally, there are problems created by both IRBs without pediatric expertise (such IRBs should not be reviewing pediatric studies), as well as IRBs populated with pediatric experts who have either little knowledge of the drug development process (including FDA requirements for studies) or have covert or overt antagonism for studies conducted by industry. Despite these problems, great strides are being made. Proof of this is provided by the number of pediatric studies being done, the knowledge being gained, the appearance of new label information, and by the rapid development of new methods and experienced personnel. Also, all the industry, regulatory, and investigative infrastructure required to conduct ethical, efficient, and valid pediatric trials is being developed. The future is indeed bright for pediatric clinical trials.

Figure 12.4.35 lists only some of the many references available on this subject. This is a rapidly changing area that requires constant updating of information but these references offer a start.

- Cost of studies ($500,000 - $5 MM per study)
- Cost & Viability of formulations ($750,000)
- Delay in adult product approval
- More complex labeling
- Lack of experience at company
- Inadequate investigators (number and research experience)
- Design & Conduct problems
  - Placebos
  - Patient recruitment
  - Multiple subpopulations, neonate vs child
  - IRB experience

FIG. 12.4.34. Challenges in Pediatric Research

U.S. HHS, FDA, CDER/CBER, ICH, Guidance for Industry, E11 Clinical Investigation of Medicinal Products in the Pediatric Population, December 2000

Murphy S. Regulatory frame work for pediatric drug development in the US. Drug Information Association presentation, June 16, 2004

Murphy D. The role of the FDA in Pediatric research. Glaser Pediatrics Research Network presentation, July 16, 2004.

Birenbaum D. Best practices for children Act: an FDA implementation perspective presentation, FDA, March 27, 2003.

Lee P. Bridging studies in drug development (for different scenarios) presentation, Pharmaceutical Sciences World Congress, June 2004.

Crescenzi TL. Statistical update (Pediatrics), FDA, March 3, 2003.

Scheiner MS. Paediatric clinical trials redressing the imbalance. Nature Reviews Drug Discovery 2003;2 (12):949-61.

Korieth K. Pediatric trials come of age. In State of the Clinical Trials Industry. Lamberti MJ (ed) Thomson Center watch, Boston MA p 85-96, 2004.

FIG. 12.4.35. References

References

1. Murphy S. Regulatory framework for pediatric drug development in the US. Drug Information Association presentation, June 16, 2004.
2. Murphy D. The role of the FDA in pediatric research. Glaser Pediatrics Research Network presentation, July 16, 2004.
3. MacLeod SM. DIA Annual Meeting, June 17, 2005.
4. Pediatrics and Child Health 2003;8 Suppl. A.
5. Heinrich J. Pediatric drug research. Substantial increase in studies of drugs for children, but some challenges remain. U.S. General Accounting Office; May 8, 2001. GAO-01-705T.
6. Scheiner MS. Paediatric clinical trials redressing the imbalance. Nat Rev Drug Discov 2003;2(12):949-961.
7. Korieth K. Pediatric trials come of age. In: State of the Clinical Trials Industry. Lamberti MJ (ed). Thomson CenterWatch, Boston, MA, 2004, 85-96.
8. American Academy of Pediatrics. 1995 Guidelines ofr the ethical conduct of studies to evaluate drugs in pediatric populatrions. Available at www.aap.org.
9. Goldkind SF. Special ethical protections for pediatric research participants. Available at www.fda.gov/oc/opt/presentations/PennnandTemple.html.
10. Birenbaum D. Best practices for children Act: an FDA implementation perspective presentation. FDA, March 27, 2003.
11. U.S. HHS, FDA, CDER/CBER, ICH, Guidance for Industry, E11 Clinical Investigation of Medicinal Products in the Pediatric Population, December 2000.
12. Lee P. Bridging studies in drug development (for different scenarios). Pharmaceutical Sciences World Congress presentation, June 2004.
13. Murphy S. Div. Ped. Drug Dev., CDER, FDA, Presentation DIA Annual Meeting, June 18, 2005.
14. Murphy S. Presentation NICHD Symposium, May 2005.

5
Psychiatry

Michael W. Jann, John J. Brennan, and Roland Garritsen VanderHoop

Introduction	338
Patient Populations in Psychiatry	339
Acute versus Chronic Trials—Design and Conduct Issues	340
Pharmacology	344
Research Challenges	346
Summary	348
References	349

Introduction

Drug development for the treatment of psychiatric disorders represents a unique challenge to the pharmaceutical industry. The *Diagnostic and Statistical Manual of Mental Disorders*, 4th edition (DSM-IV), lists more than 200 different types of mental illnesses found in the population [1]. Psychotropic medications play a vital role for many of these diseases. It is well accepted by clinicians that psychotropic drugs do not provide a "cure" for these disorders. At best and at this time, these medications can significantly improve or reduce psychiatric symptomatology and return the patient their previous level of activity prior to their illness. Successful therapies must continue to be explored for psychiatric disorders.

It has been estimated that epidemiological prevalence of various psychiatric illnesses in U.S. population ranges from low single digits (e.g., schizophrenia with 1.0%) to much higher, as for instance with depressive disorders (19 million people, 9.5% of adults) and anxiety disorders (19.1 million, 13.3% of adults 18–54 years old) [2]. As substances, tobacco, and alcohol abuse are included. The overall economic impact of psychiatric disorders places a tremendous strain upon our society affecting the patient, their caregivers, and the health care system with respect to direct costs while the effects on indirect costs, like loss of productivity, are probably much higher still. NIMH estimates the economic burden on the United States to be $150 billion per year, excluding substance abuse [2]. Substance abuse adds $246 billion in total economic costs to the United States (1992 statistics) [3]. Psychotropic medications that can improve the quality of life of patients, reduce caregiver burden, and decrease societal expenditures is the common goal for all.

Product development in psychiatry poses unique research challenges in many ways, which will be discussed in this section, such as the subjective nature of disease presentation, high degree of intersubject and intrasubject variability in diseases, overlap in symptomology with the diseases, concurrent disease presentation especially anxiety and depression as an example, assessment tools that are subjective as expected, the many assessment tools to choose from, many treatment choices for most diseases to use as comparators, the unfortunate stigma associated with psychiatric disease, and the very

12.5. Psychiatry

high prevalence of the diseases, especially depression and various anxiety syndromes. This section of the book will address a variety of patient populations and disease state issues, including the clinical assessment challenges. Pharmacologic issues will be covered, including dosing, administration, pharmacokinetics, and pharmacogenomics. We will be discussing research controversies and some dos and don'ts. The closing material addresses the research focuses and current pipeline in psychiatric research.

Patient Populations in Psychiatry

For drug research, psychiatric conditions are exceptionally prevalent and difficult to treat in the United States and around the world. They create a huge population base to draw from for studies, offering a fertile area for drug development given the disease burden on society and the medical need, providing product opportunity even though many options exist because of limited efficacy, and offering a large financial market for safe and more effective products. Figure 12.5.1 presents seven *disease categories* for mental illness from the National Institutes of Mental Health summarizing the breadth of mental disorders to differentiate, as well as recognizable common occurrence in society. These diseases are overlapping in their some of their symptomology making diagnosis and monitoring a challenge and co-present simultaneously as diseases further challenging therapy and drug research. Anxiety, for example, is composed of five major disorders, panic, phobias, obsessive-compulsive disorder, post-traumatic stress disorder (PTSD), and generalized anxiety. Even these can be further subdivided, for example, phobias into social, spatial, and specific phobias. Substance abuse is common in our industrialized society and often involves concurrent mental disorders. Furthermore, the lack of elucidation of actual disease pathology in most cases, symptomatic presentation, environmental relationships with the disease, variable patient response to disease, and the subjective nature of assessment of drug response all complicate product development.

- **Depressive disorders:**
  - Major depression
  - Bipolar depression
  - Dysthymic disorder
- **Anxiety Disorders:**
  - Panic
  - Phobias
  - PTSD
  - Obsessive-Compulsive
  - Generalized anxiety
- **Schizophrenia**
- **Eating Disorders:**
  - Anorexia
  - Bulimia
  - Binge eating
- **Autism**
- **Dementia (Alzheimer's)**
- **Substance abuse:**
  - Drugs (Rx & Illegal)
  - Alcohol
  - Nicotine

FIG. 12.5.1. Mental Disorders

- Mentally ill patients
- Many different psychiatric diseases
- Treatment-resistant patients
- Co-morbid diseases
- Substance abuse
- Age:
  - adult
  - children/adolescents
  - geriatric

FIG. 12.5.2. Special Populations

The psychiatric *patient population* presents an interesting challenge to the pharmaceutical industry (Fig. 12.5.2). By their nature, these patients are mentally ill and the question of patient competency is always an important consideration. Depending upon the illness, the patient may or may not be able to provide an informed decision to clinicians regarding participation in a clinical drug trial. Whether it is because there is considerable overlap in disease characteristics within the large number of psychiatric disorders or whether it is because the same disease mechanism sometimes underlies multiple disorders, as a consequence one medication can be approved by regulatory agencies for multiple diseases (e.g., paroxetine, Paxil®, for major depressive disorder, obsessive compulsive disorder (OCD), and social phobia, or olanzapine, Zyprexa®, for schizophrenia and bipolar disorder). Conversely, patients may not be adequately treated or respond to a single medication, may require a combination of multiple drugs, and over the longitudinal course of the disease may become "refractory" to any pharmacotherapeutic interventions. Including refractory patients in a study, who have shown a lack of response to treatments from different drug classes in a clinical trial, may lead to inadequate efficacy results with a compound that actually may be efficacious.

Psychiatric patients often have comorbid diseases such as diabetes, hypertension, cardiac diseases, and other illnesses present in the general population. When a psychotic patient neglects their diabetes, both illnesses may exacerbate and require immediate treatment. Substance, tobacco, and alcohol abuse are common among psychiatric patients. Depending upon the specifics of the substance abused, it can have effects upon the brain and can complicate the clinical presentation of the psychiatric disorder and impact treatment modalities. In clinical drug trials, patients can potentially be included with a history of substance abuse as long as a current diagnosis is not present and there are no signs of withdrawal. The substance abuse may further complicate clinical trials related to the reliability of the patient as a historian, compliance to study procedures, and extra drug use during a trial.

The patient's age must be taken into consideration as the requirements and regulations for clinical trials in age-specific groups differ. Studies involving geriatric patients (e.g., Alzheimer disease) and children and adolescents usually have a caregiver and family member involved, and legal guardianship could impact the enrollment of these subjects. Geriatric patients may provide "verbal" assent but not be able to legibly sign the written informed consent due to their illness. Children and adolescents have school and other activities, and enrollment into clinical trial may negatively interfere with these activities. As it is often difficult to conduct, especially placebo-controlled clinical trials within these populations, many psychotropic medications do not have FDA-approved indications for children and adolescents. This poses a problem, because physicians will often prescribe the treatments notwithstanding this lack of information. Finally, the lay public can question the prescribing of psychotropic medications especially in children and adolescents where adequate scientific data is lacking, generate controversy, and present clinicians with difficult treatment choices. The latest controversy regarding the potential induction of suicides by the administration of selective serotonin reuptake inhibitors (antidepressives) like paroxetine (Paxil®) and sertraline (Zoloft®) underscores this point.

The *Diagnostic and Statistical Manual* (DSM) was designed by psychiatrists to present standardized diagnostic criteria for various psychiatric disorders to health care professionals, health care organizations, and regulatory agencies for clinical drug research and evaluation (Fig. 12.5.3). The DSM continues to be revised, and the current version is DMS-IV as of 2005. It is important when conducting drug development in psychiatric disorders to always employ the most recent edition keeping in mind that future revisions could occur during a clinical drug trial and that diagnostic criteria may change for a disorder. Also, the DSM is standard in the United States, but studies outside the United States may use different diagnostic classifications. Even within the United States, the DSM-IV criteria for all psychiatric disorders may not exactly match other established references. For example, the DSM-IV criteria slightly differ from the National Institute for Neurological and Communication Disorders and Stroke (NINCDS) criteria for Alzheimer disease [4]. Clinical drug trials with antidementia drugs usually include diagnostic criteria from both DSM-IV and NINCDS for patients to be eligible for enrollment.

The DSM-IV attempts to group clusters of symptoms commonly associated with a specific psychiatric disorder. For example, the diagnostic criteria for schizophrenia include the symptom clusters of positive, negative, cognitive, and affective symptoms. However, the positive symptoms of psychosis that include delusions and hallucinations can be found in many other psychiatric and medical disorders (e.g., bipolar mania, psychotic depression, substance-induced psychosis, intensive care unit psychosis). Likewise, the negative symptoms (anhedonia, apathy, and others) overlap with major depressive disorder. Finally, cognitive symptoms can mimic the ones associated with Alzheimer disease (interesting note: schizophrenia used to be called dementia praecox or "very early" dementia). Obviously, it can greatly affect the success of a clinical trial whether or not the diagnosis can be made with certainty. When many symptoms overlap between different psychiatric disorders, clinicians have to look for other information, like risk factors or the disease's longitudinal course over time, in order to make reasonable delineations between the psychiatric disorders. Finally, even the time when symptoms occur can be dependent upon environmental factors. For example, seasonal affective disorder occurs usually during the winter months compared with the summer months. Psychiatric disorders are dynamic in nature and a patient's behavior can constantly change, which makes an accurate diagnosis challenging and can impact enrollment for clinical drug trials as well as improvement or deterioration during the trial.

Acute versus Chronic Trials—Design and Conduct Issues

Regulatory agencies can approve drugs (indications) for psychiatric disorders for the *acute* stage of an illness (intervention) or for more *chronic* or maintenance treatment (Fig. 12.5.4). Designs for the studies necessary for each will be very different

---

- Use of the Diagnostic and Statistical Manual (DSM) of Mental Disorders, 4th Edition (IV) for diagnostic criteria of various diseases
- DSM-IV criteria may differ than other established guidelines (e.g. Alzheimer's Disease)
- Symptoms can overlap between various disorders (e.g. psychosis is present in many disorders)
- Symptoms can be dependent upon environmental factors (e.g. seasons of the year).

FIG. 12.5.3. DSM and Disease Issues

- Approved indications may differ (e.g. acute bipolar mania versus maintenance therapy).
- How to define an acute versus a chronic time period for trial design?
- Disease states are of a dynamic nature and fluctuate over time.
- Diagnosis may change over time.

FIG. 12.5.4. Psychiatry – Acute vs. Chronic Trials

both with respect to duration of treatment as to the outcome variables. An example of this issue would be medications approved for the treatment of bipolar disorder. Valproic acid (Depakene®), an antiepileptic agent, is approved for acute mania [5]. All atypical antipsychotics are also approved for acute mania [6–11]. The time period for the atypical antipsychotic clinical trials for acute mania were 3 weeks, similar to the valproic acid trial [4]. On the other hand, olanzapine (Zyprexa®) and quetiapine (Seroquel®) are approved for the chronic illness of bipolar disorder [6, 8]. Quetiapine's approval for a chronic therapeutic indication involved longer clinical trials of 12 weeks [8]. Finally, lamotrigine (Lamictal®) was approved for maintenance of bipolar patients in a 52-week trial [11]. Therefore, depending upon the sponsor's seeking an acute or maintenance indication for their compound, a specific time for the trial's duration must be considered. These time periods are not precisely defined by any diagnostic criteria found in DSM-IV but are dependent on what is considered clinically necessary and may vary between indications.

Lamotrigine is also approved for bipolar depression and not acute mania [11]. Olanzapine alone and its combination with fluoxetine (Prozac®) was recently evaluated for bipolar depression making it the only agent assessed in a placebo-controlled manner for the various spectrums of bipolar illness [6, 12]. These drug approvals for the different spectrums of bipolar illness also reflects the dynamic nature of the disease that can fluctuate over time in any given bipolar patient. This change in disease over time could be a significant confounding feature in assessing drug efficacy and the necessary monitoring parameters if the patient's condition changes over time.

In order to satisfy the demands of regulatory authorities, clinical drug trials in psychiatry have to face numerous *considerations on design* (Fig. 12.5.5). Like in most other areas of drug development, the placebo controlled trial is the gold standard, especially for drugs aimed at the acute-illness phase. Although scientifically valid, placebo use in psychiatric disorders can be controversial. An ethical question is raised regarding reasonable health care and patient safety. Also, the United States and Europe have different standards for use of placebos, with Europe generally frowning on their use in clinical trials. Pharmacogenetic samples have become a commonplace occurrence to be included in clinical trials. Many pharmacogenomic differences in patients deal with metabolism, and the number of single nucleotide polymorphisms with hepatic enzymes is relatively high with psychiatric dugs. This type of information is carefully reviewed by institutional review boards (IRBs). Informed consent documents must clearly inform subjects what their sample is being tested for and how long samples can be retained by the sponsor. If new testing is to be conducted, their permission needs to be obtained prior to these new tests. The FDA (2005) has requested pharmacogenetic evaluations be completed during clinical trials to determine if specific patient groups are more prone to increased adverse events than other groups.

In most psychiatric studies, the primary and secondary outcome parameters are clinical rating scales [13, 14] that attempt to quantify the severity of a disease state, which is necessary if one wants to assess the possible positive or negative effects of a new treatment on the condition. The rating scales are therefore vital in the evaluation of drug efficacy. In order for them to be accepted by regulatory authorities and the scientific community, the scales have to be validated, which means they have been tested rigorously. These instruments translate relatively subjective observations into objective measurements. Prior to the clinical trial's initiation, investigator meetings are held to validate investigator evaluations with a specific set of rating scales. However, despite training sessions and experience with clinical rating scales, the human element in the investigator and study subject interaction still is a source of intrinsic variability.

Both the use of "as needed" (PRN) medication and stable concurrent medication can be important factors when evaluating a new medication. PRN medication, which may have to be allowed to handle acute exacerbations of disease, could influence the patient's evaluation on a specific day especially when it's the day where the clinical rating is to be conducted. Trials usually have investigators move the evaluation by another 24 hours when a PRN medication is given within a day of the scheduled evaluation.

The effects of concomitant medications or standardized treatment guidelines can also influence evaluation of a new drug during clinical trial. In Alzheimer's clinical trials, for instance, cholesterase inhibitors have become a standardized treatment for patients with dementia. New antidementia medications are usually added on to these existing drugs. The control group consists of the placebo plus the cholinesterase inhibitor. As a consequence, the clinical trial will only be able to assess the additional benefit of the new drug on top of the standard treatment, which may have implications for sample size.

Five more *study design considerations* are offered in Fig. 12.5.6 Safety monitoring always is a major component of any clinical trial. Given the patient has a mental disorder with a variety of presenting symptoms and possibly more sensitivity

- Study design methodology differs between the different psychiatric disorders:
  - Placebo use
  - Pharmacogenetic sample collections
  - Validated clinical rating scales for efficacy and safety assessments
  - As needed (PRN) medication use during the trials:
    - How to determine usage?
  - Role of concurrent medication therapy:
    - Bipolar trials – Mood Stabilizers
    - Alzheimer's Disease – Cholinesterase Inhibitors

FIG. 12.5.5. Study Design Methodology -1

- Study design methodology differs between the different psychiatric disorders (continued):
  - Safety monitoring can differ:
    - Frequency over time,
    - Number of assessments, and etc...
  - Dose – fixed versus flexible
  - How to determine efficacy or relapse "a priori" by clinical rating scales
  - Role of family members or caregivers
  - Informed consent issues

FIG. 12.5.6. Study Design Methodology -2

to personal and environmental changes, safety monitoring can be a bigger challenge in psychiatry. The frequency over time and the number of overall assessments need to be set up in advance. Guidance to patients and investigators is needed to help differentiate between disease symptoms versus drug effects versus drug-related adverse effects. Family members play a role in assessing safety and looking for adverse events, which needs to be built into a trial and is fairly unique to psychiatry and pediatrics or elderly.

Clinical drug trials in psychiatry can use either fixed-dose group assignment or a flexible-dose group [15–19]. Both methodologies have their strengths and weaknesses. Fixed-doses groups (low, medium, high) are often used to define dose-dependent relationships for efficacy and adverse events. However, drug tolerability may be low with the possibility of a high dropout rate in the higher dose group, which may interfere with the ability to show efficacy in that group. With a flexible-dose group assignment, investigators can make dose adjustments to maximize efficacy and minimize adverse events. However, an accurate correlation between dose and efficacy or tolerability can be difficult to discern. It is important to identify upfront what particular answer the trial is looking for. If it is to establish whether or not the drug works, a flexible dosing regimen, which allows optimization for efficacy and tolerability, in general will be more effective. If the aim of the trial is to determine the lowest effective or highest tolerated dose, a fixed-dose regimen may be the preferred design.

How does a sponsor actually define efficacy for a new drug appropriately? The change from baseline to the study's end point is the usual time frame to determine drug efficacy. The difference needs to be clinically meaningful, though, as well as just statistically significant. Any change can be demonstrated, provided the sample size is large enough, but if the improvement is too small compared with the possible risks and costs, it may render the drug less useful. It is essential to establish upfront what effect size and its variability in the trial is expected to be found, because these factors will determine the required sample size of the study, together with the desired power (the probability to find an existing effect, or "ß," usually put between 0.8 and 0.9) and significance (the probability of the result being a chance finding, or "α," usually put at $p < 0.05$).

The role of family members or caregivers of study patients can be vital components in a clinical trial. Often, it is a family member that plays a pivotal role in helping a patient cope with their disease and therapy, as well as affecting the general environment for a patient. This person ensures patient compliance with study medications and transportation to appointments with the investigators. These individuals can also quickly alert investigators to adverse side effects, provide vital information on patient progress and even subtle effects from the investigation drug. Caregiver presence is even required in some clinical studies (e.g., Alzheimer's trials).

As previously mentioned, *clinical rating scales* that determine a drug's efficacy must be validated and are the most important aspect during investigational programs (Fig. 12.5.7). Each clinical rating scale consists of multiple items. Each item needs to be representative of a disease manifestation and present commonly (but not always) with the disease. Often, the items are grouped into domains wherein symptoms are grouped together with a common theme such as cognitive or behavioral. Validation requires a scale to be used repeatedly and compared with existing standard scales; individual items are even evaluated and their contribution to the whole. The sensitivity of the rating scale is evaluated and documented in the validation process, too. For example in schizophrenia trials, the Brief Psychiatric Rating Scale (BPRS) is an accepted validated evaluation instrument by regulatory agencies. The BPRS scale has 18 items that assess various components of psychotic symptoms. Several items have "anchor" points indicating key aspects where investigators must have consistent interpretations of the patient's psychotic symptoms.

Inter-rater variability is an inherent problem. It can be minimized and evaluations standardized at training sessions during investigator meetings, during which investigators and study coordinators are familiarized with the appropriate way to conduct the scale by experts in the field. Another way to reduce the variability is by ensuring the same investigator

- Interrater variability:
  - Investigator meetings to standardize clinical assessments
  - Clinical and/or study coordinators
  - Investigators consistency
  - Establish "anchor" points for various rating scales
  - Environment for testing
- Frequency of assessments can differ:
  - Safety versus efficacy

FIG. 12.5.7. Clinical Ratings Scales – 1

rates a patient from beginning to end of the study. Training sessions between the established experts and investigators are routinely used by a company because they can establish a consensus among investigators for the consistent interpretation at the anchor points for the selected rating scale items. The environment where the rating scale is employed by the monitor with the patient is an important variable as well. It is important to allow enough time for completion (but not too much time to be consistent across patients), limit explanations of the items only to what is agreed upon (let patients answer it), avoid a noisy overly busy clinic setting (use a separate room or area, respect the patient's privacy in answering such psychiatric-related questions, and allow an agitated patient or hurried patient to settle down before answering the items, unless this is part of the symptomology or protocol).

The frequency of clinical efficacy assessments during an investigational study depends on the disease state and the duration of the trial. Acute manic bipolar and schizophrenic trials, for instance, require at least a weekly evaluation due to the rapidly changing nature of the disease. An effective medication would be expected to demonstrate improvement within this study time period of 3–4 weeks. On the other hand, clinical assessments of cognition in Alzheimer's trials are conducted about every 10–12 weeks during either a 6-month or 1-year time period. The goal for an Alzheimer's medication would be to demonstrate either prevention of cognitive decline or if possible improvement versus the control group over the 6-month or 1-year time period. Changes in significant cognition, unlike psychotic symptoms, cannot be found in a short time of weeks [19, 20].

Some *examples* of well-established *clinical rating scales* used for various psychiatric disorders follow below for four separate and distinct psychiatric diseases (Fig. 12.5.8). The norm in psychiatry is to develop disease-specific scales as you would expect. Also, several scales usually are available to choose from for each disease; a common practice in research is to use more than one scale in a clinical trial. For FDA approval, the regulatory agency requires that positive benefits be statistically determined by two primary efficacy scales.

- Schizophrenia—Brief Psychiatric Rating Scale (BPRS), Positive and Negative Syndrome Scale (PANSS).
- Bipolar Mania—Young Manic Rating Scale (YMRS).
- Alzheimer Disease—Alzheimer's Disease Assessment Scale – cognitive section (ADAS-cog), Severe Impairment Battery (SIB).
- Depression—Hamilton Depression Scale (HAM-D), Montgomery-Asberg Depression Rating Scale (MADRS).

As previously mentioned, the BPRS is commonly used for schizophrenia clinical trials. Although the BPRS was not originally designed for the assessment of drugs for the treatment of schizophrenia, it has become a standard instrument through continued use and established "anchor" points of how raters can interpret clinical psychopathology that translates into an objective measure. As previously mentioned, the BPRS consists of 18 items that include hallucinatory behavior, hostility, conceptual disorganization, blunted affect, suspiciousness, and other target symptoms found in psychotic patients. Each item is rated from 1 = not present to 7 = most severe. The total number of points indicates the severity of the person's illness at the time of assessment. The higher number of points indicates the increase in psychopathology severity found in the patient. As the points decrease over time (usually 4–6 weeks in a schizophrenia trial), clinical improvement is noted. Schizophrenia trials set an *a priori* improvement of 20–30% decrease in total BPRS scores when evaluating the efficacy potential of new antipsychotic drugs. Similar methods are used for these other rating instruments depending upon the disease state selected for evaluation.

The Clinical Global Impression Scale (CGIS, or CGI) is a global measure of efficacy and change determined by the clinical investigator [13] and can be used for many different conditions as it assesses change in a general way. Regulatory agencies respect the investigator's clinical experience, and a global evaluation can often detect changes in a subject's *overall* symptomatology that the other specific primary scales may miss. However, this scale is rarely acceptable as a primary outcome measure, because it is not specific to the disease studied.

Clinical trials typically include *secondary evaluations* to assess other potential changes in disease symptoms, in addition to the primary efficacy assessments with disease state specific scales (Fig. 12.5.9). Schizophrenia patients can have cognitive impairment. While antipsychotic efficacy depends upon improvement in psychotic symptomatology for regulatory agency approval, cognitive symptoms changes can be important information on the drug's overall effect on a patient. Depending upon the disease, other symptoms including mood, behavior, and anxiety can be secondary measures used in clinical studies. These cognitive, mood, behavior, or anxiety symptoms are actually a group of possible symptoms that are evaluated in a battery of assessments.

If more than one clinical rating scale is used to assess an investigational drug's efficacy, it is essential to assign primary

- Different for each disease state
  - Clinician rating scales, disease focused:
    - Schizophrenia – BPRS, PANSS, SANS.
    - Bipolar Disorder – YMRS.
    - Alzheimer's Disease – ADAS-cog, CDR/CIBIC-Plus.
    - Depression – HAMD, MADRS.
    - Use of BPRS as an example
- Clinical Global Impression Scale (CGIS):
  - Global symptom assessment of physician

FIG. 12.5.8. Clinical Ratings Scales – 2

- Primary versus secondary efficacy scales:
  - Disease state specific scales (primary)
- Secondary efficacy scales:
  - Cognitive assessments excluding dementia studies
  - Behavioral assessments
  - Mood components
- Other secondary scales:
  - Patient self-rating scales (Zung Depression Scale)
  - Caregiver reports and assessments
  - Safety and tolerability assessment

FIG. 12.5.9. Clinical Ratings Scales – 3

and secondary status. Even though a number of scales may be used, the primary efficacy variable is the one that is used to determine whether or not a study is successful and therefore has to be carefully selected. The scales described thus far have been administered by the investigator or study coordinator evaluating the study subject. In order to obtain the entire clinical presentation of the study subject, however, some clinical trials will use patient self-rating evaluations such as the Zung Depression Scale in depression studies [13]. It is obvious that some psychiatric diseases like depression are more amenable to patient self-ratings than others, like for instance Alzheimer disease, in which the cognitive functions of the patient may be impaired. Alternatively, in Alzheimer's clinical trials, caregivers can provide reliable information on the patient's behavior and functional abilities at home, information that could probably not be assessed in the research facility. Together, the combined use of multiple rating instruments can provide an accurate and more complete patient assessment of the investigational drug's effects. However, multiple scale use in patient assessments adds to the time to conduct a trial by investigators and adds significant labor costs, which needs to be considered to some extent in designing and conducting the trials.

Safety and tolerability assessments for drugs are equally if not more important than efficacy assessments. They are normally conducted at the same time as the efficacy assessments, but in chronic trials and long-term studies they may occur in the absence of efficacy assessments. The essential difference between safety and tolerability is that safety pertains to actual risks to the health of a subject (e.g., death and disability), whereas tolerability is related to signs and symptoms that are temporary in nature, even though they may be severe (like vomiting, constipation, headache, etc.). A benefit to risk evaluation is an integral part of the regulatory assessment of a submission, and regulatory agencies can delay or deny an Investigational New Drug (IND) application if the drug's safety concerns outweigh its efficacy profile.

Pharmacology

Determining the *appropriate dose* of a study medication, selected for clinical trials, can be the most important aspect of the drug's development (Fig. 12.5.10). It also greatly affects the potential approval from regulatory agencies. If the dose selected is too low, the optimal efficacy may not be adequately demonstrated. Conversely, if the dose selected is too high, efficacy may be achieved at the cost of an unacceptable safety profile. How the drug dose is selected in psychiatry for clinical trials is based on many factors that include *in vitro* drug receptor binding affinity profiles, behavioral pharmacological models in animals, and phase I clinical studies. Despite the presence of these data, it still requires a combination of "art" and "science" to estimate the dose range to use in the phase II and III clinical trials.

Optimizing the clinical dose-response to a medication in psychiatry involves the use of prospective, randomized, double-blind, fixed-dose treatment groups with adequate time and statistical power between groups to detect a significant difference versus placebo [21]. Instead of a fixed-dose paradigm, other studies have employed a flexible dosing regimen. Drug doses are individually adjusted based on patient response and attempts to minimize adverse events. Due to the wide interpatient variability with the pharmacokinetic disposition of many psychiatric drugs (including polymorphisms in cytochrome P450 metabolic enzymes), a flexible dosing regimen may actually obscure the "true" dose-response relationship between the drug and efficacy for the patient. Some studies have used a combination of a fixed dose and a flexible dose together in order to increase the capability to define the drug's best doses for efficacy [22]. This design method can lengthen the study duration and increase study costs and time frame for completion. Further, this method may still not adequately define the drug's dose range for efficacy.

Clinical trials predominantly use an oral formulation of the study medication. This factor alone dictates a certain level of patient compliance that must be adhered to in order to have a successful trial. In acute schizophrenia or bipolar trials, patients may be uncooperative, agitated, or even psychotic, which can

- How to select the dose?
  - Based upon how information from animal data and
  - *In-vitro* tests such neuroreceptor binding affinity studies
- Fixed versus flexible dosing design
- Oral tablets or capsules versus other administration routes (e.g., IM, Liquid)
- Dosing frequency – how to select based upon pharmacokinetics?
- Compliance to treatment must be assessed.

FIG. 12.5.10. Dosing and Administration

decrease compliance toward medication thus possibly affecting study completion rates. Other formulations can be used such as intramuscular (IM) drug administration (e.g., IM olanzapine) and may have advantages in acutely ill and/or uncooperative patients [23]. These formulations have the benefit of assured compliance, but the recruitment and study completion with these types of patients can present challenges to the investigator, research team, and facility.

Finally, dosing frequency of the study medication is important for patient compliance and drug administration time by the nursing staff. Medications given once-a-day offer many advantages compared with multiple daily dosing. Most psychiatric medications that are dosed once-a-day have a pharmacokinetic profile with an elimination half-life of at least 18 hours. However, there is evidence that the plasma pharmacokinetic profile of some psychiatric drugs (e.g., quetiapine) may not reflect their actual brain pharmacokinetics. Depending upon receptor–drug binding affinity profile, the drug's actual pharmacokinetic half-life at the site of the receptor in the brain may be longer than its plasma pharmacokinetic elimination half-life. This could mean that the medication may be dosed once-a-day [24]. New technologies like brain positron emission tomography (PET) studies examining drug doses and their brain and plasma pharmacokinetic profiles may be supportive to determine the correct dosing frequency [24].

The roles of *pharmacokinetics (PK) and pharmacodynamics (PD)* are undervalued in psychiatry and may provide valuable information for dose selection and dosing frequency for clinical investigation (Fig. 12.5.11). As was stated before, dose selection is based on preclinical and phase I PK/PD studies. The maximum tolerated dose (MTD) determined in phase I studies can be an important guiding principle in the drug development as it helps estimate the optimal dose range for efficacy trials [25], although for certain drugs (for instance, antipsychotics and opioids) patients can tolerate far higher doses than volunteers. Phase I studies assist to provide information for phase II studies that the doses are safe and potentially efficacious if patients are used in phase I, in the dose range evaluated. A "universal" consensus has not yet been reached on what the definition MTD is in humans. It has been accepted that MTD can be the maximum dose administered during a trial that produces mild to moderate nonlethal toxic effects in a significant percentage of individuals. Most phase I studies are conducted in healthy volunteers and may not predict the effects of a drug in a patient population. The determination of MTD in a small group of patients from the disease population can provide more information on the dose range for the investigational drug. This concept of utilizing patients instead of healthy volunteers for early safety and tolerance studies has been called a "bridging study" [25]. A bridging study uses a placebo-controlled, multiple-dose scheme with ascending doses in consecutive panels of 6–8 patients each [26]. The first panel is the MTD dose of 50% less than the MTD determined in healthy volunteers. Other panels can be 25% less than MTD of healthy subjects, 25% greater than the MTD, and 50% greater than the MTD. These studies should be conducted on an inpatient basis for safety concerns. The bridging study can define or redefine the dose range for phase II–III studies [27, 28].

In recent years, PK has played a prominent role in psychiatry for the development of various dosage formulations in marketed antipsychotic and antidepressant medications. When the new-generation atypical antipsychotics were FDA approved, these agents were initially available only as an oral tablet or capsule limiting their use to relatively compliant patients. Risperidone (Risperdal®) was developed into a liquid and a long-acting depot injection formulation [29]. The long-acting depot risperidone formulation differs from previous depot formulations (e.g., haloperidol and fluphenazine) in that the risperidone molecule is encased in a matrix of glycolic acid–lactate polymer. The older depot formulations consisted of a long-chain fatty acid. The depot risperidone is administered every 2 weeks. From a single dose, hydration, drug diffusion and final polymer erosion occurs over a time period of 9 weeks [29]. Oral risperidone supplementation is recommended for the first month as plasma drug concentrations increase from the depot formulation.

A rapid dissolving tablet and an injectable formulation of olanzapine for treatment of acutely ill patients were developed. Similarly, an injectable form of ziprasidone (Geodon®) was introduced [30]. Quetiapine (Seroquel®) has a short elimination half-life of about 6 hours, and a slow-release tablet formulation is under development. Quetiapine's molecular structure may prevent its development into an injectable formulation for acute or depot use. Instead of a liquid formulation, quetiapine is being developed as a sealed pouch of granules or "sachets" that can be dissolved in water to create a flavorless solution [31].

New formulations for antidepressants have also been developed and marketed to enhance patient compliance. These new formulations have focused on rapid-dissolving tablets,

- Undervalued in psychiatry
- Extraordinarily important regarding formulation development:
  - Oral - rapid dissolving tablets, sublingual, or sustained/extended release
  - Long-acting injections (e.g., RisperidoneConsta®)
  - Transdermal
- Dose ranging studies – effective dose range, maximum tolerated dose (MTD)
- Drug interaction studies:
  - CYP2D6 and CYP3A4 inhibitors (less metabolism)
  - CYP3A4 inducers (more metabolism)
  - E.g., ziprasidone study effects upon QTc prolongation

FIG. 12.5.11. Pharmacokinetics & Pharmacodynamics

enteric-coated, controlled-release, and extended-release oral administration [31]. Chiral enantiomers have recently played a role in psychopharmacology. Citalopram is marketed as a racemic mixture of *R* and *S* enantiomers. The *S*-enantiomer escitalopram is twice as potent as the racemate and 100-fold more potent that the *R*-enantiomer [31]. Extensive PK and PD studies were conducted on escitalopram and both formulations are available. Selegiline is a selective monoamine oxidase inhibitor (MAO) type B at an oral dosage of 10 mg/day and has been used for the treatment of Parkinson disease [32]. A selegiline transdermal formulation has been evaluated and is currently under FDA review for approval for the treatment of depression [33]. Transdermal selegiline was shown to have a 50-fold greater bioavailability than oral formulation and provided a sustained PD effect utilizing its PK of slow rate delivery through the skin (mean T_{max} 24 hours and mean elimination half-life of 9 hours) [34]. Animal studies have demonstrated a consistent PD effect of MAO-B inhibition in different brain regions [35].

PK drug interaction studies with a new investigational drug are required by regulatory agencies based on our information on CYP isozymes (cytochrome P450 hepatic enzymes responsible for metabolism). A typical panel of interaction studies includes studies with known CYP inhibitors (CYP2D6 paroxetine; CYP3A4 ketoconazole) and CYP inducers (CYP3A4 carbamazepine). These studies can be either single or multiple dose administration typically conducted in healthy volunteers. Actual psychiatric patients are usually on other prescribed medications, which can influence PK disposition of the investigational agent. PD can also play a prominent role in drug development. For example, ziprasidone's effects upon QTc prolongation was a major issue in its approval process and required an in-depth PK/PD study in actual patients to clearly define its effect on QTc. Interestingly, a drug–drug interaction was selected as the study method to evaluate this parameter [36]. Comparator medications were also evaluated with an interesting result for thioridazine [37]. The most significant QTc prolongation occurred with thioridazine and not with ziprasidone, which resulted in a "black box" warning for thioridazine and its marketed active metabolite mesoridazine.

Finally, PK studies can generate controversy. A generic formulation of clozapine was evaluated with Clozaril® in a bioequivalence study with 21 patients [38]. It was reported that when patients were switched to the generic formulation, cognitive performance was diminished with a lower T_{max} and average serum concentrations. These inconsistencies were suggested to be differences in drug absorption rates between the two products. This study has been largely criticized, and the FDA stated that studies disputing the equivalence of Clozaril and generic clozapine were not conducted in a manner that could adequately assess efficacy or safety [39].

The pharmaceutical industry has invested substantial financial resources to anonymously obtain *genetic information* from patients with various psychiatric diseases (Fig. 12.5.12).

- Genetic information is important in psychiatry for both efficacy and especially toxicity for drugs.
- Companies have invested in obtaining (anonymous) genetic information from subjects for diseases.
- Use of SNP' sassays to determine distribution of specific alleles for detection of specific effects (response versus adverse events).
- CYP profiling is conducted in trials.
- Common & major impact on drug dosing and patient responses, but underutilized.
- Future role to be determined.

FIG. 12.5.12. Pharmacogenomics

The use of single nucleotide polymorphism (SNP) assays to determine the distribution of specific alleles to potentially detect specific effects (especially adverse events) has received intense investigation. Recently, the FDA has requested this type of analysis from the industry. Patient CYP profiling is a part of these analyses. Its future role continues to be determined. However, the ultimate goals would be the identification of receptor genes that could influence drug response and metabolic profiles to determine potential drug interactions and other safety parameters resulting in individualized therapeutic decisions for patients [40]. Pharmacogenomics may play a role in the drug discovery process where researchers may identify novel genomic targets for existing drugs, develop high-throughput assays for new targets, screen new compound libraries, and eventually develop specific compounds based on genomic findings [41].

Research Challenges

Clinical research has a set of guidelines and regulations enforced by state and federal agencies including the FDA, Office of Human Research Protection (OHRP), National Institutes of Health (NIH), and other departments within the government. Even with all such guidelines to help frame research, *controversies* still exist in the design and conduct of clinical studies in psychiatric patients (Fig. 12.5.13). Ethical issues are always present in clinical research, however, seven basic universal requirements have been proposed [42]. Fulfillment of these requirements is sufficient to make all clinical research ethical. Research in psychiatric patients presents different challenges to the investigators where three basic issues arise [42, 43]. These issues are (1) Can mentally ill patients give proper informed consent? (2) What are the ethical safeguards? (3) What are competing scientific and ethical imperatives in current experimental practices?

Each issue will be briefly discussed. Acutely ill patients could have greater difficulties making a decision to participate in a trial. Schizophrenic patients have a more pronounced

- Patient competency
- Placebo use
- Study design versus treatment guidelines
- Drug dose used in clinical trials may not reflect what practitioners are prescribing when the drug is approved.
- New adverse events not found during trials are discovered after drug approval.

FIG. 12.5.13. Research Controversies

problem than patients with depression [44]. Once treated, these deficits in cognition of schizophrenics greatly improve and decisional abilities resemble other comparative groups. Severity in psychopathology was found to be a poor indicator of the inability of patients to assess the risks versus benefits. Experienced investigators can obtain proper informed consent from psychotic patients. Informed consents that are of poor quality, overly reliant on technical terms, and difficult to read are barriers to clinical trials independent of the disease state. Informed consent by mentally ill patients may present difficulties, but these problems may not be greater than other patients with medically vulnerable diseases.

Ethical safeguards for patients include the institutional review board of the institution, which oversees clinical research. The FDA and other regulatory agencies can make inspections of clinical studies, and one of the main areas of review is the informed consent document. Surrogate decision-making by legal guardians and family members of patients also provides an additional safeguard for patients (especially in dementia clinical studies). The balance between science and clinical routine practice has been clearly delineated, but at times, these "lines" can be obscured. The use of placebos always generates discussion among practitioners, investigators, and regulatory agencies [45, 46]. The placebo response rates for the various psychiatric disease states are high: schizophrenia 30%; bipolar mania 20%; depression 30%; and generalized anxiety disorders 40–50%. Interestingly, placebo response rates tend to increase when more research attention is focused on the disease and more treatments become available (example: OCD, initial studies had 0% response rate on placebo, at present the response is around 20–30%). Based on these figures, it would be difficult for regulatory agencies to remove the use of placebos from psychiatric research. Study design is another issue especially where psychiatric medication discontinuation and symptom-provoking studies are involved [47]. Adequate patient safeguards with family and/or caregiver support is necessary and clear definitions of symptom relapse are imperative in these types of studies. Clinical research trials in children and adolescents can be more challenging to conduct than in adult populations [48]. Many psychotropic medications do not have a regulatory approval indication for this population. Yet, clinicians prescribe these

- **Do's:**
  - Have close scrutiny of patient inclusion criteria.
  - Be flexibility in time between screening visit & randomization visit.
  - Use prescribed PRN medications for acute problems.
  - Call sponsor for any potential deviations from protocol versus routine medical practice.
- **Don't's:**
  - Enroll patients if questionable competency (e.g. suicide, psychosis) or compliance.
  - Hesitate to drop patients from studies due to safety concerns.
  - Engage new monitors without proper study training.

FIG. 12.5.14. Dos & Don'ts of Research

medications for these patients. The need for research in children and adolescents is great and its requirements similar to other populations.

Good clinical practice and research guidelines should be used by investigators. At this point, investigators should also use their own experience and "common sense" when conducting clinical trials with psychiatric patients. However, some major considerations of these issues are presented here as *Do's and Don't's* (Fig. 12.5.14).

Dos

- Close scrutiny of patient inclusion criteria: symptoms that are present may occur in more than one psychiatric disease (e.g., psychosis, delusions, hallucinations).
- Use PRN medications only as described in the protocol for acute problems during a trial.
- Flexibility of time between screening visit and randomization visit: When the effect of the previous medication is washed out (usually discontinued), symptoms may increase to a point where the patient may not be appropriate to enter into the study. The sponsor should include a flexible time period to allow the investigator to enroll the patient when the situation arises.
- The investigator should call the sponsoring company for any potential deviations from the study protocol and not rely on routine medical practice. An example would be in the flexibility time between screening and randomization visits. An excessive time interval may lead to changes in disease severity or complications that should exclude a patient, which would be missed.

Don'ts

Most of these items are self-explanatory as follows.

- Do not Enroll patients if there's a questionable competency issue (e.g., suicide, psychosis, cognition, language) or compliance problem.

- Do not Hesitate to drop a patient from a study due to safety concerns (e.g., suicide, psychosis, serious adverse event).
- Do not Involve a new monitor to evaluate drug efficacy with the rating scale who has not been similarly trained in the protocol and the scale as exisiting monitors.

Summary

The *top five areas* for research are schizophrenia; bipolar disorders; alzheimer disease; depression; and anxiety disorders (Fig 12.5.15). These disorders represent a significant impact upon patient care, caregiver burden, and the financial status of health care systems. For example, it has been estimated that the total direct and indirect economic costs of Alzheimer disease alone upon our society is $100 billion dollars [49]. The estimated total costs for the remaining four disorders are about $100 billion to $120 billion dollars. Antipsychotic drugs initially developed for schizophrenia have also been indicated for acute and chronic bipolar mania. Antidepressants have found new indications in various anxiety disorders. Both antipsychotics and antidepressants are prescribed by clinicians for the behavioral disturbances in Alzheimer's patients.

At any given time, there are well over 50 different types of compounds in various stages of clinical development for the different psychiatric disorders. Three different reports show the significant quantity of CNS research ongoing; 1,610 active INDs in neuropharmacology in 2001 according to FDA [50], 216 CNS products out of 1,878 listed by Medical Advertising News in 2004 [51], and 68 CNS clinical projects by seven top pharmaceutical companies according to CenterWatch [50]. R&D Directions lists 33 products for Alzheimer's, 30 for depression, and 14 for schizophrenia. Some examples of new compounds in the pipeline are found in the figure.

Carbamazepine and topiramate are antiepileptic agents already FDA approved for the treatment of various seizure disorders. Carbamazepine has been used clinically for bipolar mania for more than 20 years with its use supported by published studies. However, it is the extended-release formulation that has been evaluated in clinical trials with a New Drug Application (NDA) submitted to the FDA [52]. Topiramate remains under development. The other compounds listed are in different stages of phase I through III development.

Every clinical trial result that demonstrates a significant advance in medical science must be published in the *journal literature* (Fig. 12.5.16). Psychiatry does not differ from other medical specialties in that significant clinical trial results seek to be published in the most prestigious medical journals, such as the *New England Journal of Medicine* or *Journal of the American Medical Association* [5, 20]. However, like other medical specialties, a core set of journals within its specialty offers the industry a variety of publications in which to disseminate study results. There are more than 25 psychiatric journals where clinical research and trials are published. Some of these key journals are listed in Figure 12.5.16 for clinical trials and pharmacologic studies.

In conclusion, Drug development in psychiatric disorders will continue to be challenging to design and conduct for the pharmaceutical industry, patients, caregivers and family members, and investigators (Fig. 12.5.17). Psychiatric disorders with their comorbidities and overlapping symptomatology can make drug efficacy assessments difficult. Treatment goals for acute illness differ from chronic or maintenance clinical study trials.

- **Publications for Research in Psychiatry:**
  - Archives of General Psychiatry
  - American Journal of Psychiatry
  - Journal of Clinical Psychopharmacology
  - Journal of Clinical Psychiatry
  - Nature Reviews Drug Discovery
  - Neuropsychopharmacology
  - Psychopharmacology

FIG. 12.5.16. Key Journal References

- Drug development for psychiatric/CNS disorders will continue to be challenging.
- Psychiatric disorders with their comorbidities and overlapping symptomatology makes drug efficacy assessments difficult.
- Patient competency will be a key assessment for PIs.
- Subjective ratings scales remain gold standard in patient assessments for efficacy.
- Goals for acute versus chronic trials differ.
- Appropriate doses to maximize efficacy and minimize adverse events continues even after regulatory approval.
- FDA has guidances for studies for most mental disorders.

- **Depression:**
  Gepirone-ER, Transdermal Selegiline, Duloxetine, Neurokinin-1 antagonists
- **Schizophrenia:**
  Asenapine, Bifeprunox, Iloperidone
- **Bipolar Disorders:**
  Carbamazepine, Topiramate
- **Alzheimer's Disease:**
  Huperzine A, Nefiracetam, Neramexane
- **Anxiety Disorders:**
  Pagoclone, Eplivanserin, Ocinaplon

FIG. 12.5.15. Top Research Areas & Drugs

FIG. 12.5.17. Conclusions

- Cummings JL, Zhong K. Treatments for behavioural disorders in neurodegenerative diseases: drug development strategies. *Nat Rev Drug Discovery* 2006;5(1):64-74.
- Spedding M, Jay T, DeSilva JC, Perret L. A pathophysiological paradigm for the therapy of psychiatric disease. *Nat Rev Drug Discovery* 2005;4(6):467-476.
- Stroup TS, Alves WM, Hamer RM, Lieberman JA. Clinical trials for antipsychotic drugs: design conventions, dilemmas and innovations. *Nat Rev Drug Discovery* 2006;5(2):133-146.
- Wong M-L, Licinio J. From Monoamines to genomic targets: a paradigm shift for drug discovery in depression. *Nature Drug Discovery* 2004;3(2);136-51

FIG. 12.5.18. Readings

Patient competency is a major issue to be dealt with in psychiatric research. Patient assessment with rating scales, although challenging, is generally well established and well accepted by regulatory authorities and the psychiatric community. Identification of appropriate doses to maximize therapeutic efficacy and minimize adverse events continues to be redefined even after regulatory approval. FDA provides useful clinical guidelines for the clinical research and study design for some mental disorders (e.g., anxiety, depression, substance abuse). In general, consultation with the regulatory authorities on study design even before conduct of a clinical trial will usually lower the number of questions and improve acceptance later during the review process. Four references are provided for general reading (Fig. 12.5.18).

References

1. American Psychiatric Association. Diagnostic and Statistical Manual of Mental Disorders, Fourth Edition, Text Revision. Washington DC, American Psychiatric Association, 2000.
2. Anonymous. National Institutes of Mental Health. The numbers count. Mental disorders in America. 2001. Available at www.nimh.nih.gov.
3. Anonymous. National Institutes on Drug Abuse. NIDA Infofacts: costs to society. 1992 data set. Available at www.nida.nih.gov.
4. McKhann G, Drachman D, Folsetin M, et al. Clinical diagnosis of Alzheimer's disease: report of the NINCDA-ARDA work group under the auspices of the Department of Health and Human Services Task Force on Alzheimer's Disease. Neurology 1984; 34: 939-944.
5. Bowden CL, Brugger AM, Swann AC, et al. Efficacy of divalproex vs lithium and placebo in the treatment of mania. The Depakote Mania Study Group. JAMA 1994; 271: 918-924.
6. Zyprexa® package insert, revised August 2004.
7. Risperdal® package insert, revised December 2003.
8. Seroquel® package insert, revised July 2004.
9. Geodon® package insert, revised August 2004.
10. Abilify® package insert, revised May 2004.
11. Lamtical® package insert, revised August 2004.
12. Tohen M, Vieta E, Calabrese J, et al. Efficacy of olanzapine and olanzapine-fluoxetine combination in the treatment of bipolar I depression. Arch Gen Psychiatry 2003; 60: 1079-1088.
13. Guy W. ECDEU Assessment Manual for Psychopharmacology, revised 1976. US Department of Health and Human Services, Publication No. 91-338.
14. Sajatovic M, Ramirez LF. Rating Scales in Mental Health. Lexi-Comp Inc., Hudson, OH. 2001.
15. Beasley CM, Tollefson G, Tran P, et al. Olanzapine versus placebo and haloperidol. Neuropsychopharmacology 1996; 14: 111-123.
16. Small JG, Hirsch SR, Arvanitis LA, et al. Quetiapine in patients with schizophrenia. Arch Gen Psychiatry 1997; 54: 549-557.
17. Daniel DG, Zimbroff DL, Potkin SG, et al. Ziprasidone 80 mg/day and 160 mg/day in the acute excerbation of schizophrenia and schizoaffective disorder: a 6-week and placebo-controlled trial. Neuropsychopharmacology 1999; 20: 491-505.
18. Kane JM, Carson WH, Saha AR, et al. Efficacy and safety of aripiprazole and haloperidol versus placebo in patients with schizophrenia and schizoaffective disorder. J Clin Psychiatry 2002; 63: 763-771.
19. Corey-Bloom J, Anand R, Veach, et al. A randomized trial evaluating the efficacy and safety of ENA 713 (rivastigmine tartate), a new acetylcholinesterase inhibitor, in patients with mild to moderate severe Alzheimer's Disease. Int J Geriatr Psychopharmacol 1999; 1: 55-65.
20. Reisberg B, Doody R, Stoffler A, Schmidt F, Ferris S, Mobius HJ for the Memantine Study Group. Memantine in moderate to severe Alzheimer's Disease. N Engl J Med 2003; 348: 1333-1341.
21. Kinon BJ, Ahl J, Stauffer VL, Hill AL, Buckley PF. Dose response and atypical antipsychotics in schizophrenia. CNS Drugs 2004; 18: 597-616.
22. Davidson JRT, Bose A, Korotzer A, Zheng H. Escitalopram in the treatment of generalized anxiety disorder: double-blind, placebo-controlled, flexible dose study. Depression and Anxiety 2004; 19: 234-240.
23. Meehan K, Zhang F, David S, et al. A double-blind, randomized comparison of the efficacy and safety of intramuscular olanzapine, lorazepam, or placebo in treating acutely agitated patients diagnosed with bipolar mania. J Clin Psychopharmacol 2001; 21: 389-397.
24. Kapur S, Zipursky R, Jones C, Shammi CS, Remington G, Seeman P. A positron emission tomography study of quetiapine in schizophrenia. Arch Gen Psychiatry 2000; 57: 553-559.
25. Cutler NR, Sramek JJ, Greenblatt DJ, et al. Defining the maximum tolerated dose: investigator, academic and regulatory perspectives. J Clin Pharmacol 1997; 37: 767-783.
26. Cutler NR, Sramek JJ. Optimizing the dose for Alzheimer's disease therapeutics: bridging and dynabridge methodologies. In. Research and Practice in Alzheimer's Disease, vol. 2. Springer, New York, 1999, p. 265-269.
27. Sramek JJ, Kirkesseli S, Paccaly-Moulin A, et al. A bridging study of fanaserin in schizophrenic patients. Psychopharmacol Bull 1998; 34: 811-818.
28. Sramek JJ, Anand R, Wardle TS, et al. Safety/tolerability trial of SDZ ENA 713 in patients with probable Alzheimer's disease. Life Sci 1996; 58: 1201-1207.
29. Kane JM, Boyd MA, Casey DE, Jarboe KS, Keith S, McEvoy JP. Treatment of schizophrenia with long-acting injectable medications. J Clin Psychiatry 2003; 64: 1250-1257.

30. Kelleher JP, Centorrino F, Albert MJ, Baldessarini RJ. Advances in atypical antipsychotics for the treatment of schizophrenia. CNS Drugs 2002; 16: 249-261.
31. Norman TR, Olver JS. New formulations of existing antidepressants. CNS Drugs 2004; 18: 506-520.
32. Chrisp P, Mammen GJ, Sorkin EM. Selegiline: a review of its pharmacology, symptomatic benefits and protection potential in Parkinson's Disease. Drugs Aging 1991; 1: 228-248.
33. Bodkin JA, Amsterdam JD. Transdermal selegiline in major depression: a double-blind, placebo-controlled, parallel group study in outpatients. Am J Psychiatry 2002; 159: 1869-1875.
34. Barrett JS, Hochadel TJ, Morales RJ, et al. Pharmacokinetics and safety of a selegiline transdermal system relative to single-dose oral administration in the elderly. Am J Ther 1996; 3: 688-698.
35. Wecker L, James S, Copeland N, Pacheco MA. Transdermal selegiline: targeted effects on monoamine oxidases in the brain. Biol Psychiatry 2003; 54: 1099-1104.
36. Glassman AH, Bigger JT. Antipsychotic drugs: prolonged QTc interval, Torsade de Pointes, and sudden death. Am J Psychiatry 2001; 158: 1774-1778.
37. Harrigan EP, Milceli JJ, Anziano R, et al. A randomized evaluation of the effects of six antipsychotic agents on QTC, in the absence and presence of metabolic inhibition. J Clin Psychopharmacol 2004; 24: 62-69.
38. Lam YWF, Ereshefsky L, Toney GB, Gonzales C. Brand vs generic clozapine: bioavailability comparison and interchangeability study. J Clin Psychiatry 2001; 62 (suppl. 5): 18-22.
39. Clozapine. US Food and Drug Administration Center for Drug Evaluation and Research [online]. Available at http:www.fda.gov.cder/infopage/clozapine/clozapine.htm.
40. Arranz MJ, Collier D, Kerwin RW. Pharmacogenetics for the individualization of psychiatric treatment. Am J Pharmacogenomics 2001; 1: 3-10.
41. Wong ML, Licinio J. From monoamines to genomic targets: a pardigm shift for drug discovery in depression. Nature Drug Discovery 2004; 3: 135-151.
42. Emanuel EJ, Wendler D, Grady C. What makes clinical research ethical? JAMA 2000; 283: 2701-2711.
43. Roberts LW, Roberts B. Psychiatric research ethics: an overview of evolving guidelines and current ethical dilemmas in the study of mental illness. Biol Psychiatry 1999; 46: 1025-1038.
44. Casey DE. Clinical trial design issues in schizophrenic research. J Clin Psychiatry 2001; 62 (suppl. 9): 17-20.
45. Collaborative Working Group on Clinical Trials Evaluations. Clinical development of atypical antipsychotics: research design and evaluation. J Clin Psychiatry 1998; 59 (suppl. 12): 10-16.
46. Rosenstein DL. IRB review of psychiatric medication discontinuation and symptom-provoking studies. Biol Psychiatry 1999; 46: 1039-1043.
47. Jeste DV, Palmer BW, Harris MJ. Neuroleptic discontinuation in clinical and research settings: scientific issues and ethical dilemmas. Biol Psychiatry 1999; 46: 1050-1059.
48. Vittiello B, Jensen PS, Hoagwood K. Integrating science and ethics in child and adolescent psychiatry research. Biol Psychiatry 1999; 46: 1044-1049.
49. Ernst RL, Hay JW. The US economic and social costs of Alzheimer's disease revisited. Am J Public Health 1994; 84: 1261-1264.
50. Lamberti MJ. An industry in evolution, 4th ed. Thomson CenterWatch, Boston, MA, 2003, pg 43, 47.
51. Anonymous. From pipeline to market 2004. Med Ad News supplement, July 2004, pp.35-74.
52. Ketter TA, Kalali AH, Weisler RH for the SPD417 Study Group. A 6-month multicenter, open-label evaluation of beaded extended release carbamazepine capsule monotherapy in bipolar disorder patients with manic or mixed episodes. J Clin Psychiatry 2004; 65: 668-673.

Appendix 1
Acronyms of the Pharmaceutical Industry

AA	Administrative Assistant or Amino Acid
AAHRPP	Association for the Accreditation of Human Research Protection Programs
AAP	American Academy of Pediatrics
AAGR	Average Accumulated Growth Rate
AAPS	American Association of Pharmaceutical Scientists
AC	Administrative Coordinator or Alternating Current
ACCME	Accreditation Council for Continuing Medical Education
ACP	American College of Physicians
ACPE	American Council for Pharmaceutical Education
AD	ADvertising
ADL	Activities of Daily Living
ADME	Adsorption, Distribution, Metabolism, Excretion
ADR	Adverse Drug Reaction
AE	Adverse Event
AERs	Adverse Event Reactions or Adverse Event Reports
AERS	Adverse Event Reporting System (FDA)
AHA	American Hospital Association
AHC	Academic Health Centers
AHCPR	Agency for Health Care Policy and Research
AHEC	Area Health Education Centers
AHFS	American Hospital Formulary Service
AHRQ	Agency for Healthcare Research and Quality
AMA	American Medical Association
AMCP	Academy of Managed Care Pharmacy
AMP	Average Manufacturer's Price
AMWA	American Medical Writer's Association
ANCOVA	Analysis of COVAriance
ANDA	Abbreviated New Drug Application
ANOVA	ANalysis Of Variance
AP	Asia-Pacific (market)
APC	Ambulatory Payment Classification (government drug pricing)
API	Active Pharmaceutical Ingredient
AREA	Academic Research Enhancement Award (in NIH)
ARIS	Adverse Reaction Information System
ARS	Adverse ReactionS
ASAP	As Soon As Possible
ASCPT	American Society of Clinical Pharmacology and Therapeutics
ASHP	American Society of Health-Systems Pharmacists
ASP	Average Sales Price

AUC	Area Under Concentration curve of blood levels (measure of absorption)
AWP	Average Wholesale Price
AZ	AstraZeneca LP
B	Black or Billion
BA	BioAvailability
BB	BlockBuster (drug)
BBB	Better Business Bureau or Blood Brain Barrier
BCBS	Blue Cross and Blue Shield (health insurance)
BCG	Boston Consulting Group
BCPS	Board Certified Pharmaceutical Specialist
BCS	Biopharmaceutics Classification System
BE	BioEquivalence
BI	Boehringer Ingelheim Pharmaceuticals, Inc., or Biogen-Idec, Inc.
BIC	Best-In-Class
BID	Twice a day treatment
BIO	Biotechnology Industry Organization
BLA	Biologics License Application
BM	BioMarkers or Bowel Movement
BMR	Basal Metabolic Rate
BMS	Bristol-Myers Squibb company
BMT	Bone Marrow Transplant
BOD	Board Of Directors or Burden Of Disease
BPCA	Best Pharmaceuticals for Children Act (2002)
BRC	Business Reply Card
BRMAC	Biologics Response Modifier Advisory Committee
BRMs	Biologic Response Modifiers
BRT	Botanical Review Team (at FDA)
BSA	Body Surface Area
BSN	Bachelors of Science in Nursing
BT	Biotechnology
BU	Business Unit
BW	Body Weight or Black & White
BWI	Bacteriostatic Water for Injection
BWP	Biotechnology Working Group (in Europe)
C	Concentration
C14	Carbon 14 (radioactive material)
CA	CAncer or Corporate Accounts
CAGR	Compound Annual Growth Rate
CALGB	Cancer And Leukemia Group B
CAM	Corporate Accounts Manager
CANDAs	Computer Assisted New Drug Applications
CAP	Competitive Acquisition Program (CMS) or CAPsule
CAS#	Chemical Abstracts Service registry number
CB	Cost Benefit
CBA	Cost Benefit Analysis
CBO	Congressional Budget Office
CBER	Center for Biologics Evaluation and Research
CC	Clinical Coordinator or Copy
CCA	Comparative Cost Adjustment (in Medicare program)
CCN	Comprehensive Cancer Network
CCU	Coronary Care Unit
CD	Complement Determining region
CDA	Confidentiality Disclosure Agreement

CCC	Comprehensive Cancer Center
CDC	Center for Disease Control (in PHS)
CDER	Center for Drug Evaluation and Research (in FDA)
CDISC	Clinical Data Interchange Standards Consortium
CDM	Clinical Data Management or Managers
CDP	Clinical Development Plan
CDR	Complement Determining Region (of Mabs)
CDRH	Center for Devices and Radiological Health (in FDA)
CE	Continuing Education or Cost Effectiveness
CEA	Cost Effectiveness Analysis
Ce	Concentration of a drug for Effective activity
CEO	Chief Executive Officer
CERTS	Centers for Education and Research on Therapeutics
CEU	Continuing Education Unit
CFO	Chief Financial Officer
CFR	Code of Federal Regulations
CFR 312	Code of Federal Regulations for drug development (IND, NDA +)
CHMP	Committee for Medicinal Products for Human Use (in Europe)
CHO	Carbohydrate or Chinese Hamster Ovary cells
CI	Confidence Interval (in statistics, variations around a mean) or Curies (measure of radioactivity)
CIOMS	Council for International Organizations of Medical Sciences
CIO	Chief Information Officer
CL	CLearance (of a drug from the body)
CMA	Continuous Marketing Application (in Europe) or Cost Minimization Analysis
C_{max}	Maximum Concentration (of a drug)
C_{min}	Minimum Concentration (of a drug)
CMC	Chemistry, Manufacturing and Controls (section for NDA)
CME	Continuing Medical Education
CMO	Chief Medical Officer or Contract Manufacturing Organization
CMR	Comprehensive Medication Review
CMS	Centers for Medicare and Medicaid Services
CMTP	Center for Medical Technology Policy
CNDA	Computer assisted New Drug Application
COB	Chairman of the Board
COBRA	Consolidated Omnibus Budget Reconciliation Act (1985)
COE	Center of Excellence
COGS	Cost of Goods Sold
COI	CO-Investigator or Centers Of Influence or Cost Of Illness
COMP	Committee on Orphan Medicinal Products (Europe)
COO	Chief Operating Officer
COP	Clinical Operations Plan
COPE	Commitment, Opportunity, Promise, & Evidence (in launch)
Cp	Concentration in Plasma (of a drug or chemical)
Css	Concentration at Steady State
CPA	Certified Public Accountant
CPI	Consumer Price Index
CPM	Capacity Planning and Management
CPMP	Committee for Proprietary Medicinal Products (in Europe)
CPOE	Computerized Physician Order Entry
CPSC	Consumer Products Safety Commission
CPT	Journal of Clinical Pharmacology and Therapeutics
CPT Code	Common Procedure Terminology codes
CQI	Continuous Quality Improvement
CR	Complete Response or Clinical Research
CRA	Clinical Research Associate

CRC	Clinical Research Center or Clinical Research Coordinator
CrCl	CReatinine CLearance
CRM	Customer Relationship Management
CRO	Clinical or Contract Research Organization
CSRs	Clinical Study Review (groups at NIH)
CRT	Controlled Randomized Trial
CS	Clinical Safety or Customer Services
CSA	Controlled Substances Act or Consulting Services Agreement
CSDD	Center for the Study of Drug Development (at Tufts University)
CSF	Critical Success Factors or Colony Stimulating Factor
CSO	Contract Service Organization or Contract Sales Organization or Consumer Safety Officer (in FDA)
CSO	Chief Scientific Officer
CSR	Clinical Study Reports
CT	Cell Therapy or Clinical Trial or ChemoTherapy
CTA	Clinical Trials Application (European IND equivalent)
CTC	Common Toxicity Criteria (National Cancer Institute)
CTD	Common Technical Document (for FDA & EMEA submissions)
CTM	Clinical Trials Materials or Clinical Trials Management
CTO	Chief Technical Officer
CU	Cost Utility
CUA	Cost Utility Analysis
CY	Calendar Year
CYP	CYtochrome P450 (isoenzymes in liver metabolizing many drugs)
CYT	CYTokine
D	Dose
D&A	Dosage and Administration
D&O	Directors and Officers
DAWN	Drug Abuse Warning Network
DB	Double-Blind
DBPCRT	Double-Blind Placebo Controlled Randomized Trial
DDI	Drug Drug Interactions
DDMAC	Division of Drug Marketing, Advertising, and Communications (in FDA)
DDS	Doctor of Dental Sciences
DEA	Drug Enforcement Agency
DHHS	Department of Health and Human Services
DI	Drug Information or Drug Interaction
DIA	Drug Information Association
DIC	Drug Information Center
Div	Dose by IntraVenous route
DLE	Drug Literature Evaluation
D5LR	Dextrose 5% in Lactated Ringers
DLT	Dose Limiting Toxicity
DM	District Manager (in sales) or Data Management or Document Management
DMC	Data Monitoring Committee
DME	Durable Medical Equipment
DMF	Drug Master File
DMO	Disease Management Organization
DMP	Data Management Plan
D5NS	Dextrose 5% in Normal Saline
DO	Doctor of Osteopathy
DOD	Department of Defense
Dpo	Dose by oral route
DRG	Diagnosis Related Group (numbering system for diseases)
DRR	Drug Regimen Review

Acronyms of the Pharmaceutical Industry

DSI	Division of Scientific Investigation (in FDA)
DSM	Disease State Management or Diagnostic & Statistical Manual (for mental disorders)
DSMC	Data Safety and Monitoring Committee
DTC	Direct-To-Consumer advertising
DTP	Direct-To-Patient advertising
DUE	Drug Use Evaluation
DUR	Drug Utilization Review
DVM	Doctor of Veterinary Medicine
DW	Distilled Water
D5W	Dextrose 5% in Water
DX	Diagnosis
Es (6+6)	Essentials of Leadership
EAC	Estimated Acquisition Cost
EAP	Expanded Access Program
EBM	Evidence-Based Medicine
EC	European Community or European Council or eClinical
ECOG	Eastern Cooperative Oncology Group
eCTD	Electronic Common Technical Document
ED	Emergency Department (of a hospital) or Erectile Dysfunction
ED50	Effective Dose in 50% of animals or subjects
EDC	Electronic Data Capture
EC	Executive Committee or European Commission or Ethics Committee or Enteric Coated
ECHO	Efficacy, Cost, Humanistic, and Outcome model (for evaluations of drug's impact on health care as well as patients)
EDQM	European Directorate for the Quality of Medicines
EENT	Eye, Ears, Nose, and Throat
EEO	Equal Employment Opportunity
EFPIA	European Federation of Pharmaceutical Industries & Associations
EHR	Electronic Health Record
EIR	Establishment Inspection Report (for FDA inspections)
ELA	Establishment License Application (for biological manufacturing)
ELIPS	Electronic Labeling Information Processing System (at FDA using XML system)
E&M	Evaluation and Management (CPT codes for Medicare billing)
E_{max}	Effective MAXimum (concentration of a drug)
EMEA	European Medicines Evaluation Agency (also known as European agency for Evaluation of Medicinal Products)
ELISA	Enzyme Linked Immuno Sorbent Assay
EOP2	End of Phase 2 (meeting with FDA)
EPA	Environmental Protection Agency
EPS	Earnings per Share or Extra-Pyramidal Symptoms
ER	Emergency Room or Extended Release
ERS	Electronic Regulatory Submission
ESRD	End Stage Renal Disease (program of Medicare for kidney failure)
EU	European Union
EVP	Executive Vice-President
E&Y	Ernst and Young, LLC (consulting group)
EWOC	Escalation (in dosing) With Overdose Control
EWP	Expert Working Party (in Europe)
F	Female
Fab	Fragment of an AntiBody
FAQ	Frequently Asked Question
FASB	Financial Accounting Standards Board
FAX	Facsimile (copy sent over the telephone line)

Fc	Fragment of Constant region of an antibody
FDA	Food and Drug Administration
FDAMA	FDA Modernization Act (1997)
FDC Act	Food, Drug and Cosmetics Act (1907)
FDLI	Food Drug Law Institute
FD&C	Food, Drug and Cosmetic act
FDP	Federal Demonstration Project
FFS	Fee For Service
FHX	Family History
FICA	Federal Insurance Corporation of America
FIFO	First In, First Out (method to assess inventory costs)
FIM	First In Man (PK phase 1 studies)
FIPCO	Fully Integrated Pharmaceutical Company
FMEA	Failure Mode and Effects Analysis
FMV	Fair Market Value
FOB	Free on Board (inventory of product at company)
FOI	Freedom of Information
FPFV	First Patient First Visit
FPL	Federal Poverty Level or Final Printed Labeling
FR	Federal Register or Final Report
FTC	Federal Trade Commission
FTE	Full Time Equivalent (one staff person)
F/U	Follow-Up
FY	Fiscal Year
FYI	For Your Information
Fx	Fracture
GAO	General Accounting Office (of U.S. Congress)
G&A	General and Administrative costs
GC	Gas Chromatography (in drug analysis)
GCD	Global Clinical Director
GCP	Good Clinical Practices or Global Clinical Plan
GDP	Gross Domestic Product or Good Document Practices
GDR	Generic (drug) Dispensing Rate
Genotox	Genetic toxicity
GF	Growth Factor
GLP	Good Laboratory Practices
GM	Genetically Modified or General Manager
GMO	Genetically Modified Organism
GMP	Good Manufacturing Practices
GNP	Gross National Product
GPA	Generic drug Pharmaceutical Association
GPO	Group Purchase Organization
GRAS	Generally Regarded As Safe
GRP	Good Regulatory Practices or Good Review Practices
GSK	Glaxo-Smith-Kline, Inc.
GTP	Good Tissue Practices
HAMA	Human Antibody Murine Antibody
HCFA	Health Care Financing Administration
HCP	Health Care Professional
HCPCS	Healthcare Common Procedural Coding System (J-codes, HCFA)
HCU	Health Care Utilization
HE	Health Economics
HED	Human Equivalent Dose

HEDIS	Health plan Employer Data and Information Set
HGF	Hematopoietic Growth Factor
HDMA	Healthcare Distribution Management Association
HHS	Health and Human Services
HIPAA	Health Insurance Portability and Accountability Act
HIV	Human Immune Virus
HMO	Health Maintenance Organization
H_o	Null Hypothesis
HPLC	High Pressure Liquid Chromatography
HR	Human Resources or Home Run
HRQOL	Health Related Quality of Life
HRSA	Health Resources and Services Administration (in CMS)
HSA	Health Savings Account
HTS	High Throughput Screening
HX	History
IB	Investigator's Brochure
IC_{50}	Inhibitory Concentration at 50%
ICC	Interstate Commerce Commission
ICD-9	International Classification of Disease, 9th revision
ICH	International Conference on Harmonization
ICH E3	ICH guidance for final report writing
ICU	Intensive Care Unit (in a hospital)
ICSR	Individual Case Safety Report
ID	IntraDermal injection or IDentification
IDE	Investigational Device Exemption
IEC	Independent or Institutional Ethics Committee
IFN	InterFeroN
Ig	ImmunoGlobulin
IHS	Integrated Health System or Indian Health Service
IIS	Investigator Initiated Study
ILSI	International Life Sciences Institute
IM	IntraMuscular injection or Information Management
IN	IntraNasal
IND	Investigational New Drug application
INC	INCorporated
ILs	InterLeukins
IMS	Institute for Medical Sciences (drug marketing and sales data)
INS	INsurance or Immigration and Naturalization Service
IOM	Institute Of Medicine
IP	IntraPeritoneal or IntraPulmonary or Intellectual Property rights
IPC	In-Process Control (changes in manufacturing)
IPEC	International Pharmaceutical Excipients Council
IPO	Initial Public Offering (for sale of stock)
IPPS	Inpatient Prospective Payment System (for Medicare drugs)
IR	Investor Relations or Immediate Release
IRB	Institutional Review Board
IRS	Internal Revenue Service
IS	Information Systems or Investigative Sites
ISMP	Institute for Safe Medication Practices
ISO 9000	Quality program(s) for processes in a company
IT	Information Technology or IntraThecal injection
ITC	International Trade Commission
ITT	Intent-To-Treat (group of patients in a clinical trial)
IU	International Unit

IV	IntraVenous injection or IntraVentricular injection
IVRS	Interactive Voice Response System
JAMA	Journal of the American Medical Association
JCAHO	Joint Commission for Accreditation of Healthcare Organizations
J-code	Health care (government) procedural coding system numbers for billing
JCPP	Joint Commission of Pharmacy Practitioners
JD	Juris Doctor (lawyer)
J&J	Johnson and Johnson, Inc
JPMA	Japan Pharmaceutical Manufacturers Association
Jx	Juncture
K	Thousands (as in dollars or items)
KOL	Key Opinion Leaders
KPI	Key Performance Indicators
LA	Long Acting (formulation) or Licensing and Acquisitions
LBM	Lean Body Mass
LC	Liquid Chromatography (in drug analysis)
LCM	Life Cycle Management
LD_{50}	Lethal Dose in 50% (of animals in toxicology studies)
LE	Latest Estimate (of sales or expenses or budget, etc)
LOS	Length of Stay (in hospital)
LLC	Limited Liability Corporation
LP	Limited Partnership
LPLV	Last Patient Last Visit
LR	Lactated Ringers (solution) or Legal & Regulatory (Departments)
LRP	Long Range Plan
LSE	Last Subject Enrolled
LTC	Long Term Care
LTCF	Long Term Care Facility
LTCI	Long Term Care Insurance
Ltd	LimiTeD (as in a company business organization in Europe)
LVP	Large Volume Parenterals
M	Male or Million
MA	Medical Affairs or Medicare Advantage (service regions for Medicare Part D drug program) or Marketing Authorization (in European Union)
MAA	Medicines Authorization Application (to European regulatory agency)
M&A	Mergers and Acquisitions
Mab	Monoclonal AntiBody
MBA	Master's in Business Administration
MC	Medical Communications or Masters of Ceremony
MCB	Master Cell Bank
MCO	Managed Care Organization
MCWB	Master Cell Working Bank
MD	Medical Doctor
MDI	Metered Dose Inhaler
MDR	Multiple Drug Resistance
ME	Molecular Engineering
MEC	Minimum Effective Concentration
MedDRA	Medical Dictionary for Regulatory Activities (terms for coding adverse events in reports)
MEDPAC	MEDicaid Payment Advisory Commission
MERS	Model Errors Reporting System
MHLW	Ministry of Health, Labor, and Welfare (in Japan)

MI	Medical Information or MIle
MIC	Minimum Inhibitory Concentration
MIP	Management Incentive Plan (bonus plan)
MM	Millions (as in dollars)
MMA	Medicare prescription drug, improvement and Modernization Act (2003)
MMR	Morbidity and Mortality Reports (of the CDC)
MMS	Medicare & Medicaid Services (U. S. government health agency)
MMT	Medicare Management of Therapy program
MO	Modus Operandi (the methods by which work us done)
MOA	Mechanism of Action
MOS	Medication Outcomes Study
Motif	Structural feature of a product
MOU	Memoranda-Of-Understanding (from FDA for new regulations outside of written guidances or regulations)
MPH	Master's in Public Health
MPK	Metabolism and PharmacoKinetics
MR	Market Research or Medical Record
MRA	Medical Reimbursement Account or Mutual Recognition Agreements (in Europe among member states)
MRI	Magnetic Resonance Imaging
MRFG	Mutual Recognition Facilitation Group (in Europe for EMEA)
MRP	Mutual Recognition Procedure (in EU) or Medication Related Problem
MRSD	Maximum Recommended Starting Dose
MS	Mass Spectrophotometry (in drug analysis) or Modeling & Simulation (pharmacokinetic research in drug development)
MSC	Maximum Safe Concentration
MSA	Medical Savings Account or Metropolitan Statistical Area
MSL	Medical Science Liaison
MSM	Medical Science Manager
MTC	Maximum Tolerated Concentration
MTD	Maximum Tolerated Dose
MTM	Medication Therapy Management
MTMP	Medication Therapy Management Program
MTMS	Medication Therapy Management Services
MUE	Morbidity Utilization Evaluation
MUS	Medication Use Safety
MW	Medical Writing
n	Number (of subjects in a study)
NA	National Accounts or Nucleic Acid or Not Available
NAB	National Advisory Board
NABP	National Association of Boards of Pharmacy
NAFTA	North American Free Trade Agreement
NAMCS	National Ambulatory Medication Care Survey
NARD	National Association of Retail Druggists
NAS	National Academy of Sciences
NBAC	National Bioethics Advisory Commission
NBD	New Business Development
NC	Non-Clinical (data or study)
NCCN	National Comprehensive Cancer Network
NCD	National Coverage Determination (by CMS)
NCE	New Clinical Entity
NCI	National Cancer Institute
NCQA	National Committee for Quality Assurance
NDA	New Drug Application

NDC	National Drug Code
NDTI	National Disease and Therapeutic Index
NEJM	New England Journal of Medicine
NF	National Formulary
NFS	Not For Sale
NH	Nursing Home
NHCS	National Health Care Survey (by CDC)
NHE	National Health Expenditures
NHLBI	National Heart Lung and Blood Institute (NIH)
NHP	Natural Health Products
NIAID	National Institute of Allergy and Infectious Disease (NIH)
NICE	National Institute for Clinical and Economics (United Kingdom)
NIDP	Notice of Initiation of Disqualification Proceedings (from FDA)
NIH	National Institutes of Health
NIOSH	National Institute for Occupational Safety and Health
NL	NormaL
NME	New Molecular Entity
NNT	Number Needed to Treat
NOAEL	NO Adverse Effect Level (for dosing in animal studies)
NOV	Notice of Violation (letter from the FDA DDMAC)
NP	Nurse Practitioner
NPC	National Pharmaceutical Council
NPI	National Provider Identifier (standardized number for all providers from CMS)
NPS	National Provider System (CMS system for NPIs)
NPSF	National Patient Safety Foundation
NPV	Net Present Value
NQF	National Quality Forum
NRC	Nuclear Regulatory Commission
NS	Normal Saline
NSC#	Number from National Service Center of NCI for cancer chemotherapy products
NSABP	National Surgical Adjuvant Breast and Bowel Project
NSF	National Science Foundation
OAI	Official Action Indicated (FDA letter to company)
OBRA	Omnibus Budget Reconciliation Act (1990)
ODA	Orphan Drug Act (1983)
ODAC	Oncology Drug Advisory Committee (for FDA)
ODE	Office of Drug Evaluation (in FDA)
ODS	Office of Drug Safety (in FDA)
OECD	Office for Economic Co-operation and Development
OER	Office of Extramural Research
OGD	Office of Generic Drugs (in FDA)
OHRP	Office of Human Research Protections (in CMS)
OIG	Office of Inspector General (in Justice Department)
Oligos	Oligonucleotides
OMB	Office of Management and Budget (in U.S. Congress)
OOP	Out-Of-Pocket (expenses paid by patients outside of Medicare coverage)
OPD	Out-Patient Departments (of hospitals)
OPL	OPinion Leader
OPPS	Outpatient Prospective Payment System (for Medicare drugs)
OPs	OPerations or OutPatients
OPT	Office of Pediatric Therapeutics (in FDA)
OR	Operating Room or Odds Ratio
ORA	Office of Regulatory Affairs (in FDA)
OS	OutSourcing

OSR	Office of Sponsored Research
OTA	Office of Technology Assessment (in U.S. government)
OTC	Over-The-Counter
ORT	Office of Radiological Therapeutics (in FDA)
P	Probability (of an occurrence of an event in statistics)
3Ps	Patients, Providers, Payers
4Ps	Product, Place, Price, Promotion (in marketing products)
8Ps	People, Pipeline, Processes, Profits, Principles, Performance, Portfolio, Products (drug development needs and framework)
P.1	Phase 1 clinical trial
P.2	Phase 2 clinical trial
P.3	Phase 3 pivotal clinical trial
P.4	Phase 4 postmarketing clinical trials
PA	Physician's Assistant or Prior Authorization
PAI	PreApproval Inspection
PAP	Prior Authorization Program or Patient Assistance Programs
P&G	Proctor and Gamble company
PAS	Prior Approval Supplement
PBM	Pharmacy Benefits Management (organization)
PBPC	Peripheral Blood Progenitor Cell
PC	Placebo Control or PhysicoChemical or Pre-Clinical
PCC	Poison Control Center
PCT	Placebo-Controlled Trial
PD	PharmacoDynamics or Progressive Disease
PDA	Parenteral Drug Association or Personal Digital Assistant
PDF	Portable Document Format
PdIT	Pediatric Implementation Team (FDA cross-functional team in pediatrics evaluating PPSRs and WRs)
PDL	Preferred Drug List (for Medicare, Medicaid, or Insurers)
PDP	Prescription Drug Plan (part of Medicare drug plan, Part D)
PDRM	Preventable Drug Related Morbidities
PDUFA	Prescription Drug User Fee Act I (1992), II (1997), and III (2002)
PE	PharmacoEconomics or PharmacoEpidemiology or Physical Exam or Process Engineering
PE ratio	Profits to Earnings ratio
PEG	PEGylation or PolyEthylene Glycol
PG	PharmacoGenomics or PharmacoGenetics
PGDE	Pharmacologically Guided Dose Escalation
pH	Hydrogen concentration measurement (acidity of a solution)
PHRP	Partnership for Human Research Protection
PHC	PHarmaCeutics
PHI	Public Health Insurance
PhRMA	Pharmaceutical Research and Manufacturers Association
PHS	Public Health Service
PI	Package Insert or Principal Investigator
PIM	Exchange of Product Information (EMEA requirement for labeling System, based on XML)
PIP	Pediatric Investigational Plan (FDA requirement)
PIPE	Private Investment in Public Equity
PK	PharmacoKinetics or Protein Kinases
P&L	Profits and Loss
PLA	Product License Application
PM	Project Management or Portfolio Management
PMC	Post-Marketing Commitments (to FDA for phase 4 trials or PMS)
PMPM	Per Member Per Month
PMS	Post-Marketing Surveillance (for adverse event documentation)

PO	Per Os (Oral administration) or Purchase Order
POA	Plan Of Action
POC	Point-Of-Care
POS	Point Of Service or Probability Of Success
PP	Product Portfolio or Product Plans or Public Policy or Project Planning
PPI	Patient Package Insert or Producer Price Index
PPM	Per Patient per Month or Portfolio Planning Management
PPO	Preferred Provider Organization
PPPM	Prescriptions per Patient Per Month
PPS	Prospective Payment System (for Medicare drugs)
PPSR	Proposed Pediatric Study Request (to FDA)
PR	Public Relations
PRC	Protocol Review Committee
PREA	Pediatric Research Equity Act (2003)
PRN	Pro Re Nata (as needed, in prescription labels)
PRO	Patient Reported Outcomes
PS	Professional Services or Post Script
PSR	Professional Sales Representative
PSUR	Periodic Safety Update Report (for FDA or EMEA)
P&T	Pharmacy and Therapeutics Committee
PT	Part-Time or PharmacoTherapy or Physical Therapy
PTO	Patent and Trademark Office (in U.S.)
PUMA	Pediatric Use Marketing Authorization (in Europe)
PVG	PharmacoViGilance (for adverse event reporting)
PWG	Protocol Working Group
QA	Quality Assurance
QALY	Quality Adjusted Life-Year
QC	Quality Control
QD	Daily (once per day treatment)
QID	Four times a day treatment
QIO	Quality Improvement Organizations (related to Medicare, part D)
QOD	Every other day treatment
QOL	Quality Of Life
QoL	Quality of Life
QPQ	Quid Pro Quo
QS	Quantity Sufficient
QSAR	Quantitative Structure Activity Relationship
QTC	Quantitative Total Concentration
QWBA	Quantitative Whole Body Autoradiography
®	Registered name for a product
r^2	coefficient of determination (in statistics, amount of variation)
RA	Regulatory Affairs or Regulatory Authorities
RAC	Recombinant DNA Advisory Committee
RAPS	Regulatory Affairs Professional Society
RBRVS	Resource-Based Relative Value Scale (for payments for health care)
RBZ	RiBoZyme
RCA	Root Cause Analysis
RCT	Randomized Controlled Trials
R&D	Research and Development
RDC	Remote Data Capture
RDI	Relative Dose Intensity
RFA	Request For Applications
RFID	Radio Frequency IDentification (packaging of product, allowing tracking)

RFP	Request for Proposals
RIA	RadioImmunoAssay
RM	Regional Manager (in sales) or Risk Management (regarding adverse drug reactions or decisions)
RMP	Risk Management Program or Practices
RN	Registered Nurse
RO1	basic Research grant at NIH
RO3	small Research grant at NIH
ROI	Return on Investment
ROV	Real Option Value
ROW	Rest of World
RPH	Registered PHarmacist
RR	Recovery Room or Respiratory Rate or Relative Risk or Rest and Relaxation or Response Rate
RRR	Relative Risk Reduction
RT	Randomized Trial or Radiation Therapy
RTF	Refuse To File (FDA refuses to file NDA due to a deficiency)
RUC	Resource Utilization Committee (of AMA for Medicare programs)
Rx	Prescription or Pharmacist
S	Single
SAE	Serious Adverse Event
SAF	Standard Analytical Files (data from Medicare)
SAG	Scientific Advisory Groups (in Europe for EMEA)
SAL	Scientific Affairs Liaison
SAR	Structure Activity Relationship
SBIR	Small Business Incentive Research (grant or loan from U.S. government)
SBL	Small Business Loan
SC	SubCutaneous
SD	Stable Disease or Standard Deviation
SDTM	Study Data Tabulation Model
S&M	Sales and Marketing
SE	Side Effects or Surrogate End points or Standard Error
SEM	Standard Error of the Mean
SEER	Surveillance, Epidemiology, and End Results (database of NCIinNIH)
SEC	Security and Exchange Commission
SF-36	Short Form questionnaire of 36 questions (to assess general psychiatric and health well being of patients in many disease areas)
SFAS	Statement of Financial Accounting Standards (in company reports)
SG&A	Selling, General, & Administrative (costs of operations)
SHX	Social History
SIAC	Special Interest Area Community (of DIA)
SIP	Sickness Impact Profile
SKU	Sales configuration Unit or Shelf Keeping Unit
SL	SubLingual
SMART	Specific, Measurable, Accurate, Reasonable, Timely (Objectives)
SMDA	Safe Medical Devices Act (1990)
SMO	Site Management Organization (clinical research) or Sales Management Organization
SN	SigNs (of disease in a patient)
SNDA	Supplemental New Drug Application
SNF	Skilled Nursing Facility
SNP	Single Nucleotide Polymorphism
SO	Safety Officer
SOAP	Symptoms, Objective findings, Assessment, Plan (for medical chart notes)
SOP	Standard Operating Procedure
S&P 500	Standards & Poors 500 (top companies in stock exchange)
SPC	Supplementary Patent Certificate (in Europe)

SPL	Structured Product Labeling (FDA requirement for NDA Labeling, using XML system)
SR	Sustained Release or Statistical Report
SSA	Social Security Administration or Act
SSN	Social Security Number
ST	Stem Cell
Svc	SerViCe
SVP	Senior Vice-President or Small Volume Parenterals
SWFI	Sterile Water for Injection
SWOG	SouthWest Oncology Group
SWOT	Strength, Weakness, Opportunity, Threat analysis
SX	Surgery or Symptoms
TA	Therapeutic Area
T 1/2	Half-life
TBD	To Be Determined
TE	Time and Event (schedule) or Trial End points
T&E	Travel and Entertainment (expenses)
Telecom	Telephone Communications
TI	Therapeutic Index
TID	Three times a day treatment
TL	Thought Leader
TM	Trademark (for a product name)
™	TradeMark (for a product name)
Tmax	Time to MAXimum concentration
Tox	Toxicology studies in animals
TPD	Therapeutic Products Directorate (Canadian health agency)
TPP	Target Product Profile or Therapeutic Product Programme (in Canada)
TX	Treatment or Therapy
UHC	University Health-Systems Consortia
ULN	Under Lower limits of Normal
UN	United Nations (international organization)
UR	Utilization Review
USAN	United States Approved Name
USDA	U. S. Department of Agriculture
USP	United States Pharmacopeia
USPTO	United States Patent and Trademark Office
USRDS	United States Registry of Disease Statistics
UV	UltraViolet radiation
VA	Veterans Administration
VAI	Voluntary Action Indicated (FDA letter asking for compliance)
VC	Venture Capital (for investment in a company)
Vd	Volume of Distribution
VHA	Volunteer Hospitals Association of America or Veterans Health Administration
VP	Vice-president
W	White
WAC	Wholesale Acquisition Cost
WFI	Water for Injection
WHI	Women's Health Initiative
WHO	World Health Organization
WL	Warning Letter (from the FDA DDMAC)
WNL	Within Normal Limits
WR	Written Request (from FDA)

WW	World Wide
WWW	World Wide Web
X	Average (in statistics)
χ^2	Chi-squared (statistical test)
XL	eXtended reLease (Formulation) or eXtra Luxury
XML	Extensible Markup Language
XO	cross-over (in study design)
XR	eXtended Release
356h	FDA form for Application to market a new drug, biologic, or an antibiotic drug for human use
1571	FDA form for cover sheet for an IND
1572	FDA form 1572 for principal investigators in clinical trials
3397	FDA form for user fees
3500	FDA form for safety information and adverse experience reporting (MedWatch)

Appendix 2
Glossary of Pharmaceutical Industry Terms

Accelerated approval	FDA approval in a shorter time frame for new drugs that provide meaningful therapeutic benefit over existing treatment for serious or life-threatening illnesses, using surrogate end point or restricted use provisions.
Action plans	The specific activities in a marketing plan by the sales or other staff performed in the sales and marketing of a product, including time frame and persons responsible.
Addressable population	Group of patients or institutions that would likely be treated by a product.
Ad hoc letters	Product letters written *de novo* for new questions.
Advertising	Promotion of a product by a company including all visual materials, e.g., printed documents in press (journals, newspapers), printed sales pieces, consumer ads (television, radio, Internet).
Advocacy group	A group of people that represent in the medical arena a disease and promote its cure and support of the patients, including soliciting funds, educating the public, and lobbying efforts.
Alternative medicines	Products that are not drugs or biological products and are used to treat diseases, e.g., vitamins, minerals, and natural products from the environment.
Analyst	A person employed by an investment company who assesses a company for its value to the public; a person at a company who usually works in market research or information technology assessing data and information.
Asset	Cash, marketable securities, receivables, property, inventory, and equipment of a company.
Audience	Group of people or institutions targeted by a program or marketing.
Audit	Evaluation of a product, system, application, finances, or company by an outside group, often a regulatory agency, including usually a visit to establish compliance with SOPs or plans.
Best-in-class	The product approved for use in a specific disease, therapeutic area, or pharmacologic class that is the best based on its usage, highest percentage among competitors.
Biologicals	Antibodies, blood products, enzymes, gene therapies, growth factors, oligonucleotides, peptides, proteins (other), tissues, and vaccines.
Biomarker	Quantifiable physiological or biochemical marker that is sensitive to intervention (drug treatment). Biomarker might or might not be relevant for monitoring clinical outcome, usually used in early drug development.
Biopharmaceutical	Term used as an alternative to biological.
Biostatistics	Group at a company responsible for the information provided that designs, creates, and reports the statistical results of clinical trials.
Blockbuster	A product that sells $1 billion per year.
Bonus	Cash paid to a person or company for achieving a target in their objectives, usually annual payout.
Brand	A product with its official name or trade name that is registered and associated with a particular company, therapeutic area, and type of product within its field, along with the marketplace in which it is used and sold.

Brand awareness	Recall by a customer of the product and its main use or benefit.
Bridging study	In drug development, when the formulation changes, a study is done to demonstrate that the drug's effects are the same with new and old formulations.
Business case	A plan and rationale for a new business venture.
Business plan	The plans for the marketing and sales of a product including and integrating strategy, goals, rationale, and action plans, along with budgets, staffing, programs, and potential revenues.
Call center	Telephone service wherein calls are centralized to a company or office.
Capitalization	Value of a company in the market (stock value), effected by profitability (productivity, expenses, price of products, assets).
Category X drug	Drug with teratogenic effects in animal or human studies.
Causality	Judgment of adverse events as produced or not by a product.
CenterWatch	A company that tracks performance and writes reports about the pharmaceutical industry.
Certificate of medicinal product	In Europe, a confirmation of marketing authorization and GMP being followed by EU companies for export of products especially to developing countries.
Champion, Product	A person with expertise that strongly supports use of a product.
Change control	Part of any process that details how changes are to be made.
Clinical data interchange	Group to develop data standards to streamline access, use, and analyses.
Clinical hold	FDA stops temporarily a clinical trial during its conduct prior to product approval because of some problem, requesting the sponsor to respond to the problem.
Clinical safety	Group at a company dealing with adverse event reporting in patients from trials or postmarketing.
Clinical supply	The drug product used for clinical trials during investigational trials.
Clinical Trials Application	In Europe, this application requests approval to conduct clinical trials after preclinical work (equals IND).
Close	The final part in time or action of a sales call by a sales person with a customer, wherein they make the hopeful arrangement or commitment to prescribe or buy a product.
Closed formulary	List of drugs for patient use by patients in a health care system, wherein no other drug can be used, requiring only drugs on the list to be prescribed.
Coding	Designation of an effect or adverse event based on structured dictionary.
Comarketing	Agreement between two companies to jointly market and sell products.
Common Procedure Terminology codes	Abbreviations and numeric codes for disease categorization for billing activities.
Common technical document	The technical information for filing a new drug application that has been harmonized for worldwide use.
Compassionate use	Permission for patient to use product prior to approval or at no cost because of their medical need and poor financial status.
Competitive intelligence	Information collected and evaluated by one company about the products and actions of another company who markets products in the same arena.
Complaint, Product	A problem with a product reported by a patient or provider to a company that involves its integrity or its formulation.
Comp plan	Compensation plan to a group, including, e.g., salary, bonus, benefits, and awards.
Complement determining region	Area on monoclonal antibodies that binds to a specific site on cell surface that is an antigen.
Conjoint analysis	Market research wherein products are compared and customers or providers are surveyed.
Consolidation	A merger between two companies (or an acquisition of one by another) within an industry.
Consent form	A form requesting a patient's agreement to participate in a study.
Copyright clearance	Requests to a publisher or author for use of their materials.
Consolidated Omnibus Budget Reconciliation Act (1985)	Health care financing legislation by federal government.
Contract	Legal agreement between two parties for services or products.
Cost–benefit	Ratio of cost to benefits in healthcare with both expressed in monetary terms.
Cost-effectiveness	Ratio of cost to benefits in healthcare with cost expressed per unit of change in a measure, e.g., mm Hg of blood pressure, or in life years saved.

Cost-minimization	Drug Costs of items compared side-by-side.
Cost–utility	Ratio of cost to benefits in healthcare with benefits expressed in quality adjusted life years saved.
Culture	Style of work, including for example the dress, communication styles, approachability of senior staff, meetings (attendance, structure, and conduct), and governance.
Cure rate	A term in oncology studies and therapy that indicates the complete response (all the tumor's physical and symptoms being resolved 100%) and/or partial response (predefined major improvement in the cancer) over a specified time period observed until return of the cancer.
Customer	The person or institution who purchases, dispenses, administers, uses, or consumes a product.
Cytochrome P450	Isoenzymes in liver for metabolism of drugs.
Data monitoring committee	Independent (from a company) group of scientists and individuals with the responsibility to oversee the data collected in a clinical trial to determine safety and efficacy to stop a trial early as necessary.
Detailing	The sales process for a product by a sales person with customers.
Discount card	A card used in health care systems to help purchase prescription drugs by patients at a reduced cost.
Discovery	Basic research that identifies a drug targets and product a hits or leads in the laboratory.
Dividend	Payment to stockholder for profitability of a company in a year.
Direct-To-Consumer	Advertising by a company to patients or consumers directly via (Direct-To-Patient) media, mail, TV, or other means.
Duel eligible	Indigent patients who are eligible for payment for medical coverage by both the federal Medicare and state Medicaid systems.
Drug development	Clinical research plans (goals, budgets, staffing, and deadlines), studies, outcomes, and regulatory applications for a drug or biological.
Drug interaction	Two or more drugs being used and resulting in some interaction between them and some added beneficial or adverse outcome.
E-mail	Communication of memos over the Internet (electronic mail).
End points	The specific disease parameters being studied in a drug study, particularly focused on the end result being evaluated.
Effectiveness	Overall benefit a product produces including efficacy, QOL, and cost.
Efficacy	A product's degree of activity in producing a desired effect.
Engineering, Package	Group responsible for the design and creation of a product's outer container system, boxes or bottles, etc.
Engineering, Process	Any development or updating of any process at a company, especially in manufacturing.
Ethical company	Pharmaceutical company that develops products (R&D + S&M + L&R).
Expert working party	Group of experts in Europe for various therapeutics or regulatory categories, e.g., biotechnology, pediatrics, vaccines.
Expectedness	Character of an adverse event related to previously occurring and documented vs. new unexpected events (information found in the labeling or investigator's brochure).
Extended release	Formulation extending the time frame for product delivery.
Fast-track approval	FDA approval in shorter time frame (6 months or less) to expedite development and review of drugs/biologics intended to treat serious or life-threatening conditions and with potential to address unmet medical needs.
Filing	Submission of materials to a regulatory authority.
First-in-class	The first product approved for use by regulatory authorities in a specific class of compounds, usually based on its pharmacology or therapeutic use.
Forecasting	Evaluation of the potential sales by finance or progress in research by planners.
Formulary	List of products approved for use in an institution.
Formulation	The final form of a product, e.g., tablet, injection, liquid, lyophilized.
Fulfillment	Providing materials or a product from a company to a customer.
Gatekeeper	Persons or institutions in the health care system that are involved in decisions for access to and use of a product.
Global	International focus for a plan or organization.

Glossary of Pharmaceutical Industry Terms

Generic	Equivalency of a product in its structure and actions.
Grants	Financial support for research or educational projects from a company to a customer.
Half-life	Time for one-half of a product to be eliminated from the body.
Hatch–Waxman act	Drug Price Competition and Patent Restoration Act of 1984 (Public Law 98-417) for generic drug approvals.
Hit	A compound being screened in an early phase of drug discovery process that demonstrates some activity in the pharmacology test system.
Harmonization (ICH)	International effort to streamline drug development through standardized procedures and records.
Honorarium	Payment to a person for services rendered, usually educational or advisory role.
Hotline	Telephone system (toll free) for patient inquiries to a company.
Immunotoxicity	Toxicity studies evaluating immunologic effects of a product that are adverse events.
Information management	The systems, processes, and equipment at a company for all information coordination, processing, storage, and use.
Inquiry	A search for information by a person or agency.
Incidence	Frequency of occurrence of an event in a study or over a certain time.
Informed consent	Permission given by a patient to be a subject in a clinical trial wherein they are given full information about the study, e.g., treatments, alternatives, adverse events, and benefits.
In-licensing	Acquisition of a product from outside of a company.
Inspection (FDA) classes	FDA findings after an inspection of the company or its vendors or investigators; NAI, no action indicated; VAI, voluntary action indicated; OAI, official action indicated.
Interactive voice response system	Assignment of patients to drug treatments for a study by an automated telephone system.
Investigator's brochure	Company's extensive document that summarizes a drug's safety, efficacy, and all the research, based on the company's studies and published literature, and is provided to all investigators studying the company's drug.
Label	Approved information in the product package insert and on the package.
Launch	Process and activities of initial marketing and sales and support services for a product associated with its regulatory approval, usually the 2 years before and 2 years after approval.
Lead	A compound in the discovery phase of research that demonstrates activity in a biological system or animal.
Lead optimization	First compound in a biologic area or new pharmacologic category in which its characteristics are being improved.
The Leapfrog Group	Coalition of companies to coordinate efforts in health care coverage for their employees.
License	Approval document to market a product from a regulatory authority.
Licensing and acquisitions	Product acquisition from a source outside the company.
Life cycle	Phases of product evolution from research to clinical research to launch to marketing (growth phase) to marketing (mature) to generic.
Market	A market can be a disease area, a therapeutic class, or a pharmacologic class, wherein products are used and purchased and collectively includes all the products, people involved, and health care systems, as well as all related activities.
Market cap	Dollar value of a company based on number of stock shares times the stock price.
Marketing authorization application	The application in Europe to the EMEA for marketing approval.
Medical communications	Information and/or education about a drug and its use being provided by a company to external or internal customers.
Medical marketing	Educational materials as part of marketing materials for a product.
Medicare	U.S. government payment for healthcare, primarily in citizens at or over 65 years old.
Medicare, Part A	Medicare coverage primarily for hospital payments, also some nursing care, home care, and hospice.

Medicare, Part B	Medicare coverage for primarily physician services and other outpatient service providers; some drug coverage, such as oral cancer drugs, epoeitin for ESRD, and beta-interferon for multiple sclerosis.
Medicare, Part C	Medicare managed care programs.
Medicare, Part D	Medicare prescription drug benefit.
Market research	Science of assessment by various methods of product performance in a market or the market itself (disease or products).
MedWatch Form 5400	Official standard form to record adverse events and report to FDA.
Metabolism	Studies of a drug's elimination from the body primarily via liver.
Niche market	A specific disease or indication for a product that is narrow in scope with commensurately small market size (number of patients and possibly sales potential).
Off-label	Use of a product not included in the product labeling (nonapproved indication).
On-label	Use of a product that is approved by a regulatory authority.
Open formulary	List of drugs to be used for patients in a health care system, permitting any drug approved for use to be prescribed.
Options	Stock shares given to employees as a bonus that need to be purchased.
Orange Book	FDA-approved drug products with therapeutic and bioequivalence evaluations.
Orphan drugs	FDA program to promote the development of products that demonstrate promise for the diagnosis and/or treatment of rare disease, defined as prevalence in the United States of less than 200,000 cases.
Out-of-pocket	Expenses (payments) by patients outside of Medicare coverage for health care costs.
Out-licensing	Process for a product being sold to another company.
Out-sourcing	Use of service companies by a company that can perform various support activities, rather than adding full-time staff, e.g., clinical research, market research, regulatory affairs, drug distribution.
Package insert	The product information approved by the FDA and accompanying the product with its packaging (pharmacology, clinical data, indication, dose, warnings, precautions, adverse events, formulation, administration).
Patent	Office of patents certification that a discovery is a unique product or process.
Patient package insert	The product information designed for the patient to guide in its use and for understanding of the product's use, benefits, and risks. It accompanies the product, provided in the box or by the pharmacist.
Payor	Individual, group, or company that pays for products.
Pegylation	Attachment of polyethylene glycol molecule to a drug or biological product to alter its pharmacokinetics, usually extending its half-life.
Peri-Approval	Around the time of a product approval (1–2 years before and after)
Pharmaceutical Executive	Journal for pharmaceutical industry, especially for general industry and management issues.
Pharmaceutics	The discipline that creates formulations for products and studies their physical and chemical properties, as well as stability studies.
Pharmacoeconomics	The study of net economic impact of selection and use of pharmaceuticals on total cost of delivering health care; a more global perspective balancing more complete benefits and costs; health care utilization (efficiency) with a focus on "value," defined as a benefit for money spent.
Pharmacogenomic	Study of genes and gene fragments for their actions in disease progression and mitigation.
Pharmacokinetics	The study of the absorption, distribution, blood or tissue levels, metabolism, and elimination of drugs, biologicals, and chemicals.
Pharmacology	Mechanisms of action and drug effects studies.
Pharmacotherapy	The use of drugs and biologicals to mitigate disease pathology and/or its signs and symptoms.
Pharmacovigilance	The systems and processes to identify, evaluate, and report on adverse events with a company products.
Phase 1	Clinical study for safety and pharmacology for first use in normal humans.

Phase 2	Clinical study in patients with target disease.
Phase 2a	Early study in patients with target disease, safety and drug effect.
Phase 2b	Later study in patients with target disease with dosing determination and some efficacy.
Phase 3	Clinical study (large) to establish efficacy and safety for product registration (pivotal trial).
Phase 3b	Post–phase 3 study for expanded safety, dosing, or subpopulations.
Phase 4	Clinical study after product approval for marketing within labeling.
Pipeline	Products in research at a company.
Planning, Operational	The work guidelines to execute a promotion plan, including specific activities and responsibilities.
Planning, Strategic	The rationale and overarching ideas in the plans of product promotion or research.
Planning, Tactical	The specific activities in a marketing plan to be done to promote a product.
Platform	In discovery (basic research), a technology for identifying, validating, or characterizing a molecule or product candidate, such as monoclonal antibodies, x-ray crystallography, polymerase chain reaction, RNA blockade, or informatics.
Polymorphism	A single mutation in a gene at one nucleotide locus that potentially changes gene expression with a modified protein that may possess different properties.
Portfolio of products	List of different products that a company possesses.
Portfolio management	Coordination and leadership of the planning, timing, budgeting, and especially prioritization for all the molecules in the R&D effort at a company.
Potency	Amount by weight of a product to produce an effect.
Postmarketing surveillance	The various studies and research performed to evaluate a product after it has been approved for use and marketed, especially for adverse events.
Preclinical	Studies in animals or *in vitro* prior to human studies.
Premium	Gift to a customer, e.g., book, pen, globe, trip, as part of product promotion.
Prescription drug card	A government or health care system that provides a card to patients for drug purchases by prescription.
Prevalence	Frequency of an event in a population at one point in time.
Product candidate	A drug or biological product that has advanced through research and early clinical development successfully and moves on to become a likely marketable, safe, and effective product but it is still in early clinical research.
Product complaint	Problem with a final product, e.g., formulation or container.
Product development	Clinical research plans and outcomes for a product.
Product profile	Description of a product's properties.
Project management	Coordination and leadership of a team of cross-departmental people to manage the progress of molecules as they advance from research to launch, including budget, staffing, and deadlines.
Profitability	Net financial value of a company, sales, income, assets, and productivity minus costs including staffing, expenses, and losses.
Promotion	Activities by salespeople in the sales process with customers, using advertising or educational materials.
Provider	Physician, pharmacist, nurse, or other caregiver for patients.
Quality assurance	Company process wherein systems, processes, procedures, staffing, and actual outcomes are assessed to correct, maintain, or improve quality.
Quality control	In-process and final testing to establish and assess conformance of products and processes to specifications.
Query	A check into or of data, processes, or systems to assess accuracy.
Quid pro quo	Giving something of equal value back in a transaction.
Recall	The return of a product to a company because of some defect.
Reconciliation	Resolution of an error or missing information.
Records retention	Time frame and methods for storage and retaining documents.
Registry, Patient	A program of listing patients with a common disease or treatment for usual purposes of research subjects for studies, for adverse event reporting, or for indigent patients needing product support.

Regulatory affairs	Group at a company responsible as the official interface with regulatory agencies that approve products and materials.
Reimbursement	Payment for a product by patients or payers.
Reproductive tox	Toxicology studies for pregnancy, teratogenicity, and fertility.
Revenue	Income to a company including sales, royalties, interest, and other payments, e.g., milestone payments for research.
Risk management	The effort, processes, systems, and people that attempt to reduce the risk of problems or holdups in product development in any department at companies.
Rolling submissions	FDA drug/biologic approvals that allow for incremental submission of reviewable modules of an NDA/BLA to reduce overall FDA review time.
Safe harbor	In contracting between customers and a company and in educational programming by a company, federal drug laws and regulations need to be followed especially with FDAMA, which creates areas for collaboration that are safe from negative legal action.
Safety	Adverse event profile of a product.
Sales call	An appointment by a salesperson with a customer.
Sales reps	Salesperson.
Serious	With adverse events, a drug reaction that causes hospitalization, death, permanent disability, cancer, a life-threatening event, or an overdose.
Share-of-voice	The percentage of a market (product usage or sales) in a pharmacologic or therapeutic area that one product possesses.
Shelf keeping unit	Area for a product on the shelf of a pharmacy.
Signal	A repeated occurrence of an adverse event suggesting a rare event being detected; or a biological marker that can occur early and repeatedly in the course of a disease, suggesting a product action, such as toxicity, efficacy, or pharmacogenomic change.
Site	Institution or office where a study is being done.
Site visit	A visit to a location where a study is being done.
Six Sigma quality	Process and system in all industries to ascertain and improve the quality (to the highest level) of the targeted group or process at a company.
Slim Jim	An advertising piece for a salesperson that has a small-enough size to fit in a pocket and includes the features and benefits of a product.
Sponsor	A company or group that is supporting (financially) and guiding a project, study, or an NDA/CTD.
Stability indicating assay	An analytical test of the drug product that can differentiate between the active drug and metabolites.
Stability testing	Assessment of a product's integrity, that is, sustaining concentration, lack of degradation.
Standard letter	Medical information letters that are used frequently.
Stockholder	Person or institution that owns shares of stock in a company.
Stock options	Shares of stock given to individuals based on performance that need to be purchased at a prior lower cost but may have a higher value at a later date.
Structured product labeling	FDA requirement for NDA labeling, using XML software system.
Study coordinator	Person who is responsible for all the activities and records of a study at a site or group of sites.
Subinvestigator	An additional investigator at one site in a multicenter clinical study.
Supply chain	The companies and groups involved in distribution of a product, including at least manufacturer, wholesaler, retailer, and consumer.
Surrogate end points	Study parameters for a disease under investigation that are not end results of the disease but are representative of disease changes and drug action.
Sustained release	Delivery of a product gradually over time.
Takeaway	The core message or information in advertising, sales, or any discussion that you want the person to remember.
Therapeutic window	Regarding dosing of pharmaceuticals, the difference between the effective dose and the toxic dose.

Tiered formulary	A drug list subdivided into levels with different criteria and co-pays by the patient for patient access to the drugs.
Title XVIII	Social Security Act, 1965.
Title XIX	Medicare legislation.
Tufts Center for Drug Development	A university-based group at Tufts in Boston that tracks, evaluates, and reports on industry activities, including R&D and S&M.
Tumor rounds	The review of terminal cancer cases by the faculty and staff at a teaching medical center.
Validation	Formal assessment of a process or system to establish its reliability of its desired performance, including an analytical test, plan, protocol, and data collected.
Value	Relative benefit of a product or service to a person, group, or company, based on its activity, cost, and problems.
Value chain	A series of companies, or services, or groups, or individuals that work in a serial manner, create an outcome, and possess value to an organization; it may be, for example, a sales process for a product, manufacturing a product, or developing a product in R&D.
Vector	A carrier for a drug or genetic material from one organism to another one.
Vendor	Group, company, or individual that provides a service or product to another company.
Venture capital	Financial support from an individual or company to a new company for a new endeavor by them.
Web site	The Internet site of a company or person wherein they can be contacted or other services provided via computer systems.
Weeds and seeds	Pharmacognosy, the pharmacy discipline for botanical pharmaceuticals.
Written request	FDA document requesting specific studies to be done by a company for pediatric approval.
XML	Extensible Markup Language: Automated statistical reporting language for reports, data transfer, and archiving.
1571 FDA Form	FDA form for adverse event reporting.
1572 FDA Form	FDA form for the principal investigator in a study to complete.

Index

A

Abbreviated New Drug Applications, 160, 161, 164, 165, 167, 168, 175
Abelcet®, as amphotericin B drug, 204
Absorption, distribution, metabolism, and excretion studies, 21, 96, 101–103, 108, 123–125, 128, 132, 133, 135, 136, 139, 140, 146, 182, 183, 204, 207, 211, 268
Academy of Managed Care Pharmacy (AMCP), 260
Accelerated approval, regulatory development strategies, 158
Acquired immunodeficiency syndrome, 86–87
Active comparator trials, 110
Active Pharmaceutical Ingredient, 205, 206, 209, 210, 216, 217, 219, 220
Acute coronary syndrome (ACS), 283
Acute lymphocytic leukemia (ALL), 117, 118
Acute therapy trials, 284, 285
ADME. *See* Absorption, distribution, metabolism, and excretion studies
ADME areas as development issues, 135, 136
Adult Treatment Panel (ATP), 281
Advair®, 8, 71
Adverse Event Reporting System, 251
Adverse events, 48, 53, 58, 135, 183, 248
Advertising, product review, 167, 168
Advisory committee meetings in FDA, 153
AERS. *See* Adverse Event Reporting System
AEs. *See* Adverse events
African-American Heart-Failure Trial (A-HeFT), 286, 287
Agency for Health Care Research and Quality, 154
AHRQ. *See* Agency for Health Care Research and Quality
AIDS. *See* Acquired immunodeficiency syndrome
Allegra®, as terfenadine, 204
Allometric scaling, 123, 124, 134, 136, 137
Alternative formulation evaluation, 138, 139
Ambisome®, as amphotericin B drug, 203, 204
American Heart Association, 11, 276, 281, 283
American Heart Association's Advanced Cardiac Life Support (ACLS), 281–283
American Society of Clinical Oncology (ASCO), 11, 248
American Society of Health-System Pharmacists, 258
Amgen Inc., 28, 29
Analytical development methods in drug manufacturing, 214, 215

ANDA. *See* Abbreviated New Drug Applications
ANDA elements, 165
Angiotensin converting enzyme (ACE), 130, 276–282, 284–289
Angiotensin II receptor blockers (ARBs), 276, 277, 279, 284, 289
Antihypertensive and Lipid Lowering Treatment to Prevent Heart Attack Trial (ALLHAT) study, 282
API. *See* Active Pharmaceutical Ingredient
Aranesp®, 10
Area under the concentration-time curve (AUC), 125–128, 126, 127, 134, 138–144
Arimidex®, 67, 309
Aromatase inhibitor, 67
Aspirin, 277, 279, 280
ASSENT-2 trial, 284
Asthma Quality of Life Questionnaire (AQLQ), 116
Asthma treatment, drug usage, 203
Atorvastatin, 275, 278, 280, 281, 283, 286, 289
Atrial fibrillation (AF) development, 282
Atrial Fibrillation Suppression Trial (AFIST), 282
Azidothymidine, for reverse transcriptase, 87
AZT. *See* Azidothymidine

B

Beta-Blocker Evaluation of Survival Trial (BEST) study, 286, 287
Beta-interferons, 10
Bioanalytical method development, 132
Bioavailability (BA), 124, 126, 138–140, 146, 207, 208
Bioavailability and bioequivalence (BA/BE) studies, 138, 139, 146
Biologics License Applications (BLA), 21, 24, 34, 35, 40, 49, 67, 77, 117, 147, 157, 160, 161, 163–170, 175, 176, 192, 243, 270
Biological active ingredients in drugs, 217
Biologics, 96
Biomarker development, 145
Biopharmaceutics Classification System (BCS), 124
Biostatistics group duties, 186
Biotech products manufacturing, 217
Biotechnology industry
  cancer and, 10
  financing, 27
  pharma industry and, 29
  top companies, 12

Index

Biotechnology Industry Organization (BIO), 66, 352
BLA. *See* Biologic License Applications
Black box warning, 170
"Blockbuster" products, 70–71
Boston Consulting Group (BCG), 13, 44
Brevundimonas aeruginosa, 205
Budgeting of clinical product development, 193

C

Campbell method, 136
Cancer diagnosis and biotechnology, 9, 10
Cancer patient, genetic variations. *See* Acute lymphocytic leukemia (ALL)
Cardiology research state, 275–277
Cardiology
    diagnostic and treatment devices used in, 277–278
    standard of care in, 281
Cardiovascular drugs, clinical trial types in, 281–282
Cardiovascular endpoints evolution, 277
Cardiovascular products, 71
Carvedilol, 279
Case control design, epidemiology studies, 119, 120
Case report form, 184–186, 191, 193, 195
CBER. *See* Center for Biologicals Evaluation and Research
CDC. *See* Centers for Disease Control and Prevention
CDER. *See* Center for Drug Evaluation and Research
CDM. *See* Clinical data management
CDM group duties, 185
CDRH. *See* Center for Devices and Radiological Health
Center for Biologics Evaluation and Research (CBER), 96, 102, 150, 153, 162, 257, 270
Center for Devices and Radiological Health, 150, 151
Center for Drug Evaluation and Research, 96, 102, 150, 153, 162, 166, 246, 251
Center for Drug Evaluation and Research's Office of Drug Safety, 251
Centers for Disease Control and Prevention, 66, 150, 154, 173
Centers for Medicare & Medicaid Services, 6, 66, 154, 243, 259, 265
Central IRB use, 110
Central nervous system products, 71
Cerezyme® (imiglucerase), 114
Cerivastatin, 289
CFR. *See* Code of Federal Regulations
"Changes Being Effected" supplement, 170
Chimeric proteins, 97
Chinese hamster ovary (CHO), for clinical production, 84
Chronic diseases, 8, 9, 296
    expenditure on, 9
    prevention, 9
    survey on, 8
Chronic obstructive pulmonary disease (COPD), 116
Chronic therapy trials, 285, 286
CI-981, 88–89
Claritin® drug, 159
Clinical data management, 182, 185, 186, 191, 193
Clinical development issues, 197
    IND role and controversies, 198
Clinical development plan (CDP), 37, 38, 183
Clinical development, of product, 260
Clinical drug development, 107
    phase 2 studies, 108, 109, 268
    phase 3 studies characteristics, 109–111
        heterogeneous patient sample, 110
        scale-up in manufacturing, 109
    phase 3b studies, in drug approval, 110
    phase 4 studies, 267, 268
        clinical drug development, 111, 112
        objectives, 111
    phase studies 1, 107, 108
    phases and FDA, 111
Clinical product development, 193
Clinical regulations and guidelines, 184
Clinical research goals, 183
Clinical research organizations, 14, 21, 28, 79, 178, 181, 193–197
Clinical team participation, 180
Clinical trial and background, 178, 179
Clinical trial conduct
    conduct and design, 184
    consent form and Investigator's brochure, 187
    product profile, 181
    project management and team, 182, 183
    reports, 192
    site selection and training, 188
    standard terms, 185, 186
Clinical trial operations, 178
    background, 179
    objectives and designs, 181
    planning and activities, 180
Clopidogrel, 280
Clot busting agents, 9
CME. *See* Continuing medical education
CMS. *See* Centers for Medicare & Medicaid Services
Code of Federal Regulations, 102, 103, 153, 159, 182–185, 245, 246
Codeine, 130
Cohort design, epidemiology studies, 119, 120
Common Technical Document, 21, 35, 40, 63, 64, 156, 161, 164, 178, 179, 182–184, 186, 192, 202, 223, 240, 243
Compassionate use, clinical trials, 199
"Compassionate use" studies, 118
Compliance and quality assurance, 171–177
Compliance assurance, categories, 171
Compound safety margin, 134
Concerta®, as controlled release product, 204
Congressional acts, regulations and guidances, 152
Consent form of clinical trials, 187
Consumer Product Safety Commission, 154, 155
Continuing medical education, 69, 238, 255, 256, 258, 259
    for regional clinicians, 69, 255, 256, 259
Contract research organizations, 14, 21, 28, 59, 178, 181, 193–197
    principles in management and characteristics, 197, 198
    principles of, 194
    selection of, 196
    site management organizations, 195
Coronary heart disease (CHD), 282, 283, 286
Cost-benefit analysis, pharmacoeconomic study, 115
Cost-effectiveness analysis, pharmacoeconomic study, 113–115
Cost-minimization analysis, pharmacoeconomic study, 115
Cost-utility analysis, pharmacoeconomic study, 115

CPSC. *See* Consumer Product Safety Commission
Crestor®, for high cholesterol, 8, 72
CRF. *See* Case report form
CRFs audits, 186
Critical skills, regulatory affairs, 148
CROs. *See* Clinical research organizations
CTD. *See* Common Technical Document
CTD modules, 164
Customer contact center, in medical and professional affairs department, 245–250
Cyclosporine in initial drug formulation, 204
Cymbalta®, in gastric irritation, 203
CYP2C9 isoenzyme polymorphisms, 288
CYP2D6 isoenzyme and propafenone, 288
CYP450 drug metabolizing enzyme, 200
CYP450 drug-drug interaction probes, 128, 129
CYP450 isozymes for drugs metabolism, 129, 130
Cytochrome P450, 200

D

Data management plan, clinical trial conduct, 187
Data monitoring board, 190
Data monitoring committee, 190
Data safety monitoring committee 190. *See also* Data monitoring committee
DDMAC. *See* Division of Drug Marketing, Advertising and Communications, role
Department of Defense, 6, 68, 354
Depo-Provera®, as injectable drug product, 203
Depocyt®, as injectable drug product, 203
Diabetes treatment and *Cost-effectiveness analysis,* 113, 114
Diltiazem®, for hypertension, 72
Disease dependent drug product profile, 203
Disease state management in pharmaceutical industries, 248, 249
Disease-specific QOL instruments, 116
Division of Drug Marketing, Advertising and Communications, role, 150, 151, 247, 257
Division of Scientific investigation, role, 150, 151
DMB. *See* Data monitoring board
DMC. *See* Data monitoring committee
DOD. *See* Department of Defense
Dofetilide, 280, 281, 289
Dose proportionality, 127, 128, 139, 140
Dose-ranging studies. *See* Phase 2 studies, clinical drug development
Doxil®, as injectable drug product, 203
Drug absolute oral bioavailability, 126
Drug absorption, food-effect studies, 124
Drug approval packages in FDA, 153
Drug bioavailability (BA), 126, 127
Drug bioequivalence study, 126, 127
Drug clinical studies, 161, 162
    phase 1 studies, 107, 108
    phase 2 studies, 108, 109
    phase 3 studies, 109–111
    phase 4 studies, 111, 112
Drug clinical trial phases, 219
Drug commercialization
    postmarket safety surveillance, 250, 251
    publications and scientific meetings, 248

Drug coverage decisions, factors in, 266
Drug development
    absorption of the drug, 124
    data & possible research questions, 278
    drug's elimination, 125
    goal for, 181
    in vitro–in vivo correlation (IVIVC), 123, 124
    metabolism and pharmacokinetics role in, 146
    metabolism studies, 125
    modeling and simulation (M&S) implementation in, 145
    preclinical MPK studies aim, 133, 134
    process & initial steps in, 202
    process in API – 1, API – 2 and drug product, 216, 217
    timeline and medical affairs department, 242
Drug development plan, 182
Drug discovery process
    goals and needs of, 90–93
    MPK role, 133
    new tools of, 94
    problems with, 90
    products for, 95–98
    steps in, 86
    success factors for, 105
    targets for, 93–95
Drug dosage forms, 202–204, 209, 210
Drug elimination, 125
Drug exposure, 125, 126
Drug formulation development
    active ingredients and roles, 204, 205
    BA & PK – 1 and pharmacokinetics, 207
    BA & PK – 2, 208
    bioavailability and preformulation studies, 206
    inhaled products – 1 and 2, 210
    injectable products – 1, 2 and ophthalmics, 211, 212
    oral products – 1 and 2, 209
    rectal and vaginal products, 214
    topical products –1 and 2, 213
Drug formulator role, 205
Drug laws and regulations
    helping public, 150–156
    public health protecting and advancing, 149
Drug manufacturing
    analytical development – 1, 214
    analytical development – 2 and biological products, 215, 216
    scale-up issues – 1 and 2, 219, 220
Drug packaging and stability studies, 218
Drug pharmacokinetics
    age and gender effects on, 143, 144
    food effects on, 140, 141
Drug physicochemical properties, 213
Drug pipeline products information, 247
Drug preservatives usage, 211, 212
Drug Price Competition and Patent Term Restoration Act, 1984, 151, 160
Drug product commercialization
    approaches and practices in, 228
    creating next blockbuster in, 225
    hurdles in, 225
    labeling and sales representatives in, 246, 247

market and product shaping, 222, 223
medical affairs department role in launch and marketing, 244–245
name selection, 234
packaging, labeling and stability studies, 218
pricing decision, 233
product launch phases and commercial team role, 238, 239
profile and characteristics, 202, 203
and R&D division interaction, 230
responsibilities in, 237, 238
Drug products regulation and approval, 152
Drug quality test methods, 214, 215
Drug reformulation, advantages, 204
Drug registration process, 161
Drug, postapproval safety surveillance, 169
Drug, vehicle and skin interactions, 213
Drug-drug interaction study, 141
Drugs and biological products, R & D in
challenges, 24, 25
clinical development, 36–38
decision making in, 52
global product plan, 35
health care delivery changes and, 10
health care trends and, 10
Portfolio Planning (PP) PICTRS in, 63
portfolio planning management, 29, 43–50
product development, phases in, 34
productivity, FDA statistics, 16
spending by pharma companies, 15
spending by research phase, 15
success factors, 30, 31
Drugs inducing CYP3A, 129
Drugs with short and long half-lives, 207
Drugs, *in vitro* transport studies, 124
Drugs, research & development
and business effectiveness criteria, 69–70
company and university relationships for, 67–68
company/business for, 79–82
data/information outcomes by, 76–79
drug failures in, 73
new products outcomes for, 66
products failures in, 72
DSI. *See* Division of Scientific investigation, role
DSMC. *See* Data safety monitoring committee
Duragesic®, as transdermal drug product, 203

E
EDC. *See* Electronic data capture
EIR. *See* Establishment inspection report
Electronic data capture, 185, 191
Eligard® as subcutaneous injectable drug, 203, 207
EMEA. *See* European Agency for Evaluation of Medicinal Products; European Agency for Evaluation of Medicines
"Emergency use" provision, 118, 119
Enbrel® (etanercept), 54, 115
in rheumatoid arthritis treatment, 10, 72, 221, 238
End-of-phase 2 meeting, 161
EORTC QLQ-C30, fatigue scale, 116
Epidemiology studies, 119, 120
Epogen® (erythropoietin), 115

"Equivalence" trial. *See* Noninferiority trial
Ery-Tab®, in gastric irritation, 203
Establishment inspection report, 175
European Agency for Evaluation of Medicines, 23, 24, 45, 52, 67, 77, 155, 156, 179, 184, 186, 243, 245
European Medical Evaluations Agency (EMEA), 23
Excipient compatibility, in drug preformulation, 206
Expedited Review Requested, 170
Extended release drug products, usage, 203
External customers and internal customers in medical affairs, 243–244

F
Fast-track program, regulatory development strategies, 157
FDA. *See* Food and Drug Administration
FDAMA. *See* Food and Drug Administration Modernization Act
Federal Food and Drugs Act, 1906, 151
Federal Food, Drug and Cosmetic Act, 1938, 151, 259
Federal Trade Commission, 154, 155
Fexofenadine as terfenadine, 204
5-fluorouracil (5-FU), 204
FIFCO. *See* Fully integrated pharmaceutical company & medical affairs interactions
FIPCO. *See* Fully integrated pharmaceutical company
First-in-man (FIM) studies, single/multiple doses, 137
Food and Drug Administration, 66, 96, 107, 148, 151, 240, 245
advisory committee and hearings, 166, 167
communications, 150, 153, 168
consent decree, 176
definitions of drugs and biologics and responsibilities, 152
drug approval, 158
electronic record, 174
enforcement activities, 176
form 1572, 173
inspection-1 and 2, 175
judgement, 177
mission statement, 149
objectives, 162
patent and exclusivity interests, 160
power and regulation, 171
pre-IND meeting, 161, 162
product review, 165
purpose of drug advertising and promotion, 174
regulation laws, 151
regulatory opportunities, 157, 158
responsibilities, 175, 176
review process, 162, 165
role of, 262
Food and Drug Administration Modernization Act, 1997, 151, 152, 157, 245, 247
Food and Drug Administration regulation
content and format of the package insert, 167
Food and Drug Administration role, public health, 149, 150
Food safety issues, FDA, 150
Forced degradation studies, in drug preformulation, 206
Form FDA 483, 175
Fortovase®, as saquinavir drug, 204
Fosamax®, in osteoporosis treatment, 206
Frequently asked question (FAQs), 246
FTC. *See* Federal Trade Commission

Fully integrated pharmaceutical company & medical affairs interactions, 241, 242
Fully integrated pharmaceutical company (FIPCO), operations and divisions
    development and marketing division, 21
    global operations, 23
    manufacturing division, 23
    research division, 20
    sales department, 22
Functional Assessment of Cancer Therapy (FACT), 115, 116

G

Gantt Style Chart, 61
Gaucher disease, 67, 114, 115, 158
GCP. *See* Good clinical practice
Gene silencing, 94
General investigation plan, IND, 163
Generally regarded as safe, 206
Generic QOL instruments, 116
Genetic polymorphism, 130
Genomics study, 93
Genotropin®, 114
Gleevec®, 53
Glial-derived neurotrophic factor (GDNF), 90
Global clinical research, 192
Global organizations and medical affairs department, 241
Global regulatory harmonization, 155
Global Utilization of Streptokinase and Tissue Plasminogen Activator for Occluded Arteries (GUSTO-1), 285
GLPs. *See* Good laboratory practices, role
Glycoprotein IIb IIIa receptor drugs, 277
GMP inspection, types, 175
GMPs. *See* Good manufacturing practices
Good clinical practices, 102, 110, 171, 173–175, 183–185, 188, 190, 192, 193, 195, 331
Good laboratory practices, role, 102, 132–134, 171, 172, 174, 175
Good manufacturing practices, role, 102, 171, 172, 174, 175
Government agencies role, 113, 154
GRAS. *See* Generally regarded as safe
Guidance documents in FDA, 152, 153, 159, 160

H

H. pylori, 7
HCPs. *See* Health care professionals
Health and Human Services, 154, 252, 256, 331
Health care
    changes in, 6
    cost in USA, 5, 7, 9
    delivery, changes in, 10
    drug use and, 8
    improvements in, 9
    science and, 9
    trends, impact on drug R & D, 6
    utilization and impact, 7
Health care professionals, 68, 249, 256
Health Insurance Portability and Accountability Act roles, 245, 252
Health maintenance organization, 198
Heart failure trials, special populations studies, 286–290
Heart Protection Study (HPS), 286
Heartburn treatment, drug usage, 203

Heparin, 277
Hepatic impairment study design, 142, 143
Herceptin®, 53
    in breast cancer treatment, 53, 67, 71, 74, 118, 208, 234, 309
HHS. *See* Health and Human Services
High-pressure liquid chromatography, 132, 215, 216
Highthroughput screening (HTS), 50, 133
HIPPA. *See* Health Insurance Portability and Accountability Act
HMG CoA reductase discovery, 277
HMG-CoA reductase, and L isomer, 88–89
HMO. *See* Health maintenance organization
Homogeneous patient populations use in phase 2 studies, 108, 109
Hospital formulary approval in drug commercialization, 248
HPLC. *See* High-pressure liquid chromatography
Human immunodeficiency virus (HIV), 108
Humatrope®, 114
Hydroxy-3-methylglutaryl coenzyme A reductase inhibitors (HMG CoA RIs), 286, 289

I

IB. *See* Investigator's brochure
Ibutilide, 289
ICH. *See* International Conference on Harmonization
ICH parties, 155
ICH process, topics, 156
ICH regulatory harmonization process, 156
IEC. *See* Independent ethics committees
Immunonephalometric test, 278
Implantable defibrillators, 9
IND application contents
    investigator brochure, 162, 163
    study protocol and application submission, 163
IND safety report, 185
Independent ethics committees, 185, 189, 329, 330
INDs. *See* Investigational New Drug applications
Information technology support, 190, 191
Inhalation drug products, 203
    advantages and uses, 210
Injectable drug products usage, 203, 211, 212
Institute of Medicine (IOM), 10
Institutional review board (IRB), 109–113, 118, 163, 173, 184–187, 189, 190, 193, 269, 325, 326, 330, 334, 336
Inter and intra subject variability in drug bioavailability, 206, 207
International Conference on Harmonization, 110, 146, 155, 156, 161, 164, 173, 183, 184, 187, 188, 190, 192, 193, 214, 322, 323, 329–331, 336
International Council on Harmonization/ Good Clinical Practice (ICH/GCP) guidelines, 110
Investigational New Drugs, 5, 14, 24, 103, 107, 152, 160, 202, 205
Investigator's brochure, 78, 186, 187
Investigator-initiated studies in drug development process, 112, 113
Investigators. *See* Thought leaders
IRB. *See* institutional review board
IRB reviews documents, 190

K

Kefauver–Harris Drug Amendments, 1962, 151
Key opinion leader, 245, 253, 254, 259, 260, 267
Knockout and transgenic animals, 94
KOL. *See* Key opinion leader

Index

L

L&R. *See* Legal and regulatory department, in medical affairs department
LCM. *See* Life cycle management
Leapfrog group, 10
Legal and regulatory department, in medical affairs department, 241–243
Life cycle management, 235
Lipitor®, for hyperlipidemia, 8, 71, 88
Long half-life drugs, 207
Low-density lipoprotein cholesterol (LDLC), 286
Low-molecular-weight heparins (LMWHs), 289, 290
Lunelle™ as contraceptive injection, 207, 211
Lupron Depot®, as injectable drug product, 203
Lymphocyte function–associated antigen 1 (LFA-1), 87
Lymphoid myeloma cells, for antibody protection, 84
Lyophilized drug products usage, 203

M

Managed care organizations (MCO), 117
Market research plans in drug commercialization processes
    goal and principles of, 233, 234
    strategies in, 231, 232
Martindale: The Extra Pharmacopoeia publication, 258
Mass balance studies, 137, 138
MC. *See* Medical communication
MDIs. *See* Metered dose inhalers
MDR1 (P-glycoprotein), 129
"Me-Too" drug development, 279, 280
MedDRA. *See* Medical Dictionary for Regulatory Activities; Medical Dictionary for Regulatory Authorities
Medical affairs (MA) department, in pharmaceutical product development, 240–244
Medical affairs and R&D, 241
    types of, 268
Medical communication
    and CME assessment issues, 259
    and events and products risk types, 262–265
    and formulary submission, 260
    and product promotion & FDA enforcement, 261
    professionals in, 257
    and risk management pre-& postapproval, 262
    role of, 260, 261
Medical communications groups responsibilities and compendium, 256, 257
Medical Dictionary for Regulatory Activities, 250
Medical Dictionary for Regulatory Authorities, 156, 250
Medical field liaisons
    functions, 252, 253
    professional influence, 254–256
Medical field professionals, in pharmaceutical industries, 241
Medical information services
    customer support services, 249–250
    functions and roles, 241, 245
Medicare Modernization Act, 8
MedWatch in safety surveillance, 251, 252
MedWatch mechanisms, 170
Metabolism and excretion in drug formulation development, 208
Metabolism and pharmacokinetics (MPK), 123, 125, 133–135, 141, 142, 145, 146
Metered dose inhalers, 203
MHLW. *See* Ministry of Health, Labor and Welfare
Minimum inhibitory concentration (MIC), 123, 124, 136
Ministry of Health, Labor and Welfare, 155, 156, 179, 184, 245
Mixed effect modeling, 131
Modified or delayed release drug products, examples and usage, 203
Modules of CTD, 164
Moexipril, 280, 281
Multifaceted analysis, regulatory development strategies, 157
Multiple dose pharmacokinetics, 127
Multiple sclerosis products, 71
Myocardial infarction (MI), 282, 283

N

National Cancer Center Network (NCCN), 11, 310
National Cholesterol Education Program (NCEP), 280, 281
National Health Expenditures, 5, 6
National Health Interview Survey (2002), 8, 9
National Institutes of Health, 7, 14, 28, 79, 80, 93, 118, 154, 179, 226, 227, 328, 330–333, 346
Nature Reviews Drug Discovery, 94
NCEs. *See* New chemical entities
NDA. *See* New Drug Applications
NDA and BLA preapproval, 175
NDA submission in Phase 3 studies, 109
NDA, BLA applications, 163
NDA, regulatory expectations in, 145, 146
Neoral®, as cyclosporine drug, 204
Neulasta®, for neutropenia correction, 67, 71, 72, 79, 155, 211
Neupogen®, 67, 71, 79, 211, 309
New chemical entities, 120, 125, 128, 134, 205
    reasons of failures, 133, 134
New Drug Applications, 16–18, 21, 35, 49, 50, 57–59, 68, 77, 80, 107, 109, 111, 117, 121, 145, 151, 157, 160–170, 175–179, 181–184, 186, 191, 192, 194–196, 202, 205, 223, 240, 243, 245, 268, 271, 278, 297, 306, 348
New Drug Approvals (NDAs), 5
New medical entities, 16, 71
New molecular entities, 5, 14, 16, 17, 55, 56, 60, 71, 158, 226, 295, 324
New product development, goal, 179
New therapeutic modalities, 278–279
NHE. *See* National Health Expenditures
Niaspan®, as extended release product, 203
Nico-Derm®, as transdermal products, 203
NIH. *See* National Institutes of Health
NMEs. *See* New molecular entities
NMEs. *See* New medical entities
No action indicated (NIO), 175
No observed adverse effect level (NOAEL), 123, 124, 134–137
NOI. *See* No action indicated
Non-clinical development, 86
    and pharmacokinetics and pharmacodynamics, 98–100
Non-small cell lung carcinoma, 158
Noninferiority trial, 110
Nonischemic cardiomyopathy, 287
Norditropin®, 114
North American Trade Agreement, 19
NSCLC. *See* Non-small cell lung carcinoma
Nutropin AQ®, 114

O

OAI. *See* Official action indicated
Observational studies potential biases, 120
ODS. *See* Office of Drug Safety, role
Office of Drug Safety, role, 150, 151
Office of Generic Drugs, 150
Office of Inspector General, 25, 154
Office of Regulatory Affairs, role, 150, 151
Official action indicated, 175
OGD. *See* Office of Generic Drugs
OIG. *See* Office of Inspector General
Omeprazole, 130
Oncogene inhibitor, for breast cancer, 67
Oncology products, 71
Ophthalmic drug products, 212
Opportunity, promise, proof, and conviction in commercialization (OPPC), 235, 236
Optimal registration strategy, development, 159
ORA. *See* Office of Regulatory Affairs, role
Oral and modified or delayed release drug products, 203
Oral dosage drugs
    forms and uses, 203
    tests for, 215
Orally administered drug products, 208–209
Orange Book edition, 258
Organization evaluation, of industries, 66
Orphan Drug Act, 1983, 151
Orphan drugs, regulatory development strategies, 158
Over-the-counter (OTC), 11, 155, 159

P

Package insert, 36, 76, 77, 180, 181, 184, 185, 187, 197, 243, 245
    goals and safety information, 182
    product review, 167
PAI. *See* Preapproval inspections
Patient selection, enrollment and retention, 189
Payer relationships, 67
PDAs. *See* Personal data assistants
PDUFA. *See* Prescription Drug User Fee Act, 1992
PE drug development challenges, 117
PE studies types
    cost-benefit analysis, 115
    cost-effectiveness analysis, 113, 114
    cost-minimization analysis, 115
    cost-utility analysis, 115
Pediatric PK/PD study, 144
Pediatric Research Equity Act, 2003, 152
PEG-Intron®, as injectable drug product, 67, 203, 211
Pegasys®, as injectable drug product in hepatitis C treatment, 203, 211
Pegylation, 54
Personal data assistants, 185
Personnel in clinical trials operations, 180, 181
Pharmaceutical companies
    blockbuster products and, 12
    categories of, 11
    commercialization processes
        approaches, 223
        assessment parameters, 228
        benchmarks in, 229
        challenges, 226
        commercial input and product development decisions, 234, 235
        goals and objectives, 224, 225
        phases in product launch and commercial team role, 238, 239
        product development steps and strategies, 230, 231
        R&D role, 223
        responsibilities, 237
        strategies, 226, 227
    data warehouses in, 50
    expenditure on R & D, 13, 15
    FDA and, 17, 18
    financing of, 26, 27
    fully integrated, 19–24
    goals of, 13
    merging within, 13
    organization of, 19–24
    portfolio management by, 43–50
    product development in, 34–43
    projects in pipeline, 15
    sales figures, 11
    speed to market time, 19
    success parameters, 13
    support companies for, 11
    top ten, 13
Pharmaceutical Executive, 13
Pharmaceutical industries
    and external customers interactions, 243–244
    role of international organizations, 241
Pharmaceutical Research and Manufacturers Association (PhRMA), 16, 66, 255, 256
Pharmacoeconomics, 265–267
    definition, 113
    study, 94
Pharmacogenetic studies, use and disadvantages, 118
Pharmacogenetics & Pharmacogenomics, 129, 130
Pharmacokinetic, 92, 98–100, 123–128, 131–136, 207, 208, 247, 304, 306, 312, 313, 318, 324, 329, 332–335, 345, 346
Pharmacokinetic drug-drug interactions (PK-DDI), 128, 129
Pharmacokinetic-chemical structure-activity relationship (PK-SAR), 133
Pharmacokinetics, 134
Pharmacological-chemical structure-activity relationship (PD-SAR), 133
PhRMA. *See* Pharmaceutical Research and Manufacturers Association
PI. *See* Package insert; Principal investigator
PIPE, 27
PK. *See* Pharmacokinetic
Placebo-controlled trials, 110, 199
Podium policy in FDA, 153
Population-based PK/PD modeling, 131
Portfolio planning management (PPM)
    advantages and definition, 43
    analysis, 54–60
    budget for, 47, 48
    decision gates in, 50, 51
    elements of, 44, 45
    importance of, 63
    life cycle management in, 54

Index

managers concerns, 46
participants of, 45, 46
project management, 60–62
success criteria in, 46, 47
technology for, 50
Post-approval commitments, FDA, 169
Post-approval NDA and BLA activities, 168
Post-marketing commitments, for clinical studies, 269–271
Postapproval maintenance, FDA, 168–170
Postmarketing surveillance
programs, 262
studies, 111, 112
Pravachol®, 67, 72
Preapproval inspections, 220, 221
Preclinical development. *See* Non-clinical development
Prescription Drug User Fee Act, 1992, 151
Prescription drugs
expenditure on, 6
sales of, 11
Prevacid®, 71, 203
Principal investigator, 112, 184, 188, 190, 197, 335
Priority review, regulatory development strategies, 157
Procardia LA as extended release product, 203
Procardia XL®, as controlled release drug product, 204
Procardia®, in hypertension, 203
Procrit®, 10
Product development paradigm
peoples involved in, 41
pipelining in, 41
processes in, 40
products & portfolio in, 42
profits in, 40
Product discovery techniques, 74
Product killing, in companies, 72
Product outcomes, in research and development of drugs, 70–73
Product review, 165–167
advertising, 167, 168
FDA communication, 168
package insert, 167
Professional service unit, in pharmaceutical product development, 240–244
Project team and management, clinical trial conduct, 182, 183
Propafenone, 288
Proprietary Medicinal Products Committee, 288
Prostaglandins, 9
Prostate-specific antigen, for prostate cancer, 89
Protein pegylation, 72
Proteomics study, 93–94
Protocols of clinical trials, 186, 187
Protopic®, in atopic dermatitis treatment, 205
PSA. *See* Prostate-specific antigen
Psoriasis, 87
Public health accurate, 150–156
Public health advancing, 150
Public Health Service Act, 1944, 151
Public health, protecting
FDA approval, 150
FDA role, 149
Public/patient outcomes, in research and development, 66–70
Pulmonary capillary wedge pressure (PCWP), 285

Q

QA. *See* Quality assurance
Quality adjusted life year (QALY), 115
Quality assurance, 35, 61, 172, 182, 186, 196, 244, 248
Quality of life (QOL) studies, 115–117
Quantitative structure-activity relationships (QSAR), 97
Quantitative whole-body autoradiography (QWBA), 124, 137
Quinidine, 129

R

Radio-frequency identification, 218
Randomization code, clinical trial conduct, 187
Raptiva®, for psoriasis, 87, 99
Rectal drug products, 213, 214
Regulatory affairs, discipline, 148
Regulatory authorities
product review, 165, 166
submissions, 160–165
Regulatory development strategies
accelerated approval, 158
fast-track program, 157
orphan drugs, 158
priority review, 157
rolling submission, 157, 158
Rx to OTC switch, 159, 160
Regulatory science controversy, 176, 177
Renal impairment study design, 142, 143
Request for proposal, 196
Research-Technology Management, 62
RFID. *See* Radio-frequency identification
RFP. *See* Request for proposal
Rheumatoid arthritis, 10
Rifampin therapy, 289
Risperdal® Consta™, as injectable drug product, 211
Rolling submission, regulatory development strategies, 157, 158
RTF. *See* refuse-to-file
Rx to OTC switch, regulatory development strategies, 159, 160

S

SAE. *See* Serious adverse event
Saizen®, 114
Saquinavir in initial drug formulation, 204
SBP. *See* Systolic blood pressures
SCHIP. *See* State Children's Health Insurance Program
SDS-PAGE. *See* SDS-polyacrylamide gel electrophoresis
SDS-polyacrylamide gel electrophoresis, 216
Selegiline, 139
Self-injector pens, for insulin, 72
Serevent®, as dry powder inhalers, 203
Serious adverse event, 53, 184, 185
SF-36, generic QOL instrument, 116
Short half-life in drugs, 207
Simvastatin, 275, 280, 281, 286, 289
Single nucleotide polymorphisms (SNPs), 52
Singulair®, 71
Site visit report, clinical trial conduct, 187
SMOs. *See* Site management organizations
sNDA. *See* Supplemental NDA
Solubility studies, in drug preformulation, 206
SOPs. *See* Standard operating procedures

Spiriva®, as dry powder inhalers, 203
St. George's Respiratory Questionnaire for COPD, 116
Standard operating procedures, 40, 87, 172, 179–181, 186, 188, 196, 243, 331
State Children's Health Insurance Program (SCHIP), 6, 154
Stock price impact, 70
Supplemental NDA, 161
Surgical anesthesia, drug usage, 203
Systolic blood pressures, 192, 287

T

T-CSDD. *See* Tufts Center for the Study of Drug Development
Telithromycin clinical development, DDI studies in, 142
Telmisartan, 280
Tentative approval, FDA, 160, 165
Therapeutic dosage regimen determination, 131
Therapeutic index (TI), 131
Therapeutic window, 131
Thiopurine, 130
Thomson-CenterWatch (2004), 52
Thought leaders, 35, 38, 55, 67, 68, 188, 194
Thrombolytic therapy, 285
Ticlopidine, 280
TNKase®, 71, 72
TOBI®, in cystic fibrosis treatment, 203, 205
Topical drug products, 213
Torsades de pointes (TdP), 288, 289
Toxicokinetics, 134, 135
Training of investigators, 188
Transdermal drug products, 203
"Treatment INDs" provision, 119
Tufts Center for the Study of Drug Development (T-CSDD), 13, 17, 52, 66, 193
Tumor necrosis factor (TNF), in sepsis, 72

U

Ultra-Lente insulin, as injectable drug product, 203
United States Pharmacopeia Drug Information®, 258
United States Pharmacopeia, in drug quality tests, 215, 252, 257, 258, 264, 328
United States Pharmacopeia–National Formulary (USP-NF), 258

V

VAI. *See* Voluntary action indicated
VALIANT study, 284
"Value" in economic analyses, 113
Vasodilation in the Management of Acute CHF (VMAC) study, 285
Veterans Administration Cooperative Vasodilator-Heart Failure Trial(V-HeFT), 286, 287
Vioxx®, 25
Voluntary action indicated, 175

W

Warfarin, 112
Warning letters in FDA, 153, 176
Waxman–Hatch legislation, 151
Wellbutrin XL®, as controlled release drug product, 204

X

Xeloda® as prodrug, 204
Xenical®, 114
Ximelagatran, 279
Xopenox®, as nebulizer, 203

Z

Zemplar®, in secondary hyperparathyroidism treatment, 211
Zocor®, 8
Zyban®, as extended release drug product, 203

Printed in the United States of America.